THE CONSUMER'S ENERGY HANDBOOK

Contents

Introduction   v

Part 1: Energy Facts and Information

1   All About Energy Technology   3
2   A Look at an Energy-Efficient Residence   44
3   Homes and Buildings   48
4   Waste Heat Recovery: More Power from Fuels   51
5   Why Heat Pumps?   53   HP·111
6   Emerging Age of Solar Energy   55   59
7   Buying Solar   104
8   Possible Effects of Solar Energy   123
9   State Solar Legislation   135
10   How a Nuclear Plant Works   144
11   Shipping Nuclear Fuel   146
12   Safeguarding Nuclear Material   148
13   Transportation   150
14   Tomorrow's Cars   153
15   Experimental Electric Cars   155
16   New Electric Light Bulb   157
17   Basic Energy Conservation Code   158
18   U.S. Department of Energy   170

Part 2: Energy-Saving Tips

19   Saving Energy in the Home and Automobile   203
20   Automatic Thermostat Controls   215
21   Selecting the Right Heating System   217
22   How to Understand Your Utility Bill   223
23   How to Survive a Tough Winter   227
24   How to Buy Insulation   231
25   How to Weatherize Your Home   234
26   Firewood: How to Choose It and Use It   272
27   Landscaping to Conserve Energy   276
28   Energy Activities for Children   278

Part 3: Who to Contact

29   Associations and Organizations   289
30   State Energy Offices   292
31   Insulation Contractors   296

32 National Manufacturers and Distributors of Solar Equipment 301
33 National Solar Heating and Cooling Information Center 333

Appendix: National Energy Act 334

Glossary 339

Bibliography 353

Index 357

THE CONSUMER'S ENERGY HANDBOOK

Part 1
Energy Facts and Information

1 All About Energy Technology

The United States, which has 6 percent of the world's population and consumes 35 percent of the global production of energy and minerals, is confronted with an energy shortage. There are many reasons for this. First of all, our total energy consumption has risen rapidly as the population and the per capita demand have increased. Our domestic supplies of natural gas and oil are running out, and dependence on foreign sources could create international political and financial risks. Moreover, the production of energy in the United States is now affected by environmental regulations protecting air, water, land, recreational, and aesthetic resources. Finally, we are not developing new sources of energy or new energy production systems fast enough to keep up with the increasing demand.

The term *energy technology* refers to the application of all scientific and engineering knowledge to the complex problems of locating, recovering, developing, storing, and distributing various forms of energy in a clean, convenient, and economical manner. In an energy-intensive country such as the United States, it is essential to explore new methods of improving the technology so that future needs can be met at a reasonable cost and without damaging the environment. Many government, university, and industrial organizations are involved in this effort.

ANNUAL CYCLE ENERGY SYSTEM (ACES)

Because much of the United States is involved in seasonal changes (i.e., moderate temperature variation between winter, spring, summer, and fall), there is a good balance between the heat required to provide residences with hot water all year round and warm air in winter, and the cooling required to air-condition them in summer. The Annual Cycle Energy System (ACES) is being developed at the ERDA Holifield (Oak Ridge) National Laboratory to take advantage of this annual weather cycle to conserve energy in the heating and cooling of residential and commercial buildings. It is estimated that over 50 percent of the energy now used for space heating and cooling and for providing hot water can be saved with the use of ACES (see Figure 1-1).

The principal component of the ACES is an insulated tank of water that serves as an "energy storage bin." For well-insulated homes within the applicable zone, this bin requires 2 cubic feet of water for each square foot of living space. Thus, a home with 1500 square feet of living space would require a 3000-cubic-foot tank of water, which would be equal one-fourth to one-third the size of the basement. The bin could be located in the basement of an existing house or could be built under the driveway, carport, or patio of a house under construction.

During the winter, heat is obtained from the bin by a heat pump, which draws the heat from the water in much the same manner that a conventional home heat pump draws heat from air. Heat drawn from the water is used to warm the house and to provide hot water. This removal of the heat from the water gradually turns the water into ice over a period of months. In the summer, the chilled water from the bin is used to provide air conditioning without the operation of the heat pump compressor. The air-conditioning action causes a gradual melting of the ice over a period of months and thus stores heat for use in the winter.

Cost and Savings

The purchase price of an ACES will, of course, depend upon the size unit required for a given residence. For most homes, the estimated cost will be about $1700 greater than conventional central heating and cooling systems. Installing the system into an existing home would cost approximately $3000, depending upon the type of construction and previous heating system of the home. The energy and money savings will also depend on the size of the residence and cost of electricity in the area. It has been calculated that a residence containing 1500 square feet will consume about 19,400 kilowatt hours per year in electric heating, air conditioning, and water heating by conventional means. Equipped with ACES, the same home would consume only 7400 kilowatt hours for the same functions. In an area where the electricity cost is 3.5 cents per kilowatt hour, ACES would provide an operating cost saving of approximately $400 per year.

Organization

The ACES is being developed by the U.S. Department of Housing and Urban Development and by the ERDA Office

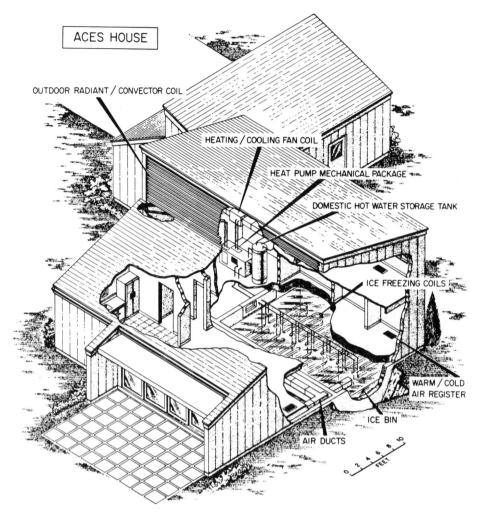

Figure 1-1. Annual cycle energy system.

of Conservation, Division of Buildings and Industry. The research effort is being carried out by the Energy and Reactor Divisions of Holifield (Oak Ridge) National Laboratory, which is operated for ERDA by Union Carbide Corporation's Nuclear Division.

BATTERIES

Large-Scale Storage of Electricity in Batteries

Widely dispersed battery stations may play a subtle but crucial part in the United States' future efforts to conserve scarce fuel resources. Under certain conditions, batteries can help to meet major energy needs without damaging the environment, keep energy costs within reasonable limits, and make sure that energy is available wherever and whenever it is required without encouraging waste.

The potential use of very powerful and long-lived batteries in electrical vehicles has attracted a fair amount of public interest. However, the possible role of batteries in bulk energy storage facilities within the nation's electric power networks is probably not so well understood.

By the beginning of the 1980s, banks of batteries near Princeton, New Jersey, will be integrated experimentally with an operating utility grid for the first time. The same small, unimposing building—officially designated the *Battery Energy Storage Test facility* (BEST)—will also be the site for continuing, realistic evaluation of various advanced battery systems, which may make bulk chemical storage of electricity commercially feasible by around 1985.

What Is a Battery?

Batteries do not generate energy or consume fuel in the usual sense. A battery is simply a device that stores energy in chemicals that react to produce a flow of electrons (an electrical current) when the basic constituents—two dissimilar terminals immersed in an electrolyte—are connected externally by a piece of conducting material, such as wire.

In *primary* batteries (e.g., the dry cell used in most flashlights), the chemical reaction is not readily reversible. Therefore, once a limited amount of electricity has been drained from the system, the battery is "dead" for good. On the other hand, *secondary* batteries (such as those used in automobiles) can be recharged by feeding an electric current back into them and thus restoring the original chemical potential. The secondary battery is the focus of most studies now being carried out by ERDA, the utility industry's Electric Power Research Institute (EPRI), and a number of individual manufacturers. The goal of this research is to develop a battery system that will store sufficient energy in small volume and undergo thousands of reasonably long charge-discharge cycles over a lifetime of a decade or two. In addition, it must operate at an overall cost that is comparable to alternate approaches for providing electric power to homes, businesses, and industry.

New Sources and Old Problems

If the factors affecting national energy policy were less complex, the investment of tens of millions of dollars annually in research on large-scale battery systems might be inappropriate. It happens, however, that chemical storage is a logical way to accentuate the usefulness of new basic energy sources—including photovoltaic conversion of solar energy, wind power, and geothermal energy—which we will need to exploit more fully in the future.

In the shorter run, batteries are a practical means of providing electricity at times when demand is low, thereby making it possible to use generating facilities only during peak hours.

Demands Are Great

An ideal utility battery faces far stiffer challenges than the one in a conventional automobile. The former must be capable of operating steadily for hours on end, allowing itself to be drained of most of its power, and then rapidly recharged four or five times each week. It should be able to repeat this cycle a couple of thousand times or more.

Most of the battery research that has been—or is being—sponsored by ERDA aims at obvious cumulative benefits. Reducing the effects of internal corrosion, eliminating impurities from the electrolytes, and solving the high-temperature sealing problems should extend the useful life of batteries. New combinations of materials could offer higher energy capacity, because potential voltages vary from different chemical reactions. Improvements in either materials or design might result in higher energy density—that is, more watt-hours of storage capacity per pound. Better manufacturing techniques and standardization of components might replace the job-shop approach with something closer to automated production lines; this would undoubtedly affect the most sensitive factor of all in regard to bulk battery systems—their cost.

Net Costs Are the Key

Although energy storage promises to conserve our scarcest domestic fuel resources (oil and natural gas), it presents an economic picture that is far from clear. Some energy is always lost when one form is converted into another; and this rule applies to chemical storage of electricity as well as to the generation of electricity from chemical fuels. What comes out of a battery is always less than what went into it. The net efficiency of baseload generators combined with advanced batteries should usually be greater than the oil- and gas-burning combustion turbines that have customarily been used as peaking units. Even so, savings on this account alone may not always be dramatic. The economic case for dispersed energy storage can be further strengthened by consideration of the possible postponement or elimination of the need for new transmission and distribution facilities, as well as lowered requirements for generating capacity.

Like generating capacity, power lines must ordinarily be planned to handle peak loads rather than the average. If battery storage stations of adequate capacity are installed near load centers, less transmission capacity will be needed to connect the ultimate users of electricity with the ultimate source—central power plants. In urban areas, where underground lines are becoming an expensive practical requirement, the savings through the use of storage facilities could be appreciable. Service might be more secure too, because some emergency reserve power would usually be available during unanticipated outages of plants or lines.

At the generating centers themselves, a certain amount of wear and tear on the equipment itself can be avoided by the addition of battery storage to a utility network. Besides base-load and peaking units, electric utilities normally use some "intermediate" generating units, which are regularly cycled up and down in output to match supply and demand as demand varies, whether by the season, day, or hour. Such cycling adversely affects the efficiency and life of mechanical equipment (even within plants designed expressly for the intermediate purpose). In some cases, the effects of cycling may be another substantial economic factor in determining the point at which battery storage could begin to pay for itself and turn a profit.

Batteries Have Competition

There are, of course, other means of storing electrical energy. ERDA is examining them too—including giant flywheels, compressed air storage, electrolysis of water to yield hydrogen fuel, and energy storage in the form of heat.

Presently, the most common form of utility storage is "pumped hydropower," in which electrical pumps raise water during off-peak periods to an elevated reservoir behind a dam. On command, the water can fall back through gravity-operated turbines when additional generating capacity is needed to meet short-term demand peaks. Because acceptable sites for new pumped-storage dams are limited, ERDA is investigating the feasibility of creating similar systems that would combine water storage at the surface with excavated reservoirs from 2000 to as much as 5000 feet underground.

All in all, battery storage looks like as good a bet as any as a part of the total energy system for the mid to late 1980s or early 1990s.

Simple, Compact, Quiet, Nonpolluting

In some ways, batteries are a nearly ideal power source. Battery power systems can be built in modules like filing cabinets to fit any space requirement, and they can be readily expanded if the need arises. Battery banks respond immediately to changes in power level dictated by a utility's load.

Batteries take up relatively little room. In fact, an operating station of the future could probably fit into about one-quarter of a city block.

There is very little operating noise and practically no emissions or waste disposal to worry about. Fuel delivery and storage are nonexistent. Operating crews are likely to be unnecessary for normal operation.

Nevertheless, as is usually the case in the energy field, battery storage is not a magic solution in itself. A variety of conservation measures to reduce the wasteful use of electricity (and every other form of energy) will be elements of American national policy for a long time to come. In addition to battery power, other forms of "load management"—e.g., shifts in working hours, lower rates of off-peak use, etc.—will try different means simultaneously to smooth out the peaks and valleys of electrical demand.

In the long run, utilities will probably take many approaches; the exact "mix" will vary from one area to another because of local factors. But batteries will probably be used increasingly in electrical supply systems of the future. And the function of the BEST facility is to help decide more precisely what their role should be.

FUELS FROM BIOMASS

Photosynthesis—the biological process that provides food and fiber from the plants—is supplying humans with a renewable source of energy. The energy is released when plants (*biomass*) are broken down by biological or thermochemical processes into fuels and other energy products.

Technology development efforts are divided into two main groups: production and harvesting of biomass, and conversion of biomass to fuels and petrochemical substitutes. The major products that can be obtained from biomass are liquid fuels (methanol, ethanol, gasoline); gaseous fuels (methane, hydrogen, ammonia, medium-Btu gas for onsite application, upgraded synthetic natural gas fed into natural gas pipelines); petrochemical substitutes (alcohols, ketones, turpentines, resins); heat; process steam; and electricity. Research projects range from burning wood to produce heat and electricity, to obtaining crude oil directly from a species of plant that was developed by Dr. Melvin Calvin of the University of California, Berkeley.

Trees and Woody Plants

Biomass currently provides almost 2 percent of the nation's supply of total energy consumption. The primary biomass resource is wood from trees in the forest or from residues of logging operations.

About 10 percent of the nation's potential grassland and forests could be made available for wood energy farms. These farms would probably consist of selected, rapidly growing types of trees planted close together, harvested in rotation, and trained to regenerate by "coppicing" or sprouting from stumps. The Department of Energy (DOE) is currently testing this type of forestry and funding research projects aimed at increasing production and decreasing costs of growing and harvesting woody plants.

Agricultural and Herbaceous (Nonwoody)

Herbaceous crops, such as sugar cane, can be grown closer together and have more frequent harvesting seasons than woody plants. Experiments with sugar cane and sweet sorghum have already shown that the amount of energy yield from these plants can be doubled by spacing crops closer together and choosing better species of plants.

Other major research efforts include additional work with sugar cane and sweet sorghum, a study of sugar cane and tropical grass production in Puerto Rico, and programs to identify and develop new species of herbaceous plants that have high energy content, growth rate, resistance to disease, and production rate.

Aquatic Biomass

Several types of aquatic resources, both freshwater and marine biomass, are being studied by DOE. These include red, brown, and blue-green algae, and water hyacinth, duckweed, kelp, and other species. Some of the algae and freshwater weeds have good potential for large-scale growth and conversion to energy.

Studies are underway to identify the most promising plants and energy-farm sites, and to select various growing systems and analyze their overall cost-effectiveness and potential.

Experimental work is continuing at California Institute of Technology on the growth of giant kelp; at Woods Hole Oceanographic Institute on freshwater and marine species; and at the University of California, Berkeley, on freshwater algae.

Residues

About 100 million dry tons of biomass residues—leavings from field crops, forest and mill operation, and animal manures—are produced in the United States every year. Biomass residues are currently used in animal feeds, fiberboard products, fertilizer, and as soil additives and conditioners. Their cost varies greatly according to type, application, and location. Because residues are the most readily available biomass products, they are used as feedstocks for most of the conversion processes now under study by DOE.

Conversion Technologies

Several different techniques for converting biomass to clean fuels and petrochemical substitutes are being investigated:

direct combustion; gasification and liquefaction; anaerobic digestion; and fermentation.

Direct Combustion. Large-scale direct burning appears to be the most immediately available way of using biomass to produce energy. The overall efficiency of burning biomass to produce steam is about 80 percent, and for power generation, 20 to 30 percent, depending upon moisture content and plant design. Coal combustion plants typically operate with a 34 to 40 percent efficiency.

Gasification and Liquefaction. Gasification and liquefaction involve thermal decomposition of biomass feedstocks to produce gaseous and liquid fuels. Biomass, pretreated with chemicals and subjected to high temperatures and pressure, can be converted to a medium Btu gas. The gas, in turn, can be upgraded to synthetic gas, methane, or gasoline.

Anaerobic Digestion. Anaerobic digestion takes place when treated biomass is consumed by bacteria in an oxygen-free atmosphere. The process, which produces methane and carbon dioxide, is commonly used in wastewater treatment facilities.

DOE experimental studies have shown that animal manure can be economically converted to methane in an anaerobic digester. Experimental and pilot test units are operating on cattle feedlots and dairy farms throughout the country.

Fermentation. Fermentation uses yeast to convert sugars to ethanol and petrochemical substitutes. The process is well established, but is not used widely in the United States, since it is cheaper to obtain those same products from petroleum.

If low-cost sugars can be found, fermentation could become an economically competitive source of ethanol. The most promising fermentation feedstock appears to be cellulosic materials, which can be transformed into sugar by a chemical process called *hydrolysis,* using either acid or enzymes to break down the cellulose into sugars and other by-products.

Fermentation can also be used to produce alcohol from surplus grain, which is then mixed with gasoline to create "gasohol." Although there are no technical problems involved with the process, it is more expensive than petroleum-derived gasoline.

Environmental and Other Concerns

The environmental impacts of biomass-based energy production are expected to be small. However, some basic concerns must be addressed, including: the potential erosion and depletion of organic content of the soil from overuse, or removal of too many residues; possible release of particles into the air, resulting from the direct combustion process; and disposal of by-products and residuals produced in many of the conversion processes.

Other factors that will have a strong impact on production of crops for energy use include the economics associated with using land for growing energy crops as opposed to food and fiber crops, and the need for large amounts of water to be used for growing feedstocks and for use in the conversion processes themselves.

BREEDER REACTORS

To a world faced with escalating fuel prices and rapidly shrinking fuel reserves, a device that produces electric power and creates, at the same time, more fuel than it burns should seem a godsend. Such a machine does in fact exist, its welcome has been mixed.

This machine is the "breeder reactor," which is a second-generation or advanced nuclear fission reactor. Like its predecessor, the nuclear converter reactor, the breeder reactor will obtain energy from the fission reaction—the splitting of a fissionable element such as the heavy-nucleus uranium or plutonium. The breeder reactor, therefore, shares the same suspicion that has dogged the rest of the nuclear projects—that is, will it explode like an atomic bomb?

The "magic" of the breeder is, in principle, fairly simple to understand. To produce energy, a fissionable nucleus is caused to split, releasing energy and at the same time two or three neutrons. In turn, one of these neutrons must cause another fission in order to continue a chain reaction, which provides the power. The other neutrons can be used to create fissionable nuclei out of material that is not otherwise usable as nuclear fuel; in particular, it can create useful fuel out of nonfissionable uranium or thorium. There is no controversy over this. If more than one fissionable nucleus is created this way for each one that is split, then the reactor is called a *breeder* for it "breeds" more fuel than it uses.

The controversy that does exist falls into the same categories—economic, environmental, and safety—as does the controversy over the conventional reactor. The arguments put forward in each category, however, are quite different. The following sections will briefly describe the ways in which the breeder works and will examine the contrasting statements and opinions on its economic feasibility and necessity, its effect on the environment, and its safety-related issues.

Resources

A good bit of the controversy that surrounds the breeder reactor concerns the amount of resources. The United States' known resources (reserves) of inexpensive uranium are not infinite or even very large. If the fuel is recycled (it is not at the present), these reserves will be used up during the 30-year average lifetime of reactors currently operating, under construction, on order, or announced. Without fuel recycling, reserves will only fuel about 60 percent of that generating capacity. If the United States were to develop and mine no more uranium than these reserves (an unlikely prospect), the nuclear age would be a short one. These reserve estimates deal with uranium 235, a rare form of uranium that is needed to fuel the conventional reactors. It makes up only 0.7 percent of the uranium in uranium ore. Uranium 238 accounts for almost all the rest. What the breeder reactor does is convert uranium 238 into plutonium, thus using almost all of the energy available in the uranium fuel instead of the 1 and 2 percent conventional reactors use.

In 1976, ERDA estimated that the reserves of uranium for breeder use, at 1975 levels of consumption, would be enough to meet all of this country's energy needs for about 1800 years. More astounding, however, is that because the breeder reactor uses fuel so efficiently, uranium price is no longer a major contributor to the price of electricity. This opens up more dilute and higher-priced ores to consideration.

There is no contesting the breeder's ability to greatly magnify the energy potential of United States uranium resources. However, the important questions are: how quickly will United States' resources of inexpensive uranium be used up? How much will the price of electricity increase? The answers will go a long way toward determining the breeder's economic viability.

In addition, the economic viability depends, for instance, on whether the spent fuel is reprocessed and the uranium and plutonium recycled, and on the amount of uranium 235 left in the "tails," discarded from the enrichment process. It also depends greatly on comparisons of capital costs and on assumed electricity demand and the fraction of that demand that is to be nuclear.

Table 1-1 gives the complete ERDA uranium estimate. The total amount, including less certain categories such as "possible" and "speculative," is for 3,500,000 tons–almost four times the "reserve" figure. In an ERDA study, the cumulative uranium requirement from 1976 to 2000 (assuming fuel recycling) is 1 million tons; without recycling, it is 1.3 million tons. Although this is half again as much as the reserve, it is only one-third of the potential resources. Other studies have produced essentially similar results.

In summary, it seems fair to say that the potential inexpensive uranium resources are sufficient to take a healthy electric industry (5.2 percent per year growth) with a healthy

nuclear component (about 40 percent of the total generating capacity in the year 2000) to 2000 and beyond, perhaps to 2020 or 2040. The resources will last longer if electrical and nuclear growth are lower than projected. It should be noted, however, that if this growth occurs, all of the estimated economically available uranium resources would be committed by the mid-1990s, and no more plants could be assured of fuel supplies after that time. If the nuclear components of electricity generation are to continue to grow in the next century, development of the breeder will be necessary.

Technology

The "breeding" of nuclear fuel consists of converting a nonfissionable nucleus such as uranium 238 into a fissionable nucleus such as plutonium 239. For example, in the fissioning of uranium 235, several neutrons are released along with energy and the radioactive fission products. On the average, about 2.5 neutrons are released per fission (which means, of course, sometimes 1, and more often 2, 3, or more, since a fraction of a neutron has no meaning). One of these neutrons is needed to maintain the power-producing chain reaction. The others can be used to form new fissile fuel. The process, in nuclear notation, is

$$U^{238} + n \underset{\text{Rapid}}{\underline{\quad}} U^{239} \underset{\text{Rapid}}{\underline{\quad\text{-----}\quad}} Np^{239} \underset{\text{Slow}}{\underline{\quad}} Pu^{239}$$
$$\text{Rapid} \qquad \text{Rapid} \qquad \text{Slow}$$
$$\text{decay} \qquad \text{decay} \qquad \text{decay}$$

(*Rapid decay* means that the change takes place in less than a month, while *slow decay* takes hundreds or thousands of years. The letter n means a neutron.)

A neutron strikes a uranium 238 nucleus (called a *fertile* nucleus) and changes it into uranium 239, which then changes, through a sequence of spontaneous nuclear reactions, into plutonium 239, a long-lived material that can serve as a fuel reactor.

This process takes place, of course, in conventional reactors. What is important is the ratio of fissioned nuclei to the fertile nuclei converted to fuel. In conventional reactors,

Table 1-1. United States uranium resources, January 1, 1977.

| $/POUND URANIUM COST CATEGORY | RESERVES | POTENTIAL | | | |
		PROBABLE	POSSIBLE	SPECULATIVE	TOTAL
$10	250,000	275,000	115,000	100,000	740,000
$10–$15 Increment	160,000	310,000	375,000	90,000	935,000
$15	410,000	585,000	490,000	190,000	1,675,000
$15–$30 Increment	270,000	505,000	630,000	290,000	1,695,000
$30	680,000	1,090,000	1,120,000	480,000	3,370,000
$30–$50 Increment	160,000	280,000	290,000	70,000	800,000
$50	840,000	1,370,000	1,410,000	550,000	4,170,000
By-product 1975–2000[a]	140,000	–	–	–	140,000
	980,000	1,370,000	1,410,000	550,000	4,310,000

[a]By-product of phosphate and copper production.

this "conversion ratio" is 0.4 to 0.5, which means that one fertile nucleus is converted for every two fissionable nuclei destroyed. In the breeder reactors, this ratio is expected to be 1.1 to 1.3. Thus, in the optimistic case, about four fissionable nuclei are formed for every three destroyed.

A relevant measure of merit of a breeder is the "doubling time," the time needed for it to not only reproduce fuel for its own use, but also to produce enough to fuel another similar reactor. In the first experimental breeder, this figure may be as high as 50 years; advanced breeders have been projected to have doubling times as short as 10 to 15 years—conveniently, the recent doubling time of electric energy consumption. Thus, perhaps breeder reactors can be "turned" (restructured) so that fuel and power production match electric energy demand.

To convert uranium 238 to plutonium 239, "fast" or high energy neutrons are needed, and so the conventional water moderator and cooling fluid cannot be used. Instead molten sodium or perhaps helium will be used.

The fuel assembly and control rods are similar to those in a conventional reactor. Around the fuel region, however, is a "blanket" of uranium 238 assemblies in which the plutonium is formed.

Molten sodium circulates through the core, cooling it and carrying the heat energy away. Since sodium becomes radioactive under the intense neutron bombardment inside the reactor, it must not be allowed to contaminate the rest of the system. A second cooling loop, also sodium, picks up the heat from the first loop and is used to heat the steam.

By using sodium, steam temperatures as high as 1050°F can be reached, and so the conversion of this heat energy to mechanical energy at the steam turbine (and then to electrical energy at the generator) can proceed at fairly high efficiency. The expected efficiency of 40 percent for the breeder reactor will approximate that of the best modern coal-fired plants.

Sodium, however, is a very active metal—for instance, it burns in contact with air or water. It offers both advantages and disadvantages as a coolant. It will have to be kept in liquid form and will require development of appropriate pipes, valves, pumps, etc. Since it has a high boiling point, however, it will be used at low pressures rather than at the high pressures needed for steam. This will make for simplicity and safety.

Much has been learned from experience with small experimental breeders in the United States and from breeder programs in other countries. The more difficult problems remaining are not technological; they are the resource availability problems, which were discussed previously, and economic and social problems, which will be discussed next.

Economics

Proponents of the breeder reactor attribute huge savings to its introduction—from $20 to $50 billion between 1970 and 2020 according to an ERDA study. On the other hand, the reactor's critics see no savings and predict increased costs.

Resolution of these differences depends on such factors as the cost of uranium (which will depend on source availability and total electrical demand) and on the capital cost of the breeder.

The major economic advantage of the breeder is its relative insensitivity to the cost of uranium. As an example, a $10-per-pound rise in the price of uranium 308 increases the cost of electricity from a conventional reactor about 0.15 mills per kilowatt hour (or about 0.05 out of 3 or 4 cents per kilowatt hour). However, the same price rise would only increase breeder-produced electricity 0.015 mills or 0.0015 cents. Therefore, if one anticipates a long-range price squeeze on uranium, then the breeder could save a significant amount of money.

Most of the cost of the breeder is in its construction and, in particular, in the interest paid on the capital investment. Maintenance and fuel costs make up but a small percentage of the total. The amount of capital estimated by the General Accounting Office (GAO) to be needed for a breeder program is staggering: $150 billion before the year 2000. The costs for other types of electricity generation will also be high. The conventional reactors cost about $1 billion per 1000-megawatt breeder, and coal-fired plants about $800 million. It is the expectation of the breeder proponents, of course that, as the price of coal, oil, and uranium increases, utilities will be willing to initially invest more in a plant in return for low 30- to 40-year fuel costs.

The estimate that attributes high dollar savings to the introduction of the breeder makes the following assumption. Electrical demand will grow (until at least the year 2000) at about 5.2 percent per year (compared with 7 percent growth rate in 1960–1970); by the year 2000, about two-thirds of the primary energy will be used to generate electricity, and half of this generation will be nuclear. The cost of uranium will rise steadily. The cost of breeder reactors will drop as more is learned about them and construction and fabrication practices become standardized.

Breeder critics have performed their own cost-benefit analyses with contrasting results. In a report on misplaced federal energy priorities, it is concluded that by 2020, the research and development for the breeder program will cost 10 times as much as the savings. In fact, the "potential" resources will be found and developed; electricity demand will slacken (as it did during the recession years of 1974 and 1975) as the price of electricity continues to rise and more conservation efforts begin to bear fruit; and the breeder costs will not decrease along some "learning curve," but will stay constant or increase. This last argument is supported by experience with present-day reactors; that is, capital costs of the conventional reactor increased from about $200 to $1000 per kilowatt in less than a decade, and only about half of this increase can be attributed to inflation.

There is a related worry over the immense amount of capital that will be needed for the breeder program. The energy industry is the most capital-intensive of all industries, to a large extent employing dollars rather than people. The electric utilities are the most capital-intensive of the energy industries, and the breeder will follow this pattern—and in all likelihood, so will fusion, solar energy farms, or solar satellites. Therefore, it is argued that the dollars invested in the breeder will not only deprive other energy options of research and development money, but will generate much less employment than they would if invested in less centralized energy options (conservation practices, solar home heating, etc.) or in non-energy-related activities.

There is clearly no common ground of agreement on breeder reactor economics. Readers will have to make their own assessment of the different assumptions and projections. In arriving at a final cost-benefit ratio, however, the environmental and social costs will also have to be considered.

Environmental Effects

The breeder offers environmental advantages that go beyond the nuclear vs. coal advantages. Its higher thermal efficiency leads to even smaller thermal waste. It uses uranium so efficiently that it requires even less mining than the conventional reactors.

The decision to commercially develop the breeder will in some ways intensify nuclear disadvantages. The breeder will represent the full maturation of the nuclear program; fission energy will be an important part of the energy mix for a few centuries; and amounts of radioactive wastes, potential radioactive pollution from reprocessing plants, etc., will continue to increase. The major concern, however, will be plutonium; which will be used in much greater amounts.

Safety

The environmental problems posed by the breeder reactor are of two types, technical and social. They are viewed, of course, quite differently by the pro- and anti-breeder camps. Those opposing the breeder assert that use of highly concentrated plutonium 239 (about 20 percent)—instead of the 3 to 4 percent in enriched uranium 235 used in conventional reactors—makes the chance of nuclear explosion greater. Since there are several critical masses of plutonium in a breeder, the concern is that a meltdown might collect enough of it in one place to cause a small nuclear explosion within the plant. Occurrence of a core meltdown is expected to be extremely rare, of course, but if a meltdown does occur it may be more serious in a breeder than in a conventional reactor.

To most critics, however, the most serious problem posed by the breeder reactor is not the accident threat but the threat of deliberate misuse of plutonium. This synthetic material is not only a good fuel for nuclear reactors, it is also the raw material for the nuclear bomb.

This is not a new discovery. The fission reaction in the reactor and in a bomb are basically similar. In the former, it is slowed and controlled. Any material that can fuel a reactor can, in theory, be used in a bomb. In practice, of the three fissile elements—uranium 235, uranium 233 (which can be bred from thorium), and plutonium 239—only

plutonium 239 is likely to be of use to a clandestine bomb-maker because it can be chemically separated from the other fuel materials.

Nevertheless, as it goes to the reactor and travels between the reprocessing plant and the fuel fabrication plant, plutonium could be the target of a hijacking operation. Likewise, without proper safeguards, a breeder reactor or plutonium-containing fuel sold to another country could be the equivalent of an invitation into the nuclear club. Refinement of the raw material is the only really difficult step in the fabrication of a nuclear weapon. India created its bomb out of plutonium reprocessed from research reactor fuel, and several undergraduate science students in the United States have designed nuclear bombs that experts judged to be probably workable. (Other experts suggested that the students would probably have been killed in the process of making the bombs.)

The "plutonium economy," as this anticipated new nuclear age is being called, will present problems that are different from any we now face. First, much greater security will be required over the vulnerable parts of the nuclear fuel cycle. Plutonium is presently formed in special reactors and separated out for weapons used as a part of the United States defense program. Since no commercial reprocessing plant is currently in operation, the plutonium remains in the stored spent fuel, amply protected from misuse (except with great difficulty) by the radioactive fission products mixed in with it.

In the future, there will be an increase in plutonium in transit and in storage if it is recycled as planned. Significant quantities of plutonium—about 200 kilograms per year per 1000 megawatts—are bred from uranium 238 in present reactors. Thus, the 42,000 megawatts of nuclear generation now in operation produce 4 to 5 tons of plutonium per year. The breeder yields about the same net plutonium production, but maintains about five times as much in the fuel cycle as does a conventional reactor. With the breeder program in full bloom, however, the output plutonium (net production plus recycled fuel) will be measured in thousands of tons. An Atomic Energy Commission (AEC) projection for the year 2020, with half of the reactors being breeders, is for 30,000 tons of plutonium to be produced per year. It takes only 4 to 10 kilograms (13 to 22 pounds) of plutonium to make a nuclear bomb, so scarcity will not be the problem. The potential danger of plutonium proliferation is yet to be solved. Stepped-up security and concentration of reactors and processing plants in a single, easily guarded area is one proposal. A more imaginative one is to forego the uranium 238 to plutonium 239 breeding reaction entirely and to use instead thorium 232 as the fertile element, converting it in a breeder to uranium 233. Thorium is a least as abundant as uranium. Uranium 233, if carefully controlled and guarded (it is also potential bomb material), could be blended with uranium 238 to thus frustrate any would-be bombmakers. Whatever the solution, an agreed-to and workable international plan to keep plutonium from criminal individuals or irresponsible nations must be established before the United States enters into a large-scale breeder program.

COAL

Coal, the first of the fossil fuels to become industrially important, is still, on both national and international scales, a most abundant resource. During very early experimentation, it was realized that useful, clean fuels could be made from it. Kerosene—still called "coal oil" in the rural United States—was produced from coal in the mid-nineteenth century, and "town gas" or "water gas" produced from coal provided the fuel for urban lighting and cooking until the late 1800s and early 1900s. The discovery of large quantities of easily obtainable oil and natural gas led to the replacement of coal and coal-based fuels in the twentieth century. Germany, forced by the Allied blockade and European geology to find a substitute for oil during World War II, demonstrated on a large scale the technical feasibility of obtaining liquid and gaseous fuel from coal. The return of cheap oil and gas to the marketplace after the war led again to a general demise in the importance of "King Coal."

A major research and development effort is now underway to produce cost-competitive gaseous and liquid fuels from coal. This resurgence in interest seems likely to continue as long as there are coal supplies. The new impetus is provided by the realization that the end of the petroleum era is near. National and international oil resources, depleted by the rapidly increasing consumption of the past 50 years, will barely last beyond the turn of the century. In addition, increasing scarcity and the near-monopoly enjoyed by the OPEC nations have caused the price of oil products to skyrocket. As a result, coal and coal-derived fuels can once again compete for the energy dollar.

Resources

Coal is our most abundant resource. Listed in Table 1-2 are the "ultimately recoverable" resources—that is, the total resource, discovered and potential, which is thought to be economically exploitable with present technology. As can be seen, the coal resource is 10 times larger than the others.

In 1975 United States coal production was 646 million tons. *At that rate of use,* American coal resources will last about 600 years. The rate of use will increase, of course, especially if new fuels are successfully produced from coal. Even if coal use increases exponentially by only 2 percent per year, coal supplies will last only 100 to 150 years.

Technology

Coal is almost entirely carbon and contains only small amounts of hydrogen, oxygen, sulfur, and nitrogen. Synthetic fuels are hydrocarbons—molecules made up of hydrogen and carbon. Therefore, the basic conversion task (or gasification of coal) is to add hydrogen to the carbon atoms.

The following numbers give some idea of how much hydrogen is needed. In coal, there are 16 carbon atoms to each hydrogen atom; in heavy fuel oil, the carbon-to-hydrogen ratio is 6 to 1; in gasoline, there is about 1 carbon atom to every 2 or 3 hydrogen atoms; and in natural gas (methane), the ratio is 1 to 4. Thus, to make any of these fuels from coal, a rather large amount of hydrogen must be supplied.

The four primary ingredients of fuel are carbon, hydrogen, oxygen, and heat. Coal supplies the carbon and some of the hydrogen. Additional hydrogen is usually provided either by breaking down natural gas or by reacting coal with steam. Oxygen is supplied either in a pure form or as air, and the heat is supplied by burning coal or a by-product gas.

The basic chemical reaction for coal gasification is:

Coal + Water = Heat → Carbon monoxide + Hydrogen

The "power gas," a mixture of carbon monoxide and hydrogen, is combustible but has a low heat (low Btu) value—about one-third that of methane. It can be burned onsite, but long-distance pipeline shipping is uneconomical.

To make a natural gas substitute, the power gas undergoes a second "methanation" step in which the carbon monoxide and hydrogen are combined in the presence of a catalyst to make methane, the chief component of natural gas. There is an intermediate step in which the carbon monoxide reacts with steam to produce carbon dioxide and hydrogen. The carbon dioxide is then removed along with another gas, hydrogen sulfide, which is formed by combining hydrogen with the sulfur impurities in coal. Thus, sulfur, the most damaging pollutant in coal is not present in the gaseous fuel produced from it. Furthermore, the hydrogen sulfide can be easily converted to elemental sulfur and disposed of or sold.

The steps just outlined are basic to all gasification processes. The processes differ in how the heat is obtained and in such details as operating temperatures and pressures, and whether pure hydrogen and/or oxygen is required.

Synthetic Oils and New Solid Fuels. Two other types of fuel are being developed from coal—liquid fuels, and a solid fuel with higher energy content and less sulfur and ash impurities.

Table 1-2. Ultimately recoverable resources.

RESOURCE	AMOUNT RECOVERABLE	POTENTIAL ENERGY (CALORIES)
Coal	390 billion tons	2380×10^{15}
Oil	112–189 billion barrels	$168–284 \times 10^{15}$
Natural gas	761–1094 trillion cubic feet	$198–284 \times 10^{15}$

Producing a liquid fuel from coal encounters the same problem that occurred in gasification. The 16 to 1 ratio of carbon to hydrogen needs to be lowered, at least to the 6 to 1 ratio of fuel oil. This reduction can be accomplished by one of three basic processes: *hydrogenation,* the addition of pure hydrogen; *pyrolysis,* the heating of coal in the absence of oxygen; and *catalytic conversions* between carbon monoxide and hydrogen.

Hydrogenation produces a heavy fuel oil that is usable in power plants. Pyrolysis produces three fuels: high-Btu or pipeline gas, a synthetic crude oil (syncrude), and char, a carbon residue that can be burned itself (as long as it does not contain too much sulfur). Hydrogen is added in the hydrogenation process, whereas, in the pyrolysis process, the coal molecules—which contain some hydrocarbons—are reformed into the hydrocarbon molecules, and the excess carbon is removed. In the catalytic conversion process, the gasifying reaction described earlier takes place, producing carbon monoxide and hydrogen, which are combined under pressure and in the presence of a catalyst to make liquid fuels.

The process of producing a solid fuel from coal is called *solvent refining;* the product is *solvent refined coal (SRC).* In the process, crushed coal is mixed with a hydrocarbon solvent, such as light oil, at high temperature and pressure. The coal dissolves, and the ash and much of the sulfur can then be filtered out. The fuel, which is like solid tar, can be pulverized or heated and melted, and thus piped if desired. Its heating value is far greater per ton than that of coal and is a much cleaner fuel.

A summary of the ranges of conversion efficiencies to be expected follows:

Process	Efficiency (percent)
Solvent refined solids	65 to 70
Liquefaction	62 to 69
Low-Btu gasification	65 to 95
High-Btu gasification	54 to 68

While low-Btu gas and solvent-refined coal have the highest conversion efficiencies, it must be remembered that these fuels will be used, for the most part, in the inefficient (about 30 percent) conversion of heat to electrical energy. Overall, the reliance on coal-based synthetic fuels will lower the efficiency with which we use our primary energy resources. Given the dwindling supplies of oil and natural gas, however, this inefficient use of coal seems necessary.

Economics

That coal can be used to supplement dwindling domestic supplies of oil and natural gas is beyond question. Whether it will be used in this way is another question.

As with any new technology, the start-up costs are phenomenal. Without government incentives, private companies will not make the investments necessary to build synthetic fuel plants (over $1 billion each), especially when the market for the fuels is an unknown quantity. However, with the price of oil steadily on the increase as the supply dwindles, the economic picture for developing coal's synthetic fuels brightens.

Environmental Problems

The most serious problem faced by the synthetic fuel plants, the gasifiers in particular, is the need for water. Not only is water one of the raw materials in the conversion process (it provides the hydrogen), but it also carries out its usual cooling role. The amount of water needed will be 1.5 to 3 pounds (or about 2 quarts) for each pound of coal processed. This is about twice the needs of an electric power plant of the same energy output.

Unfortunately, most of the coal that is targeted for gasification is either in the arid Southwest, or in the equally arid north central region (the Dakotas, Montana, and Wyoming). Some hard decisions will have to be made between agricultural, industrial, and synthetic fuel needs for water in these areas, where agriculture is the dominant factor.

A large amount of solid waste will be generated along with the gas. Presumably, this will be dumped back into the mine. The air emissions will require both particulate and sulfur oxide removal units since the potential air emissions are large. As reported earlier, the process itself allows the sulfur to be removed as hydrogen sulfide.

Land Use. Most of the coal for the gasification plants presently under construction will be strip-mined. (To provide the tons of coal per day needed by a typical processing plant from a coal seam 10 feet thick—typical of southwestern coal deposits—almost 400 acres per year will have to be stripped.) This is an economic necessity since surface-mined coal is about 20 percent less expensive than deep-mined coal. The problems associated with strip mining will be aggravated because irrigation will probably be required for land reclamation in the arid Southwest and North Central coal areas.

Carcinogens. A new worry for the synthetic fuel industries has recently surfaced. Because of the high temperatures used in these coal conversion processes, molecules known as *polycyclic aromatic hydrocarbons (PAH)* are formed. There is ample evidence that many of these PAHs are carcinogenic—that is, cancer-causing.

The evidence comes not only from observations of workers in coal liquefaction plants, but also from studies of workers in existing goal gasification plants. The process for producing oil distilled from oil shale also produces PAHs. In one study, it was reported that a worker exposed to shale oil was 50 times more likely to get skin cancer than one exposed to oil from a Pennsylvania well. Coal-produced synthetics seem, from the evidence, to be more dangerous to health than are petroleum products.

Summary

The technologies for coal gasification and liquefaction, and for producing clean, high-Btu solid forms of fuel have been demonstrated in laboratory and small pilot plant processes. These synthetic fuels are not presently economically competitive with other primary energy forms, particularly oil or natural gas.

If the federal government wishes to assure the development of large-scale demonstration plants, it must depart from its usual role of supporting only research and pilot plant development and underwrite the construction and operation of some large-scale demonstration plants.

CONVENTIONAL REACTORS

The technology of conventional nuclear reactors was already understood in principle by the early 1940s, and the first operating commercial plant went on-line in 1957.

However, conventional reactors are still relatively new to the energy scene, and their proponents expect them to play a major energy role in the 1980s producing as much as 20 percent of our electric energy by 1985. They are also the most controversial component of the different types of energy sources, bitterly opposed by some and just as strongly championed by others.

Most of our present reactors are "converter reactors," which convert fissionable fuel into energy. The "breeder reactors," on the other hand, can produce new fuel as well as electric power.

Resources

Present-day converter reactors obtain energy from the fissioning of a rare form of uranium, uranium 235. This isotope makes up less than 1 percent of common uranium ore, most of which is uranium 238. Uranium 235 releases energy when the nucleus undergoes fission. The fissioning of 1 pound of uranium could produce 9 billion calories of energy—3 million times the energy released by the burning of a pound of coal. As will be seen, however, the actual comparison is much less dramatic.

Uranium is found in small deposits in most rock. To be mined, however, the deposit must be concentrated. Commercial ore, on the average, contains 4 or 5 pounds of uranium oxide (the commonly occurring uranium compound). Of these 4 or 5 pounds, only 0.02 to 0.03 pounds are the sought-after uranium 235.

To produce uranium fuel, ore that is rich enough to mine must be located first. The raw ore is then "milled," concentrated to a product called *yellowcake*, which is 80 percent uranium oxide. A 1000-megawatt reactor uses 170 tons of yellowcake per year at 80 percent efficiency. If such a reactor were to operate at 70 percent efficiency, it would use 130 tons of yellowcake per year.

ERDA estimated uranium reserves (identified resources) are for 430,000 tons available at under $15 per pound and 640,000 tons plus 140,000 tons of by-product uranium oxide at under $30 per pound. (The by-product is uranium that could be produced during the mining and refining of other materials.) Potential resources (not yet discovered and developed) are estimated to be 2,050,000 tons at under $15 per pound and 3,700,000 tons at under $30 per pound.

These tons of uranium represent an enormous amount of energy. However, reactors use only a small percentage of their fuel. A practical assessment of the energy equivalent of these resources can be obtained using the 130 tons annual requirement of yellowcake in a 1000-megawatt reactor. For instance, we can calculate that the 45,500 megawatts of reactors presently operating will use up the 420,000 tons of $15 per pound reserves in 73 years.

This calculation suggests that the resource lifetime of uranium may be comparable to that of oil or natural gas, which is supported by studies that take into account the expected nuclear growth. ERDA estimates that all of the $30-per-pound reserves will be used up during the 30-year operation lifetime of the reactors presently in operation, under construction, or on order (a total generating capacity of about 235,000-megawatts). Even this estimate counts on the recycling of plutonium; without this process, the under-$30 reserves will only fuel about half this amount of generating capacity.

So far, we have considered only the reserves. If the potential resources are as large as projected, the lifetime becomes longer. The whole subject of uranium resources is controversial, and it enters strongly into the controversy over the breeder reactor. In summary, uranium supplies seem to be sufficient to fuel the planned reactor program into the next century; however, without new supplies or a new technology (such as the breeder reactor), uranium resources will provide only temporary relief for the energy shortage.

Technology

This section presents an overall view of the step-by-step movement of uranium from mine to reactor, etc. Some additional technological details will be developed in "Safety" (p. 15).

Since the raw ore contains only 2 percent or 3 percent of uranium, 100,000 or more tons of ore must be mined to provide 150 to 200 tons of uranium oxide, which is about the annual requirement of a 1000-megawatt nuclear reactor. The ore is refined and concentrated in the milling operation. It is then converted to a gas, uranium hexafluoride, which is the form in which enrichment takes place. In the enrichment plant, the amount of uranium 235 is increased from the original 0.7 percent to 3 or 4 percent. This enriched uranium is then assembled into the fuel elements, while the uranium that has been depleted in uranium 235 (which is largely uranium 238) is stored. It is from this stockpile that the breeder reactor may draw.

After a year's operation, the fuel is removed from the reactor. The present practice is to replace one-third of the reactor fuel each year. The initial loading of the 1000-mega-watt reactor would therefore take three times the amounts of yellowcake, uranium, etc.

Economics

In 1970 and 1971, the capital costs of generating plants were commonly quoted as $250 and $200 per kilowatt for nuclear plants and coal-fired plants, respectively. In 1975, Southern California Edison estimated that in 10 years nuclear plants would be delivered at $940 per kilowatt and coal-fired plants at $860 per kilowatt. While these costs (which mean that a 1000-megawatt plant now requires almost a billion dollars to build) may not be exact national averages, they do correctly reflect the remarkable inflation that has taken place. In addition, wages have increased, the duration of the construction has lengthened, materials are more expensive, and new safety devices and antipollution equipment are required. Moreover, a major factor is the cost of the money itself, that is, the large increase in interest rates. The operating and maintenance costs and fuel costs must also be figured in. If the interest cost is placed somewhere in the future—1986, for instance—it is very difficult to produce an acceptable projection because there are so many variables. Table 1-3 gives representative 1975 costs and projections of costs in 1985 for the atomic industry. In this comparison, the nuclear reactor produces the least expensive electricity.

However, these figures are vigorously questioned by nuclear power opponents. There is not even agreement on the historical data, let alone on the projects.

The 1975 data in Table 1-3 show nuclear-produced electricity averaging 13.9 mill per kilowatt-hour and coal-produced electricity (without scrubbers) averaging 18.1. The overall industry averages reported in *Electrical World* (November 15, 1975) for a 2-year period are: nuclear, 18 mill; coal, 13.8 mill. It seems to make a difference if only those plants that have both nuclear and coal-fired plants (as in Table 1-3) are surveyed or if total industry data are used.

The projections are even more open to controversy because values for interest rates, inflation, fuel costs, etc., must all be assumed. Much of the controversy has centered on the selection of an appropriate "capacity factor," which measures the ratio of the electric energy actually produced to the total rated capacity of the plant. So far, the larger plants (over 800 megawatts) are less reliable whether they are coal-fired or nuclear. If a lower capacity factor is assumed for nuclear, the mill/kilowatt-hour nuclear advantage weakens.

In summary, nuclear and coal-fired plants are economically competitive. Which one has the advantage depends on many factors, including air-pollution control requirements, cost of fuel, and cost of money. Probably the only thing that we can be certain of is that the cost of electric energy produced by either plant will increase rapidly over the next decade.

Environmental Effects

If nuclear reactors come to dominate electric power production, they will ease several perplexing environmental problems. They do not produce the air pollutants associated with coal or other fossil-fuel plants. In addition, the amount of ore to be mined per year—10,000 tons of uranium vs. 2,500,000 tons of coal for a 1000-megawatt plant—would help to reduce the concern over strip mining. However, reactors have their own environmental impact.

Radioactive Pollution. Radioactivity is the by-product of the fission process. Fission fragments are radioactive—they emit energetic particles of radiation, which can penetrate tissue or other matter. The amount of radioactive material that accumulates during a year's operation of a reactor is enormous. As an example, if half of the strontium 90 (one of the more hazardous radioactive waste products) that is accumulated in one year of operation by a 1000-megawatt plant were to be sprinkled over the country, it could contaminate all the freshwater runoff in the continental United States to six times the official "maximum permissible concentration" of that contaminant.

Such selective deposition is not possible, of course, and radioactive pollution is rigorously controlled at the reactors. The regulatory guidelines are quite strict, and a person could live quite safely (as far as routine emission of radioactivity

Table 1-3. Economic comparison of two 100-megawatt generating units (mill/kilowatt-hour).

	YEAR	OIL	COAL-FIRED SCRUBBERS	WITH TALL STACKS	NUCLEAR
Capital	1975	4.1	6.1	4.5	9.5
	1985	14.7	20.8	18.1	27.4
Fuel	1975	20.9	13.3	12.3	2.9
	1985	32.4	25.0	22.7	6.9
Operating and Maintenance	1975	1.2	2.7	1.3	1.5
	1985	2.4	5.0	2.6	2.5
Total	1975	26.2	22.1	18.1	13.9
	1985	49.5	50.8	43.4	36.8

Source: V.S. Boyer, Philadelphia Electric Co.

is concerned) right at the plant boundary line. The exposure caused by the routine operation of even the several hundred reactors projected for the future will still be relatively small.

Unlike control at the reactors, the control over releases from reprocessing plants is not as satisfactory. In some instances, releases of some gaseous radioactive products have been as large as 20 percent of the allowed maximum. Clearly, more stringent safety measures are needed here as more and larger reprocessing plants go into operation.

Radioactive Waste. The leftovers, the "ashes" of the conversion process, are radioactive—and they remain so for a long time. The crucial parameter of a radioactive substance is its "half-life," the time during which half of a sample of the material will "decay," or change into something else. Radioactive species with short half-lives are intensely dangerous for short times, whereas longer half-lives mean that the radiation is less intense and is spread out over a longer time.

In assessing the danger to humans, the biological properties of a contaminant have to be considered, in addition to the half-life. Several of the fission products, notably radioactive forms of strontium and cesium, have dangerous combinations of half-lives (about 30 years) and biological properties. Thus, they must be kept completely out of the ecosystem for hundreds of years.

It is this necessity for complete isolation that is the challenge of radioactive waste disposal. It will have to be stored and monitored for several generations. There is no precedent in our history for such a task. It is an ethical matter because we are, in effect, leaving this material to our descendants.

The technological solution seems to be easier. One plan being considered calls for flash-drying the high-level radioactive liquid waste to a ceramic material with a large reduction in volume. These "hot rocks" would then be stored for at least a few hundred years in some dry and geologically isolated spot—a salt mine, for instance.

Since the annual amount of dried high-level radioactive waste from a 1000-megawatt plant occupies only about 60 cubic feet of space (a cube measuring a little less than 4 feet on each side), the requirements of storage space will not be overwhelming. Of course, the waste cannot be stored this compactly because it is physically hot as well as radioactively "hot."

The nuclear critics point out that no agreed-to official plan exists and that the waste is not yet being processed in this way. Instead, it is rapidly piling up in temporary storage. The resolution must come quickly if the reactor program is to grow at the projected rate. The problem must be faced, however, even without considering electricity generation, because of the waste produced in the manufacture of nuclear weapons.

Safety

The most emotion-charged issue in the nuclear controversy is safety. Safety involves the safe transport of radioactive fuels and waste material, the protection of nuclear fuels from theft by would-be bombmakers, and the safety of the nuclear plants themselves.

Although there is a growing concern over the increasing volume of radioactive materials on our roads and railways—52 shipments per day by 1980—the great care exercised in the transportation of nuclear material so far has prevented serious accidents, and this problem seems manageable. However, the danger of theft seems more threatening. It is a threat that may grow if plutonium recycling and the breeder program are instituted.

Nuclear Accidents. It is not a nuclear explosion that is feared because the uranium 235 is not present in sufficient concentration to make a "critical mass" and produce a nuclear explosion. The greatest danger perceived is that a reactor will lose its cooling water, melt down, and either melt through its container or fracture it by a steam explosion. If this happens, the lethal contents may be released to the countryside.

How could such an accident occur? The nucleus (uranium 235, in most cases) absorbs a neutron and undergoes fission. Energy is released along with, on the average, 2.5 additional neutrons (that is sometimes 1, sometimes 2 or 3). These neutrons sustain the reaction. At least one must be used to cause another fission; however, if two fissions are caused, a chain reaction builds up.

Control is obtained through a series of "control rods," which are similar to the fuel rods that make up the reactor core, and are interspersed among the fuel rods. The control rods are filled with a neutron-absorbing material, and the number of neutrons available for fission is regulated by moving them in and out of the core. The reactor can be shut down, or "scrammed," by running all of the control rods into the core.

The reactor remains very hot for some time after shutdown and cooling is thus extremely important. Without a cooling system, a reactor would melt in less than a minute. To prevent this, there is, in addition to the regular cooling system (the water that carries the heat energy to the steam generator), a backup "emergency core cooling system" designed to dump an enormous amount of water into the reactor if the regular system fails. Much of the controversy over reactor safety has focuses on this emergency system, which has never been fully tested. (The opponents of nuclear power point to Three Mile Island in Pennsylvania and say, indeed, there are obvious safety hazards.)

In general, reactors are designed with several layers of safety systems, and their record so fair is impressive. There have been no radiation-related fatalities or injuries at a commercial reactor. In fact, this very lack of experience contributes to the uncertainty of some opponents.

It is extremely difficult to arrive at a consensus assessment of the accident hazard. Nuclear proponents point to the record of almost 1000 reactor years of worldwide commercial operations without a fatal accident and to studies that give a probability of 1 in 100 million per reactor year that an accident causing 1200 or more deaths annually will occur.

The critics pointing to the assumptions and estimates in the calculation and the lack of actual experience, question both the probability calculations and their implications. They cite the many minor accidents that have occurred due to human and equipment malfunction that were not foreseen by regulations and elaborate safety systems. They also point out that 1000 reactors operating for 10 years lower the safety margin offered by the odds to 1 in 10,000.

The prudent course of action would seem to be to recognize the great danger of this form of energy conversion, keep reactors away from the cities, continue to enforce stringent safety and security regulations, and look to new ways to make them even safer.

Summary

Nuclear energy is like a genie released from a lamp, offering great power at a time when traditional sources of energy and power are disappearing. It is a genie needing rigorous control.

The genie has been released, and the country's nuclear generating capacity may grow to 200 to 300 thousand megawatts by 1990. But the road to a nuclear future is not smooth. More uranium will have to be found, or the second generation, the breeder reactors, will have to provide the fuel. The ultimate cost comparison with coal as an energy source is still uncertain. It is clear that the capital requirements (now at $1 billion per 1000-megawatt plant) are enormous. And there are hazards involved, as discussed previously.

FLYWHEELS: STORING ENERGY AS MOTION

A child's yo-yo, the propulsion mechanism of a toy friction car, a potter's wheel—all are commonplace examples of flywheels. A flywheel is a device that stores mechanical energy, absorbing it by spinning faster and giving it up by slowing down. Larger flywheels being developed today with new materials and in new shapes promise increased energy-storing capacity. These new high-performance flywheels, designed to do bigger tasks, such as powering cars and providing supplemental power at electric power plants, could help the United States to stretch its limited supplies of oil and natural gas.

Past Uses of Flywheels

A familiar form of the flywheel is a flat "pancake" that spins around a perpendicular axis, such as a potter's wheel. For thousands of years, potters' wheels have been used to store and smooth the irregular spinning and pedaling energy of potters trying to form circular vessels. About a hundred years ago, flywheels spun up by an external steam turbine were used to propel naval torpedoes over long distances. In the 1950s, a Swiss company developing a small flywheel-powered railroad engine for switchyard work saw the possibility of using flywheels in public transportation. The company developed 35-passenger buses that were used in

Switzerland and the Congo. About every half-mile, the flywheel would be spun for 1 to 2 minutes by a motor that was powered by electricity from overhead trolley wires. Although the buses operated successfully, they were replaced after about 10 years by cheaper, more convenient diesel buses that could operate independent of any charging apparatus.

In industrial equipment and vehicle engines, a flywheel is used to smooth out, or "level," variations in power from an intermittent source. For example, the internal combustion engine in a conventional car runs on energy from a series of small explosions in the cylinders. A flywheel on the main shaft smooths out the bursts of energy from the explosion.

Pros and Cons

Compared to batteries, which store energy chemically and transform that energy directly into electricity, flywheels have certain advantages. They can be charged and discharged quickly, and they have an extremely large number of charge/discharge cycles. However, storing energy in flywheels is expensive because materials and fabrication costs are high, and the devices store no more energy per pound than batteries. And there is the hazard associated with a wheel suddenly flying apart when small flaws lead to rapid failures.

Reinventing the Wheel

Capitalizing on the inherent advantages of flywheels will require "reinventing the wheel." This will entail improving the wheel itself, as well as the other components that make up a flywheel system (for example, bearings and suspension systems, housing, and motor/generators and their controls). In addition, integrated systems for specific applications must be developed.

In the past, most flywheels have been made of steel. More recently, titanium alloys have also been used. Newer composite materials—many developed for the American space program—are being investigated. They are made by embedding into a plastic or resin high-strength fibers such as fiberglass, graphite, and Kevlar, a new organic fiber material used to add strength to automobile tires.

Improving the flywheel so that it can serve as a modern and efficient storage device involves trade-offs in weight, volume, and cost. Metal flywheels, although heavier than those made of composite materials, are more compact and cost less. Both metal and composite materials will probably be used in the future. Metal will be the likeliest choice for stationary flywheel applications where weight is not a concern, and composite materials will be used for large mobile applications where lighter weight is important. Flywheel materials can be fabricated, not only in the familiar pancake shape, but in other shapes as well. The amount of energy stored depends on the flywheel's shape and weight, and, most important, the spinning speed. The energy stored is proportional to the square of the flywheel speed—that is, a wheel spinning at three times the speed of a second identical wheel stores nine times the energy of the second. Because composite materials have higher strengths than most metals, they can be spun faster and hence can store more energy than metals of equal weight.

High-speed flywheels, however, are not without problems. One problem is friction, which slows the spinning flywheel. Thus, flywheels are usually housed in vacuum chambers to reduce the air drag and allow the wheel to spin longer. Also, the bearings supporting moving parts must be durable and able to minimize the energy losses in friction.

Another consequence of high speeds is the large centrifugal forces that tend to make the wheel fly apart. Thus, spinning speeds are kept well below the maximum. Wheels of composite materials are safer than others because, if they fail, they disintegrate into a substance resembling cotton candy. This has been verified in laboratory testing sponsored by DOE. On the other hand, when a metal flywheel fails, it almost explodes, usually throwing off high-speed projectiles. Housings, which can be a part of or separate from the vacuum housing, are provided to contain the pieces in the event of failure. The probability of failure can be reduced by constructing wheels with two or more segments, for separate segments are unlikely to fail simultaneously.

New Uses

A number of new applications for flywheels are under investigation. During the mid-1970s, the U.S. Department of Transportation (DOT) supported the demonstration of a flywheel propulsion system on two New York City subway cars. As a subway car is brought to a stop, a considerable amount of energy (heat of friction in the brakes) is normally lost. With the flywheel propulsion system, the forward motion of the car is transferred to the flywheel, slowing down the car and storing the energy of deceleration. Later, this energy is "regenerated" to help accelerate the car, replacing some of the electric power previously needed. The flywheel system cuts subway power costs by 20 to 30 percent and reduces heat in the subway tunnels. However, ways must be found to enable financially pressed transit companies to purchase flywheel propulsion systems.

The regenerative flywheel/battery combination is a clean (no combustion products), reliable, and silent (no explosions in the cylinders) propulsion system of great practical value for vehicles. It is especially suited for city driving, it does not idle, which wastes energy when the vehicle is stopped at a light or in heavy traffic, and it can be restarted quickly.

Flywheels for vehicle propulsion could have a tremendous impact on oil consumption. If, by the year 2000, only 15 percent of American automobiles are powered by hybrid flywheel propulsion systems, the oil savings could reach 300 to 500 million barrels per year. The savings would result from reduced use of conventional fuel-burning engines, use of off-peak electricity (from coal, nuclear, or hydro plants) to spin up flywheels, and use of the energy normally lost during braking for spin up.

Besides vehicle propulsion, flywheels may be used to store energy captured by wind turbines. Since wind blows intermittently, any wind energy that exceeds the demand for power can be stored for use when the wind is not blowing. By the early 1980s, flywheels might be used in areas where electricity rates are higher during peak daylight hours. Lower-cost electricity would spin up flywheels at night to provide power during the day.

Flywheels for use in power plants will have to be able to store far more energy than those used for applications discussed previously. Utility scale flywheels may have to be made of composite materials, since metal flywheels of the required size would be very heavy and uneconomical. Even made of lighter composites, these flywheels will probably weigh 150,000 pounds and measure 15 feet in diameter and 10 to 20 feet thick. The flywheels would be spun up at night with equipment that would otherwise be idle. Being able to store the excess energy generated during periods of low demand for use during periods of high demand—a process called *load-leveling*—means that the plant can always operate at its highest, most efficient rate and so avoid wasting energy. So much new technology will have to be developed that it will be 1985 at the earliest before load-leveling with flywheels can be demonstrated.

FUEL CELLS: A NEW KIND OF POWER PLANT

The fuel-cell concept itself is not new, such cells have already provided power for moon landings and, between 1971 and 1973, provided electric power to 50 apartment houses, commercial establishments, and small industrial buildings. What is new is an effort to capitalize on the fuel cell's inherent flexibility, safety, and efficiency by putting together a generating system that can use a variety of fuels to meet today's utility-scale power needs economically.

Fuel Cell Operation

In the combustion process used in conventional power plants, power generation requires three steps: (1) fuel and air combine releasing their stored chemical energy as heat energy which (2) is used to make steam, and (3) the mechanical energy of steam turning a turbine is converted to electric energy.

In the electrochemical process that takes place in a fuel cell, the chemical energy that bonds atoms of hydrogen (from a hydrocarbon fuel such as coal, oil, or gas) and oxygen (from air) is converted directly to electrical energy.

A fuel cell is a sandwich consisting of an anode (a negatively charged terminal), an electrolyte (a subsystem that becomes electricity-conducting), and a cathode (a positively charged terminal), which is much like a battery. Hydrogen-rich fuel is fed down the anode side of the cell, where the hydrogen loses its electrons, leaving the anode with a negative charge. Air is fed down the cathode side, where its oxygen picks up the electrons, leaving the cathode with a positive charge. The excess electrons at the anode flow toward the cathode, creating electric power. Meanwhile, hydrogen ions produced at the anode (when electrons are lost) and oxygen ions from the cathode migrate together in the electrolyte. When these ions combine, they form water, which leaves the cells as steam because of the heat of the cell processes.

Basic Fuel Cell System

The fuel-cell generator system has three parts: a fuel processor, a fuel cell stack, and a power inverter. Hydrocarbon fuel and steam (recycled from the fuel-cell operation) are first fed into the fuel processor and converted to hydrogen and carbon dioxide. This hydrogen-rich mixture is then fed into the fuel-cell stack, where the fuel cells are piled one on top of another. As electricity is produced, exhaust from the stack includes carbon dioxide, nitrogen, and water condensed from steam.

The dc-power electricity from fuel-cell electrochemistry must be converted to ac power for utility applications. This conversion takes place in the third part of the fuel-cell system, the power inverter, which can convert large amounts of dc power to ac at nearly 96 percent efficiency.

The three-segment fuel-cell system can already produce electricity, from fuel to utility grid, at an efficiency of 38-40 percent, which is comparable to the best conventional combustion plant. By 1985, this efficiency is expected to be 50 to 55 percent. In addition, emission from the generating process is 10 times cleaner than the U.S. Environmental Protection Agency (EPA) requires.

Advantages

In a time when power plant costs are high, construction lead times are long, and future power demand is uncertain, the flexibility of fuel-cell systems is a major attraction. Like the fuel cells themselves, all three segments of a fuel-cell system are modular and can be connected in parallel to meet additional power demand. Standardized fuel processors, fuel cell stacks, and power inverters could be factory-built and trucked to the plant site. This modularity could mean lower cost, shorter plant construction lead time, and greater flexibility in the size of the plant.

Fuel-cell systems, in various sizes, could serve at any point in a utility system, from a central power station to a user site (e.g., home). Siting close to point of use can reduce transmission losses (now about 8 to 10 percent of power transmitted) and cut the costs for transmission lines, which add about $50 to $100 per kilowatt to the cost of present plant construction. Onsite use allows additional fuel savings through constructive use of system heat usually discarded in central generating systems.

Fuel-cell system modularity also affects system efficiency. A large-scale combustion plant is inefficient at part-load operation because some of its expensively built capacity is not being used. Fuel-cell plants, however, operate equally efficiently at part or full load. Fuel cells could also provide

"instant response," generating electricity from the moment they are turned on.

Besides physical and operating flexibility, fuel flexibility is a major advantage of fuel-cell systems. Fuel processors already accept a variety of hydrocarbon fuels, including light distillates, natural gas, methanol, and high-, medium-, and low-Btu gases. By 1985, fuel processors should be able to accept synthetic fuel products from the nation's more plentiful coal supplies. Also expected during 1985 is development of a more advanced technology for integrating fuel-cell systems directly with coal gasification units.

The greater efficiency of the electrochemical process gives fuel-cell power plants advantages in operating costs. With the installation of 20,000 megawatts of fuel-cell power, savings of $1 billion per year in electrical generating costs could be expected by 1985. This includes a yearly fuel saving equivalent to more than 100 million barrels of oil. To achieve these benefits, further research and development must be done to lower the installed cost and to increase the reliablity and durability of these fuel-cell systems.

Such environmental considerations as low water requirements, limited emissions, and quiet operation make fuel-cell plants an attractive power option. Where fossil-fuel and nuclear plants require large quantities of water for cooling, fuel cells, which generate less heat, will be air-cooled by low-speed fans. Because fuel cells can use a variety of hydrocarbon fuels, they share, with conventional generating processes, the environmental problems currently associated with extracting and processing fossil fuels. However, since fuel cells do not involve a combustion process, emissions from their operation are significantly lower than emissions from conventional plants and well within EPA requirements. And fuel-cell plants operate with very little noise, making them environmentally "good neighbors."

Application

The range of sizes, the modularity, and the environmental advantages make the fuel-cell system a candidate for power generation in a variety of utility applications:

- upgrading old urban plants, using existing sites more efficiently with decreased environmental impact.
- supplying new generating capacity where environmental considerations restrict combustion plants (especially when transmission right-of-way is limited and plants must be sited close to population areas)
- complementing the peak load capacity of existing power systems, where quick response and part-power efficiency are required
- supplying power for small and medium-sized municipal and rural utilities under 100 megawatts, a range in which other types of power plants cannot operate as efficiently.

Because they can produce electricity efficiently on both small and large scales, fuel-cell systems are also candidates for onsite power generation. Eighty percent of the commercial and multiunit residential buildings built in the United States annually have a maximum power rating under 100 kilowatts. Onsite fuel cells could save 25 to 30 percent of the fuel required to supply electricity to such buildings. Recovering the heat of by-products for space and water heating could further stretch fuel resources. An apartment house study showed that fuel cells, coupled with heat pumps and thermal storage, would be able to provide all of the building's electrical and thermal requirements with less fuel than was previously used to supply conventional heating alone.

GEOTHERMAL ENERGY

Geothermal energy, which is the internal heat of earth, is one of the most immense energy resources available to humans. However, it has not been studied extensively, primarily because it exists at such great depths in most areas that it has been difficult to reach. But it is generally agreed that vast and virtually inexhaustible quantities of heat would be available if one could drill far enough down into the earth at any location.

Origin and Nature

The heat energy of the earth's core is evident most dramatically in volcanic eruptions. The sight of hot molten rock pouring from the earth and the sensation, even at great distances, of its intense heat, bring a frightening realization of the fantastic quantity of energy involved.

The direct tapping and use of the heat of this molten rock, which geologists call *magma,* is a fascinating possibility, but one that will require many years of study and development to bring to fruition. However, even if only a small portion of the total internal heat of the earth could be harnessed for useful purposes, there would probably be enough for centuries.

Geothermal energy is abundantly evident in many other forms, some of which are quite as spectacular as—although less terrifying than—a volcano. Geysers, such as Old Faithful, result from the interaction of hot pressurized geothermal waters and underground cavities. Fumaroles (holes that emit vapor), hot springs, and bubbling mudholes are other examples of the earth's internal heat breaking through to the surface.

Although in many areas of the world there is not perceptible evidence of geothermal heat at the surface, there is still a measurable underground flow of heat that is quite variable and depends upon geologic conditions. For example, temperatures in mines and wells usually rise with increased depth at an average rate of 35°C per kilometer of depth. At the base of the outer layer of the earth, which geologists call the *continental crust,* temperatures range from 200° to 1000°C. At the molten center of the earth, temperatures are believed to range from 3500° to 4500°C. This internal heat is considered to be mainly the result of the natural decay of long-lived radioactive materials and thus virtually inexhaustible.

While these temperatures in the earth's crust represent average conditions, there are also significant concentrations of heat at much shallower depths—less than 3 kilometers (2 miles)—in certain locations. The energy stored in these "hot spots" can exist in solid rock, or in water and steam that fills cracks and pores in the rocks. The water and steam are sometimes at relatively high pressures, which may go above 135 atmospheres (2000 pounds per square inch), and at temperatures of up to 300°C (570°F).

The geothermal energy in such a relatively shallow region is often observed as a natural hot springs or geyser, and it can be tapped by drilling wells at proper points. Since water under high pressure may exist as a liquid, even at temperatures far above its normal boiling point at the earth's surface, many such underground geothermal concentrations consist of liquid water rather than steam.

When the hot water in such a reservoir comes to the surface through a well, the high pressure is normally reduced, and some of the water spontaneously boils, or "flashes," into steam. In this way, a typical geothermal well yields a mixture of hot water and steam; the proportions depend on the temperatures and pressures at the particular location. These areas are called "wet-steam" fields. Some of the major wet-steam fields are in Mexico, the Salton Sea—Imperial Valley area of California, and the Yellowstone region of Wyoming. Others are in New Zealand, Japan, the Philippines, Indonesia, Taiwan, and Turkey.

The Salton Sea—Imperial Valley region is a good example of a potentially enormous source of geothermal energy in a wet-steam field that, if developed, could provide very important economic benefits for the West Coast. This valley covers approximately 1000 square kilometers (400 square miles), and temperatures of 300°C (570°F) have been verified at depths of only 1500 meters.

If this energy is to be usefully exploited, some serious technical problems as well as environmental, economic, and legal difficulties must be overcome (see p. 000). But if large quantities of very hot salty water could be brought to the surface, they could be partially flashed into steam at pressures that could be used to generate electricity. The steam would then be condensed into fresh water for many valuable uses in the arid Salton Sea—Imperial Valley Area. The remaining heavily mineralized water could either be returned to the ground or, under certain conditions, further distilled to produce useful minerals. The development of the geothermal resources in this region could thus have a threefold benefit—the yielding of electricity, fresh water and chemicals.

A relatively small number of geothermal heat concentrations produce superheated steam with almost no water. The steam from these vapor-dominated or dry-steam fields can be conveniently piped directly to steam turbines to generate electricity as described. Only a very few of these fields are known to exist; the main ones are located at Larderello in Italy and in The Geysers region in California.

In the western United States, many active geothermal areas are located near hot springs. Huge concentrations of hot water in large geologic basins are also known to lie under the Gulf Coast area at considerable depths. In this region, many wells have encountered large quantities of hot water under abnormally high pressures, but generally at relatively lower temperatures than those found in the wet-steam fields in the western states.

Many of the hot rock masses underlying the United States do not come into contact with water systems at all and so they are dry. It is generally considered that the total heat contained in these dry deposits is much greater than that which exists in wet-steam or dry-steam reservoirs. The problem of tapping these tremendous dry heat resources, which have never been used, is discussed later.

There is not enough information available to accurately estimate how much total geothermal heat of all kinds exists in the United States. From the information at hand, the amount has been calculated, to a depth of 6 miles, to be the equivalent of burning 1 quadrillion (1,000,000,000,000,000) tons of coal. If made available at the surface, most of this energy could be used at least for direct heating and would heat every American home for thousands of centuries. It is not yet known how much of this energy exists at high enough temperatures to be of practical use in generating electricity.

History

Hot springs, where underground geothermal water flows naturally to the surface, have been known and used since ancient times. The Romans developed such watering places for recreational and medical purposes all around the Mediterranean as well as throughout their empire. Medical spas were also in existence in ancient Japan and elsewhere in the Far East. Today, hot springs are popular as resorts both in the United States and at other locations throughout the world.

Most geothermal waters contain a variety of dissolved minerals, which are often considered to have medicinal value. In some cases, they might even be recovered economically. For example, the first use of geothermal wells for commercial purposes was recorded in 1812 in Italy; it was not used for heating but for the production of boric acid.

As pointed out previously, some geothermal springs or wells yield steam in addition to hot water, a very few produce dry steam alone. Useful applications for these various types have depended on the composition, the temperatures and pressures involved, and the geographical locations.

Direct Use of Heated Water

In the 1930s, Iceland pioneered the use of geothermal hot water for household and commercial heating. Near the capital city, Reykjavik, an elaborate network of pipes and conduits carries hot water from more than 100 geothermal wells to 90 percent of its homes. The deepest well is about 2000 meters, and the temperatures of the basic geothermal reservoirs range from just under the boiling point to 150°C. The hot water is carried for distances of up to 16 kilometers and is delivered at about 80°C.

An additional 25,000 people in towns outside the Reykjavik area will also soon receive geothermal water for a variety of direct heating applications. In addition to house heating, the water is widely used in Iceland for heating baths, swimming pools, and greenhouses, which furnish fresh vegetables throughout the year.

In the United States, geothermal space heating has been successfully carried out for many years in southern Oregon. At Klamath Falls, for example, geothermal water from 350 wells supplies heat, either directly or through heat-exchanger systems, to pure municipal water, which is used for heating buildings. In other parts of Oregon, geothermal waters are used for greenhouses, baths, farm buildings, schools, and resorts. Similar projects have been undertaken in Boise, Idaho, in towns in California, and in some farms and villages in other western states.

In Japan, there has been extensive exploration for geothermal resources, which have been employed for a variety of direct heating applications during the last century. The Japanese employed geothermal heating of greenhouses in the early 1920s, a technique that is still widely used to produce many vegetables and tropical fruits. Although space heating has not been developed as widely there as in Iceland and the USSR, Japan has thousands of natural hot mineral-water resorts, baths, and therapeutic spas, and probably leads the rest of the word in the use of geothermal resources for such purposes. There are also several industrial applications of geothermal energy in Japan, including sulfur recovery, commercial baking, salt recovery from seawater, and experimental fish farming.

In the USSR, there has been considerable development of geothermal heating systems over a wide geographical area. At several locations, centralized municipal geothermal heating installations furnish hot water and heat houses for communities of 15,000 to 18,000 people. Geothermal energy is also used in the Soviet Union in an oil refinery and to heat greenhouses, seedbeds, and baths. Although precise information about the total consumption of geothermal energy in direct heating applications in the USSR is not known, it is believed to be the equivalent of over 1 million tons of fuel oil per year.

In Hungary, geothermal reservoirs have been found under large areas and have been widely used for space heating. For example, some 1200 housing units and associated municipal and commercial buildings are heated by hot geothermal water in the city of Szeged at a cost well below that of conventional fuels. Geothermal heat is also used extensively in Hungary for agricultural purposes such as heating greenhouses and farm buildings and drying crops. Such applications are expected to increase during the 1980s.

New Zealand is another country that has used direct geothermal heating extensively. At one place, natural steam from a number of wells is used to heat pure water for the generation of high-quality steam, which is then used directly in industrial mill processes. This steam is also used to dry timber, operate log-handling equipment, and generate electricity.

In the New Zealand city of Rotorua (population 30,000), more than a thousand hot-water wells supply heat to commercial establishments, houses, schools, hospitals, and hotels. There is also a hotel in the city with a geothermal air-conditioning system, which has a very low operating cost. In addition, direct geothermal heating is used in New Zealand agriculture.

Many other countries have used direct geothermal heat successfully and economically. Such applications are limited, however, to the immediate geographical region of the wells.

From the standpoint of distribution, a much more flexible way of using geothermal sources is through the generation of electricity. Until the present time, however, this has been practical with only a relatively few geothermal deposits.

Generating Electricity

Generation of electricity from naturally occurring steam was first undertaken in Italy at Larderello in 1904. As early as 1913, a 250-kilowatt station was established, and electricity has been generated geothermally at this location almost continuously ever since. The present level of generation is about 365 megawatts. There are 13 individual plants included in the Larderello geothermal area, which is now being used nearly to capacity.

Geothermal electricity-generating stations of various sizes have also been established in various other foreign countries including New Zealand, Japan, Iceland, the USSR, and Mexico. The largest such installation—and the only one in the United States—is located in California at the geothermal steam field known as The Geysers.

The development of electric power generation has proceeded rapidly at The Geysers, which is the only operating field other than Larderello that has dry steam rather than hot water. The first 12-megawatt turbine was installed in 1960, and since then additional plants have brought the present capacity to over 400 megawatts. Based on a steady expansion of this field, the final total capacity now planned is almost 1200 megawatts, which will make it the largest complex of its kind in the world and sufficient to meet the power needs of a large metropolitan area.

There are various estimates of the total sustainable geothermal power available at The Geysers field. These are based on different interpretations of data concerning the extent of the field, the yield and spacing of wells, and the total fluid stored in the reservoirs. These estimates range up to 4000 megawatts.

Over 100 wells have been drilled in The Geysers area, which is about 3 by 13 kilometers; the deepest well is more than 2300 meters. Temperatures of the reservoirs from which the heat is drawn are about 255°C. The basic source appears to be a mass of heated rocks at a depth of 5 to 8 kilometers and covering an area of about 250 to 1300 square kilometers.

Geothermal Technology

Geothermal technology covers a wide range of different scientific and engineering fields. First, new geothermal resources must be identified and evaluated. This process uses geophysical exploration instruments and techniques that are similar to those used in prospecting for petroleum or minerals.

Next, these heat concentrations must be reached, and satisfactory methods must be devised for bringing the heat to the earth's surface. This involves drilling geothermal wells. Facilities must then be designed and built to use the heat in a practical and economical manner without adverse environmental effects.

In connection with the use of heat, especially in the generation of electricity, it is important to understand the limitations imposed by some of the basic principles of *thermodynamics,* which is the science that deals with the relationship between heat and work. In discussing how geothermal heat can best be used, it is convenient to divide geothermal resources into three major classes:

1. Those producing steam alone—the dry-steam field
2. Those yielding mixtures of steam and hot water in widely varying proportions, depending on different temperatures and pressures caused by geologic factors
3. Those consisting of hot rock that has not come into contact with water and is hence in a dry condition

The geothermal sources that produce only steam are relatively rare. This is unfortunate because the technology to use dry steam has been reasonably well developed.

Based on present knowledge, it seems that wet-steam areas are perhaps 20 times more plentiful than dry-steam fields. The technology for using wet-steam fields varies widely according to such factors as proportions of steam and water, and temperatures, impurities, pressure, and depth of the wells.

The technology of identifying and exploiting dry hot rock is in a very early state. However, the total amount of heat available in such deposits is frequently estimated to be at least 10 times the total heat contained in the wet- and dry-steam fields combined.

Finally, some new and imaginative concepts have been recently proposed to improve the efficiency and economics of geothermal heat utilization. These are discussed briefly at the end of this section.

Prospecting for Heat

Geothermal heat sources must be accurately located and analyzed before they can be used. Hence, it is very important that more reliable and economical geothermal prospecting techniques be developed. Prospecting for geothermal heat sources has much in common with prospecting for petroleum deposits, and many of the same sophisticated prospecting techniques developed by the petroleum industry can be used in locating and identifying geothermal resources.

In the past, geothermal exploration, using petroleum prospecting methods, has generally been confined to locations in the vicinity of hot springs. However, such techniques have not been widely employed to discover and evaluate possible new geothermal fields prior to drilling, nor have new specialized methods been well developed.

One of the most important first steps is to outline broad regions where the outward flow of heat near the earth's surface is significantly greater than average. However, temperature measurements right at the surface can, for various reasons, be misleading; it is therefore more reliable to measure how fast the temperature is increasing with depth at distances of 30 to 90 meters underground.

Aerial surveys, including those using infrared rays, can be useful but may also be affected by factors other than geothermal heat deposits. Many other heat-prospecting approaches are now being investigated more thoroughly, including measurement of the electrical conductivity (which is related to temperature) of rock masses at various depths; magnetic, electromagnetic, and gravity measurements; and seismic methods. Chemical analysis of geothermal waters can also give information about the nature of heat concentrations.

Sometimes, an explosive charge is dropped down a test well. How sound waves created by the explosion travel depends on the types of rocks and soil in the vicinity. Measurements of these waves help geologists to determine the location of geologic structures that may contain petroleum, mineral deposits, or geothermal reservoirs.

Drilling Technology

At the present time, well-drilling techniques in the United States are adequate for depths up to approximately 6000 meters and temperatures around 200°C. Costs increase sharply beyond these limits, and the accuracy of information about conditions at the bottom of the well—"well logging"—deteriorates rapidly. If it is true, as many geothermal experts agree, that vast heat concentrations lie at considerable depths under most of the United States, then it is important that the technology of deep drilling at high temperatures be improved. Perhaps then geothermal resources can eventually be made economically available to the whole nation. In addition, it will be necessary to develop better high-temperature drilling fluids and new drilling techniques, such as electric melting penetrators and turbine drills. Based on the available evidence, a deep-drilling research program will be essential to the accurate evaluation and utilization of new geothermal resources.

Thermodynamics—Heat and Work

Energy—the capacity to do useful work—can exist in several different forms. Heat, first discovered by early humans as fire, is still the most important source for many applications. Another form is kinetic energy—the energy of mechanical motion—as illustrated by a moving vehicle, a pile driver, or a steam turbine.

In the early days of history, wood was the chief substance burned to produce heat. But within the last two centuries, the largest amounts of heat energy produced by civilized humans have come from burning the fossil fuels—coal, oil, and gas.

Heat energy has always been used directly for household and industrial purposes, and these applications amount to 30 percent of all the energy consumed in the United States. Since the discovery of the steam engine and later the gasoline engine, large amounts of heat have also been used to power machines and provide means of transportation. Around 1900, heat energy began to be used extensively to generate electricity. In the last half century, the demand for electricity has almost doubled during each decade. More than 98 percent of American farms have electricity, and the country now consumes about 33 percent of all the electric power generated in the world—although we have only 6 percent of its population.

As its name suggests, *thermodynamics* is the science that involves the interaction between heat and work. One fundamental principle is that a heat source cannot be used to produce mechanical work unless the heat can flow to a region of lower temperature, i.e., a *heat sink*. Furthermore, the amount of useful work that can be obtained in any kind of heat-using machine or engine depends largely on the temperature of the heat source relative to that of the heat sink.

A heat source can be compared in some ways to a body of water that is confined at a specific level, such as a lake or reservoir. No matter how much water is available, it will not produce useful work unless it flows to a lower level. Furthermore, the higher the initial level, the greater the amount of work that will be produced as the water flows to the lower level. Also, the amount of mechanical work that can be extracted from a source of heat, no matter how large it may be, is limited by the difference between the temperature of the heat source and that of the region to which it flows. This is true for all kinds of engines, turbines, or other machines that convert heat into work.

In an automobile engine, for example, only 10 to 20 percent of the energy of the burning gasoline is transformed into work. None of the original energy actually disappears, but the large fraction that is not converted to work is discharged to the environment as waste heat in the radiator and exhaust.

On the other hand, if heat energy is used to produce high-temperature steam, 30 to 40 percent can be converted to work by turbine in order to generate electricity as in steam-electric power plants. In such cases, the remaining energy also goes to the environment—usually into large bodies of water or through cooling towers to the atmosphere.

Most geothermal sources provide heat energy that is contained in hot water and/or steam at temperatures of 150 to 250°C, which are much lower than those used in modern steam-electric plants. The overall efficiencies of power plants using such geothermal energy are hence correspondingly less than those of fossil or nuclear plants. In instances where geothermal waters provide heat at a temperature in the lower portion of the preceding range, ordinary steam turbines do not operate properly, and new technologies must be developed, such as the vapor-turbine cycle.

Dry-Steam Wells

The technology of using dry steam from geothermal wells, which is done at The Geysers and in Larderello, Italy, is well-developed and is analogous to that of modern steam-electric plants except for a different range of pressure and temperature.

By the time the steam from typical natural steam wells reaches the turbines, it is at a pressure of 7 atmospheres (100 pounds/square inch) and at a temperature of 200°C (400°F) in contrast to pressures of 200 atmospheres (3000 pounds per square inch) and temperatures over 540°C (1000°F) for modern power-generating stations using fossil fuels. Because of these lower initial temperatures, the geothermal plants have a lower efficiency—about 14 percent—compared with 40 percent for the latest fossil-fuel plants. But after the geothermal steam leaves the turbines, which must be of special design, it is processed in the same way as in fossil plants. First, it is condensed, and then the water containing the waste heat is fed to cooling towers or disposed of in other ways depending on the local conditions.

Wet-Steam Wells

Much more common than the dry-steam wells are wells yielding a mixture of hot water and steam in different proportions depending on a number of factors, mainly pressure and temperature. The steam can often be separated and used to generate electricity; the remaining hot water can be employed for many different purposes. The technology of using geothermal water for various direct heat applications is relatively simple, except that, in many installations, the dissolved minerals foul the piping and heat-exchange equipment.

Installations have been established where some of the hot geothermal water, after separation from the steam, can be distilled by its own heat using low pressures. An extension of these ideas has been proposed that would have a threefold purpose. Steam from a wet-steam well would be separated and used to generate electricity. The hot water would be fed into a desalination plant, and a portion would be used to produce fresh water. The remaining water, with an increased concentration of minerals, would be evaporated to yield those minerals. The design is still being studied.

Secondary-Fluid Cycle

The secondary-fluid cycle is a relatively new process in which the hot water is brought up from the geothermal reservoir and kept at relatively high pressure throughout the process so that it does not boil and produce steam. Through a heat exchanger, it gives up most of its heat and causes a secondary fluid, such as isobutane or Freon, which has a lower boiling point, to vaporize. The vapor expands through a turbine to drive an electric generator and is then condensed to a liquid before being returned to the heat exchanger to start the cycle all over again.

The original geothermal well-water, still at high pressure but having lost most of its heat energy, is forced back into the ground. This procedure tends to maintain the underground pressure at a useful level and prevent subsidence (lowering) of the local ground level. At least some of the returned water may come into contact again with the basic geothermal heat source.

There are many advantages to the secondary-fluid concept. Among these are: smaller turbines; reduction of the heat losses that occur in an unpressurized well when some of the hot well-water flashes into steam on the way to the surface; and reduced atmospheric pollution under certain conditions. A plant of this kind, now in operation in the USSR, used Freon as the secondary fluid. Similar plants are planned in Japan.

Dry Hot Rock

As mentioned previously, there is evidence that the total geothermal heat energy in dry hot rock is much greater than that associated with hot water and steam systems. However, this resource has not yet been used, primarily because there is no technology concerning it.

ERDA's Los Alamos Scientific Laboratory in New Mexico is working on a new approach to this problem. The method, based on the hydrofracturing technique already used extensively in the petroleum industry, has not yet been clearly demonstrated as being practical for the kind of rock formations that contain geothermal heat deposits.

The first step is to pump water under very high pressure down into a dry hole to open large vertical cracks in hot rock, such as granite, which overlies an area of high heat flow from the earth's interior. A second hole is drilled down to intersect the crack. Subsequently, cold water is forced down through the crack, becomes heated, and then returns to the surface to provide energy for power plants or other applications.

Experiments in New Mexico indicate that the granite can indeed be fractured successfully, and that the crack will retain the water tightly. It is believed, but has not yet been demonstrated, that the hot rock will tend to contract as it cools so that the initial crack will be extended to make new hot rock available for the extraction of additional heat.

Another method is to create large artificial cracks with explosives and then to circulate water from the surface through the extended areas of the cracks in order to extract the heat. This technique must be further investigated because there are several major problems, such as the effect of blast waves on surface facilities, the economics of creating sufficient fresh rock surface to extract heat in useful quantities, and the limitation on how fast the rock will conduct heat to cracks where it is being withdrawn. In all, development of techniques to use dry hot rock has just begun, but the results so far are promising.

Other New Techniques

Other concepts have been proposed to advance the technology of geothermal energy use. One proposal is to make a direct tap into molten rock or magma, possibly near a volcano, and to install a collector or heat exchanger to bring the heat to a power plant on the surface. With temperatures of 1000°C (1800°F) or more, the heat would be available in almost limitless quantities. It is estimated that a cubic mile of such material contains enough heat to run several 1000-megawatt electric plants for a century. However, many formidable technical problems must be solved. These include the selection of the proper materials for the heat exchanger, which must last for a reasonable period of time at the high temperatures involved, and the technology needed to bring about the required flow of the molten rock around the collector so as to maintain the necessary transmission of heat.

Another proposal, which does not appear to require such radical extensions of present technologies, has been formulated at ERDA's Lawrence Livermore Laboratory in California for application to the hot waters of the Salton Sea area in California, which contains heavy concentrations of minerals. The basic idea is called the *total flow concept*. The object is to convert some of the thermal energy of a pressurized-steam hot-water mixture into kinetic energy by forcing it through a converging-diverging nozzle. The resulting high-velocity jet output would be used to drive a modified hydraulic impulse turbine. Theoretically, this method should produce 60 percent more power than other systems, operational or proposed, for this type of application. The advantage is that the total flow is used, which allows recovery of some energy in the liquid that would otherwise be lost. Much of the necessary technology already exists, and preliminary estimates indicate that the capital and operating costs will be attractive.

Environmental Effects

Generally speaking, the practical use of geothermal energy has some special advantages from an environmental standpoint when compared with other methods of energy production or generation. Very important is that the environment effects associated with geothermal power generation are generally restricted to the immediate vicinity of the plants. This is in marked contrast to coal, oil, and nuclear installations, all of which involve a wide variety of environmental impacts far removed from the generating plants themselves.

In the case of coal, for instance, the mining and transportation activities, as well as the final disposition of the ash and other combustion products, are usually at some distance from the plant. For oil-fired plants, distant activities include extracting and transporting the crude oil, the refining processes, the transmission of fuel to the plant, and the disposal of undesirable combustion products. In the case of nuclear plants, some of the essential activities undertaken at widely separated locations include mining of the ore, a complex series of industrial processes, and the ultimate disposal of the potentially hazardous spent fuel.

At some geothermal installations, the dissolved minerals in the wastewater can be separated out, which yields fresh water that can be important in arid regions. This situation is found, for example, in the Salton Sea–Imperial Valley region of southern California.

Nevertheless, use of geothermal energy resources presents special environmental considerations that must be carefully evaluated for each installation. The more important of these are:

- Prior to the construction of a geothermal facility, the noise and other objectionable aspects of well-drilling may be important depending on the specific location. Noise may also continue to be a problem after operations begin.

- It is necessary in certain instances to provide safe disposal methods for toxic and saline liquid wastes involved in some geothermal hot-water fields. Reinjection of such fluids back into the geologic formation is a possible solution, but it has not been fully investigated.

- Geothermal fluids, both liquid and steam, often carry substantial amounts of dissolved gases that are released upon loss of fluid pressure or condensation of the steam. Commonly encountered gases are carbon dioxide, hydrogen sulfide, methane, and ammonia. Although carbon dioxide is relatively harmless when discharged into the atmosphere, hydrogen sulfide and methane can be dangerous and/or objectionable depending on concentration and must be properly controlled. Ammonia occurs only infrequently and, particularly in low concentrations, is not regarded as a hazard.

- Withdrawal of substantial volumes of fluids from geological formations over relatively long periods of time has caused extensive differential land subsidence in the Wairakei geothermal field in New Zealand and during petroleum extraction in the United States. Land subsidence, which in the petroleum industry has been controlled by repressurization with water, is irreversible and can be damaging to irrigation, flood control, and sewer systems, as well as to structures such as railroad beds, airport runways, and foundations.

- It is even possible to trigger earthquakes by the high-pressure injection of fluids into geologic formations if they are in direct contact with seismically active faults. Present evidence indicates that such consequences can be avoided by eliminating the possibility of any hydraulic contact with a seismic fault.

- Subsurface groundwater contamination can be caused by improper well-drilling and cementing procedures.

- Under atmospheric conditions, local fogging may result from the release of large amounts of water vapor from a geothermal facility where flashing of the hot pressurized water occurs.

- For applications related to direct heating, such as for household, industrial, or agricultural purposes, the environmental impact is related mainly to the ultimate disposal of the water, which usually contains many dissolved minerals.

- In geothermal power plants using relatively low-temperature steam, more waste heat is dissipated to the environment than from fossil-fuel or nuclear power plants of the same generating capacity that operate at much higher temperatures. This is because the overall efficiency of transforming heat into mechanical (and, in turn, electrical) energy is limited by the difference between the initial steam temperature and the final condenser temperature. For example, the electrical power output of the plants at The Geysers averages about 14 percent of the total heat energy of the input steam. The remaining 86 percent must be dissipated to the environment in some acceptable way.

- For those geothermal electrical generation systems designed to start with hot water instead of steam—as in the secondary-fluid cycle described previously—the overall efficiency may be even less because of still lower initial temperatures. Hence, the fraction of the extracted geothermal heat passed on to the environment via the condenser cooling water may be over 90 percent. However, the heat still remaining in the original geothermal water will be returned to the ground as the water is reinjected; this procedure has the additional environmental advantage of preventing subsidence.

Despite these environmental considerations, which may or may not be problems in any particular installation, the use of geothermal energy resources appears to have environmental advantages that generally outweigh the drawbacks. In addition, the technology to deal with these environmental effects exists although each such facility must be considered as a special case.

Economic Considerations

The most basic economic fact about geothermal energy is that, once tapped, the "fuel"—the natural heat of the earth—is free. However, the overall production costs associated with various kinds of geothermal energy installations are generally difficult to compile for comparative purposes because the background data and assumptions are not always available. Different tax rates and amortization periods may be involved as well as a number of hidden costs and other variables.

In direct heating applications in various foreign countries, such as Hungary, Iceland, New Zealand, and the USSR, it seems to have been clearly shown that the use of geothermal heat in agriculture, industry, and household heating is definitely less expensive than the use of other fuels for the same purposes. In Iceland, for example, the direct use of hot geothermal water for heating buildings is cheaper than providing the heat by other means, including hydroelectric power.

In the past, natural gas and coal have been competitive in cost with geothermal energy for direct heat applications. However, gas is becoming increasingly scarce and costly, and coal produces serious environmental effects that are expensive to control in order to meet today's pollution standards.

At present, there are so few geothermal electric power plants that it is difficult to draw meaningful conclusions regarding the economics of these plants compared with fossil-fuel, nuclear, and hydroelectric plants. However, the best available information indicates that the overall cost, including capital amortization, of geothermally produced electricity—such as at The Geysers—is less than that for fossil and nuclear plants, but higher than for hydroelectric installations.

One economic advantage of geothermal power installations at certain locations is that they are practical even in relatively small sizes that are not economically feasible for other kinds of power plants. As an illustration, in Iceland many small noncondensing geothermal turbines are economically operated to generate electricity for villages and farms.

Some of the basic costs associated with providing geothermal energy, such as drilling costs, are well known. Other important cost elements, such as those related to the life expectancy of the "fuel," routine maintenance, and environmental control, are not known and are difficult to estimate; however, in geothermal desalination plants, they are estimated to be well below those of other methods.

With the development of multipurpose geothermal plants for several combined end products, such as electricity, hot-water uses, desalination, and mineral production, the economic benefits can be expected to be correspondingly greater than for single-purpose facilities. Today it is estimated that an electric plant using natural geothermal steam can be built at a cost far less than that of a modern fossil-fuel or nuclear plant of the same capacity. On the other hand, a series of proposed demonstration geothermal plants that will be constructed to test some of the new complex designs, such as the secondary-fluid cycle, will probably cost initially at least as much as fossil or nuclear installations of comparable size. Eventual capital investment costs for such facilities cannot be forecast at this time.

Future Prospects

Geothermal energy could, if more extensively developed, play a significant part in the future increase of the United States' total energy resources to meet rapidly mounting demands. But quantitative forecasts by knowledgeable experts differ widely, mainly because of insufficient data and uncertainties regarding the size and success of future research efforts.

With regard to power generation, estimates of installed geothermal electrical capacity in the United States by 1985 range from a conservative 5000 megawatts, assuming only technology presently available or under development, to a potential 20,000 to 30,000 megawatts based on an energetic and successful research and development program by both industry and the federal government. If this potential could be realized, it would represent about 3 percent of the total electrical generating capacity of the country projected for 1985.

In addition to generating electricity, geothermal heat will undoubtedly have increasingly important applications for direct use in such diverse fields as space and industrial process heating, agriculture, refrigeration, and the production of fresh water and certain minerals. Programs to promote these kinds of direct geothermal heating applications are being pursued all over the world.

Research and Development Programs

The most important factor in rapidly bringing geothermal energy into a more useful roll in the United States is the research, development, and engineering programs in this field that will be undertaken in the near future. There are many broad areas in which new technological frontiers must be aggressively explored and progress achieved. The most important of these are:

- Improving our knowledge of the location, size, and type of our geothermal resources. This demands significant advancement in drilling technology and exploration and prospecting techniques.
- Investigating the feasibility of using explosives or hydrofacturing to artificially stimulate geothermal sources, such as hot dry rock deposits.
- Advancing the technology of generating electricity with geothermal hot waters.
- Expanding the direct use of geothermal waters as heat for a wide variety of applications.

Legal and Regulatory Aspects

The multitude of laws and regulations that govern the exploration and use of geothermal resources must be carefully analyzed. Three-quarters of the known geothermal energy deposits lie below land owned by the federal government, which must assume a leading role in the overall program of geothermal development. Recent legislation in this field has extended the authority of the U.S. Department of the Interior to establish more precise standards for what are called *known geothermal resource areas (KGRAs)*. The secretary of the interior is empowered to formulate and carry out regulations pertaining to the competitive leasing of such lands. The exercise of such authority will have an important bearing on the future effective use of our geothermal resources.

In accordance with these statutes, federal lands were opened to private industry early in 1974 for the first time for geothermal exploration. It is clear that a vigorous movement exists to mobilize many different kinds of private and government efforts to exploit geothermal resources and to search for more in spite of the risks and technological problems involved.

Present federal policy emphasizes a strong and continuous working relationship with the private geothermal energy industry. It is hoped that industry will assume an increasingly active role and accept a greater share of the risk as the whole national research program in this field begins to pay off. The hope is that private industry development will facilitate a rapid transfer of research results and achieve a substantial acceleration of the development of United States geothermal resources. Perhaps this development activity will become a kind of "geothermal rush," like the famous gold rush of 1849.

To whatever extent these developments occur, they will represent significant steps toward better use of this almost limitless domestic source of energy and a resulting increased conservation of previous fossil fuels. It is therefore vital that government and private industry continue their cooperative efforts to develop geothermal energy for the benefit of the American public.

HYDROELECTRIC POWER: USE OF A RENEWABLE RESOURCE

Falling water in streams and rivers has been harnessed to produce electric energy for almost 100 years. As it falls, the water passes through turbines, which turn to generate electricity.

In the mid-1930s, hydroelectric power ("hydro") comprised about 30 percent of the nation's total generating capacity. As fossil fuels came to be relied on more and more, however, and as many of the best sites for hydro were developed, the relative contribution of hydro to the nation's capacity decreased to 20 percent in the early 1960s and to 15 percent in 1978.

Although the relative contribution of hydro has decreased, hydroelectric development in the United States has steadily increased, from 7800 megawatts in 1930, to 31,900 megawatts in 1960, and to 69,000 megawatts in 1978.

Hydroelectric plants depend upon water, which is a renewable resource because of the recurring cycles of rainfall, runoff, and evaporation. Hydroelectric plants do not consume water, heat the water of streams, or contribute to air pollution. These favorable characteristics, combined with increasing shortages and costs of fossil fuels, make hydroelectric power an increasingly attractive choice as a source of electric power generation.

History of Hydro Regulation

Before passage of the Federal Water Power Act in 1920, a special act of Congress was needed to build and operate a hydroelectric power project on navigable streams or lands of the United States. Congress first authorized construction of a hydroelectric plant in 1884. Demand for electric power suddenly increased during World War I, and the need for an orderly means of advancing water-power development became pronounced.

To meet this need, Congress enacted the Federal Water Power Act, which established the Federal Power Commission (FPC). The FPC was given the responsibility of licensing nonfederal hydroelectric power projects that affect navigable waters, occupy United States public lands, use water or water power at a government dam, or affect the interests of interstate commerce. The act also required the FPC to license projects that were "best adapted to a comprehensive plan for improving or developing a waterway or waterways."

In its first 2 years, the FPC received 321 applications involving the construction of about 15,000 megawatts of new generating capacity, more than twice the capacity of then-existing water power installations in the country. In 1935, the Federal Water Power Act was incorporated into the Federal Power Act, which extended the FPC's authority to regulate the interstate aspects of the electric power industry.

Hydroelectric projects built by the United States government are authorized by Congress and do not require FPC licenses. Most federal water projects are multipurpose projects, which serve other primary purposes besides generating electric power, such as navigation, flood control, or irrigation.

The FPC ceased to exist when the U.S. Department of Energy (DOE) was activated on October 1, 1977. The newly created Federal Energy Regulatory Commission (FERC) now carries on many of the former FPC's functions, including its licensing of nonfederal hydroelectric power projects.

Role of FERC

FERC issues licenses for hydroelectric power projects for periods of up to 50 years. In addition to safety and engineering requirements, the commission often adds special license provisions on flood control, navigation, soil erosion, water quality, municipal water supply, recreation, protection of archaeological and historical sites, and protection of fish and wildlife.

When the project license of a privately owned utility expires, the Commission may issue a new license to the original licensee or to a new licensee, or it may recommend takeover by the United States when it determines that this would serve the public interest. Licensed, publicly owned projects are not subject to federal takeover. If a federal agency recommends takeover of a project, the commission postpones its decision for 2 years to provide time for Congress to consider the takeover recommendation.

Current Development

Approximately 1400 hydroelectric power plants are currently operating in the United States, with 59,000 megawatts of conventional generating capacity and 10,000 megawatts of pumped storage capacity. Approximately 49 percent of this capacity is federally owned, 28 percent privately owned, and 23 percent nonfederal publicly owned. About 90 percent of the nonfederal capacity operates under about 500 FERC licenses.

Forty-seven states have hydroelectric power projects in operation, the exceptions being Delaware, Louisiana, and Mississippi. Almost half of the nation's total hydro capacity exists in Washington, Oregon, and California. However, New York is second only to Washington in total hydro capacity within a single state.

The largest, privately owned, conventional hydro project currently in operation is Susquehanna Power Company's Conowingo project on the Susquehanna River near Conowingo, Maryland, with 476.7 megawatts of generating capacity. The largest, nonfederal, publicly owned conventional project is the Power Authority of the State of New York's Robert Moses project at Niagara Falls, New York, with 1950 megawatts of capacity.

The largest conventional hydro project owned by the federal government is the Grand Coulee project on the Columbia River near Coulee Dam, Washington, with a capacity of 4063 megawatts. The project is currently being expanded to include an additional 2100 megawatts of generating capacity.

Total investment in a new hydroelectric plant varies greatly due to size, location, type, land costs, and the costs of relocating highways, railroads, and other facilities. Investment costs per kilowatt for conventional hydroelectric plants may or may not be higher than for steam-electric plants depending on site characteristics. However, production expenses are much lower because no fuel is required, and operating and maintenance costs are much less.

Approximately 22 applications to build major new hydro projects (over 2000 horsepower) were pending at the FERC as of March 1978. These applications together involved more than 5100 megawatts of new capacity. Also pending at the FERC were outstanding preliminary permits and applications for preliminary permits to study the feasibility of installing about 16,000 megawatts of new capacity. The cost of these potential developments in total is estimated at $6.7 billion.

Pumped Storage

Hydroelectric power plants are particularly well suited for providing peaking (when demand for electricity is high) and reserve capacity to complement the output of large fossil-fueled and nuclear steam-electric plants. Pumped-storage hydro projects are uniquely suited for this purpose, and consequently they are being planned and built at an increasing rate. Pumped-storage projects use the same principles as conventional developments, but can reuse water repeatedly by pumping it from a lower to an upper reservoir during off-peak periods, where it is stored to generate power during peak periods.

Pumped-storage projects are particularly effective at sites having high heads (large elevations). The Cabin Creek project in Colorado has the highest head of current pumped-storage projects—approximately 1200 feet.

The largest, privately owned pumped-storage project currently in operation is Consumers Power Company's Ludington project on Lake Michigan near Ludington, Michigan, with 1009 megawatts of generating capacity. The largest, nonfederal, publicly owned, pumped-storage project is the Power Authority of the State of New York's Blenheim-Gilboa project on Schoharie Creek near Schoharie, New York, with 1000 megawatts of capacity. The largest federal pumped-storage project is the 424-megawatt San Luis pumped-storage project on San Luis Creek near San Luis, California.

The FERC recently issued a license to the Virginia Electric and Power Company (VEPCO) to build the largest pumped-storage project in the world. VEPCO's 2100-megawatt Bath County pumped-storage project is under construction on Back Creek and Little Back Creek near Mountain Groves, Virginia, and will cost an estimated $783 million.

Dam Safety

The FERC staff closely supervises the construction and operation of licensed projects to make sure that all license conditions are being met, with an emphasis on dam safety. Projects are inspected monthly while under construction, and designs of dams, powerhouses, and other structures are analyzed for safety and adequacy.

After they are operating, licensed projects are inspected annually to assess their structural integrity, to assure the safety of people and property located downstream and upstream, and to see that license conditions regarding reservoir levels, discharges, recreational developments, and fish and wildlife are being met. Projects with dams higher than 35 feet, or with total storage capacity of more than 2000 acre-feet, must be inspected once every 5 years by a qualified independent engineering consultant.

FERC has instructed all licensees to prepare emergency action plans to provide an early warning system if there is a sudden release of water due to a dam failure or an accident to the dam. The emergency plans are required to include operational procedures that may be taken in such an event, such as reducing reservoir levels, reducing downstream flows, and procedures for notifying nearby residents and also appropriate officials.

Recreation and Environmental Protection

Any application to build a major hydroelectric power project must include an environmental report describing the effect the project would have on water quality, land use, fish and wildlife, scenic values, recreation, archaeological sites, etc. Licenses issued by FERC contain conditions for protecting these values, and for mitigating any potential adverse effects that the project may have.

Another important part of a license application and of any license subsequently issued by FERC is a recreational-use plan for project waters and adjacent lands. People spend an estimated 73 million user-days annually at hydroelectric projects participating in such activities as fishing, boating, swimming, picnicking, camping, hiking, hunting, and guided tours of the generating facilities.

Small Dams

The United States government is actively encouraging the development of small dam sites throughout the nation for electric generation. A study by the U.S. Army Corps of Engineers identified approximately 47,000 dams throughout the United States that are 25 feet or higher and at which no electricity is generated and found that a total of 54,000 megawatts of generating capacity could be developed at these dam sites. The greatest potential was found to exist in New England, followed by the Mississippi Valley and the upper Northwest.

HYDROGEN FUEL

Plentiful, clean, high in energy content, adaptable to power generation and to industrial, residential, and transportation uses—this could be a description of the perfect fuel. In fact, it describes hydrogen, the lightest and one of the most abundant chemical elements, which is found in water and in all of the earth's organic matter. Pure hydrogen is a clean fuel: its only combustion product when burned with oxygen is water. Even when burned in air, it yields almost no pollutants.

Hydrogen's energy content per cubic foot is less than one-third that of natural gas. But its energy content per pound, almost three times that of gasoline, is the highest of any fuel known.

When hydrogen gas is cooled to a liquid, it takes up less than 1/700 as much space; hydrogen thus becomes a natural for space propulsion, which requires high-energy, low-weight fuel. The space shuttle and the rockets propelling the Apollo missions to the moon burn liquid hydrogen. Hydrogen's high-energy content could also make it desirable for more ordinary transportation, as well as for home and industrial use, helping to shift us away from dependence on scarcer fossil fuels.

Hydrogen's major drawback is that it is extremely rare in its elemental form. Although abundant, hydrogen is almost invariably locked (bonded) into chemical compounds. Two common examples are water, which covers 70 percent of the earth's surface, and all organic matter. Releasing the hydrogen stored in these materials requires expending significant amounts of energy.

Since energy must be invested in hydrogen before energy can be gotten out—that is, before hydrogen is useful as a fuel—hydrogen is considered a means of storing energy. The heat or electrical energy required to separate it from the elements to which it is bonded is in effect stored in hydrogen until that fuel is burned.

Old Uses . . .

Hydrogen, discovered in 1766, served as a buoyant gas in balloons and as an agent for extracting metals from raw materials. Late in the nineteenth century, people began to burn "town gas" or "manufactured gas," a half-hydrogen, half-carbon-monoxide fuel made from coal. Networks for distributing town gas still exist in several countries, including Brazil and Germany.

Currently, several million tons of hydrogen are produced annually in the United States, primarily for use in petroleum refining and in making ammonia and methyl alcohol, two major industrial chemicals. Most hydrogen is produced by reacting natural gas or light oil with steam at a high temperature. Small amounts of very pure hydrogen are produced by electrolysis: an electric current passing through water splits it into its two components, hydrogen and oxygen. This more expensive material is used in such special applications as food processing which require a higher purity gas than can be inexpensively produced from natural gas or light oil.

. . . And New

A much bigger role is now envisioned for hydrogen as a storage medium. Energy storage has come to be recognized as vital to using energy resources wisely. Storage can make generating electricity both more efficient and more economical. It can also help to make the best use of variable solar energy. And hydrogen could be used for both of these storage applications.

The demand for electricity produced at power plants is variable—higher during the day than at night; higher during weekdays (when businesses are in operation) than on weekends; and, in many parts of the country, higher during summer (when cooling needs are up) than during the rest of the year. Peak demands, which threaten to exceed a power plant's generating capacity, alternate with demands so low that much of that generating capacity is idle. Being able to store the excess energy that could be produced during low demand times to supplement the electrical output during periods of peak demand—a process called *load leveling*—would enable power plants to operate more efficiently and economically. With this process, much more of the power the plants provide can be produced by their large-capacity nuclear and coal-fired equipment. Inefficient older generating equipment and petroleum- or natural-gas—fired turbines do not have to be brought on-line to keep up with peak demands. And the expense of building additional generating capacity to cover those peaks might be avoided.

Electricity itself is difficult to store economically, but it can be converted into a more easily storable form, such as hydrogen. A utility could produce hydrogen with excess electricity during off-peak times, store it, and reconvert it to electricity, probably in fuel cells, for use during peak demand times.

Practical use of solar energy also requires a way of storing energy for backup use when there is no sunlight. Hydrogen produced with electricity during high-sunlight hours could be burned later to produce supplemental heat for solar heating systems or to drive electrical generating systems.

As an energy carrier and a storage medium, hydrogen has several potential advantages over electricity and devices that convert and store electrical energy. Hydrogen can be transported long distances by pipelines instead of by expensive overhead transmission lines, which require rights-of-way. Hydrogen can probably be stored underground as cheaply as natural gas is, providing longer-term storage (from week to week and season to season) than storage devices like flywheels and super-conducting magnets can. And, compared to "going electric," it is relatively simple to modify present automobiles, home furnaces, power plants, and industrial plants so that they can burn hydrogen instead of fossil fuels. Hydrogen production in effect converts energy from sources as varied as coal, the sun, and nuclear materials to a uniform, widely useful fuel.

Taking advantage of hydrogen's versatility, however, will require developing less expensive, more efficient production, storage, and distribution methods, for at present hydrogen is much more expensive than the fossil fuels it might replace. For instance, it costs about $5 to produce the hydrogen equal in heating value to 1000 cubic feet of natural gas, which costs today's user about $2.25 at intrastate rates. If new technology is developed and if fossil-fuel prices continue to increase—and both are likely—then hydrogen may become a practical and affordable solution to part of the national energy problem.

Production Methods, the Major Need

Cheaper and more efficient methods of hydrogen production are the greatest need. Since natural gas is the scarcest of the fossil fuels, its use in making hydrogen will probably give way to using abundant coal. Hydrogen is necessarily produced as an intermediate step in plants making synthetic gas or oil from coal. However, it will probably be just as easy to make those synthetic fuels as to make hydrogen and more convenient to transport and burn them in existing facilities. Making hydrogen from fossil fuels makes sense primarily for those applications requiring hydrogen's lightweight, clean-burning qualities.

To minimize the depletion of fossil-fuel reserves used as both the source of hydrogen and the source of the power needed to produce it, the U.S. Department of Energy (DOE) is working on producing hydrogen from water by electrolytic and thermochemical methods that can use solar and nuclear energy.

Since water is a poor conductor of electricity, making an electrolytic cell to separate hydrogen from water requires adding an electrolyte, a substance that does conduct electricity. Essentially, an electrolytic cell is made of two electrodes suspended in an electrolyte like potassium hydroxide, a corrosive chemical. The electrodes are made of expensive, corrosion-resistant metals such as platinum or palladium. When water is pumped into the cell and electric current is passed through it, hydrogen gas collects at one electrode and oxygen at the other. An asbestos cloth separator between the two electrodes permits current to pass but prevents the two gases from mixing.

Since generating electricity is a rather inefficient process (60 to 65 percent of the fuel energy consumed is lost), it is important that systems using high-quality electric energy be very efficient users of that power. Two approaches to improving present electrolytic cells, which convert only about 65 percent of the electric energy they use into hydrogen, are being studied.

The first approach is to improve the performance of the potassium hydroxide cell by increasing the temperature at which electrolysis occurs and to reduce the cost by using less expensive materials, especially for the electrodes. Since asbestos fails at about 80°C, other separator materials are also being tested as possible substitutes. A porous mat made of thin fibers of a synthetic polymer or an oxide material is one possibility. It might allow operating temperatures as high as 150°C, thus increasing conversion efficiency to 90 percent.

The second approach is to use a thin sheet of an acidic fluorocarbon plastic as the electrolyte. Tiny particles of a platinum alloy pressed onto both sides of the solid electrolyte serve as electrodes. The raw material, water, is the only liquid used. DOE researchers have used this technology to convert electricity to hydrogen with 90 percent efficiency in a small laboratory module. Tests are underway on a 2.5-square-foot module. The goal is to be field-testing a 10 square foot module (a size that would be suitable for commercial hydrogen production) by the early 1980s.

Electricity needed to electrolyze water for hydrogen production can come from today's fossil or nuclear power plants, or it could be produced with solar energy. At one kind of solar power plant, a field of solar collectors would concentrate the sun's rays on a boiler atop a tall "power tower," creating the high-temperature steam needed to generate electricity. Another type of solar power system would generate electricity directly from arrays of photovoltaic cells.

The sun's energy could also be used to produce hydrogen using a process that would shortcut the electricity-producing step. Preliminary studies of *photolysis,* the process by which light decomposes materials, suggest that electrodes made of semiconducting materials can absorb sunlight and split water at the electrode surface. A solar-powered "electrolyzer" based on this process would need little outside electricity.

Producing Hydrogen from Heat Alone

Production of hydrogen using thermochemical cycles is a newer concept than electrolytic production, and hence has not been as well studied. A thermochemical cycle is a series of linked chemical reactions that produce hydrogen and oxygen from water and heat; all other chemical compounds produced are completely recycled. Although thermochemical cycles are unlikely to be developed for large-scale use in this century, they hold special promise. Because they could use the heat from solar and nuclear plants directly, they are potentially more efficient and cheaper than electrolysis, which requires the energy-wasting step of converting heat into electricity. Thermochemical cycles could produce hydrogen directly from heat energy with 40 to 75 percent efficiency, compared to electrolysis with an overall heat → electricity → hydrogen conversion efficiency of 35 to 55 percent.

Many thermochemical cycles are theoretically possible, but the most efficient ones require very high temperatures, 800°C (1472°F) and above. For efficient, inexpensive hydrogen production, researchers are looking for cycles that involve only a few, rapid reactions; that use inexpensive, easily available, and noncorrosive chemicals; that require few extra processes (such as separation of a solid from a liquid); and that lose little energy in linking the reactions into a complete cycle.

Cycles involving sulfur or halides are especially promising. The DOE is presently focusing on two sulfur-based cycles. One cycle involves reacting sulfur dioxide, water, and iodine to produce sulfuric acid and hydrogen iodide, which then decompose to hydrogen and oxygen, plus sulfur dioxide and iodine. The other cycle, involving sulfurous acid, is a "hybrid" production method that couples a high-temperature thermochemical step with an electrolytic step. The electrolysis replaces a difficult thermochemical step, and as a result reduces materials handling, the number of reactions, and the number of extra separation processes. Although the hybrid process requires some electricity in addition to heat, the quantities of electricity are much smaller than those necessary for electrolysis alone.

The temperatures of 800°C (1472°F) and above, which are required to make thermochemical processes efficient, can be supplied by nuclear reactors or solar concentrators. The leading reactor design for production of hydrogen is the high-temperature gas-cooled reactor. An experimental gas-cooled reactor has reached temperatures up to 1100°C (1922°F). Similar reactors have been used to produce electric power although not at full commercial scale. The concept of using high-temperature nuclear reactors to provide the heat for thermochemical cycles is still in its infancy and is being developed mainly in Italy, the United States, Germany, and Japan.

Concentrated solar energy might also be coupled with thermochemical cycles to split water. The necessary high temperatures have already been achieved in demonstrations with concentrating mirrors that focus the sun's rays on a reaction vessel.

Whether hydrogen is produced by electrolytic or thermochemical methods, whether the source of energy is a coal-fired or nuclear plant, a solar power tower, or a field of solar electric cells, the resulting fuel is an energy carrier much more convenient and versatile than the coal, uranium, or solar energy from which it derives.

Storing Hydrogen Poses Unique Problems

Hydrogen's low energy content per cubic foot as a gas and the very low temperature needed to liquefy it (−267°C, at normal pressure) cause unique storage problems. A number of solutions are in various stages of development.

NASA has developed procedures for storing liquid hydrogen under normal pressure in refrigerated metal tanks, a method also used in industrial plants, and has built a 1-million-gallon tank, the world's largest refrigerated tank, at the Kennedy Space Center. However, the process of converting hydrogen gas to a liquid and keeping it refrigerated is expensive, consuming the equivalent of 25 to 30 percent of hydrogen's energy value.

Even larger amounts of hydrogen could perhaps be stored as a gas under high pressure in natural cavities such as depleted oil and gas fields, mines, and caverns. This high-pressure storage would require less energy than refrigerating liquid hydrogen.

Storing large amounts of hydrogen gas at very high pressure in metal tanks is another possibility, but it appears to be much more expensive to build tanks to withstand high pressure than to use existing natural storage sites.

A new and promising means of storing hydrogen exists in hybride compounds. Certain metals and their alloys, such as magnesium-nickel, magnesium-copper, and iron-titanium, can absorb hydrogen gas to form chemical compounds referred to as hydrides. When heated, these compounds release the hydrogen again. This storage method avoids having to contain large volumes of hydrogen as a gas or to maintain the special pressures and temperatures needed to store hydrogen as a compressed gas or as a liquid. Although they can store twice the hydrogen in the same space required by liquid hydrogen tanks, hydrides store relatively little energy per pound. Research underway to find lighter hydrides or hydrides with even larger storage capacities could lead to convenient use of hydrogen in vehicles.

Transport by Pipelines

Small amounts of liquid or gaseous hydrogen are now routinely transported on land or water in special storage vessels. The large amounts envisioned in the future can be transported only by pipelines or ocean tanker. Systems of pipelines have transported gas mixtures containing hydrogen for many years. However, improved equipment will be necessary to handle pure hydrogen gas at high pressures because the pipe materials commonly used become brittle under such conditions. Stainless steel, while it does not become brittle, is prohibitively expensive for long-distance pipelines. If pure hydrogen gas is to be transported in pipelines, inexpensive, embrittlement-resistant materials, or new techniques for designing pipelines with existing materials, must be developed. Adding certain impurities, or "inhibitors," to hydrogen could also eliminate embrittlement.

Phasing in of Hydrogen

Hydrogen may come into wider use, as early as the 1980s, the timing depending largely on economics. The first new application may be as a raw material rather than as a fuel. Substituting hydrogen, produced from water or coal, in the manufacture of various chemicals now made from natural gas would stretch out dwindling natural gas supplies.

Adding hydrogen to the natural gas used for fuel can also stretch supplies. There appear to be no major technical problems in using mixtures containing 10 to 20 percent hydrogen, although due to the price of hydrogen, consumer gas bills would increase.

Using hydrogen for load-leveling in coal-fired, nuclear, or solar power plants might become a reality in this century.

Power plants could use their excess electricity to produce hydrogen electrolytically during off-peak time and then reconvert the hydrogen, in fuel cells or gas-turbine-driven generators, to electricity for use during peak times. Although this electricity → hydrogen → electricity conversion is only 40 to 45 percent efficient with today's equipment, efficiencies of 60 to 65 percent would be possible with new electrolyzers and fuel cells being developed. Supplementing power-plant generating capacity with such peak power systems based on hydrogen could save scarce fossil fuels. In addition to its regular use for rocket propulsion, hydrogen has also been used experimentally as a fuel for airplanes, naval vessels, and motor vehicles. Hydrogen's light weight is an obvious plus for use as airplane fuel, although the large storage space needed (four times the space that jet fuel requires) would require modifications in airplane design. However, the reduced weight of the fueled aircraft would reduce fuel requirements by about one-third.

A conventional automobile engine can operate on hydrogen with relatively simple changes in the carburetor. Although they require twice the space, liquid hydrogen tanks for cars may have an advantage over hydride storage because they weigh only one-tenth as much. However, since hydrogen is extremely flammable, techniques to prevent static sparks in hydrogen storage and handling areas need to be developed. Since hydrides release their hydrogen slowly when heated, there is little risk of explosion from damage to a hydride storage tank. Hydrides may thus be transported more safely. Hydride storage has already been used successfully in a city bus operating between Provo and Orem, Utah.

Technology aside, hydrogen may be slow to take over as a substitute for gasoline because the vast existing network for distribution of gasoline will have to be modified to handle hydrogen. Furthermore, several million automobile drivers and service-station attendants will have to be educated in the safe use of hydrogen. Hydrogen is fundamentally no more dangerous than gasoline or natural gas, but it has different properties that require different precautions. For example, since hydrogen burns with an invisible flame, a flame colorant may be required. With the exception of nitrous oxide, exhaust emissions from hydrogen-fueled vehicles should be environmentally acceptable (no lead, sulfur, smoke, or odor; very little carbon monoxide; few hydrocarbons).

Hydrogen as a Universal Fuel

Widespread use of hydrogen as an all-purpose fuel would be the last stage in the evolution of hydrogen energy systems. The heat produced by solar and nuclear plants could provide the energy to generate hydrogen. Instead of using that hydrogen at the power plant to generate electricity, it could be piped to urban centers, where it would fuel vehicles and provide heat for homes and for industrial processes. Electricity, which now accounts for about one-quarter of United States energy production, would have a different role than it does now. Most of the electricity needed for specialized uses, like running motors and light, could be generated in dispersed, hydrogen-fueled substations, eliminating the need for massive cross-country transmission lines.

If such a "hydrogen economy" becomes a reality, it will likely occur well into the twenty-first century. Considerable new technology must be developed, and other factors can stall introduction of that technology—including concern about safety and the environment, and the large investments that will have to be made in new facilities if hydrogen is to become a common fuel.

Thus, it will probably be a long time, if ever, before the world runs completely on hydrogen. In the meantime, the hydrogen stockpiled in overwhelming quantities in every body of water could help to replace petroleum and natural gas by powering cars, providing heat to homes and industry, and serving as a chemical feedstock.

NUCLEAR FUSION

Most of the energy we use on this planet is converted from its stored (potential) form by combustion—by "burning" fuels. The important feature of this combustion reaction is that it is, in scientific language, *exothermic*—that is, it releases more energy than is required to start or keep it going.

In the search for ways to replace the primary energy of fossil fuels, we have turned to different kinds of exothermic reactions—nuclear ones. Fission, which is the energy source for nuclear power plants, is one such reaction. The fusion reaction is one of the most promising exothermic reactions—from the point of view of energy released per pound of fuel—that science has yet discovered.

It is not enough, however, for a reaction to be exothermic. To be a large-scale source of energy, a reaction must be self-sustaining, or, to use fusion terminology, something like a chain reaction must occur. It must be arranged so that some of the energy released in each event is absorbed by the surrounding fuel material causing further reactions to occur. Only under this condition will one get more energy out than is put in. For instance, we attempt to light a fire by supplying energy with a match. However, unless the fire wood has been arranged so that smaller sticks catch fire and heat larger sticks to their kindling point, the fire does not burn. It is toward the achievement of this second condition that fusion scientists are currently struggling.

The fusion reaction was first scientifically demonstrated and studied about 45 years ago. It has glimmered brightly before us for almost 30 years, giving us hope for a return to the Eden of abundant energy. To date, however, the only synthetic, self-sustaining fusion reaction has been the explosion of a hydrogen bomb—which has nothing to do with Eden.

Fusion uses "fuels" that are essentially inexhaustible. It appears that it does not pollute the environment. It does not (unless it is designed to do so) produce materials that can be sidetracked for bombmaking. In fact, solar energy alone can compete with its promise. But the achievement of

this promise on a commercial scale is surely the most difficult technical task that our species has yet undertaken.

Resources

The resources for the fusion reactions are deuterium, an isotope of hydrogen, and the light metal, lithium. Lithium is an indirect resource since the other important fusion fuel, tritium, a radioactive isotope of hydrogen, will be "bred" from it in the fusion reactor.

Deuterium, which is rare, combines with oxygen to make the laboratory curiosity "heavy water." In a natural sample of water, only one molecule in 6500 is of that variety. In other words, in 60,000 pounds of water, there is only about 1 pound of deuterium. But water is enormously abundant, and the earth's oceans, rivers, and lakes contain 10 trillion tons of deuterium. The world's total recoverable resources of coal are estimated to be only 6 to 8 trillion tons. In a fusion reaction between dueterium nuclei, the total amount of energy released is 340 million Btu per gram of deuterium or about 1.5 quadrillion (1.5×10^{15}) Btu per ton. In contrast, coal releases at most 25 million (25×10^6) Btu per ton when burned. Thus, each ton of deuterium could produce 60 million times more energy than a ton of coal.

The energy content of all this deuterium is difficult to comprehend. The total energy the world uses in a year could be obtained from 200 tons of deuterium. Even if the world consumed twice the annual amount of energy it now does, the deuterium supply would last about 50 billion years—which is longer than we can be sure the world will last. The use of deuterium as a fuel is the most attractive fusion possibility.

The easier reaction to achieve experimentally, and therefore the one presently emphasized, uses an even rarer isotope of hydrogen, tritium. Lithium, from which tritium could be made, is less abundant than deuterium. However, from its known reserves alone (and there has been little exploration for lithium), we could fuel the world at twice its present level for almost 50,000 years.

It will require energy, of course, to separate that one deuterium atom from its 6500 or so chemically identical counterparts. We do that now; heavy water is produced routinely, with the energy loss of providing a gram of deuterium amounting to less than 1 percent of the fusion energy available from that gram. Clearly, fusion energy will not be limited by fuel resources. Limits may be set, by other sources—for instance, capital or the materials needed to build the complicated plant and its machinery.

Technology

Despite more than two decades of determined effort by scientists and engineers of several countries, the demonstration of a controlled, self-sustaining fusion reaction has not been achieved.

Fusion is a nuclear reaction. In a sense, it is the opposite of fission. The release of energy in fission occurs because a large nucleus (uranium, for instance) is split into two smaller ones. In the fusion reaction, two very small nuclei combine to form a larger one. In both cases, the mass of the end product(s) is less than the mass of the original reacting nuclei, and this lost mass is converted into energy.

The early study of fusion reactions was accomplished with "particle accelerators"—cyclotrons, Van de Graaff generators, etc. In these experiments, deuterium particles were hurled against a stationary target of deuterium or tritium. Created in this fashion, however, the reactions are not self-sustaining. Even though net energy is released in each successful reaction, most of the incoming particles miss, and the emerging particles do not hit others and cause a chain reaction. On the whole, much more energy is used to cause reactions in these experiments than is released by them.

Fusion reactions do occur in the sun and in hydrogen bombs. In those cases, energy is supplied in the form of heat. If a mixture of deuterium and tritium (D + T) can be held together and brought to a temperature of 50 to 100 million °C, the fusion reaction will take place. The *ignition temperature*—as this reacting temperature is called—is about 500 million °C for a D + D mixture. Since the ignition temperature for D + T mixtures is lower, the experiments now underway concentrate on this reaction.

The enormous temperatures that are needed greatly limit confinement techniques. Ordinary vessels such as bottles, cans, and tanks cannot be used. The reacting particles must be suspended in a vacuum, free of any matter that could conduct their heat away. There are two ways to accomplish this; magnetic confinement and inertial confinement.

In magnetic confinement, the deuterium-tritium mixture is given enough energy so that the electrons are stripped from the nuclei, forming a "plasma" of charged electrons and nuclei. This plasma can be controlled by a magnetic field in much the same way that a beam of electrons is controlled in a television tube.

In inertial confinement, a solid target (a droplet or sphere) of deuterium and tritium is heated extremely rapidly so that it reaches the ignition temperature for fusion before it can expand and reduce its density. Bombardment of a small sphere of deuterium from all sides with a high-powered laser is one method that may achieve this.

Magnetic Confinement. There are three basic magnetic confinement systems under development.

1. *Toroidal-shaped chambers* ("doughnuts"), in which the plasma travels around inside an evacuated chamber. The Russian-invented Tokamak is the most successful of these and about 70 percent of the American effort is going into similar devices. Examples are the Princeton Large Torus (PLT), Oak Ridge Tokamak, Alcator at the Massachusetts Institute of Technology, Doublet IIA at the General Atomic Company in La Jolla, California, and the much larger Tokamak Fusion Test Reactor under construction at Princeton University.
2. *Magnetic mirrors.* These are linear tubes in which the magnetic field that confines the plasma is shaped so

that it turns the particles around at each end, as a mirror does to light beams. The most successful of these devices is the 2X-IIB at the Lawrence Livermore Laboratory of the University of California. Mirrors are now the principal alternative to Tokamaks.
3. *Magnetic pinch device.* In these, the interior space is filled with plasma, which is then "pinched" by a rapid compression of the magnetic field. This is accomplished by increasing the strength of the field and forcing the plasma toward the center of the tube. The Scyllac at Los Alamos is a pinch device.

In addition to the proper confinement conditions, the plasma must be heated to 50 to 100 million °C. Heating can be accomplished in three ways.

1. The plasma may be heated up induced electric currents. The plasma heats up like a resistive wire does when current flows through it. The toroidal machines rely, in part, on this type of heating.
2. A plasma can be heated by injecting it with an energetic beam of nuclear particles. Techniques of firing beams of uncharged deuterium atoms into the plasma are used in the toroidal and mirror machines.
3. The plasma acts in some respects like a gas (a gas-charged particle). Therefore, like a gas, it can be heated by compression. The magnetic pinch devices not only confine their plasmas, but heat them.

Inertial Confinement. The confinement problems can also be solved by freezing the mixture of deuterium and tritium to form a solid. The major problem left is that of heating and compressing the solid to the temperature and density needed for the reaction to occur.

Experiments with inertial confinement have proceeded along the following lines. A small frozen droplet of the fusion fuel, preferably spherical and less than a millimeter in diameter is placed in the center of an evacuated chamber and bombarded from all sides by an energy source—a laser, or a beam of charged particles (ions or electrons).

Lasers are powerful sources of light that can be accurately focused to a very small spot. In order to bombard the pellet symmetrically from all sides, the main beam is usually broken up into several smaller beams by a system of mirrors, and these beams are then all brought to bear on the target sphere. The burst of energy must last only about 10^{-9} seconds in order that the deuterium-tritium mixture is heated rapidly without expanding. As the energy hits the outside layer, the material essentially vaporizes. In rushing outward, it exerts a reaction force back against the sphere, which "implodes" the sphere, crushing it inward and greatly increasing its density. The high temperature and high density allow ignition of the fusion reaction, and the pellet explodes. Most of the energy is carried out by neutrons released in the fusion reaction.

Nuclear Fusion in Practice!

For self-sustaining fusion to occur, the plasma density, confinement time, and temperature must all be above certain values. Satisfactory values of each of these parameters have already been achieved separately, but not in the same machine at the same time.

What we are seeking is a chain reaction, of sorts, in which the energy released by one fusion event causes another. To occur, either the particles must be very close together (have a high density), or there must be sufficient time for the energetic particles to wander around until they hit other ones (a large confinement time).

Fusion Reactors

Given the current status of fusion and the scientific and engineering problems that must be solved before the scientific feasibility of fusion can be demonstrated, it may seem premature to consider what a commercial power-producing plant would look like. It is important, however, to some of the problems that will remain after the scientific success is achieved.

The magnetic confinement machines will all be large, it appears, with generating capacities in the thousands-of-megawatts range. The major components will be the chamber itself and a surrounding thick blanket of lithium—probably in a molten form—which will absorb the neutrons and convert their energy to heat. The lithium will also be the source of the tritium and Helium³ which will be created by reactions between the neutrons and the lithium nuclei and then separated from the blanket and used as fuel. Heat extracted from the hot lithium will be used to create steam to turn a steam turbine. It is expected that high temperatures will be achieved and that efficiencies will approach 40 percent.

There is some hope for even greater efficiencies. In one fusion reaction, the end products are charged. It may be possible to separately collect these particles and the electrons accompanying them, and to directly convert some of the kinetic energy into electrical energy. Efficiencies in the 80 percent range may be feasible.

Unless some unforeseen breakthrough occurs, however, the first fusion reactors will be huge machines, practical only as sources of large amounts of electrical energy. (They could also be used for chemical processing or to breed fissionable fuel.) They will certainly increase the already important centralization of energy production.

Laser fusion may depart from this trend. The energy released in each explosion will be relatively small. They will be pulsed, one small explosion following another. It may be that generators with capacities as low as 100 megawatts will be feasible.

A laser fusion generator will be quite different than the magnetic machines. The reactor vessel will be a fairly large sphere, strong enough to withstand the repeated small explosions as the pellets are bombarded. Lithium will be introduced in some manner around the inside surface of this vessel. It will perform the same function as in the magnetic devices, creating new fuel, absorbing the energy from the neutrons, and converting it to heat. Ample challenge still

remains, moreover, in the search for a means to make inexpensive little target pellets (the electricity generated from each explosion is worth only a few cents) and in constructing a vessel that is not weakened by the neutron bombardment and/or damaged by the explosions.

Environmental Effects and Safety

Fusion reactors seem to offer significant advantages over fission reactors in their possible effects on the environment and on the society that uses them. The threat of an accident will be greatly reduced. Fusion reactors will not contain the huge amounts of radioactive material characteristic of fission reactors. Tritium will be produced, but the entire system will be designed to recapture and consume this material, and its stockpile, even if released, presents much less of a threat than the fission products. Radioactivity will be produced in the reactor materials by the neutrons, and, while this may be a problem, it probably can be reduced by proper choice of construction materials.

There is no "critical mass" involved in fusion. Any malfunction would destroy the plasma and stop the reaction. Although there will be some "after heat" remaining in the reactor structure, it will be much less than that in a hot-fission reactor core, and it will not pose the meltdown problem.

The radioactive waste problem would be ameliorated. Some storage of discarded materials may be necessary, but in the long run, large-volume storage problems would be avoided.

Successful fusion power generation could offer relief from the plutonium nuclear-bomb threat. Unless neutrons from fusion reactions were deliberately used to breed plutonium from uranium 238 (as has been proposed), no potential bombmaking materials would be produced or transported, a very significant advantage.

Because of the enormous amounts of energy that fusion fuels contain, their production should cause little disruption of the environment. Removing one hydrogen atom in 6500 from the ocean will have no measureable effect, and the amount of lithium needed is miniscule when compared to the projected needs of coal or even uranium. The fusion reaction would be, it appears, as environmentally benign as any technology except solar energy.

OCEAN THERMAL ENERGY CONVERSION

More than 70 percent of the solar energy reaching the earth falls on the oceans and other bodies of water. Almost a century ago, the French scientist Jacques d'Arsonval considered the possibility of tapping the thermal wealth of the oceans. In the 1920s, his countryman Georges Claude, already famous for his work in liquefying gases and developing industrial uses for acetylene, helium, and neon, made ocean thermal energy conversion (OTEC) a reality. He succeeded in producing 22 kilowatts of electric power at a facility on the Cuban coast that was supplied with water from the depths through large pipes.

Because Claude used pumping equipment that required more power than he produced, most engineers considered his OTEC project a failure. However, as it became obvious that alternate sources of energy were desirable, the idea of drawing power from ocean heat again began to receive attention. OTEC is now one of a number of important alternative energy concepts being developed by the U.S. Department of Energy (DOE).

Tapping Ocean Thermal Energy

Solar energy warms the surface water of the oceans and other large bodies of water to temperatures considerably higher than those beneath the surface. The difference in temperature is a source of heat energy, which can run a steam engine such as the one Georges Claude built. In his design, warm surface water was boiled at low pressure to drive a turbine and was then condensed by cold water pumped up from the depths. Solar energy replaced heat taken from the surface water. Claude's steam engine operated with a water-temperature difference of only 24°F.

Present OTEC designs use fluids more suitable than seawater as the working medium. D'Arsonval had suggested ammonia, and DOE is taking that approach. Even so, an OTEC plant cannot be as efficient as a high-temperature steam-electric plant. However, with the vast amount of solar heat stored in the ocean, efficiency is not as critical as it is when expensive fuels are used.

The energy potential of the oceans is tremendous. One estimate is that 180 trillion kilowatt-hours of power could be generated annually from the Gulf Stream flowing along the eastern coast of the United States. This is about 75 times the amount of power that the country will be using in 1980. Coastal areas with suitable temperature differences for OTEC plants include an appreciable share of the world's population, and the United States is especially favored in this respect.

Advantages of Ocean Thermal Energy

OTEC accounts for almost three-fourths of the solar energy the earth receives, and it has other advantages as well. Unlike direct solar energy sources, the oceans are a huge reservoir of energy conveniently stored for use at any time. This is particularly attractive in the generation of electric power since the ocean would provide a 24-hour-a-day, year-round source, rather than the intermittent supply of sunshine.

Conventional power plants produce pollution of one kind or another, and their waste products must be disposed of. OTEC is a clean source of power and can be located out of sight of land on relatively cheap real estate. Plants other than just power plants might be built in areas near raw materials. Aluminum plants are an example; bauxite ore could be delivered short distances to the OTEC plant by cheap water transport.

By-products of OTEC plants may make them even more attractive. Fresh water is becoming as vital as energy in some areas, and a 100-megawatt OTEC plant could produce millions of gallons of fresh water daily by desalting the warm water used in its boilers. OTEC plants might also create productive fisheries in the waters surrounding them. Pumping cold water from the depths also brings up nutrients that attract fish to the area. The Humboldt Current, a natural cold-water upwelling off the coast of Peru, creates a natural fishery, supplying about 20 percent of all the commercial fish in the world.

Problems to Be Solved

Problems must be solved before OTEC can supply part of the United States energy. These include the corrosion of metal components in salt water and the "biofouling" of heat exchangers and other equipment by marine organisms. The construction of huge boilers, pipes, and other parts as well as their assembly at sea will require new techniques. The stability of giant OTEC plants in heavy seas is another question, as is the feasibility of laying and maintaining long undersea transmission lines.

Just as marine life may adversely affect OTEC plants, OTEC may also be a hazard to these organisms. For example, ammonia leaking into the sea would not only corrode metal parts but also endanger marine life.

In addition, the long-range effect of removing heat from the ocean must be resolved. Although OTEC scientists generally agree that, as heat is withdrawn, the ocean water will retain additional solar heat to maintain its temperature, DOE must determine that no adverse effects exist, such as a change in ocean temperatures or even harmful local changes.

A prototype OTEC plant with about a 25-megawatt electric output will be constructed in the early 1980s. If this pilot plant is successful, a 100-megawatt demonstration plant will be build by the mid-1980s. Georges Calude's prediction of energy sufficiency from the oceans would then move a step nearer fulfillment.

OIL

It isn't worth much per gallon, or even per barrel, but for decades the value of the total world production of petroleum—often called *black gold*—has been many times that of all the gold mined during the same period. The annual world oil output is now about 20 billion barrels, an average of 760 liters (200 gallons) for every human being on earth. The United States consumes about 30 percent of this total, or more than 4160 liters (1100 gallons) per person per year.

Through its many important applications as an energy source and a lubricant, and through its unique potential as a source of thousands of valuable petrochemicals, petroleum has probably had a greater overall usefulness to humanity than any other material found in nature except air, water, and the land itself.

Petroleum now supplies directly or indirectly, a total of 44 percent of American energy requirements. Over half of the petroleum consumed in the United States is used for transportation by automobiles, airplanes, trains, and boats. Modern technology has not yet produced a satisfactory substitute for oil as an energy source for most of these transportation applications. Without an adequate and continuing supply of crude oil, or a satisfactory replacement for it, the civilized world would come to an abrupt stop.

Origin and Nature

Early scientists classified petroleum as a mineral produced within the earth or as a residue somehow transmitted here from outer space. The present consensus is that petroleum was created hundreds of millions of years ago from the decomposition of simple forms of plant or animal life. There is evidence that these organic substances were later buried by sedimentation; through the action of heat and pressure over millions of centuries, they were transformed into petroleum.

Petroleum consists of a mixture of a large number of complex chemicals called *hydrocarbons*. As the name suggests, these compounds consist largely of the basic chemical elements, carbon and hydrogen, sometimes combined with other elements such as sulfur and oxygen. Although usually a liquid, petroleum ranges from light-colored water-thin products to black, solid, pitch-like bitumen.

Petroleum deposits are usually found in sedimentary rocks, which were formed aeons ago from a consolidation of sand and other minerals in water—often the sea. Frequently, petroleum is intermingled with natural gas and salt water in consolidated (solidified) or unconsolidated (loose) sand. Petroleum is seldom, if ever, found in a large underground pool; it is usually mixed with salt water and gas in the pores of rocks that appear to be solid but are filled with minute pores. When a well is drilled into oil-bearing rock, the pressure of the gas trapped in the oil often forces part of the oil to the surface. This led to the early "oil gushers" that were spectacular proof, but also a gross waste, of the oil fields of Texas and elsewhere.

History

Petroleum has a fascinating history. In ancient Persia, the kind of petroleum known as *bitumen* was used as a bonding material instead of mortar. It was also widely used to caulk ships, including Noah's Ark, according to certain versions of the story of the Great Flood. Bitumen was even supposed to have great curative powers.

In early Christian days, the Arabs developed methods for distilling petroleum to produce various products for uses such as lighting and cleaning textiles. The Chinese also accidentally discovered oil in the third century while drilling for salt. Marco Polo, during one of his celebrated thirteenth-century trips, noted the early commercial development of the oil fields of the Baku Peninsula, in a section that was

to supply Russia's fuel during World War II. The primitive religious cults of fire worshipers in this area were based on the "eternal fires," which may have been caused by natural gas seepages ignited by lightning in prehistoric times.

In the mid-1800s, relatively limited quantities of liquid fuels were regularly produced in the United States and abroad. These were mainly for illumination and cooking and were produced by the simple distillation of raw petroleum available as ground seepages or skimmed from the surface of certain creeks or rivers. Near one such stream—Oil Creek in northwestern Pennsylvania—the most important event in United States oil history occured in 1859. At that time, the first well in which oil was found at a depth of only 21 meters (69 feet) was drilled at Titusville.

Oil in commercial quantities was also discovered and exploited in the late 1800s in other parts of the world, but the United States and Russia accounted for 90 percent of the world's crude oil output at the beginning of the twentieth century.

From 1903 to 1911, a severe petroleum shortage developed in the United States because of the rapidly expanding market for gasoline. This was due to a tenfold increase in the number of "horseless carriages." In 1913, scientific research, stimulated by the shortage, produced the "cracking" process, an ingenious chemical procedure that doubled the refinery yield of gasoline and helped to establish the abundant energy source for the automotive and aircraft industries.

Between World Wars I and II, many improvements were made in the techniques for locating, recovering, and refining petroleum. Gasoline production increased enormously in order to take advantage of the innumerable convenient applications of gasoline and diesel engines. From 1937 to 1947 (which included World War II), more petroleum was taken from the ground than had been produced in all previous history. After World War II, the rate of discovery of new wells in the United States began to fall off, but the demand for petroleum and its products continued to increase.

Recently, apart from its use as a source of many different fuels, petroleum has also been increasingly in demand by the giant petrochemical industry, which now annually turns out billions of dollars worth of synthetic chemicals and useful substances of all kinds. In addition, hundreds of millions of United States machines and engines—most notably, automobiles—depend for efficient operation on specially designed lubricants refined from petroleum.

The development and widespread exploitation of crude oil has taken place in just a little over 100 years. Particularly during the last 50 years or so, a substantial percentage of the world's known crude-oil supplies has been consumed. New developments, such as additional petroleum fields, more complete recovery from known deposits, or the discovery of satisfactory substitutes, will be needed in the near future to meet the constantly increasing demand for the end products of petroleum. The U.S. Energy Research and Development Administration (ERDA) has overall responsibility for developing the technology for accomplishing these objectives.

Finding and Producing Oil

Modern methods used to search for oil are based on sensitive magnetic, gravity, and seismic (shock) measurements. But even though these techniques involve complex technical equipment and ingenious analysis, oil prospecting is still far from an exact science. This is because petroleum itself has no characteristics that enable its presence underground to be specifically pinpointed from the earth's surface. The only sure way to determine whether oil is present in a given spot is to drill for it, which explains why well-drilling is frustrating and often financially ruinous.

Drilling deep wells requires a substantial investment in thousands of feet of well casing, tubing, and drill pipe; the derrick; a substantial power source; drill bits; associated drilling muds; and chemicals. It also requires a lot of trained personnel.

Refining Crude Oil

Petroleum refining consists of a series of complex physical and chemical processes by which crude oil is separated into its various components or "fractions." These may, in later stages of operations, be recombined or chemically altered to yield the desired products for final consumption.

Originally, refining was simply the selective distillation—boiling off at various characteristic temperatures—of the many different compounds that make up crude oil. The more complex and efficient refining methods developed in recent years involve the alteration of the hydrocarbon molecules—the basic building blocks of the various compounds. The full series of intricate modern refining chemical processes include: cracking to break down the larger hydrocarbon molecules, reforming to change the internal molecular structure, and purifying for environmental reasons.

The success of refinery technologists in improving gasoline quality made possible a dramatic increase in the use of internal combustion engines for automobiles, trucks, aircraft, and farm machinery. It also contributed greatly to the "quality of life" and high industrial productivity in the United States and around the world. At the same time, it must be recognized that this high productivity requires the consumption of tremendous amounts of energy.

Petrochemicals

The term *petrochemicals* does not usually refer to fuels, such as gasoline, kerosene, jet fuel, and heating oils, but is restricted to a large number of useful nonfuel substances and chemicals also derived from petroleum. Some of the most important of these are feedstock chemicals for synthetic rubber, many different kinds of plastics, solvents, detergents, textile fibers, fertilizers, construction materials, coatings, adhesives, and paints.

Basically, the science of petrochemistry is an extension of petroleum refining. It comprises the complex procedures by which petroleum hydrocarbon molecules, consisting mainly of hydrogen and carbon, are broken down, combined, internally altered, and rearranged into new molecules that form different substances with unique and desirable characteristics. These processes use controlled conditions of temperature, pressure, and time, and require special catalysts that stimulate and hasten the chemical reactions.

Transportation

Much of the world's petroleum production is located in areas remote from centers of population. In its movement from the oil field, to the refinery, to the consuming public, petroleum and its products require many different kinds of transportation. These include pipelines, oceangoing tankers, barges, railroad tank cars, and highway tank trucks.

Shortly after the first United States oil well was drilled in 1859, the nation's first pipeline—177 kilometers (110 miles) long—was laid in northwestern Pennsylvania. Since then, use of pipelines has grown steadily, and the United States is now criss-crossed by over 322,000 kilometers (200,000 miles) of oil pipelines. (This does not include an even larger network of gas pipelines.)

The latest development in pipeline technology is the Trans-Alaska Pipeline, which carries oil from Prudhoe Bay on the Arctic Ocean in the north of Alaska to a tanker outlet in an all-weather bay in Valdez in the south. From there, it is transported to the United States and other markets. This remarkable pipeline snakes its way above and below ground, across 1300 kilometers (800 miles) of some of the wildest and coldest country on earth, where average winter readings range as low as -57°C (-70°F). Newly invented design features speed the movement of warm oil through the huge 1.22-meter (48-inch) diameter pipe without melting or destabilizing the permafrost, the permanently frozen ground, that is a natural feature of the region.

Next to pipelines, the cheapest mode of transportation is ocean tanker. These have been used extensively since the first John D. Rockefeller sent a clipper ship full of Pennsylvania crude oil around South America's Cape Horn to California in the early 1860s. There are now over 3500 oil tankers, but there are some economic uncertainties regarding the demand for such a large number.

Future of Petroleum Use

The United States is now consuming about 17 million barrels of petroleum per day. But the demand for petroleum will undoubtedly continue to expand as more machines, heat energy, and electric power are needed for industry, agriculture, transportation, and business operations to keep up the steady pace toward a higher standard of living. Great quantities of petroleum will also continue to be required to produce lubricants and useful petrochemicals. It is predicted that American oil consumption will double in the final quarter of this century. This prospect forces the country to face some hard and sobering facts.

Oil production in the United States dropped 10 percent during 1973-1976, continuing a downward trend that began in 1970. For the first time in history, the USSR, not the United States, was the world's leading oil producer. Furthermore, oil imports, which have been increasing for many years, in 1976 reached a rate of half of our total consumption. If such sizable imports are not balanced by corresponding exports of other goods and services, serious currency and other economic problems may result. Heavy dependence on foreign sources for oil will also jeopardize national security.

Figure 1-2 presents a summary of the best available data on past and projected crude-oil production in the United States. It is significant that of the nation's total original petroleum deposits that can be recovered by known technology, about three-fourths are expected to have been consumed between the Great Depression of the 1930s and the end of this century. The United States is estimated to have now only 8 percent of the world's remaining recoverable oil, whereas the Middle East possesses 62 percent, of which 25 percent is in Saudi Arabia.

Thus, the United States must find more petroleum, process it for carefully planned uses, recover more petroleum from known domestic deposits, and at the same time explore other alternatives and options for suitable substitute energy sources.

Two of the most important areas of investigation relating to petroleum are (1) the enhanced recovery of oil remaining in the ground after active fields have been "played out" and (2) the possibility of using the oil contained in certain types of shale rock, of which there are vast quantities in the western United States. Many other alternatives are also being explored in such fields as improved coal utilization and nuclear, solar, and geothermal energy.

Ordinary oil-recovery methods extract about one-third of the oil at a specific site. Enhanced oil-recovery techniques are designed to recover some of the remaining two-thirds. The most common procedure, now used, "waterflooding," is to pump water under pressure into an oil reservoir, in order to sweep a sizable portion of the remaining oil toward the producing wells. About 40 percent of the current United States production is by this method. Other procedures use detergents, solvents, acids, and heat, depending upon conditions. Although some of these methods appear promising, the technology in most cases is far from the practical stage. Thus, it may be many years before they are used extensively.

Attempts to produce oil from oil shale involve many difficult technical, economic, and environmental problems. On the average, 1 ton of rock must be processed for every 95 liters (25 gallons) of oil recovered. In addition, present techniques require large amounts of water, which cannot be recycled because it is chemically bound with the waste material. The disposal of this water poses a gigantic environmental problem.

In summary, recovery of petroleum will continue to be of critical concern throughout the world. No reasonable amount of conservation can possibly make available the

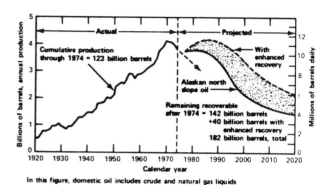

Figure 1-2. Actual and Projected United States
Domestic Oil Production.

vast amounts of petroleum required to meet all of the anticipated demands. The present high productivity and standard of living in the United States are based on energy-intensive methods of agriculture and manufacturing. Productivity—basically, the rate at which an individual worker can turn out goods and services for other people—has been directly correlated with increased energy use. Therefore, the success of ERDA's National Plan for Energy Research, Development and Demonstration is of critical importance to the American people.

OIL SHALE

Oil shale is a fine-grained rock that contains varying amounts of organic material called *kerogen*, which, upon heating, or retorting, to about 482°C (900°F), yields a synthetic oil and gas. The major deposits in the United States exist in the Colorado-Wyoming-Utah area, known as the Green River Formation, and the area of the Upper Mississippi Valley to Michigan, known as the Devonian and Mississippian deposits.

The richer shale deposits are located in the Green River Formation. The estimate of recoverable oil there is about 0.6 trillion barrels, about one-third of the 1.8 trillion barrels of oil contained in the shale. This is based on use of shale that yields from 15 to 25 gallons of oil per ton of shale. The Michigan Antrim oil-shale deposits, on the other hand, consist of predominantly low-grade oil shale (less than 15 gallons per ton). Gasification is considered the most applicable technology for this resource, with the potential yield equal to some 3 trillion barrels of oil if the gas can be recovered economically.

During World War II, a high demand for liquid fuels prompted Congress to pass the Synthetic Liquid Fuels Act of April 6, 1944. The federal government then began to develop ways to produce synthetic oil from shale. The U.S. Department of Interior's Bureau of Mines (BOM) began development of a gas-combustion retort in 1945 at the Anvil Points Facility, a research complex on the Naval Oil Shale Reserve near Rifle, Colorado. The work was performed at the laboratory and pilot-plant stages for the next several years. In 1958, the facility was shut down by BOM. A six-company industry group leased it from BOM, modernized the retort, and conducted further experiments starting in 1964.

Although oil shale is not now used commercially in the United States, most of the usable technology was developed through earlier BOM research. Much of that technology, however, is suitable only for mining relatively shallow deposits, which represent only 20 percent of the nation's oil-shale resources, and aboveground retorting.

Oil-Shale Energy Systems

Shale oil can be produced by three general retorting processes: *surface* or *aboveground in situ,* and two underground methods known as *true in situ* and *modified in situ.* In each case, the oil shale must be heated (retorted) to 900°F or higher to bring about the chemical decomposition of shale oil or gas. The process used to produce the synthetic fuel depends on the depth, assay, and geographic location of the shale formation or deposit, and the technological, economic, and environmental tradeoffs.

Surface processes require mining of oil shale by either underground or surface methods. The mined shale must be crushed and sized before retorting, followed by disposal of the spent (retorted) shale.

Two methods for surface retorting—the direct and indirect heated systems—are of industrial interest. The heat required to decompose the kerogen for the direct-heated process is supplied by creation of a combustion zone within the retort. The combustion is sustained by gases released from the heated oil shale. In the indirect-heated processes,

Table 1-4. Potential environmental concerns.

PHASE	OIL SHALE PROCESSES	PHYSICAL DISTURBANCES	POLLUTANT DISCHARGES	AFFECTED RESOURCES	
				PHYSICAL RESOURCES	SOCIOECONOMIC
Extraction through retorting	Surface retorting	Aquifer local interruption Land disconfiguration (strip mining) Roof collapse Noise (drilling, retorting) Retorted shale waste piles	Runoff or leachate (metals, organics, salts) from retorted shale pile Dust from mining, crushing, and grinding Fugitive emissions and off-gases from retort (venting to air) Contaminated retort water (metals, organics, salts) in settling ponds	Water for dust control, process cooling, and vegetation and community use Secondary recovery of minerals	Financing Labor force Community services Housing Power Equipment
	True in situ	Work site disturbance Subsidence or uplift Noise (drilling, fracturing) Aquifer local interruption Heat	Leachate (metals, organics, salts) from retorted shale into aquifer Fugitive emissions and off-gases from retorting (venting to air) Contaminated retort water (metals, organics, salts) in containment ponds	Water for community use and process cooling	
	Modified in situ	Aquifer local interruption Subsidence or uplift Noise (drilling, fracturing) Raw shale waste piles Heat	Leachate (metals, organics, salts) from retorted shale into aquifer Runoff or leachate (mainly salts) from raw shale piles Dust from mining/fracturing Fugitive emissions and off-gases from retorting (venting) Contaminated retort water (metals, organics, salts) in settling ponds	Water for community use, processing, and vegetation of raw shale	
Upgrade through end-use	All processes	Land disturbances for facilities, roads, other transportation Physical plants	Evaporation and emissions of crude-oil volatiles, during storage, upgrading, and refining Accidental spillage	Water for upgrade/ end-use stages and community use	

gases are circulated to an external reactor or combustion unit.

Environmental Concerns

Three classes of environmental concerns related to oil-shale technological development are physical disturbances, pollutant discharges, and affected resources. The major concerns for the oil-shale processes are listed in Table 1-4. The upper portion of the table covers the extraction through retorting phases of the oil-shale energy system. Some of the concerns—contaminated retort water, subsidence, and uncontrolled emissions—are common to the three processes. Others are specific to one or two processes; for example, the retorted shale-waste piles from surface retorting and leaching of wastes directly into six aquifers with the *in situ* processes.

In another study, scientists will analyze plants grown on oil-shale wastes to determine the amounts and kinds of trace elements absorbed from the wastes into the plants' systems. The studies are key parts of a DOE program to gather detailed information on potential effects of future oil-shale development designed to understand and lessen environmental problems.

PHOTOVOLTAICS

Photovoltaic devices provide a clean, simple way of converting sunlight directly into electricity. They can be used in a wide variety of applications, the first well-known extensive use being in the United States space program. Photovoltaic systems do not require complex machinery, engines, or heat-conversion methods, and they operate silently. The scientific principles and technology are established. The major impediment to wide use is the high cost associated with the manufacture of photovoltaic devices.

Solar Cells

The photovoltaic process occurs when light hits units of sensitive materials, creating an electric current. These basic units, called *solar cells,* are made of solid crystalline materials, such as silicon.

Solar cells can be connected electrically to form solar modules, which are the basic building blocks of solar electric systems. Almost any amount of power output can be supplied by combining solar modules into a solar array, controlling the voltage, and directing the power into a storage and distribution system. A solar array is rated in peak watts.

Environmental and Other Considerations

Photovoltaic power is expected to have little adverse effect on the environment. Indirect environmental impacts may occur, however, from the industrial processes that will be used in the production of millions of solar cells and in the amount of land required for large-scale photovoltaic installations.

The most important concerns at present are the high cost of photovoltaic arrays, the small production capacity of the current industry, and limited demand.

If the DOE goal of $0.50 per peak watt by 1986 is reached, electricity from solar cells can be expected to make a substantial contribution to the nation's energy supply by the turn of the century.

SOLAR THERMAL ENERGY

The sun's energy, falling to earth in a diffuse manner, can be trapped between the glass and black metal surfaces of a simple flat-plate solar collector and heat water to more than $100°F$. Such can provide heat and hot water for home and industrial use. But if the sun's rays are concentrated with mirrors or lenses, temperatures of hundreds of degrees can be produced. And fluids heated to a few hundred degrees can drive heat engines coupled to generators to produce electricity or mechanical power, which is how most of the nation's electricity is produced.

There is no technical limitation that would prevent use of solar thermal technology right now. The challenge lies in bringing down the cost of solar thermal energy systems so they can be competitive with conventional energy sources.

Two main approaches are being developed: central-receiver systems, and distributed-receiver systems for dispersed power application. In a central-receiver power plant, there are numerous movable mirrors, called *heliostats,* which focus the sun's rays on a central collection device located in a tower. This device, called a *receiver,* transfers the concentrated solar energy to a liquid or gas, which drives a turbine generator to produce electricity. The central-receiver approach is well suited to large power plants.

In the distributed-receiver system, each lens or reflector has its own thermal collector. The hot fluid from the collectors is pumped to one or more small heat engines to be transformed into electrical or mechanical energy. Distributed-receiver systems are adaptable to communities, shopping centers, office buildings, factories, power plants, and farms.

Total Energy Systems

Total energy systems generate electricity and use rejected heat for heating buildings and actuating cooling systems or for industrial processes.

Two total energy systems are sheduled for completion in 1981. One is at the Fort Hood military base near Killeen, Texas. It will provide a significant portion of the electricity, space heating, hot water, and air conditioning for a modern residence complex on the base. It will have 125,000 square feet of line-focusing collectors, which will generate 250 kilowatts of power.

The other system will provide electricity, steam, heat, and hot water for a 42,000-square-foot knitwear plant being built at Shenandoah, Georgia. The system will use parabolic dish collectors (point-focusing) to heat a working fluid to about $600°F$. This hot fluid will be used to boil water, which in turn will drive a 400-kilowatt turbine generator, providing both process steam and electricity. The system is expected to produce 60 percent of the electricity and 50 percent of the factory's steam requirements.

Distributed Systems for Irrigation

Much of the food that American produces, consumes, and exports comes from irrigated fields. Because of rising fuel costs, new ways to power irrigation pumps are being developed, including use of solar energy. For example, distributed collectors or small central-receiver systems can be used to pump irrigation water, using either thermally driven heat engines or by generating electricity for conventional electric pumps.

An experimental solar-powered irrigation system is operating on a farm near Willard, New Mexico, about 60 miles from Albuquerque. Its parabolic trough-shaped collectors concentrate sunlight receivers, which contain an oil-like fluid that remains stable at high temperatures. The concentrated sunlight heats the fluid to about 420°F. The hot fluid moves through insulated pipes to a boiler/heat exchanger, where it transfers heat to another liquid. The second liquid (Freon 113) expands into a high-pressure gas, which drives a mechanical turbine to operate a 25-horsepower irrigation pump. The pump brings up water from a 75-foot well, and the water is used to irrigate 100 acres of wheat, corn, and potatoes. After passing through the turbine, the Freon is condensed and cooled so that it can begin the cycle again.

Small Solar Electric-Power Systems

Distributed-receiver systems also can be used to generate power for small communities. These systems take a different technical approach from larger systems, using a concept that lends itself to "modularity," so that capacity can be added as demand grows.

Outlook for Solar Thermal Power

The advantage of solar thermal energy technology is that it uses a proven technology and a free, clean, and renewable energy resource. It also is very versatile, being adaptable to a wide range of system sizes and temperature requirements. Solar thermal energy systems were used as early as the late nineteenth century, but fell into disuse with the advent of low-cost electric power from fossil fuels.

Solar thermal technology has been criticized as being useful only in the sunny Southwest. However, with modest improvements in technology, solar thermal power systems could be used economically in most parts of the United States.

According to recent studies, solar thermal economics appear favorable. Currently, power from a dispersed, distributed-receiver solar power plant would cost about three or four times as much as power from a conventional plant; power from a large central receiver plant would cost four to six times as much as conventional power. These economics are encouraging when compared to economics of some other emerging energy technologies.

URANIUM: ENERGY FOR THE FUTURE

Resources Finite and Infinite

For many years, energy was abundant and cheap in the United States, encouraging Americans to both consume, and unfortunately waste, large amounts of it. Now, the fuels on which the United States depends most heavily—oil and natural gas—are the least available domestically. Plentiful coal alone cannot meet increasing demand. And no longer can energy be cheap; from now on, its price must reflect the constantly growing cost of recovering diminishing resources from the earth.

The era of fossil fuels, then, is fast approaching a close. We must develop as soon as possible technologies that will lead us into the age of infinite fuels. Already we are benefiting from one of humanity's greatest scientific achievements: the release of usable energy from the atom. Like the burning of fossil fuels in a furnace, the fissioning, or splitting, of certain atoms of uranium in a nuclear power reactor is a source of heat for steam to generate electricity. And the United States appears to have enough economically recoverable uranium to fuel, for their lifetime, the nearly 400 reactors of current design that could be in service at the turn of the century. By then, these plants would be providing about one-third of the nation's electricity.

Furthermore, when existing technologies such as fuel reprocessing and breeder reactors become commercially available, nuclear power can realize its potential as a source of energy, not only during the transition, but for as long as we need it.

Exploration and Development

As with other fuels, harnessing the energy of uranium begins with its discovery and extraction from the earth's crust. Because it is slightly radioactive, uranium can be detected by sensitive instruments from considerable distances. In fact, prospectors can even locate some ore bodies from airplanes. After a deposit has been indentified, geologic surveys, including drilling and sample-taking, determine its size and value.

Uranium ore is then recovered through surface, underground, or solution mining. On the average, 1 ton of ore currently yields about 3 pounds of uranium oxide, the form in which uranium appears in nature. At a mill, the ore is crushed, ground, and dissolved. The uranium is chemically extracted from this solution and dried to a concentrate known as "yellowcake." Later it is converted from uranium oxide powder to solid uranium hexafluoride, which, when heated to a gas is a feed material for the next—and the most difficult—step in nuclear fuel supply.

ENRICHMENT

Only 0.7 percent (7 grams in every kilogram) of natural uranium is the fissionable type, or isotope, uranium 235. The remainder, mostly uranium 238, is not readily fissionable. Such a low concentration will not efficiently sustain the fission chain reaction in today's reactors. The proportion of uranium 235, to uranium 238, therefore, must be increased.

This enrichment to about 3 percent uranium 235 takes place at one of three gaseous diffusion plants owned by the federal government. The isotopes of uranium are separated by forcing uranium hexafluoride gas through barriers perforated with holes less than 2 millionths of an inch in diameter. Since molecules containing the lighter uranium 235 will pass through at a faster rate, each barrier, or stage, produces a gas slightly enriched in uranium 235.

After more than 1200 stages, 1 kilogram of uranium hexafluoride has been separated into about 180 grams of 3 percent enriched uranium and 820 grams of depleted uranium, or "tails," made up of uranium 238 and a minimal amount of uranium 235. The enriched uranium hexafluoride gas is then converted to uranium dioxide powder and formed into 0.5 inch pellets, which will be placed into thin metal tubes, 12 to 14 feet long. As many as 50,000 of these rods form the fuel core of a large power reactor.

Economics

Do so many elaborate and sometimes costly procedures in the production of the fuel make nuclear energy too expensive for generating electricity? Actually, nuclear power plants produce electricity at substantially lower average total cost than coal or oil-burning stations.

Of course, as all fuel prices rise, so will the overall cost of generating electricity. However, the price of uranium oxide from the mill makes up only 5 to 10 percent of the cost of nuclear generation. Nuclear power, therefore, can more readily withstand significant increases in the cost of fuel while remaining economically competitive. For example, assuming all other costs in nuclear generation remain the same, it would cost 1.7 cents to generate a kilowatt-hour with uranium purchased at recently reported prices of about $40 per pound. Even if this price were doubled, the cost of a kilowatt-hour would increase by less than 0.3 cent. A similar increase in price for a ton of coal would add twice as much—0.6 cent—to the cost of a kilowatt-hour. Nearly 2 cents would be added by doubling the price of a barrel of oil.

Because uranium is an extremely compact fuel, it requires far less mining and transportation for the energy obtained. For example, to generate 191 billion kilowatt-hours, domestic nuclear plants use approximately 4800 tons of uranium oxide. Similar electrical output would require the mining, shipping, and burning of 90 million tons of coal, 325 million barrels of oil, or 2 trillion cubic feet of natural gas.

Supply and Demand

Utilities planning to build nuclear power plants must have meaningful and reliable estimates of uranium resources. The U.S. Energy Research and Development Administration (ERDA), the U.S. Geological Survey, and private industry are conducting programs to increase the knowledge in this area.

According to ERDA the United States has about 1.8 million tons of known uranium reserves and probable resources recoverable at future costs—which are not the same as selling price—of up to $30 per pound. Another 1.6 million tons are listed as possible and speculative resources. Developing a fraction of the less well-defined resources into actual reserves can make domestic uranium sufficient to meet the 2.2-million-ton, 30-to-40-year requirement of the reactors that may be operating in the United States in the year 2000. If lower-grade ores (those yielding less uranium oxide) can be mined competitively, then additional uranium could be made available.

Rather than producing more expensive uranium for today's reactors, however, several options can significantly extend the energy available from uranium resources already known to exist:

1. Reducing the concentration of uranium 235 remaining in enrichment tails—from 0.3 to 0.2 percent, for example—requires more energy and enrichment plant capacity, but increases the energy potentially available from a given amount of uranium 308 by about 20 percent.
2. Spent, or used, fuel from light-water reactors contains plutonium and unfissioned uranium 235, which can be recycled to generate about 50 percent more electricity.
3. Breeder reactors, now being developed, can use uranium 60 to 70 times more efficiently than current light-water reactors by converting otherwise useless uranium 238 into fissionable plutonium. In fact, a breeder creates more nuclear fuel than it consumes in generating electricity. Transformed to plutonium, uranium 238 now in storage in enrichment tails could eventually generate as much electricity as 580 billion tons of coal: one and one-third times the nation's demonstrated reserve.

Consumed in light-water reactors, uranium will last for decades. Converted in breeders, it will last for centuries. In whatever manner uranium is used, ensuring its availability is vital to the future supply of energy.

POWER FROM THE WIND

In areas where there are strong steady winds, modern wind-turbine generators may be able to provide significant proportions of electric power needs. Windmills have been used for centuries to pump water and grind grain, and wind electric systems have been in use since the late 1800s. But despite the antiquity of wind systems, very little is known about wind resources and how the wind can be used most efficiently.

Large Wind Systems

The most promising device for harnessing wind power appears to be the large horizontal-axis wind turbine with propeller blades. Utilities could use these machines in windy areas to supplement their power supply, and power would be distributed to the community through existing utility lines.

Vertical Axis Wind Turbines

The vertical-axis wind turbine (VAWT) accepts wind from any direction and therefore does not have to be turned into the wind to operate.

One type of VAWT, the Darrieus rotor, is considered to be a potential major competitor to the propeller systems. It was invented by G. J. M. Darrieus of France in the 1920s. With its eggbeater-shaped blades, the Darrieus has relatively high power output per given rotor weight and cost.

Small Wind Machines

Smaller wind turbines, which produce less than 100 kilowatts of electricity, are economical in some rural and remote areas where other energy sources are expensive and difficult to obtain. As the cost of conventional power resources rises and the cost of producing small wind turbines decreases, it is expected that wind power will become an economical power resource for farms, rural homes, and certain small industries and communities.

Once active and prosperous, the American wind-energy industry has declined in recent decades. Even so, it is possible today to purchase small wind generators from United States firms. Some are produced in foreign countries, and some are older American models that have been refurbished. The economics and reliability of many of these systems may be uncertain.

Innovative Wind Energy Ideas

Innovative wind-energy concepts that offer the potential for significant improvement in energy cost in the future are being investigated. Some of these concepts involve devices called *augmentors,* which increase the speed of airflow through wind machines. By doing so, they offer great potential for reducing wind system costs by allowing the use of smaller, cheaper rotor blades.

Also under investigation is a tornado-type wind turbine, which would consist of a tall cylindrical tower with an open top, slotted slide openings, and guide vanes to create a swirling tornadolike vortex flow. Outside air would be allowed to rush into the base of the tower and would be drawn upward through the vortex's low-pressure-core, spinning rotor blades, which would drive a generator.

2 A Look at an Energy-Efficient Residence

EER RESEARCH AND DEMONSTRATION PROGRAM

Developed for
U.S. Department of Housing and Urban Development
by
NAHB Research Foundation, Inc.

The energy-efficient residence (EER) research and demonstration program was conducted by the NAHB Research Foundation, Inc. under contract to the U.S. Department of Housing and Urban Development. The purpose was to develop information and data for cost effective design and construction of energy-conserving homes.

A wide range of energy-conserving options were investigated including design and orientation, insulation, windows, doors, heating/cooling systems, water heating, and appliances, as well as some more sophisticated concepts. The most promising and practical options were incorporated into a demonstration house (Figures 2-1 and 2-2). The EER demonstration house will be monitored for one year following occupancy to determine the relative cost effectiveness of the special features. The energy-efficient residence is expected to require 33 1/3 to 50 percent less energy than a typical existing home.

DESIGN AND PLANNING FEATURES

- Compact rectangular plan minimizes heating and cooling loads
- Unconditioned vestibule/storage room buffers end wall
- Vestibule "air lock" entrance isolates conditioned space
- 7-foot, 6-inch ceiling height reduces interior conditioned volume
- Family retreat closes off for comfort conditioning
- Special circulator fireplace uses outdoor combustion air
- South-facing windows provide solar heating in winter
- Roof overhang designed to shade south-facing windows in summer
- Deciduous trees provide summer shading to south side of house
- North-facing windows reduced in size to 8 percent of floor area

Figure 2-1. EER demonstration house.

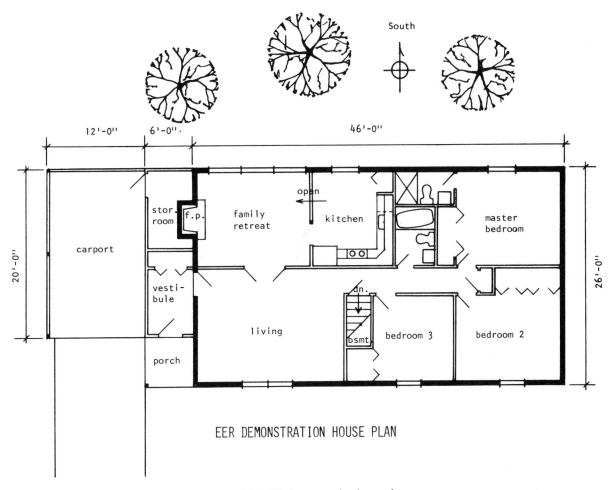

Figure 2-2. EER demonstration house plan.

EER HOUSE SPECIFICATIONS

Foundation/Floor

- Dry basement construction with gravel and sump under slab
- 6-mil polyethylene film beneath slab and behind concrete block walls, which are backfilled
- Exposed walls are stucco-finished to seal concrete block against infiltration
- 2 x 3, 24-inch on-center studs are set out from wall to accommodate insulation
- R-19 pressure-fit insulation batts on exposed walls
- R-11 pressure-fit insulation batts on below-grade walls
- 2-inch-thick plastic foam perimeter insulation at exposed slab edges
- 1-inch fiberglass sill sealer between foundation and sill plate
- R-19 band joist insulation
- All utility entrances sealed with heavy caulk
- Basement storm windows

Exterior Walls (see Figure 2-3)

- Wall height of 7 feet, 7 inches (nominal 7-foot, 6-inch ceiling) reduces interior volume

- 2 x 6 studs are spaced at 24-inch on-center with single top plate
- Bottom plate sealed to deck with construction adhesive
- R-19 unfaced pressure-fit insulation batts in walls
- Continuous 6-mil polyethylene vapor barrier behind dry wall
- R-5 plastic foam sheathing extends up between trusses, down over band joist
- Plywood box-header over openings insulated as walls
- Two-stud corner post with dry-wall backup clips to accommodate insulation at corner
- Partition posts deleted, and dry-wall clips used to accommodate insulation at partition intersections
- Surface-mounted electrical outlets with wiring in floor used to avoid penetrating wall

Doors/Windows

- Insulated steel entrance door with double glazing and magnetic weather strip
- Mechanical door closer on entrance door
- Insulated, weather-stripped inner vestibule door
- Interior doors permit closing off family retreat area for comfort conditioning

- Weather-stripped window units with double-insulating glass plus storm windows (triple glazing)
- Insulating drapes at windows control winter heat loss and summer heat gain
- Cracks around door and window frames are filled with insulation and sealed with polyethylene
- Window area reduced to 8 percent of floor area except on properly shaded south orientation

Roof/Ceiling

- Standard trusses are cantilevered over wall plate to provide clearance for insulation at exterior wall
- 12-inch-thick R-38 pressure-fit insulation batts installed from below
- Continuous 6-mil polyethylene vapor barrier below insulation
- Gable end vents provide 1 square foot of ventilation for each 300 square foot of ceiling
- 24-inch soffit overhang provides summer shading for south-facing windows
- Attic access door located in vestibule outside of conditioned area
- Surface-mounted ceiling or wall-mounted lighting fixtures to avoid penetrating ceiling

Heating/Cooling System

- Simplified duct system with low inside registers and low central return
- Special reduced-capacity heat pump with compressor installed indoors
- Controlled bypass on inside heat pump coil for improved summer dehumidification
- Heat-recovery device on compressor heats domestic water with waste heat

- Manually controlled bathroom heaters for increased comfort
- Heat-circulator fireplace unit with glass door enclosure uses outdoor combustion air
- Large south-facing windows in family retreat area contribute solar heating in winter

Water Heating/Appliances

- Heavily insulated water heater with insulated jacket set back to 120°F temperature
- Hot and cold water pipes are insulated to reduce heat loss and control condensation
- Low-water-use devices on kitchen and bathroom faucets and shower heads

Table 2-1. Comparison of Heating and Cooling Loads at Design Temperature Conditions

SOURCE	HEATING (BTU)		COOLING (BTU)	
	TYPICAL HOUSE[a]	EER HOUSE	TYPICAL HOUSE[a]	EER HOUSE
Ceiling	4,610	2,130	4,170	1,640
Walls	5,510	2,730	2,600	1,240
Windows	6,170	3,290	6,430	5,600
Doors	1,650	260	800	550
Foundation	5,050	2,370	1,140	500
Infiltration	8,580	5,250	5,990	4,400
Internal Load	–	–	3,250	3,250
Total Heating Load	31,570	16,030		
Total Cooling Load			24,380	17,180

[a]Loads for "typical house" represent a well-insulated electrically heated home that meets or exceeds current standard practice.

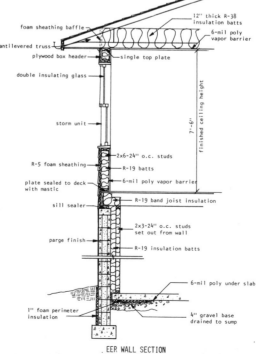

Figure 2-3. EER wall section.

Table 2-2. Projected Annual Energy Consumption in Millions of Btu's

ENERGY USE	TYPICAL HOUSE	EER HOUSE
Heating	67.2	14.5
Cooling	11.8	8.3
Hot Water	12.3	9.8
Appliances	15.2	13.0
Miscellaneous	13.5	13.0
Total Energy Use	120.0	58.6

Projected energy savings in EER house—61,400,000 Btu/year
Estimated savings on electric bills @ 3.5¢ /Kwh—$630/year

- High-efficiency refrigerator with improved insulation to save energy
- Electric range with heavily insulated standard oven plus microwave oven
- Energy-saving dishwasher uses less water and air-circulation drying
- Front-loading clothes washer uses less water and has load-size scale and selector switch
- Bathroom vent fans with effective damper, which exhaust through second damper in exterior wall
- Fluorescent lighting, four times as efficient as standard incandescent bulbs, used wherever appropriate

HOW MUCH ADDED COST TO THE BUYER IS JUSTIFIED FOR ENERGY-CONSERVING OPTIONS?

This question is best answered by determining the present value of anticipated savings on utility bills over a reasonable period of time. The present value of such future savings

Table 2-3. Additional Cost to Home Buyer for Energy-Conserving Options Justified by a First-Year Savings of $630 in Utility Bills.

ANNUAL RATE OF INCREASE IN PRICE OF ENERGY (%)	SAVINGS ON UTILITY BILLS			
	NO. YEARS ALLOWED TO RECOUP ADDED INITIAL COST OF ENERGY-CONSERVING ITEMS			
	3	5	7	10
15	$2106	$3710	$5496	$8560
12	1996	3420	4923	7337
10	1925	3238	4575	6627
8	1856	3064	4251	5991

depends primarily on the rate of increase in energy costs that is assumed and the number of years allowed to recoup the added cost of the energy-conserving options.

Given an estimated savings in utility bills for the first year, the added investment justified for energy-conserving options may be determined. Table 2-3 shows present values for future energy savings based on a savings of $630 in electric bills for the first year in the EER house, assuming a 9 percent finance cost. An annual rate of energy price increase of 10 to 12 percent and 5 to 7 years to recoup the investment are considered to be prudent assumptions for the near future.

Depending on the energy price increase assumed and the time allowed to recoup the cost, the values shown in Table 2-3 are a good indication of justifiable added cost for the energy-efficient residence from the home buyer's standpoint. Obviously, additional construction costs for the energy-conserving options must be judged accordingly.

3 Homes and Buildings

Of the total quadrillion Btu of energy used in the United States in 1979, about 30 percent was used in the residential and commercial sectors. In these sectors, the major energy-consuming unit is a building—a home or apartment building, school, office building, theater, etc.—and the major use of the energy consumed is for heating and cooling. A breakdown of the various uses of energy in all sectors of the American economy follows:

End Use	Percent
Transportation	25
Space heating*	
Residential	11
Commercial	7
Process steam†	17
Direct heat†	11
Electric motors†	8
Raw materials†	6
Water heating*	4
Air conditioning*	3
Refrigeration*	2
Cooking*	1
Lighting*	
Residential and commercial	2
Other electric	
Industrial	1
Miscellaneous	2
	100

†Industrial uses 43%.
*Used in buildings 30%.

We see that building-related uses (space heating, water heating, air conditioning, etc.) account for 17 or 18 percent of the total.

The efficiency of energy use or, conversely, the amount of energy that is wasted in a given consuming area should be determined. Such information is harder to obtain, but the intensive studies of energy consumption undertaken during the past few years have begun to provide it. Fortunately for our conservation efforts, initial studies show that there is much wasted energy—and plenty of room for improvement.

Modern buildings, with very few exceptions, were not designed to be energy-conserving, and some consume extraordinary amounts. The World Trade Center in New York City is an extreme example. It requires more power than Schenectady, New York, a city of 100,000 people.

Heat leaks through building walls and windows—*in* during the summer, *out* during the winter. In addition, buildings are overlighted; they use energy-consuming appliances; they are build of energy-intensive materials, such as glass.

CONSERVATION STRATEGIES AND SAVINGS

Energy can be saved in the construction of a building, in its operation, and by careful selection of the appliances within it.

In Building Construction

There are three major ways in which the structure of a building can influence energy use. The building materials themselves can be energy-intensive or energy-economic. The insulation, window area, and sealing determine its thermal properties. Finally, a building can be sited and oriented to make the most of natural heating and cooling.

The last of these is most applicable to private homes as their siting and orientation is the most flexible. Roof overhang can reduce the entrance of direct sunlight in the summer and allow it in the winter. Trees shade and cool; banks of earth around northern walls insulate. Such customized design has limited usefulness on a national scale. For the individual consumer, however, an energy-conserving design can have an important payoff.

Energy savings in materials can be accomplished by selecting energy-economic materials over energy-intensive ones. Architects point out that buildings are routinely overdesigned. The safety factors are much larger than reasonable; structural pieces are designed for mass production and ease of handling rather than for efficient use of materials. Estimating that 10 percent of our total energy goes into building construction, some architects have suggested that 10 percent of this energy, or 1 percent of the total, could be saved without any loss in building safety or usefulness.

Much larger savings are available, moreover, if the buildings are designed to operate with less energy. The closest thing to an energy-conservation building code is the Federal Housing Authority (FHA) standards. Up until 1971, these permitted heat losses as high as 50 Btu per square foot per hour—almost 2,000,000 Btu, or the equivalent of 15 gallons of fuel per day. New standards established in 1971 became stricter (but apply only to new buildings).

Most houses built before 1965 have about 1.5 inches of insulating materials in the ceiling and none in the walls and floors. In addition, only about half of all single-family houses currently have storm windows. Current practices call for 2 to 3 inches of insulation in the walls and 4 to 6 inches in the ceiling. A report of the U.S. Environmental Protection Agency estimated that complete retrofitting of one-third of the existing homes would save 0.5×10^{15} Btu.

The energy that can be saved through insulation, weather stripping, and storm windows to residences is accompanied by financial savings. Attic insulation, for instance, which most homeowners could install themselves, would cost about $400 in an "average" home (1600 square feet of floor space) and would pay back this investment (in lower fuel bills) in 2 to 5 years.

The energy—and therefore dollar—savings to an individual homeowner that should be expected for each of the recommended practices are summarized below.

Conservation Practice   *Savings*

- Caulk and weather-strip doors and windows — 10%
- Install storm doors and windows — 15%
- Insulate the attic (6 inches) — 20%

For air-conditioned houses, these practices will bring additional savings.

Energy savings with new homes are equally impressive. If all of the 15.4 million housing units projected to be built between 1972 and 1982 conformed to the 1971 FHA minimum insulation standards, rather than to the 1965 standards, the total savings would be almost 3×10^{15} Btu, or 10 percent of the expected 1982 residential heating and cooling consumption.

The waste of energy in commercial buildings follows a pattern similar to that in residences. Insulation and draft sealing is not significantly better. Air leaks or "infiltration" is particularly troublesome, accounting for 25 to 50 percent of the energy used in heating and cooling. In addition to leakage, there is overventilation of a more or less deliberate type. Quite often, an entire building is ventilated to the same level as is required in special areas such as bathrooms, kitchens, and conference rooms. It is estimated that reduction of excess ventilation alone could reduce fuel needs by about 15 percent.

Through Reduced Heating and Cooling Demand

The strategies discussed so far require either some awkward and expensive "retrofitting" (putting in new insulation, etc.)

or apply only to new building construction. However, significant energy savings are within reach—they can be had for a twist of the thermostat control. Not surprisingly, heat losses depend on the temperature difference across a thermal barrier, be it wall or window. Thus, the hotter the inside of the house, the easier it is to lose heat to the outside in wintertime; the cooler the inside, the easier heat leaks in from the outside in summertime.

This information can be turned into conservation figures. Lowering daytime indoor temperatures from 74° to 68°F and nighttime temperatures from 66° to 60°F will reduce wintertime heating needs by 15 percent, saving 15 out of every 100 heating dollars. If every American household had lowered its wintertime temperature 6°F in 1975, enough fuel would have been saved to provide heat for approximately 10 million additional houses during the winter.

By raising the thermostat 6°F, from 72 to 78°F, cooling costs can be cut almost in half. If all American homes had made this change in 1975, as much as 2 percent of the nation's total consumption of electricity could have been saved—exactly the increase in electrical consumption between 1974 and 1975.

Lowered air-conditioner demand has a double payback to the nation (and eventually to the consumer). The air-conditioning load usually comes at the time of daily peak demand and often seasonal peak demand. Peak electricity is expensive since the smaller, less efficient standby generators must be turned on to provide it. It is quite likely that future rates will reflect this cost and thus add economic motivation for thermostat twisting in the summer.

With Appliances

Hot-water heating and refrigerators together use 4 to 5 percent of household energy. An average dishwasher uses 14 gallons of hot water per load. One less load per week would save enough energy to heat approximately 150,000 homes during the winter.

The automatically defrosting refrigerators are also energy-greedy, using half again as much energy as those requiring manual defrosting. The automatic feature in some models can be switched off.

The appliance industry, in fact, presented a new look in fall 1976, when it displayed not only the efficiency of the various models, but also the typical annual energy use. Consumers can now base purchases not only on price, but on lifetime energy costs. Such information is especially useful on air conditioners, whose efficiencies vary by as much as a factor of three.

An old/new appliance, which seems to be designed for importance, is the heat pump. It operates much like a reversible refrigerator; it "cools the outside" in the winter, pumping the heat indoors, and can be turned around to cool the indoors in the summer. Heat pumps use 60 percent less electricity than do the usual electric resistance heaters in providing the same amount of heat.

Lighting

In the breakdown of energy consumption (p. 000), lighting represented a fairly small fraction of the total energy consumed, about 1.5 percent. Its percentage share is growing and accounts for a rather substantial fraction of electrical energy consumption: 21 percent overall of the 1974 electrical consumption and as high as 60 percent of the electrical consumption in a city such as New York.

Energy (and money) can be saved by turning out lights or by following the federal government's lead and removing each third bulb in a series of fluorescent lights. (The Federal Energy Administration estimates that such a 30 percent reduction in lighting would save average householders 4 percent of their electrical bill.) In addition, energy can be saved by switching to the more efficient fluorescent bulbs whenever possible.

The major wasteful use of lighting is in commercial buildings—offices, schools, hospitals, etc. For reasons that seem to have no sound physiological basis, the illumination levels recommended by the Illuminating Engineering Society of North America have been increased over the last few decades. In 1952, for instance, the recommended classroom lighting was 20 foot-candles (a measure of brightness). This was raised to 30 in 1957 and to 60 in 1971. In libraries, the recommended level has gone from 20 in 1952, to 40 in 1957, to 50 in 1959, and to 70 in 1971.

Thus, "no contrast" lighting has resulted in buildings in which all space is brightly lit, including walkways and even space near windows for which daylight would suffice. Estimates of possible savings in electrical energy used for lighting in new buildings range upward to 50 percent. A more realistic target of 25 percent in new buildings and 15 percent in existing buildings would reduce electrical demand by 4 percent (which is twice the rise in demand measured between 1974 and 1975).

Summary

Buildings are involved, one way or another, in the consumption of about 30 percent of our total energy. They are thus a worthwhile target for the two-pronged goal of energy conservation—reduction in use and improved efficiency of utilization.

A savings of 30 to 50 percent of the energy in the residential and commercial building sectors, and thus 14 percent of the United States total, is not impossible to achieve. The Energy and Research Development Administration goal for the year 2000 is to reduce energy demand in buildings by 7.1×10^{15} Btu. This is 4 or 5 percent of the 150×10^{15} Btu or so per year that some forecasters expect us to be using in that year—a not unreasonable goal.

Increasing the supply of usable energy by using it more wisely is the most appealing option we have. It can be accomplished relatively swiftly; it has generally positive environmental effects; it provides the least expensive energy and even saves the consumer money. While there are technological challenges to be met (for instance, a screw-in fluorescent bulb and better insulating material), the real challenges are in the institutional area, building codes, rate structures, and tax laws. Breakthroughs in the social economic arena, therefore, are as important to our energy future as are the technological breakthroughs we have described.

4 Waste Heat Recovery: More Power From Fuels

POWER PLANTS AND EFFICIENCY

Today's electric power comes from many types of generating stations—from hydroelectric plants, from uranium-fissioning nuclear reactors, and from fossil-fueled steam, diesel, or gas-turbine generators. None of these systems for transforming energy into electricity is more than 40 percent efficient. That is, none produces electric energy equal to much more than a third of the mechanical or chemical energy it consumes.

In the past, the type of generating station built has depended on factors such as proximity to energy resources (adequate water for a hydroelectric plant, for example), size of plant needed (steam plants were most economical for large utilities; diesel generators a relatively inexpensive way to supplement capacity), and fuel costs. Today, plant efficiency, making the most of scarce fuels like petroleum, is a factor of growing importance. While part of energy conservation involves developing more efficient generating systems for the future, much work is being done to increase immediately the efficiency of existing generating systems with a lot of useful life left in them. One way to increase efficiency is to make use of a system's waste heat.

WASTE-HEAT RECOVERY

Plants that generate electricity by combustion operate at temperatures well over 1000°F. Waste heat from these systems is still very hot and capable of doing more work. Utilities can recover some of this heat energy by adding another type of generator as a "bottoming cycle." Since it operates at a lower temperature, the bottoming cycle can be "fueled" by the hot exhaust from the main engine.

Adding bottoming cycles to the diesel engines and gas turbines used by small utilities and industrial plants requires relatively uncomplicated engineering and is relatively low-cost. Because the diesel has a smaller generating capacity (about 3 megawatts compared to 20 megawatts for a gas turbine), it will be the main engine used in a number of bottoming-cycle demonstrations to be operating by 1980. Results from these demonstrations will help in scaling up bottoming cycles for use on the larger gas turbines.

COMBINED SYSTEM OPERATION

A widely used work-producing system called a *Rankine cycle* has been chosen for a demonstration bottoming cycle. The Rankine cycle describes a closed loop containing a working fluid. Pressurized working fluid absorbs heat from an external source at one end of the cycle and uses that heat energy to drive a turbine and generator at the other end. The working fluid is then cooled and recycled to be pressurized and heated again.

Ordinarily, the heat to warm the working fluid in a Rankine system comes directly from burning fuel. In the proposed Rankine bottoming cycle, the heat source is the hot exhaust (500° to 1000°F) passing out of the stack of an internal combustion engine. Because diesel exhaust is not as hot as burning fuel, the Rankine bottoming cycle will probably use an "organic" working fluid instead of the more familiar steam, which requires a higher temperature range. The organic Rankine bottoming cycle has three units:

1. a heat exchanger (vaporizer), tailored to the application and installed at the exhaust stack, to capture waste heat
2. a power control unit, the minigenerating plant that converts the exhaust's heat energy into electrical energy
3. a control system

The main engine operates as usual, turning about 38 percent (in the case of a diesel engine) of the heat energy from burning fuel into horsepower to drive a generator, losing about 28 percent in cooling operations, and passing about 34 percent up the stack. With the bottoming cycle to capture exhaust heat, 7.6 percent more of the original heat energy can be converted into mechanical or electrical energy. As in the main system, some heat energy is lost in cooling and some rejected as exhaust, but exhaust temperature at the end of the bottoming cycle is much lower, about 250°F.

ECONOMICS OF WASTE-HEAT RECOVERY

A typical diesel system turns out about 8800 horsepower, which can generate about 6250 kilowatt-hours of electricity. The bottoming cycle fitted to such an engine will recover enough energy from the exhaust to produce an additional 900 horsepower or generate 600 kilowatt-hours of electricity. This increased efficiency means fuel cost savings that should offset the cost of installing the bottoming cycle in 5 years or less.

Getting more electricity from limited fuels will mean important nationwide fuel savings. If only 500 bottoming-cycle systems are installed on diesel engines, the fuel savings could approach 15,000 barrels of oil per day or 5.5 million barrels per year.

The combination of diesel engine and Rankine bottoming cycle is only one example of a combined system that can increase power production without increasing fuel use. Gas turbines exhaust 60 percent of the heat energy they use. Refitting them with bottoming-cycle systems can greatly increase their efficiencies, recovering an additional 15 percent of the initial fuel energy as shaft horsepower for generating electricity. The fuel-stretching benefits of waste-heat recovery systems are not limited to producing electricity. Bottoming cycles can also be developed to increase the efficiency of the diesels and gas turbines used for industrial processes, for pumping gas and petroleum, and for propelling ships.

5 Why Heat Pumps?

In many parts of the country, a heat pump can provide an economic and energy-conserving alternate to electric-resistance heating (furnace or baseboard) with a central air conditioner.

WHAT IS A HEAT PUMP?

A heat pump, which typically runs on electricity, works like a refrigerator or air conditioner. It transfers the heat present in air from one place at a relatively low temperature, to another place at a higher temperature.

The heat pump can be used for heating and cooling. When a heat pump is used for heating, cool air outside the building is blown over an even colder outdoor coil. Heat is transferred from the outdoor air to this coil. This heat evaporates the liquid refrigerant in the coil, which is raised above room temperature by compression and then condensed in the indoor coil. By condensing, heat is released and transferred to the indoor air. A fan blows the warm air through ducts to the rooms to be heated.

When a heat pump is used for cooling, the functions of the indoor and outdoor coils are reversed. In this case, heat is absorbed in the indoor coil (thereby cooling and dehumidifying the indoor air) to evaporate the liquid refrigerant, which is then circulated through the outdoor coil after having its temperature raised above the outdoor temperature by compression. The refrigerant vapor is then condensed by giving up its heat to the outdoor air circulated over the coil. This is the typical air-conditioning process.

ADVANTAGE OF HEAT PUMPS

The appeal of the heat pump is that, except in very cold weather, the amount of heat energy delivered to the indoor air is greater than the amount of electrical energy put into the system. The ratio of the annual heat delivered to the annual electrical energy input is called the *seasonal performance factor (SPF)*. The heat pump's SPF is typically between 1.5 and 2.0 for many regions of the country. As a result, a heat pump operating in these regions will annually use 30 to 50 percent less electrical energy than an electric-resistance heating system, which has a nominal efficiency of 100 percent (SPF = 1). The SPF of the electric furnace would typically be somewhat less than the SPF of the baseboard system because of the duct losses of the former.

PRACTICAL PROBLEMS

With the uncertain future of oil and natural gas supplies and the trend toward all-electric homes, the heat pump offers an important opportunity for energy conservation. There are, however, practical problems in comparing the use of heat pumps with electric-resistance systems:

- Since the heat pump provides both heating and cooling, the decision to use a heat pump will, in most cases, depend on whether central air conditioning and a ducting system are also needed. If summer air conditioning is not needed, it is likely that the heat pump's higher first cost will make it uneconomical on a life-cycle basis.
- At the present time, heat pumps are rated under steady-state, full-load operating conditions and are given an energy-efficiency ratio (EER) value. In practical operation as a heating device, the efficiency of a heat pump is reduced by intermittent operation, frost buildup, and the process of defrosting. Because of these losses, the actual seasonal performance of a heat pump will usually be less than the predicted seasonal performance that would be calculated from a manufacturer's performance data. The size of the discrepancy is likely to vary among products and will depend upon the climate in which the heat pump operates. A recent study at the National Bureau of Standards indicated that these factors resulted in an almost 20 percent drop in the calculated SPF on a typical residential heat pump operated in the Washington, D.C., area.

HOW TO DECIDE

Consider the case of whether a heat pump should be installed instead of an electric furnace with central air conditioning. It will be assumed that the first cost of a heat-pump system is $350 more than the first cost of the latter system and that both systems have an expected service lifetime of

15 years. It is further assumed that the present mortgage interest rate is nine percent; the current cost of electricity is 3¢ per kilowatt-hour and is likely to increase at a rate of 5 percent a year; and the average cost of maintaining the heat pump is expected to be $25 per year more than the furnace and central air conditioner due to the heat pump's increased complexity. The house that will be heated is assumed to have a 28,600-Btu/hour heating requirement at the outdoor design temperature for the region of 10°F. The number of degree-days in the heating season will be taken to be 4500, and it has been determined that the cooling cost for similar houses in the area runs about $175 a year.

For this particular example, installing a heat pump instead of an electric furnace with central air conditioning will save the home buyer, on the average, $131 per year over the 15-year life of the system.

6   The Emerging Age of Solar Energy

SOLAR ENERGY: AN OLD IDEA REBORN

Humans have always made use of the sun's energy. In fact, without the sun's energy, humanity could not survive. All our food and fuel have been made possible by the sun through the photosynthetic combination of water and atmospheric carbon dioxide in growing plants. Fossil fuels represent energy photosynthesized and stored in dead plant and animal matter millions of years ago. The earth's climate is also dependent on solar energy. The winds, tides, rivers, and ocean currents all rely on the daily heat of the sun for their motion. People have long made use of these forms of solar energy—they have used the wind and water currents to move from place to place, and they have used the sun to grow food and warm their shelter.

Primitive people learned that it was advantageous to find caves and to place wall openings in the direction of the sun's path in order to capture its warmth during winter. They also discovered the unique ability of certain materials to retain the sun's warmth and release it after the sun had set. The use of solar energy continued and, indeed, increased as a tool and craft society developed.

The earliest attempts to harness the sun's energy include the fabled burning of the Roman fleet by Archimedes in 212 B.C. Archimedes reputedly set the attacking Roman fleet afire by "burning glass composed of small square mirrors moving every way upon hinges which when placed in the sun's direct rays directed them upon the Roman fleet so as to reduce it to ashes at the distance of a bowshot."* Whether Archimedes actually did set fire to the sails of the attacking ships does little to dampen the fact that early in history, solar devices were being designed and built.

The marvelous inventions of the Renaissance included many solar devices. One of the most original was built by Salomon de Caus (1576-1626) of France, who used the sun to heat air in his solar "engine," which in turn pumped water.

The solar devices of the Renaissance were generally gadgets with little practical application. During the latter part of the eighteenth century, however, solar furnaces capable of smelting iron, copper, and other metals were constructed of polished iron, glass lenses, and mirrors. The furnaces

were in use throughout Europe and the Middle East. One such furnace, designed by the French scientist Antoine Lavoisier, attained the remarkable temperature of 1750°C (3182°F). The furnace used a 52-inch lens plus a secondary 8-inch lens to attain temperatures far exceeding those achieved up to that time—and, as it turned out, not to be achieved again for the next 100 years.

Early in the nineteenth century, numerous hot-air engines were developed; the famous Stirling two-piston air engine, although not designed to be operated by the sun, was ideally suited for such use and later was adapted for solar power. A curious assortment of solar engines were built over the next century, powering everything from printing presses and electric lights to distillation operations.

One unique variation, which occurred during the late 1800s and early 1900s, was the use of flat-plate collectors to intercept the sun's rays and power equipment. Until that time most solar devices used what is called a *focusing collector*—one that focuses the sun's rays upon a single area where the energy is collected. The flat-plate collector, however, does not focus the sun's rays to a single point, but collects the solar energy over a uniform horizontal surface. Flat-plate collectors were both less expensive and simpler to construct and operate than the focusing collector. Also, whereas focusing collectors required clear skies for operation, flat-plate collectors could function under cloudy conditions. Several solar-powered pumping facilities were constructed during the early 1900s in the United States, which utilized either focusing or flat-plate solar collection.

In 1901, A.G. Eneas installed a 33-foot-diameter focusing collector, which powered a water-pumping apparatus at a Pasadena, California, ostrich farm. The device consisted of a large umbrellalike structure open and inverted at an angle to receive the full effect of the sun's rays on the 1788 mirrors that lined the inside surface. The sun's rays were concentrated at a focal point where the boiler was located. Water within the boiler was heated to produce steam, which in turn powered a conventional compound engine and centrifugal pump. During the next 50 years, many variations of this process were designed and constructed using focusing collectors as the means of heating the working fluid that powered mechanical equipment.

*Johannes Tetzes, twelfth-century author.

Frank Shuman, an inventive engineer, favored the more economical flat-plate collector. Using 1200 square feet of collector area, his test engine produced 3.5 horsepower. The flat-plate collector, constructed in 1907 in Tacony, Pennsylvania, was used to heat water, which, in turn, boiled ether. The ether vapor was then used to drive a vertical steam engine, which pumped water. Although Shuman's solar engine did not produce anything near the 100 horsepower predicted (in part due to the polluted air and cloudy conditions of Tacony), the technique of collecting and utilizing solar energy was significantly advanced.

Despite the increased sophistication and reliability of the solar-powered devices, none of the early devices survived competition with the emerging use of cheaper fossil fuels. For, although solar energy was free and readily available, the capital investment was so high for the necessary solar collectors and associated equipment that it cost more to run a solar engine than to run a conventional type. However, the energy and environmental situation that we now face has made solar energy an idea whose time has come and gone, and come again—this time to stay.

SOLAR HEATING AND COOLING: A HISTORICAL PERSPECTIVE

Pueblo Structures

The hot-arid climate of the Southwest is characterized by high daytime temperatures and uncomfortably low nighttime temperatures. The solution best suited to such a wide temperature fluctuation is to delay the entry of heat as long possible so that it will reach the interior late in the day when it is needed. The Pueblo Indians achieved this desired thermal performance by using materials of high heat capacity, such as adobe mud and stone, which provide a "heat sink," absorbing heat from the sun during the day and reradiating it into the dwelling during the night. Also, by crowding their dwelling spaces together, side by side and one on top of another, the Pueblos achieved maximum

volume with the minimum surface area exposed to the outside heat. They thereby reduced the area exposed to the sun while increasing the mass of building as a whole, thus increasing the thermal time-lag. (See Figure 6-1.)

In hot-arid climates it is important to avoid interior heat buildup during the day. The Pueblos accomplished this by separating the cooking space from the living spaces, by reducing the number and size of windows and placing them high on walls to reduce radiation gain, by painting the dwelling white or some other light color to reflect a maximum of radiant heat, and by minimizing ventilation during the hottest part of the day.

A unique example of these principles of solar design is found in the cliff dwellings of the Sinaqua and Anasazi Indians in Arizona. Some of the best preserved cliff dwellings—now the Montezuma Castle National Monument— were built about 1100 A.D. and made use of the heat capacity of mud, rock, and other indigenous materials to absorb direct solar radiation on the south-facing vertical walls. The heat was then reradiated to the interior spaces during the evening. The dwellings were built under an extended portion of the mountain, and so the overhang blocked the high summer sun, thereby providing natural cooling for the dwelling.

The Saltbox

In the New England saltbox (Figure 6-2), the primary source of heat is inside the dwelling rather than outside and the attempt to stop heat flow is in the opposite direction from that of the Pueblo dwelling. Large stoves and fireplaces are often found at the center of the saltbox to capture heat given off by cooking and space heating. Reducing heat loss is achieved by a compact plan, a minimum surface area exposed to the outside, materials of good insulating characteristics, and the prevention of drafts and air leaks. Also, by sloping the roof on the north side of the dwelling toward the ground, shelter from the wind and a reduction of surface area exposed to the cold are achieved. This allows snow to build up on the roof and thus further insulate the dwelling from heat loss. Unlike the dwellings of hot-arid

Figure 6-1. Taos pueblo, New Mexico.

Figure 6-2. New England saltbox.

areas, the saltbox is constructed to capture as much solar radiation as possible during cold weather. Therefore, the windowed two-story portion of the house faces south, and the dwelling is painted a dark color (which absorbs more solar energy than do light colors).

American Solar Dwellings

Pueblo and colonial architecture are representative of the intuitive responses to local climatic conditions and the beneficial effects of proper building exposure to solar radiation. The intuitive approach to solar heating and cooling has laid the groundwork from which a scientific understanding of solar radiation and its corresponding climatic impact on building have recently developed.

Only since the 1930s in the United States have buildings been designed that attempt to collect, store, and distribute solar energy as a principal heat source for human comfort. The Crystal House at the 1933 Chicago World's Fair is an example of the direct "greenhouse" effect, whereby glass walls or windows were used as heat collectors. The architects, George and William Keck, incorporated the ideas used in this design into other dwellings. Their designs used large expanses of south-facing glass, which allowed the low winter sun to heat the interior masonary floors and walls of a building during the day (similar to the Pueblo structures), which in turn radiated the stored heat to the spaces during the evening. These structures were first called "solar houses" by a reporter for the *Chicago Tribune*.

The houses built by the Massachusetts Institute of Technology (MIT) as part of the Cabot Solar Energy Conversion Project between 1939 and 1956 and the Peabody-Telkes-Raymond house built in 1949 were the first fully documented solar dwelling designs in which a major portion of the heat requirement was obtained by a formal solar collector and storage system. In the latter house, all of the heat was obtained by solar collector and storage systems. In all, four solar-heated houses were built by MIT. The first two, built in 1939 and 1947, were more experimental laboratories for evaluating solar equipment than actual homes. However, the third dwelling, constructed in 1949 from the shell of the second, was designed to house a student family with one child. The solar heating system consisted of a south-facing flat-plate collector mounted on the roof with a large water-storage tank located directly behind it. Solar radiation, collected by water circulated through the flat-plate collector and distributed by copper tubing mounted in the ceiling, provided more than three-quarters of the house's heat. In sunny winter weather, the large windows along the south wall provided more heat than needed for several hours of the day and so had to be released by ventilation. During the hot summer months, however, the south-facing windows were shaded by an overhang.

Also during 1949, a solar-heated house was built in Dover, Massachusetts (Figure 6-3). The house was engineered, designed, and sponsored by Maria Telkes and her coworkers. The major objectives were (1) to prove that complete solar

Figure 6-3. Peabody-Telkes-Raymond House, Dover, Massachusetts.

heating was feasible in the Boston area despite the chance of three to five consecutive sunless days with no collection and (2) to demonstrate the merits of using the heat of fusion of Glauber salt as a means of storing heat. The dwelling used both south-facing windows and vertical collectors to capture direct and reflected solar radiation from the low winter sun. The hot air removed from the solar collectors was used to heat 5-gallon storage cans filled with Glauber salt. When heat was needed, air was circulated around the cans by small fans, and then was heated and distributed to each room. The technical feasibility of total solar heating in the Boston area and the use of the heat fusion of Glauber salt as a storage medium were both adequately demonstrated.

The fourth and final solar house built by MIT under the Cabot Foundation Project was constructed in 1956 at Lexington, Massachusetts (Figure 6-4). The house had both solar space heating and solar domestic water heating. The dwelling used a flat-plate collector tilted toward the sun and covered with two panes of glass. Water was heated by the sun as it circulated through the collector and was stored in a 1500-gallon insulated storage tank located in the basement. Heat from the water-storage tank was transferred to air by a heat exchanger and distributed throughout the house by a blower and ducts. This house was an example of the emerging sophistication of solar dwelling design. The relationship of the dwelling with the solar equipment (collector, storage, distribution) was treated as an integral problem and

Figure 6-4. MIT Solar House No. 4, Lexington, Massachusetts.

not as the separate concerns of equipment design and dwelling design.

Except for an active, but relatively unheralded community of solar researchers who built upon the MIT work, there was little application of solar technology to buildings until the 1970s. Two notable exceptions are the works of George Löf and Harry Thomason.

George Löf, a Colorado chemical engineer, constructed a solar house in Denver in 1957. The solar heating system, which is still in operation, consists of two arrays of south-facing flat-plate collectors and two vertical storage cylinders filled with granite rock. A blower draws air through the rooftop collectors to the storage cylinders and distributes the heat through floor ducts to the individual rooms. Domestic hot water is preheated by running the hot-water piping through the incoming heated air duct from the collector.

Harry Thomason, a Washington, D.C., lawyer, constructed three solar homes in the Washington area between 1959 and 1963. A similar solar energy system is used in each dwelling. The collector is made of blackened, corrugated aluminum covered with a single pane of glass placed on either a sloping roof or wall that faces south or slightly west of south. Solar radiation is collected by water trickling down the open channels of the corrugated aluminum. The heated water is transferred to a storage tank surrounded by fist-size rocks. The rocks, which are heated by the water in the tank, provide an additional storage medium and the mechanism for transferring the heat from water to air. To heat the house, air is drawn from the house through the rock bed; it picks up heat from the tank and rocks and is blown through ducts into the individual rooms. Two of Thomason's solar dwellings incorporated a compression refrigeration system for cooling. Operated at night, the system chills air, which cools the rock, which in turn distributes cool air during the day.

Early in the 1970s, the economic and environmental cost of conventional fuels brought attention back to solar energy as an alternative energy source for heating and cooling. Although the earlier solar houses at MIT and Dover have either been demolished or are operating without their solar equipment, they nonetheless proved that solar heating was possible with existing technology. Given sustained technical refinement, appropriate architectural design, and industry and marketing economics, solar energy could possibly become a major source of energy for the heating and cooling of buildings.

The solar dwellings built over the last several years range from those that require no mechanical equipment for their operation to those that generate their own electricity from solar energy and collect and store radiation for heating and cooling. In the following pages, solar dwellings built between 1972 and 1974 are described. They represent solar dwelling designs that employ different ways of collecting, storing, and distributing solar energy for space heating and cooling and domestic water heating.

Built in 1974, the David Wright residence in Santa Fe, New Mexico, illustrates a modern adaptation of the Pueblo

Figure 6-5. David Wright house, Sante Fe, New Mexico.

Indian structures. The design is extremely simple and open (Figure 6-5). On first inspection, the only collector visible is a 32-square-foot solar water heater on the ground, so some might question whether the house is actually a *solar dwelling*. However, the sun dictated the design of the house from the very beginning. The solar aspects are so well integrated that the house actually becomes both the solar collector and the heat-storage system. The south wall, constructed entirely of insulating glass, serves as the solar heat collector. Adobe in the exterior walls and beneath the brick floor provides storage by absorbing incident solar radiation. Several 55-gallon drums filled with water are buried beneath an adobe banco along the south wall to provide additional heat storage. Insulation is located around the entire outside of the adobe walls and beneath the adobe floor. The insulation minimizes the flow of heat from the walls and floor to the colder outside air and the ground. Thus, the heat is stored until the temperature inside the house drops, and then the stored heat is radiated and convected into the space. The fabric of the building is capable of storing enough heat to keep the house comfortable for three or four sunless days.

The solar house in Atascadero, California, illustrates another variation in solar collection and storage design. This house used horizontal water ponds (in plastic bags) located within the roof structure for solar collection, storage, and distribution. The house's solar concept, developed and patented by Harold Hay, is capable of both heating and cooling. During the heating cycle, insulating panels that cover the ponds are removed so that on a sunny winter day, the water bags underneath are entirely exposed to the sun's rays. The solar radiation warms the mass of water, which in turn warms the supporting steel ceiling deck; the ceiling radiates the collected solar heat to the interior of the dwelling. At night, the panels are rolled back automatically to cover the water ponds and thus act as a "thermal valve" to retain the heat collected during the day.

Cooling is provided by reversing the winter procedure and opening the insulating panels at night so that the heat absorbed by the roof ponds during the day can be dissipated by nocturnal cooling. The water in the ponds thus cooled is able to absorb heat from the house during the day. As the heat is absorbed, the interior of the house is cooled. Radiation to the sky, evaporation, and, at certain times, convection all play a part in the nocturnal cooling process. This system is most effective in regions where the summer dew-point temperature or humidity is relatively low.

The Hay system is similar in concept to a solar dwelling constructed in France in 1956. Dr. Felix Trombe, former Director of Solar Energy Research for the Laboratoire d'Energetique Soloire of the Centre Nationale de la Recherche Scientifique, and architect Jacques Michel designed a solar house that, instead of using horizontal roof ponds for solar collection and storage, used a massive vertical south-facing concrete wall with a glass wall in front of it. The warmth from the inner surface of the concrete wall, trapped by the glass cover, provides most of the heat at night during the winter.

The American modification of the Trombe/Michel concept involves the use of additional thermal storage via a rock bed located beneath the floor. By the use of blowers, the heat from the south wall can be stored in the rock bed, and thus storage capacity of the system can be increased considerably.

The Phoenix of Colorado Springs is a solar house built in the winter of 1974 during the critical shortage of natural gas. The solar system of this dwelling is significantly different from those already mentioned. Solar energy is captured by two banks of solar collectors facing due south at an angle of 55°. The aluminum collector panels are covered with two panes of glass to reduce heat loss from the collection surface. The fluid that circulates through the collector to absorb the heat is propylene glycol, which will not freeze in the extremely low temperatures of Colorado Springs. Heat from the collector is transferred to a large storage tank buried beside the house. The distribution system is composed of two heat exchangers, both located in the storage tank. The first transfers the collected heat to water, which in turn heats air for distribution through ducts to each room; the second preheats the domestic hot water before it passes through a conventional electric water heater.

Solar One, located in Newark, Delaware, is the first building in which both thermal and electrical energy are derived from the sun. It was designed and built under the auspices of the Institute of Energy Conversion of the University of Delaware, with support from the National Science Foundation and Delmarva Power and Light Co. The south-facing roof, tilted at 45° to the horizontal, supports 24 solar panels, all of which will eventually be covered with cadmium sulfide photovoltaic cells. Also six vertical collector panels, reminiscent of the Dover house of 1949, are mounted on the south-extending bays. Storage for the solar electric systems consists of automobile-type 12-volt storage batteries, which are housed in a small frame shed on the east side of the house. Eutectic salts, in small plastic containers, are used to store heat from the solar thermal collectors.

Solar One represents the first attempt to combine the onsite generation of electricity with the collection and storage of thermal energy. Economical and technical attainment of these goals would be a major step toward making dwellings energy self-sufficient.

Solar Energy in Other Countries

The growing interest in solar energy, along with increased research, development, and application, has not been confined to the United States. In particular, Japan, Israel, and Australia use solar energy for domestic hot-water heating. In addition, India, France, and the Soviet Union have been involved for many years in developing and applying solar heating and cooling technology in both commercial and residential buildings.

During the 1930s and 1940s in the United States, prior to the availability of inexpensive energy and utility services, solar domestic water heating found widespread application in Florida, Arizona, and southern California. In fact, in the Florida area alone, approximately 50,000 solar domestic water heaters were in use before natural gas and electricity replaced solar energy as a primary fuel source. The resurgence of interest in solar energy today may once again give rise to buildings heated and cooled by the sun.

There are inumerable design variations possible to achieve total or partial solar heating and cooling. The preceding survey provides just a glimpse of the range of possibilities in both dwelling and solar-system design. The appropriateness of a particular solar-dwelling design depends on such factors as climate, occupant comfort, site conditions, building characteristics, and the proposed solar system. These factors are discussed in the following sections.

SOLAR HEATING/COOLING AND DOMESTIC HOT-WATER SYSTEMS

Solar heating/cooling and domestic hot-water systems that are properly designed and integrated into a dwelling to use heat efficiently can provide a large percentage of the dwelling's space heating, cooling, and domestic hot-water requirements. Technically, it is possible to achieve close to 100 percent solar heating and cooling; however, a more realistic and economically feasible goal, given present technology, would be 70 percent solar space heating and 90 percent solar hot-water heating. On the other hand, mechanical solar cooling* although technically feasible, requires additional research and development to achieve the same levels of efficiency and cost-effectiveness as present solar heating systems. Therefore, this section will focus on solar heating and domestic hot-water systems.

In some cases, with total energy-conserving design, a renewable auxiliary energy source, and a provision for solar-produced electrical power to operate appliances and controls, a dwelling can become completely independent of nonrenewable fossil fuels. Lower utility costs and the possibility

*Mechanical solar cooling involves the use of solar-produced heat to power conventional mechanical cooling equipment. Heat pumps, absorption cycle, and Rankine cycle systems are representative of such mechanical cooling equipment.

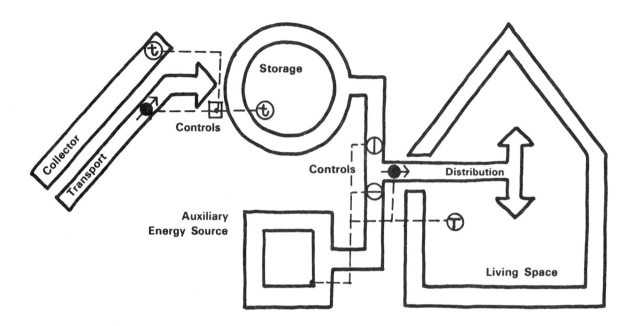

Figure 6-6. Simplified diagram of a solar heating system.

of total energy self-sufficiency are two major factors for the growing interest in solar energy.

Solar System Components

Any solar system consists of three generic components—collector, storage, and distribution—and may include three additional components—transport, auxiliary energy system, and controls. These components may vary widely in design and function. They may, in fact, be one and the same element (a masonry wall can be viewed as a collector, although relatively inefficient, which stores and then radiates or "distributes" heat directly to the building interior). They may also be arranged in numerous combinations depending on functions, component compatibility, climatic conditions, required performance, and architectural requirements.

Solar energy, also known as solar radiation, reaches the earth's surface in two ways: (1) by direct (parallel) rays and (2) by diffuse (nonparallel) sky radiation, reflected from clouds and atmospheric dust. The solar energy reaching the surfaces of buildings includes not only direct and diffuse rays but also radiation reflected from adjacent ground or building surfaces. The relative proportions of total radiation from these sources varies widely in each climate, from hot-dry climates where clear skies enable a large percentage of direct radiation to reach a building, to temperate and humid climates where up to 40 percent of the total radiation received may be diffuse, to northern climates where snow reflection from the low winter sun may result in a greater amount of incident radiation than in warmer but cloudier climates. The need for and the design of solar system components will vary in each locale as a result of the preceding differences, as well as differences in climate, time of year, and type of use—that is, space heating, cooling,

or year-round domestic water heating. Recognition of these differences is important for the proper design and/or selection of solar-system components. A simplified diagram of a solar heating system is presented in Figure 6-6.

Collector. The collector converts incident solar radiation (insolation) to usable thermal or electrical energy by absorption on a suitable surface. In nonphotovoltaic (nonelectrical) systems, the thermal energy captured is transferred to a heat-transfer medium, usually gas or liquid, within the collector. Collectors are generally classified as focusing or nonfocusing depending upon whether the sun's energy is concentrated prior to being absorbed or collected at the density received at the earth's surface.

Collectors generally use a transparent cover sheet (cover plate) to reduce convective and radiative cooling of the absorber. Glass or plastic is commonly used for the cover sheet because each has a high transmittance of short-wavelength (ultraviolet) solar radiation and a high absorptance of long-wavelength radiation, thus trapping heat that is re-emitted from the absorber. Ideally, a cover sheet and an absorber should be normal—that is, perpendicular to the sun's rays. When the angle at which the sun's rays strike the collector is less than 30°, the loss of radiation by reflection can be greater than that being collected.

An absorber can be any building material on the inside of the cover sheet. However, an efficient absorber surface will have a high solar absorptance and low emittance (that is, it will absorb solar radiation and not reradiate it). When the absorber is used to conduct heat to a liquid or gas, it must have high thermal "conductivity" as well. Most absorber surfaces are coated with a dark substance, either paint or a special chemical coating, to increase their absorption of radiation. Some coatings are designed to be selective—

that is, to maximize the rate of absorption and minimize emissivity losses.

Storage. The storage component of a solar system is a reservoir capable of storing thermal energy. Storage is required since there may be energy demand during the evening or on consecutive sunless days when collection is not occurring. Storage acquires heat when the energy delivered by the sun and captured by the collector exceeds that demanded at the point of use. The storage element may be relatively simple, such as a masonry floor, which can store and then reradiate captured heat, or it may be relatively complex such as chemical phase-change storage. Heat storage is also required for solar-assisted domestic water heating. This may be provided within the larger space-heat storage component, with a separate but smaller storage tank, or in conjunction with the storage capacity of a conventional water heater.

Distribution. The distribution component receives energy from the collector or storage component and dispenses it at points of consumption—that is, spaces within the dwelling. For example, comfort heat is usually distributed in the form of warm air by ducts or pipes within a building. Distribution of energy will depend upon the temperature available from storage. Temperatures as low as 90°F may still be useful for space heating if the baseboard convectors are increased in size or if they are used in conjunction with a heat pump or auxiliary heating system.

Because solar-produced temperatures in storage are normally 90° to 180°F, distribution ducts and radiating surfaces are normally larger than those used in conventional heating systems. Therefore, careful consideration is required in the design of heat-distribution systems throughout the dwelling.

Domestic water heating is also a part of the distribution component. Its distribution system generally consists of a heat exchanger, backup heater, piping, and controls.

Transport. Most solar systems have an energy-transport component that provides a means of moving a fluid carrying thermal energy to and from the collector and storage. The transport component also regulates the flow through the collector and storage. In liquid or gas systems, this component consists of pumps, valves, and pipes or blowers, dampers, and ducts.

Auxiliary Energy. The auxiliary energy component provides a supply of energy for use when the solar system is inoperable and during periods of extremely severe weather or extended cloudy weather when solar-produced temperatures from the collector and storage are not sufficient to satisfy the building's heating or cooling load. Presently, the experimental nature of solar heating, cooling, and domestic hot-water systems and the possibility of extended sunless days generally require that the auxiliary energy component be capable of providing the total energy demand of the house if the solar system is inoperable.

The auxiliary system can be powered by conventional fuels such as oil, gas, and electricity, by alternative fuel sources such as wood burned in fireplaces, methane gas, electricity, or by a combination of these. The component may operate independent of or in conjunction with the solar system. This is usually accomplished by locating the auxiliary energy system between the storage and distribution components to permit a full or partial operation, or by supplying the energy directly to the heat-storage component, thereby raising it to a usable temperature.

Control. The control component performs the sensing, evaluation, and response functions required to operate the system in the desired mode. For example, a thermostat senses the temperature in the house and relays it to the distribution component (pump or blower) when heat is required. The controls generally distribute information, including fail-safe instructions, throughout the system by means of electrical signals. However, the control function can be performed by automatic pneumatic controls or by the dwelling occupants who initiate manual adjustments to alter the system's operation.

Solar Collectors

There are numerous concepts for the collection of solar radiation. These concepts range from the most simple—a window—to those that are quite complex and require advanced technology for their development—for instance, a solar cell.

Historically, solar collectors have been classified as either focusing or nonfocusing. In a nonfocusing collector, the absorber surface is essentially flat, and the absorber area is equal to the aperture for incident radiation. A focusing collector, however, is one in which the absorber area is smaller than the aperture for incident radiation, and consequently there is a concentration of energy onto the absorber surface. Numerous solar collectors have been developed, which illustrate each of these concepts. Several recent collector designs have been developed that do not fit into either category, thus creating considerable confusion in collector classification. Consequently, the following description of solar-collector concepts is not separated exclusively into focusing or non-focusing.

Flat-Plate Collectors. Of the many solar heat collection concepts being developed, the relatively simple flat-plate collector has found the widest application. Its relatively low fabrication, installation, and maintenance costs have been the primary reason for its widespread use. Additionally, flat-plate collectors can be easily incorporated into a building shape provided the tilt and orientation are properly calculated.

Flat-plate collectors utilize direct as well as diffuse solar radiation. Temperatures to 250°F (121°C) can be attained by carefully designated flat-plate collectors. This is well above the moderate temperatures needed for space heating, cooling, and domestic water heating.

A flat-plate collector generally consists of an absorbing plate, often metallic, which may be flat, corrugated, or grooved; painted black to increase absorption of the sun's heat; insulated on its backside to minimize heat loss from the plate; and covered with a transparent cover sheet to trap heat within the collector and reduce convective cooling of the absorber. The captured solar heat is removed from the absorber by means of a working fluid, generally air or treated water, which is heated as it passes through or near the absorbing plate. The heated working fluid is transported to points of use or to storage, depending on energy demand.

In selecting any particular collector, one should consider thermal efficiency, the total area and orientation required, durability of materials, and initial operating cost. Following are descriptions of three types of collectors.

Open-Water Collector. At present, collectors that are factory-produced and shipped to the building site are relatively high in cost, due in part to the small volume manufactured. Collectors built from commonly available materials and fabricated on site are less expensive. Their thermal efficiency, however, may be lower than factory-produced units. An open water collector of the type used in the Thomason House (p. 58) is representative of onsite fabricated collectors, which use corrugated metal roofing panels that are painted black and covered with a transparent cover sheet. The panels thus provide open troughs in the corrugations for trickling water to be fed from a supply at the top of the roof to a collection gutter at the base, where it is then transported to storage. Heat losses that occur by evaporation in open systems are reduced in some designs by the nesting of two corrugated sheets with a small enough passage between them to force the water into contact with the top sheet. An open water collector should be carefully evaluated before use in cold climates to determine the extent of condensation and corresponding loss of efficiency.

Air-Cooled Collector. Collectors have been developed that employ air (or gas) as the transport medium between collector and storage. George Löf designed such a collector and installed it on his own home in Denver, Colorado (p. 58). Low maintenance and relative freedom from the freezing problems experienced with liquid-cooled collectors are two of the chief advantages of air collectors. In addition, the heated air can be passed directly into the dwelling space or into the storage component. Disadvantages are the inefficient transfer of heat from air to domestic hot water and the relatively large duct sizes and electrical power required for air transport between collector and storage. Although few air collectors are now readily available from manufacturers, in contrast to the more than one dozen sources of liquid-cooled collectors, it is predicted that air-cooled collectors will soon be more widely used.

Liquid-Cooled Collector. Most collectors developed since the time of the MIT experimental houses have used water or an antifreeze solution as the transport medium. The liquid is heated as it passes through the absorber plate of the collector. Then it is pumped to a storage tank, where it transfers its heat to the storage medium.

Freezing, corrosion, and leaks have been the major problems that have plagued liquid-cooled systems, which are otherwise efficient collectors and transporters of heat. These problems are generally prevented by using oil or water treated with corrosion inhibitors as the transport medium or by designing the collector to drain into storage during periods of noncollection.

Increasing the Performance of Flat-Plate Collectors. Flat-plate collectors are frequently mounted on the ground or on a building in a fixed position at prescribed angles of solar exposure—which vary according to the geographic location, collector type, and use of absorbed solar heat. The fixed mounting has advantages of structural security and design integration, but must be oriented within prescribed limits to receive a level of solar radiation commensurate to the capital investment involved in installing a solar energy system. For space heating, a tilt of latitude plus 15° and an orientation true south to southwest (afternoon air temperatures are higher and thus bias an orientation west of true south) are considered optimal. For combined space heating and cooling, collector orientation remains the same while collector tilt is changed to latitude plus 5° of these optima. There have been several proposals to improve the annual thermal performance of flat-plate collectors. One involves the use of adjustable sun-tracking flat-plate collectors, (Figure 6-7) that are continually or periodically adjusted, and another uses reflecting panels to increase thermal yield.

Mounting a flat-plate collector on adjustable or sun-tracking mechanisms can improve the annual thermal performance of the collector by as much as 70 percent. A collector can be tilted from the annual optimum (described as latitude plus 15° for space heating) to more closely

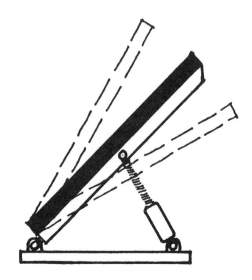

Figure 6-7. Flat-plate collector with adjustable tilt.

approximate the seasonal optima. For instance, season optima for space heating will vary 47° from December to June. A monthly adjustment can be made manually, provided there is access to the collectors; this can yield at least a 10 percent improvement over a fixed position. Adjustable tilt mechanisms can usually be accommodated most easily on ground or flat roof installations.

A collector can also be rotated on one axis to follow the daily path of the sun's orientation from east to west. This requires an automatic tracking mechanism, unless a method for manual adjustment tracking during the day is provided. With an automatic sun-tracking mechanism, an annual energy increase of 40 percent could be expected in areas where the predominant source of thermal energy is direct radiation.

The tilt and orientation of a flat-plate collector can be maintained in any optimum position throughout the day and year by use of a heliostatic mount, which maintains the collector at a perpendicular (normal) exposure to direct solar radiation. Although the control mechanism and structure mounting are complicated, a sun-tracking flat-plate collector could collect 70 percent more solar radiation than the same flat-plate collector in a fixed optimum position. The cost of such a sophisticated tracking solar collector should be compared with the cost of a larger fixed collector, which would deliver the same energy output.

Another concept for increasing the thermal yield per unit of flat-plate collector area is the use of panels that reflect additional solar radiation onto the collector, thereby increasing its thermal performance. The same panels, if so designed, can also be used to cover and insulate the collector during noncollection periods, at night, and on cloudy days. The panels may be operated manually or by sun-sensitive automatic controls. In either case, questions of maintenance and operation should be considered, particularly in areas that experience snow or ice.

Some collection arrangements reflect incoming solar radiation several times onto a focused or concentrated area, thereby using an optical gain to increase the unit collection of the absorbing surface. As a result, the area of collector-absorbing surface needed is reduced. There are numerous concepts based on such principles. One patented idea that has found application is a flat-plate collector located in a roof shed where an adjustable reflecting panel directs the radiation to an otherwise weather-protected collector. As mentioned previously, cost, fabrication, and maintenance of adjustable reflectors are major considerations.

Concentrating Collectors

Concentrating collectors use curved or multiple point-target reflectors to increase radiation on a small target area for either a tube or point absorber. Presently, concentrating collectors cost more than the flat-plate collectors and involve problems of reflective surface maintenance. For use in sunny climates, however, they provide more than double the temperature generated by flat-plate collectors. Concentrating

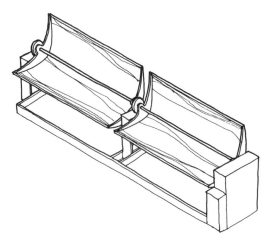

Figure 6-8. Linear concentrating collector.

collectors are best suited for areas with clear skies where a major portion of solar radiation is received in direct rays. Their inability to function on cloudy or overcast days is a significant disadvantage; however, they may be successfully used in solar cooling systems. At present, they require more development than flat-plate collectors.

Linear Concentrating Collector. A linear concentrating collector (Figure 6-8) is curved in one direction to focus radiation on a pipe or tube absorber. Heat is removed from the absorber by a working fluid circulating through the pipe and transported to point of use or storage. The absorber is generally covered with a transparent surface to reduce convective or radiative heat losses. The working fluid should have a boiling point above the expected operating temperature of the collector and also be resistant to freezing.

Linear concentrating collectors can be designed with the long axis horizontal (in an east-west direction) or at an optimum tilt (in a north-south direction). Since the change in the sun's altitude during the day is less than the change in its compass direction, horizontal linear concentrating collectors can be manually adjusted every few days to tract the average path of the sun as it changes from season to season. Tilted linear concentrators with the long axis north-south must track the sun throughout the day, as must horizontal linear concentrators designed for precise focusing.

Circular Concentrating Collector. A circular concentrating collector (Figure 6-9) is a reflector in the shape of a dish or hemisphere that is used to focus solar radiation on a point target area. An absorber located at the focal point absorbs solar-produced heat which a working fluid then transports away from the collector. The collector may be fixed and the target area (absorber) movable to accommodate the daily path of the sun as the focus point changes with the direction of incoming rays. More commonly, the entire reflector and absorber assembly is made to follow the sun. Several experimental designs have been constructed utilizing this principle. The high temperatures achievable with such collector mountings may eventually justify their use despite problems of operation, durability, design integration, and structural mounting.

Passive Collectors. The collector concepts discussed so far have been relatively independent elements that can be organized and operated quite apart from the building itself. That is, the relationship of the collector to the building it serves can vary significantly without a major alteration of system performance. There is a point when it is advantageous in terms of cost, building design, and system operation that the collector, storage, and building be physically integrated.

There are numerous collector concepts in which the relationship of the collector to the building is direct, and thus alteration of the collector design would modify in varying degrees the building's design. These concepts use the entire building or various elements of the building (walls, roof, openings) as solar components. As such, the collector and building are one and the same and therefore cannot be separated from each other. This type of collection method has come to be known as *inherent* or *passive solar collection.*

The integrated nature of passive solar collectors permits the design of a variety of collector concepts. The imaginations of the designer and the builder are the only limiting factors. Passive collectors can be as simple as a window or greenhouse.

Incidental Heat Traps. All collectors are "heat traps" in that they capture heat from direct solar or diffuse sky radiation or from adjacent ground or building surface reflection. There are numerous building components including windows, roof monitors, and greenhouses, which are not normally considered solar collectors but which can be used as such along with their other principal function. Because

they can serve various purposes—visibility, ventilation, and natural illumination, as well as incidental heat collection—they become particularly valuable components in integrated solar-dwelling design. They also illustrate that energy conservation does not preclude the use of windows, glass structures, or skylights if they are designed to maximize direct solar heat gain.

The use of south-facing windows to increase heat gain directly into a building is well known. Beginning with the Crystal House in 1933 (see p. 57), the concept was used in the United States in popular house plans, which were often referred to in the 1940s and 1950s as "solar homes." The south wall orientation is considered ideal because in northern climates, shading a south window to prevent summer overheating is easily accomplished by an overhang calculated to equinox sun angles. In such arrangements, careful attention must be given to insulating the window at night, preferably from the interior, to reduce heat loss. With an interior insulating drapery, or even better, a shutter that is much more airtight, windows do indeed function as effective heat traps and have been shown to be able to provide a sizable percent of the annual heat requirements of a building (estimates vary from 25 to 60 percent, depending on climate and use).

Windows used as solar collectors have the drawback of overheating the space they serve. In order to reduce the overheating effect, masonry surfaces such as concrete, brick, tile, or stone on floor or walls can be used for their heat-storage capacity, absorbing the heat during the day and subsequently radiating it for several hours or more. The storage effect of a particular floor or wall can be calculated as a function of the specific heat of the masonry, its volume and weight, and the expected temperature differences it will experience throughout the day. Too great a storage effect in the exposed room surfaces can have a negative effect on occupant comfort or fuel consumption if the morning "reheat" time of the materials is too long. However, properly designed, the thermal mass construction materials can play a significant role in an integrated solar dwelling.

Use of a greenhouse as an incidental heat trap is a further elaboration of the solar window concept. It exposes more glass area to solar radiation than a window but experiences greater heat loss if no provision is made for insulation. Fiberglass panels and temperature-sensitive gels, which will reduce heat loss while allowing a comparable amount of direct solar radiation to enter the greenhouse, are under development. Another idea that has found application is a patented vacuum-driven system, which fills and empties a cavity between two sheets of glass or plastic with polystyrene beads, thus substantially increasing the insulation characteristics of the transparent surface.

Again, storage such as masonry surfaces or a rock pile under the greenhouse may be necessary to avoid overheating the space. One advantage of a greenhouse as a solar collector is that it can be closed off from the rest of the house on sunless days, thereby reducing the net heated area of the dwelling. However, during cooler weather, it can be used as a supplementary room, such as an atrium, or a dayroom.

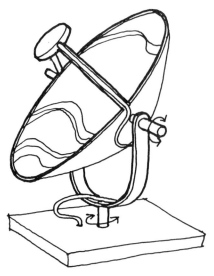

Figure 6-9. Circular concentrating collector.

Another valuable passive collector concept is the roof monitor. A roof monitor is a cupola, skylight, or clerestory shed designed to control heat gain, natural light, and/or ventilation. Roof monitors by themselves are poor devices for gaining usable heat, because the heat enters at the high point of the building and the roof exposure gains so much solar radiation in the summer that shading or insulating arrangements are required. Also, due to the low winter sun angle, the sun entering through the roof monitor would probably not reach the floor or lower wall surfaces to offer any storage effect. However, if a return air register is located at a high point in space, the trapped heat can be recirculated. This technique will prevent temperature stratification within a building by continually returning solar heat gained through the building and roof monitor to the lower occupied spaces.

Roof monitors are excellent sources of natural lighting and can be used in summer months to augment natural cooling through the "thermal chimney" effect. Roof monitors, designed with proper insulating and ventilating controls, can be used to great advantage to help lower home energy costs.

Thermosiphoning Walls/Roof. Another set of passive collector concepts makes use of heat that is built up within a wall or roof structure by siphoning or drawing it off and supplying it to a room or storage element. Thermosiphoning, a term traditionally applied to mechanical systems that use the natural rise of heated gases or liquids for heat transport, is the primary method for moving captured heat to point of use or storage. To avoid overheating in the summer, the space where the heat builds up is vented to the exterior.

One thermosiphoning concept that has found application in traditional as well as solar designs is the use of solar heat trapped in airspaces in walls and roofs. When the trapped air temperature exceeds the temperature of the internal building space, it can be drawn off by direct venting or forced-air duct arrangements.

This method of solar heat collection is marginal at best due to the small amount of heat collected, problems in the control of temperature differences between the inside of the wall or roof structure and the occupied space, and the large ducts and electrically powered fans needed to move any sizable volume of heated air. However, the concept of using solar heat that is built up within a building's walls or roof deserves consideration as a multipurpose solution to annual climate conditions.

A more effective variant of the preceding concept is one in which the external surface or internal and external wall or roof surface are transparent. The heated air trapped between the building surfaces can be used because it will usually be hotter than the temperature of the occupied space. Ducted fiberglass panels are under development for just such an application. This concept is not as effective a formal solar heat collector (i.e., flat-plate collector) but it has the advantages of admitting natural light and of providing better insulation than a plate-glass window.

The previous two thermosiphoning concepts used the walls and roof of the building for solar heat collection. In a further elaboration, the building envelope or exterior walls are also used as storage. In several examples described previously, notably the Michel/Trombe House (see p. 59), a glass wall is placed over an absorbing material—generally masonry—which is painted dark and serves as the heat storage area. The airspace between the glass wall and the absorbing material is vented to the interior at the top of the wall or ducted to rock storage elsewhere in the building. A cold air return must be located at the bottom of the collector so that a thermosiphoning arrangement can be used to facilitate air circulation. As previously mentioned, the morning reheat time of large storage media becomes a design calculation that is particularly critical with such direct storage concepts.

Solar Ponds. A solar pond is a particularly interesting passive collector concept because it can provide both heating and cooling. Also, a solar pond may be integral with the building structure—on the roof, for example—or entirely separated from the building on adjacent ground. For either situation, control must be maintained over the heating and cooling processes for efficient operation. This can be accomplished by the use of movable insulating panels to expose or conceal the pond, by filling and draining the pond according to the heating and cooling demand, or by covering the pond with a transparent roof structure.

Although use of solar ponds at present has been confined to the Southwest, it has also been proposed for use in northern climates. Solar ponds are particularly appropriate to climates where the need for cooling is the principal design condition and where summer night temperatures are substantially lower than daytime temperatures. The combination of these climatic conditions, found usually in hot-arid regions, permits the ponds and elements of the building (exterior walls) to be cooled by natural radiation to the night sky and to effect a time lag of temperature through the building envelope, in ideal cases for 8 hours or more. This permits numerous design concepts to aid natural cooling by radiation and evaporation. In hot-humid climates, high vapor pressure, cloudiness, and a small diurnal temperature range limit the cooling efficiency of solar ponds.

Roof ponds have found a more widespread application than ground ponds. The major advantage of a solar roof pond is that it does not dictate building orientation or exposure, and, as long as there are no barriers between sun and pond, it will provide a completely even heating or cooling source over the entire living area of a building. Several dwellings incorporating a solar roof pond have been designed and constructed. The Atascadero house (see p. 58) is an example of a dwelling concept using a roof pond. In this patented design, water containers are dispersed on a flat roof and covered with insulating panels. During summer days, the panels are closed. At night, the panels are removed to lose heat to the cooler night sky. Suitably cooled, the water containers then draw heat during the day from the building interior (usually through a metal deck roof). In winter, the process is reversed, exposing the containers during the day to collect and store solar thermal radiation. At night, the containers are covered by the insulating panels, providing heat to the interior.

The same process is applicable to northern climates. However, the roof pond is covered by a transparent roof structure oriented to receive maximum incident solar radiation. Heat is trapped in the attic space, thus warming the pond. Insulating panels cover the transparent surface during periods of no collection to reduce heat loss. The transparent surface could be removed during the summer to increase evaporative and convective cooling.

An alternative to water containers covered with insulating panels or enclosed in an attic space is a roof pond with circulating water between the roof and the storage tank, which is located below or within the occupied space. With this system, the need for movable insulating panels is eliminated. During the heating cycle, circulation would take place during the daytime. The water, heated by the sun, would be stored or distributed depending on the dwelling's heating requirement. To prevent cooling by evaporation, the pond must be covered by a transparent surface—glass or plastic, which can float on the surface. During the cooling cycle, the circulation of water would take place only during the night. The cooled water would be stored to draw heat during the day. The transparent cover sheet would no longer be needed for efficient cooling, since cooling would be achieved by night sky radiation, convection, and evaporation.

SOLAR HEAT STORAGE

The intermittent availability of solar radiation requires that heat be stored during times of favorable collection for later use for such purposes as space heating, cooling, and domestic water heating. Solar heat may be stored by raising the temperature of inert substances such as rocks, water, masonry, or adobe (sensible heat storage); or it may be stored by reversible chemical or physical/chemical reactions such as the dehydration of salts or phase changes (latent heat storage).

In many cases, the use of several storage methods for the storage of heat at different temperatures has been advantageous for supplying the heating and cooling demand of specific buildings. The type, cost, operation, and required size of the solar storage component will be determined by the method of solar collection, the dwelling's heating and cooling requirement, and the heat-transfer efficiency to and from the storage unit.

Sensible Heat Storage

Room Air and/or Exposed Surfaces. The solar radiation received from south-facing windows or transparent panels increases the temperature of room air and surfaces exposed to the sun's rays. As such, the room's air and exposed surfaces (e.g., walls, floor) are the solar storage component for a window wall, greenhouse, or transparent panel that collects the radiation. For most situations, the storage capacity of the air and surfaces will not be sufficient for long periods

of heating demand. In addition, in the process of "charging" the storage, the space may become overheated and possibly extremely uncomfortable for the occupants.

To use effectively the radiation stored in the air and room surfaces, careful attention must be given to minimizing the loss of heat at night or when collection is not occurring. Insulated drapes, shutters, and other such devices are necessary to reduce heat loss and increase the use of trapped heat. The room size, the window placement, the expected temperature difference, and the material composition, volume, and weight will also determine the performance of the solar storage.

A more direct application of this concept involves the placement of a glass or transparent wall over an exposed masonry surface such as concrete. The exposed surface, which serves as the heat store, is painted a darker color and located directly behind the transparent surface. The thermal conductivity and specific heat of the wall material and the expected temperature range will determine the volume of the wall. The reradiation time lag must be accurately calculated to assure proper heating of the space. It is sometimes necessary to place insulation on the room side of the storage wall to avoid overheating the space. The exposed masonry storage method is usually used in conjunction with interior and exterior vents to control the heat distribution to the space or to another storage system.

A variation on exposing a masonry surface to solar radiation is to expose containers filled with water. The exposed water containers may be placed on the roof or used as interior or exterior walls. The Atascadero house, which uses plastic bags filled with water placed on the roof, is an example of this storage method. Again, careful calculations are required to properly size the storage capacity. Also, a similar means of thermal control is necessary to assure a proper lag-time to avoid overheating.

Rock Storage. A common method of heat storage, most often associated with air-cooled flat-plate collectors, is rock storage. Pebble beds or rock piles contained in an insulated storage unit have sufficient heat capacity to provide heat for extended sunless periods. The rock storage is heated as air from the collector is forced through the rock container by a blower. Rock storage will require approximately two-and-a-half times the volume of water storage, assuming the same temperature range. For example, a rock pile with a void space of one-third of the total volume can store approximately 23 Btu/cubic foot/°F, while water can store 62.5 Btu/cubic foot/°F.

Rocks appropriate for storing solar heat are about 2 inches in diameter. A decrease in pebble size increases the airflow resistance through the storage and can affect distribution efficiency and blower and duct size. Unlike the other storage methods, rock storage does not have to be in close proximity to the collector. However, as the distance increases, the heat-transfer losses between the heated air ducts increase, and more electrical power is generally required for moving air among the collector, storage, and heated spaces.

Water Storage. Water has the highest heat capacity per pound of any ordinary material. It is also very inexpensive and therefore is an attractive storage and heat-transfer medium. However, it does require a large storage tank, which may be expensive. The tank is usually insulated to reduce conductive heat losses. Also, it is sometimes practical to compartmentalize the storage tank to control temperature gradients (different temperatures within storage tank) and to maintain an efficient heat transfer. Potential disadvantages of water storage include leakage, corrosion, and freezing.

Heat is generally transferred to and from storage by a working fluid circulated by an electric pump. The heated working fluid itself may be placed in storage or its heat transferred to the storage tank by a heat exchanger. The process of heat transfer to water is more efficient than to rock, and therefore less surface is required for the heat exchanger. With water storage, proximity of storage to the collector is not as critical as with rock storage. Also, compared to rock storage, water occupies a comparatively small volume.

Latent Heat Storage. The use of the heat of fusion or heat of vaporization, associated with changes of state or with chemical reactions, offers the possibility of storing a great deal of heat in a small volume. Although numerous physical/chemical processes have been investigated and offer numerous advantages compared with sensible heat storage, a completely reliable storage method using latent heat does not exist. However, to illustrate the basic principles of latent heat storage, several examples are presented.

Salt Hydrates. The salt hydrates are among the simplest types of chemicals used for heat storage. The heat-storage process involves a phase change—generally, liquid to solid to liquid—which is induced by and produces heat. When the temperature of salts (such as the Glauber salt used in the Dover house) is raised to a specific value dependent on the chemical composition, heat is absorbed, releasing *water of crystallization* which dissolves the salt. When the temperature drops below the crystallization temperature, the stored heat is released, and the solution recrystallizes. This phase change allows the salts to store a large amount of heat per unit volume. Unfortunately, after many phase-change cycles, the salt hydrates have a tendency to break down, thus discontinuing their heat-evolving crystallization.

Paraffin. Paraffin storage is similar to salt-hydrate storage because thermal energy is stored by heat-of-fusion. Paraffin does not have the crystallization problem common with salt hydrates. However, waxes do tend to shrink upon solidifying and thus lose contact with the heat-exchange surfaces (i.e., walls of containing vessel), thereby reducing the rate of heat transfer.

Thermal Energy Distribution. Generally speaking, there are three methods by which thermal energy from storage or collector can be distributed to point of use: gas flow, liquid flow, and radiation. Within each category, there are several techniques by which the distribution of energy to occupied spaces can be accomplished. Some involve mechanical and electrical equipment and processes, while others utilize natural convection and radiation. How solar radiation is collected and stored will usually determine the means of distribution. For example, if an air-cooled flat-plate collector is used to capture solar radiation and a rock pile is used to store the heat, distribution is usually accomplished by air.

Gas Flow Distribution

Natural Convection. Natural convection is the circulatory motion of air caused by thermal gradients without the assistance of mechanical devices. An example of convection is the motion of smoke toward room lamps—the hot air generated by the lamp rises because it is less dense, and cooler air moves in to replace it.

Natural convection is a useful means of distributing solar thermal energy because it requires no mechanical or electrical input. However, careful attention to design is required to maintain proper control of convective distribution methods. The placement of solar collector, storage, interior and exterior walls, and openings is extremely important for the successful operation of convective distribution.

The operation cycle of natural convective distribution is quite simple. Heat from the collector or storage is supplied to the habitable space. This process is controlled by the collector or storage design or by wall or floor vents. As the hot air rises to displace cooler air, convection currents similar to those causing winds occur, and the air is distributed through the space. The air is cooled, becomes dense, and falls toward the floor where it is captured by cool-air return vents, passed through the collector and storage, and once again distributed. The cycle will continue as long as there is a temperature difference between the collector/storage components and the room air. When the convection heating cycle is not desired—in the summer, for instance—the warmer air may be vented to the exterior.

Forced Air. A forced-air system relies on mechanical equipment and electrical energy for the distribution of thermal energy. Design for solar systems is much the same as for conventional forced air systems. However, because solar-produced temperatures in storage are often relatively low, distribution ducts and vents must normally be larger than those used in conventional heating/cooling systems. Therefore, to achieve maximum efficiency in a solar system, careful attention to the design of air distribution throughout a dwelling is required.

Forced-air distribution for solar systems is similar to conventional air distribution. Air from either the collector or storage is blown through ducts to the occupied spaces. The type of solar collector or storage is not the determining factor for selection of a forced-air ducted system; the system is adaptable to rock, water, or phase-change storage, air is simply blown through the storage to ducts, which supply the dwelling spaces. In the case of water storage, a heat exchanger is required to transfer heat from the liquid to air, which is distributed to dwelling spaces.

Liquid Flow Distribution

Forced Radiation. Forced radiant distribution relies on the transfer of heat to air in the occupied spaces by radiation and convection from circulating hot water through tubes. For cooling, the forced radiant system is generally used in conjunction with a refrigeration unit, which passes chilled water through a fan-coil unit located at the point of distribution. A blower is used to force air through the cooled fan-coil unit and into occupied spaces.

The piping for the radiant system may be located in the ceiling, floor, or along the wall in fin-tube baseboard units. The only significant alteration required of conventional radiant systems for use by solar systems is the enlargement of the radiating surfaces—larger fin tubes or closer spaced ceiling or floor coils—because of lower temperatures from storage.

Natural Radiation. Natural radiation is the transfer of heat by electromagnetic waves without the assistance of mechanical devices. The radiation properties of the emitting and absorbing surfaces, which are influenced by their temperature, will determine the rate of heat flow between them.

Unlike natural convection, which is dependent on differential air temperatures for distribution, natural radiation is dependent on differential surface temperatures. An example of natural radiation is the sun warming a greenhouse on a cold day. The radiant energy is transferred directly to the greenhouse surfaces and is not significantly affected by the cool temperature of the surrounding air. Natural radiation is particularly useful for collector or storage systems that are directly exposed to the occupied spaces. The captured energy can be emitted by natural radiation directly to the room's surface. The walls, floors, and ceiling of a dwelling, which are used to collect and/or store thermal radiation, will radiate directly to a room's other cooler surfaces. The Atascadero house is an example of solar heat distribution by natural conduction and radiation.

Collector Storage and Distribution Components

Solar system design assembles the collector, storage, and distribution components into a heating, cooling, and/or domestic hot-water system. Each component may be compatible with a limited number of other solar components or may be compatible with many. For example, a solar collector may be compatible with a specific storage component, which in turn may serve one or several types of distribution systems.

To illustrate the compatibility of various solar components and to describe the process of converting solar radiation into thermal energy for heating and cooling, several representative solar systems will be discussed. Each system is made up of a collector, storage, and distribution component.

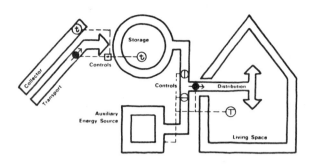

Figure 6-10. A solar system.

A Solar System. A solar system may be diagrammed as shown in Figure 6-10 above. The basic function of a solar system is to convert solar radiation into usable energy. Radiation is absorbed by a collector, placed in storage, without or with the assistance of a transport medium, and distributed to point of use—an occupied space. Each operation is maintained and monitored by automatic or manual controls. An auxiliary energy system is usually available for operation, both to supplement the output provided by the solar system and to provide for the total energy demand should the solar system become inoperable. With this relatively simple process in mind, a more detailed explanation of several solar systems is presented to illustrate variations in solar system design and operation.

Warm-Water Flat-Plate System. Solar heating using water as the heat-transfer and storage medium is the most common system in use today. More information is available about the behavior of water systems than about either air or passive systems. The basic components of a typical water system consist of: a collector; storage; a system of piping, pumps, and controls for circulating water from storage through the collector; and a distribution network for transferring stored heat to the dwelling space.

The liquid-cooled flat-plate collector has a flat absorbing surface, integrated with transfer fluid piping, which collects both direct and diffuse radiation. Energy is removed from the collector by a liquid flowing through conduits in the absorber plate. The transport fluid is pumped to storage where its heat is transferred to the storage medium (in this case, water) and then returned to the collector to absorb more heat. Generally, the transfer fluid is circulated through the collector only when the absorbing surface is hotter than storage (except in instances when snow has covered the collector surface and the heated transport fluid is circulated through the collector to melt it).

Storage consists of either a concrete or steel tank located near or beneath the building (access should be provided). The tank should be insulated to minimize heat loss. A concrete tank should be lined with a leak-proof material capable of withstanding high storage temperatures for extended periods without deterioration. Heat from the collector is transferred to storage by a heat-exchange coil passing through the storage tank. Coil length and size are dependent on expected collector operating temperatures.

The distribution system consists of a pump and pipes that deliver heated water to occupied spaces. A thermostat controls the operation of waterflow or fan-coil unit used in each room or dwelling. Baseboard heaters (convectors) require careful evaluation when not used in conjunction with a fan-coil unit. Liquid-cooled flat-plate collectors seldom deliver water above 150°F in winter operation without auxiliary energy or reflected surface focusing. Thus, most warm-water distribution systems use fan-coil units or enlarged convectors.

Energy is transported away from the collector to storage by water or a water/antifreeze solution. Liquid transport fluids should be carefully evaluated before selection. The liquid must absorb heat readily at various collector temperatures and easily give up heat to the storage medium. In addition, the liquid should not be toxic, susceptible to freezing or boiling, or corrosive to the system components.

A gas-fired conventional boiler is integrated with the solar system to provide an auxiliary energy supply should the solar system fail to function or not meet the dwelling's heating requirement. The distribution piping is run through the boiler where an energy boost may be supplied when temperatures from storage are not sufficient to heat the dwelling adequately.

Domestic hot water piping is run through the central storage tank prior to passing through a conventional water heater. Storage heat is transferred to the hot water piping, thereby either eliminating the need for additional heating or substantially reducing the energy required to raise the water to the needed distribution temperature. The domestic water-heating system may operate independently of the space-heating system. This is very useful in the summer, when space heating is not required.

The advantages of a warm-water flat-plate system include:

- It has repeatedly been proved to work well.
- Water is a cheap and efficient heat-transfer and storage medium.
- Piping, as opposed to ductwork, uses little floor space, is easily interconnected, and can be easily routed to remote places and around corners.
- The circulation of water uses less energy than the circulation of air with corresponding heat content.
- It requires much less-heat exchanger area than an air system.

The disadvantages are:

- High initial cost, particularly when expensive prefabricated collectors are employed. With the use of large areas of lower-efficiency collectors, the total system cost may be lowered considerably.
- Care must be taken to prevent corrosion, scale, or freeze-up capable of causing damage or blockage.
- Leakage anywhere in the system can cause considerable damage to the system and the dwelling.

- Contamination of the domestic hot-water supply is possible if a leak allows treated water storage to enter the domestic water system.

Warm-Air Flat-Plate System. In warm-air systems, unlike warm-water systems, air is used to transfer heat from collector to storage. The storage medium can be water; however, rock piles are used more frequently. Heat, stored in the rock pile, can easily be distributed to the dwelling space by a forced-air system.

The air-cooled flat-plate collector has a solid absorbing surface and collects both direct and diffuse radiation. Energy is removed from the collector by air flowing in ducts beneath the absorber plate. The system may be operated in four different modes:

1. heating storage from collector
2. heating house from collector
3. heating house from storage
4. heating house from auxiliary energy system

The four modes of operation are regulated by several sets of dampers. One set of dampers will direct airflow from the collector into storage or directly into the occupied spaces while another set will regulate airflow from storage to the occupied spaces. The dampers may be adjusted by manual or automatic controls. During modes 2 and 3, an energy boost may be supplied to the warm air by the auxiliary energy system before the air is distributed to the occupied space. The amount of energy boost is determined by the temperature of the air passing through the auxiliary heater and the amount of heat required at the point of use.

Storage consists of rocks about 2 inches in diameter, contained in a concrete bin in a basement area or underground beneath the building. The container is insulated on the earth sides to reduce heat loss. The storage capacity should be sufficient to provide several days of winter heating.

Because temperatures in rock storage are typically highly stratified from inlet to outlet, the airflow providing heat to storage should be from top to bottom. This insures that the temperature of air returning to the collector from the storage is as low as possible, thereby increasing collector efficiency. The airflow, when removing heat from storage, should be in the opposite direction to insure that air returning to the rooms is as warm as possible.

The hot air distributed to the rooms comes either directly from the collectors or from storage. The ducting required to conduct the air from the collectors to storage is extensive when compared to analogous piping requirements for liquid-cooled collectors. Two blowers are required to distribute air throughout the system.

Almost any type of auxiliary energy system may be used in conjunction with a solar system. The auxiliary system may be completely separate or fully integrated with the solar heating/cooling system. However, in most cases it makes economic sense to integrate the backup system with the solar system. This may mean running the distribution component from heat storage to the occupied space through the auxiliary system where an energy boost may be supplied when storage temperatures are low. Heat from storage may also be used in conjunction with heat pumps, absorption units, or Rankin engines. The heat pump, a device that transfers heat from one temperature level to another by means of an electrically driven compressor, uses the solar heat available from storage to supply necessary heat to the occupied space. The advantage of the heat pump/solar system integration is the reduction of electrical energy required by the pump because of heat supplied by solar storage. Also, the heat pump is the most efficient device presently available for extracting and transferring electricity into heat.

Piping for the domestic hot water is run through the rock pile storage bin. As a result, the domestic hot water is preheated before passing through a conventional water heater, thus reducing the water heater's energy requirement.

The advantages of warm-air flat-plate systems include:

- Capital cost tends to be lower than a water system of the same capacity.
- There is no problem with corrosion, rust, clogging, or freezing.
- Air leakage does not have the severe consequences of water leakage.
- Domestic hot water supply is not subject to contamination by leakage from heat storage, as in the water system.

The disadvantages are:

- Ductwork risers occupy usable floor space and must be aligned from floor to floor.
- Air, having a lower thermal-storage capacity than water, requires correspondingly more energy to transfer a given amount of heat from collector to storage, and from storage to occupied spaces.
- Air collectors and storage may need frequent cleaning to remove deposits of dust (filters may solve this problem).
- Air systems require a much larger heat-exchange surface than liquid systems.

Warm-Water Concentrating System. Solar systems with concentrating collectors have not been extensively used to provide space heating or cooling. The absence of such equipment from the market and the high cost and uncertain reliability of tracking or concentrating equipment in freezing rain, ice, wind, and snow have been the primary reasons for their limited use. However, they do offer advantages over flat-plate collectors—primarily the generation of high temperatures to operate heat-driven cooling systems. The warm-water concentrating system uses a linear concentrating collector.

The collector is a linear concentrator with a glass-enclosed pipe absorber. The collector captures only direct radiation and is therefore limited to climatic regions with considerable sunshine and direct radiation in winter. However, where applicable, the linear concentrating collector offers considerable

economies over flat-plate collectors since the necessary absorber area is reduced and the assembly is often simpler to construct.

The absorber pipe is a black metal tube within a glass enclosure in a vacuum to reduce convection and radiation losses. Radiation is focused on the absorber by a trough-shaped reflector surrounding the pipe.

Storage consists of a steel tank or a lined concrete block enclosure filled with water. Again, the storage unit should be insulated to minimize heat loss. As with almost all solar storage techniques, special structural support will be required if the storage tank is to be located in the dwelling.

The distribution system is by heated water to baseboard convectors. Heat is removed from storage by liquid-to-liquid heat exchanger. The heated water is pumped to baseboard convectors located throughout the building. If storage is below a present minimum temperature, the pump continues to operate with a conventional oil- or gas-fired furnace assist in the liquid distribution loop.

Collector fluid is transported by means of a pump, which causes the fluid to flow through the absorber, and into the storage heat exchanger from which energy is removed and transferred to storage. The working fluid should be a heat-transfer medium that has excellent transport properties and a boiling point above the expected operating temperature of the collector.

Domestic hot water piping is run through a heat exchanger in storage, thus preheating the water, before it proceeds to a conventional water heater, which also provides storage. The water heater may supply a boost to the water depending on its temperature.

The advantages of warm-water concentrating systems include:

- Potential for more than double the temperatures of either air or water flat-plate collectors (particularly useful for solar cooling).
- Total absorber area needed is substantially smaller than flat-plate collectors.
- Collector forms lend themselves to mass-production techniques.

The disadvantages are:

- Capital cost of collectors is greater than either air or water flat-plate collectors.
- Concentrating collectors may present problems of operation, reflecting surface durability, and structural mounting.
- Leakage at flexible absorber connections may present possible problems.
- Climatic applicability for winter space heating is limited.

Warm-Air Passive System. The passive system described here is one possible concept among many. It makes use of extensive south-facing glazing with an intermediate collection/storage wall between the glazing and the occupied space. It relies in part on the thermosiphoning principles discussed earlier.

The passive collector, made up of a massive south-facing wall of either concrete or masonry separated by an airspace from an outer wall of glass, captures direct, diffuse, and reflected solar radiation. With the use of automatic or manually operated dampers and vents, the system may operate in four modes:

1. natural ventilation—no collection
2. house heating from collector
3. storage heating from collector
4. house heating from collector and storage

When no collection or heat distribution is required, the vents and dampers may be opened to provide natural ventilation and removal of heat striking the collector. The space may be heated directly from the collector by closing the storage vent duct, thus forcing the heated air into the occupied space. Once sufficient heat has been transferred to the space, the storage vent may be opened and heated air from the collector transmitted to storage. If heat is required at a later time, the storage vent may be opened to allow stored heat to enter the occupied space.

The system employs several storage concepts. The exposed masonry wall that the radiation strikes acts as a storage element. The warmed masonry surface transmits collected heat to occupied space by radiation. The second storage element is a rock pile located beneath the occupied space. Insulation is placed between the rock pile and floor surface to avoid overheating the space. Water or containerized salts could have also been used as the storage medium.

Heat is distributed to the occupied spaces from the collector or storage component. Ducting is required to transport the heated air from the collector to storage, and a small fan may be necessary to circulate this air. Heat is distributed to the space by convection from the collector and/or storage, by radiation from the collector and surrounding surfaces, and to a small degree, by conduction from the collector and surrounding surfaces.

Domestic water heating is not directly integrated in the solar space-heating system. However, a storage tank or the domestic hot water piping may be placed in the rock-pile storage to preheat the water before it passes through a conventional electric or gas water heater.

The advantages of a warm-air passive system include:

- A system with electrical controls can be designed to operate manually in a power failure.
- Cost should be reduced through simpler technology and elimination of a separate collector.
- Collector serves multiple functions (e.g., it can be a wall or a roof).

The disadvantages are:

- May not be cost-effective relative to warm-air or water flat-plate collector systems.

- In many cases, it requires automatic or manual insulating devices, which are expensive and may require life-style modification.
- Larger unobstructed area is needed to the south of the house for a vertical passive collector than for a roof collector.
- In some climates and for some passive systems, low winter-sun angles may disturb the occupants.
- Potentially large nighttime thermal losses from collector if not properly insulated.

SOLAR-DWELLING DESIGN

Solar heating and cooling systems are substantially different from conventional systems. The major difference is the use of solar radiation as a fuel source, which requires a large collection and storage area not associated with conventional systems. As a consequence, a major portion of the system is exposed. It cannot be easily hidden, nor should it be. Rather, the exposed components of a solar system should be integrated with the architectural design to enhance the overall appearance of the dwelling.

Solar Design Factors

Solar building design is inextricably related to a number of critical factors that influence the resultant architectural expression. Some factors affect the realistic opportunity for the utilization of solar heating and cooling systems; others affect the physical capability of a designing and constructing dwellings and solar systems. The former relates to considerations that may impede or accelerate the use of solar energy systems; the latter includes considerations that will influence the physical design of the dwelling and solar system.

The major opportunity factors are:

- *Legal*—such as building codes or zoning ordinances
- *Economic*—cost effectiveness of solar heating and cooling
- *Institutional*—such as lender's attitudes toward solar energy
- *Sociological*—such as a society's environmental and energy attitudes
- *Psychological*—such as an individual's expectation

The major physical factors are:

- *Climate*—sun, wind, temperature, humidity
- *Comfort*—such as an occupant's comfort zone
- *Building characteristics*—the thermal behavior of buildings
- *Solar system*—collector, storage, and distribution component integration
- *Site conditions*—such as topography, ground cover, and vegetation

The interrelationship of these and other factors will ultimately affect the design of solar-heated/cooled buildings.

Climate

The Given Conditions. Sun, wind, temperature, humidity, and many other factors shape the climate of the earth. Basic to using the sun for heating and cooling is understanding the relationship of sun and climate. What are the elements of climate? How do they affect solar-dwelling design for a specific site or climate.

Global climatic factors such as solar radiation at the earth's surface, tilt of the earth's axis, air movement, and the influence of topography determine the climatic makeup of any area on earth. These factors will determine the temperature, humidity, solar radiation, air movement, wind, and sky conditions for any specific location. Regional patterns of climate will emerge from a commonality among these climatic influences. In addition, local climates are influenced by site topography, ground surface, and three-dimensional objects. The sum total of these climatic factors will determine the need for and the design of solar dwellings and sites. In essence, climate is the given condition within which solar dwellings are designed.

What Is Climate? Climate (from Greek, *klima*) is defined in *Webster's* as "the average course or condition of the weather at a place over a period of years. . . ." Since weather is the momentary state of the atmospheric environment (temperature, wind velocity, and precipitation) at a particular location, climate could be defined as "the sum total of all the weather that occurs at any place." Like the weather, climates are directed by the sun and are influenced by all the physical conditions of the earth—the nearness of an ocean, the presence or absence of a mountain, prevailing winds, and so on. The climates of particular localities are comparatively constant, and, despite pronounced and rapid changes, they have weather patterns that repeat themselves time and again. Cold days occasionally occur in hot climates, and hot days are not unknown in cold climates; dry climates often have rainy periods, and wet ones have extended periods of drought. Even so, every place on earth over an extended period of time exhibits its peculiar combination of heat and cold, rain and sunshine.

In response to these differing climatic conditions, housing styles in one area of the country are substantially different than those found in other areas. This same differentation will be particularly true for dwellings heated and/or cooled by the sun.

Elements of Climate

The earth's climate is shaped by thermal and gravitational forces. Regional pressure, temperature, and topographical differences influence the climatic conditions on a continental scale. The conditions of weather that shape and define local and regional climates are called the *elements of climate.* The five major elements of climate are temperature, humidity, precipitation, air movement, and solar radiation. In addition, sky condition, vegetation, and special meteorlogical events are also considered elements of climate.

A designer or builder is primarily interested in the elements of climate that affect human comfort and the design and use of buildings. This is especially true if the building is to be powered by solar energy. The information a designer would like to know includes averages, changes, and extremes of temperatures; the temperature difference between day and night (diurnal range); humidity; amount and type of incoming and outgoing radiation; direction and force of air movements; snowfall and its distribution; sky conditions; and special conditions such as hurricanes, hail, or thunderstorms.

The National Weather Service gathers climate data for each of these elements at airports and meteorological stations. Generally, the information is not collected specifically for use by builders or designers, and so may omit data relevant to solar building design. Therefore, it is often necessary to supplement published data with information obtained directly from the meteorological station. In many cases, however, the most frequently used climatic data are organized into helpful design manuals. The use of these climatic design tools eliminates the need for painstaking analysis of unedited climatic data. The careful analysis of climatic elements will identify the local climatic features that are potentially beneficial or detrimental to human comfort and to solar-dwelling and site design.

Climatic Regions of the United States. A number of systems have been proposed for classifying climatic regions (common climatic conditions within a geographic area) within the United States. For purposes of discussing the variations in solar-dwelling and system design that result because of different climatic conditions, only a broad description of climates of the United States is required. W. Koppen's classification of climates, based upon vegetation, has been the basis of numerous studies on housing design and climates. Using Koppen's criteria, four broad climatic zones are to be found in the United States: cool, temperate, hot-arid, and hot-humid.

The climatic characteristics of each of the four regions are not uniform. They may vary both between and within regions. In fact, it is not unusual for one region to exhibit at one time or another the characteristics associated with every other climatic region. However, each region has inherent weather patterns that distinguish it from others. A brief description of the four climatic zones will identify the general conditions to which solar-dwelling and site designs in those regions must be responsive.

A wide range of temperature is characteristic of *cool regions.* Temperatures of –30°F (–34.4°C) to +100°F (37.8°C) have been recorded. Hot summers and cold winters, with persistent winds year-round, generally out of the northwest and southeast, are the primary identifiable traits of cool regions. Also, the northern location most often associated with cool climates receives less solar radiation than southern locations.

An equal distribution of overheated and underheated periods is characteristic of *temperate regions.* Seasonal winds from the northwest and south along with periods of high humidity and large amounts of precipitation are common traits of temperate regions. Intermittent periods of clear sunny days are followed by extended periods of cloudy overcast days.

Hot-arid regions are characterized by clear sky, dry atmosphere, extended periods of overheating, and large diurnal temperature range. Wind direction is generally along an east-west axis with variations between day and evening.

High temperatures and consistent vapor pressure are characteristic of *hot-humid regions.* Wind velocities and direction vary throughout the year and throughout the day. Wind velocities of up to 120 mph may accompany hurricanes, which can be expected from east-southeast directions.

Climate and Solar-Dwelling Design. Although numerous atmospheric and surface conditions shape the earth's climate, four elements of climate are particularly important for solar-dwelling and system design. These are solar radiation, air temperature, humidity, and air movement. Each of these imposes specific design conditions to which the dwelling, site, and solar system should be responsive. In some cases, protection of the building and solar system from excessive climatic exposure will be the primary concern; in others, it will be to maximize climatic impact. Regardless of the area of the country, a careful analysis of these elements should be undertaken before detailed dwelling or solar-system design.

Solar Radiation. In providing radiation, the sun provides the earth with almost all of its energy. Solar radiation is electromagnetic radiation transmitted in wavelengths that vary from 0.29 to 3 microns in length (1 micron equals 0.001 millimeter). VHF and other radio waves are familiar examples of infrared or long-wave radiation; X rays, gamma rays from radioactive substances, and cosmic rays are examples of ultraviolet or short-wave radiation. The human eye perceives radiation in the region between ultraviolet and infrared—that is, between 0.36 and 0.76 microns—as visible light. Visible light is therefore only a part of solar radiation. Solar radiation and light should not be confused with each other. Recognition of this difference is of critical importance for the design of solar dwellings and systems since collector designs and building materials will respond differently to solar radiation.

Solar Constant. The intensity of radiation reaching the upper surface of the atmosphere is taken as the solar constant. The solar constant may actually vary +2 percent due to variations in the sun's energy output and +3.5 percent due to changes in the distance between the earth and the sun. Solar constant (on a plane perpendicular to the sun's rays) equals 429.2 Btu per square foot per hour, or 1353 watts per square meter.

The radiation that finally arrives at the earth's surface —called *insolation*—is less than the solar constant and arrives in either direct (parallel) or diffuse (nondirectional) rays. The solar radiation reaching a building includes not only direct and diffuse rays but also radiation reflected from adjacent ground or building surfaces. These three sources of solar radiation may be used to heat and cool buildings.

Insolation. The insolation at a particular point on the earth is affected by several factors. The angle at which solar radiation strikes the earth's surface changes because of the earth's curvature and tilt of its axis. Consequently, the radiation received per unit area perpendicular to the incoming radiation is greater than that received per unit area on a horizontal surface. Thus, solar collectors are generally placed on an angle. Radiation reaching the earth's surface is also affected by the condition of the atmosphere: its purity, vapor, dust, and smoke content. Radiation is absorbed and scattered by ozone vapors and dust particles in the atmosphere. The lower the solar altitude angle, the longer the path of radiation through the atmosphere; thus, the amount reaching the earth's surface is reduced. And obviously, the length of the daylight period, which is dependent on the day of the year, also affects the amount of radiation striking a particular location.

Each area of the earth will be affected differently by these factors, and the percentage of direct and diffuse insolation at a particular location will vary.

Tilt of the Earth's Axis. The earth rotates around its own axis; one complete rotation takes 24 hours. The axis of this rotation (the line joining the North and South Poles) is tilted at an angle 23.5° to the plane of the earth's orbit, and the direction of this axis is constant.

Maximum intensity of solar radiation is received on a plane normal—that is, perpendicular—to the direction of radiation. The equatorial regions of the earth, which are the closest to the direction of solar radiation, would always receive maximum radiation if the earth's axis were perpendicular to the plane of the orbit. However, due to the tilt of the earth's axis, the area receiving the maximum solar radiation moves north and south between the tropic of Cancer and the tropic of Capricorn. This is the primary cause of seasonal changes.

Atmospheric Conditions. The momentary and long-term state of the atmosphere has considerable effect on the type and amount of insolation at the earth's surface. Clouds will reflect a considerable portion of incoming radiation back into outer space, while water droplets, dust, smoke, or other particulate matter will absorb or scatter solar radiation. Even when the sky is clear, as much as 10 percent of the total insolation may arrive as diffuse, scattered solar radiation. The insolation at the earth's surface is greatly reduced during cloudy weather, and most of the radiation that gets through is diffuse sky radiation.

Air Temperature. Temperature is not a physical quantity. Rather, it can be thought of as a symptom or as the outward appearance of the thermal state of substance. For instance, if energy is conveyed to a substance—the human body, for example—the molecular movement within the body is increased, and it appears to be warmer. If this molecular movement spreads from the body to other substances—to air, for instance—its intensity within the body decreases, and the body appears to be cooling. This principle applies to all substances—living or inanimate. As such, air temperature is considered the basic determinant of heat loss or gain of a substance such as a human body or a building.

Degree-Days. An analysis of selected temperature differences (design conditions) between the interior and exterior of a building will establish the heating and cooling load of a building. The relationship of the outdoor air temperature to a building's heating and cooling load has led to the concept of a *degree-day* for predicting the energy consumption of a building. The degree-day concept is particularly useful for relating the energy load of a building to the amount of solar energy available to satisfy this load.

Heating Degree-days are the number of degrees that the daily mean temperature is below 65°F. A day with an average temperature of 45°F has 20 heating degree-days (65 – 45 = 20). One with an average temperature of 65°F or more has no heating degree-days. A similar degree-day concept has been recently developed for cooling. As with the heating degree-day, the point of departure for a *cooling degree-day* is 65°F, so that a mean daily temperature of 85°F would result in 20 cooling degree-days (85 – 65 = 20).

The relationship between degree-days and residential fuel consumption is linear. For example, doubling the degree-days doubles the energy requirement. However, deviation from the degree-day norm of 65°F or the design of an energy-conserving building will substantially alter the energy requirement of the building and thereby change the linearity of the degree-day/energy-consumption relationship. It remains, however, a useful concept for relating the overall need for heating and cooling of a particular area with the amount of solar radiation available for use. By plotting degree-days per month and insolation on a horizontal surface per month simultaneously, it is possible to derive a "figure of merit" that indicates, in general terms, a relative feasibility for solar systems in various parts of the country. The figure of merit is a qualitative description and takes into account only climatic data. Other factors such as solar system performance, functional characteristics, cost, and socioeconomic conditions must be examined and traded-off against climatic conditions before a realistic appraisal of solar heating or cooling is possible.

Air Temperature Patterns. Patterns of air temperature are also important for solar-dwelling and system design. The fluctuation of air temperature throughout the year and during regular day-night cycles will establish criteria for system design. Regular day-night cycles are important for systems dependent on thermal inertia for heating and cooling. Convective and radiative cooling both require appreciable day-night temperature differences to function effectively. Similarly, collector size and storage capacity will be influenced by patterns of air temperature associated with changing weather conditions. Constant patterns of air temperature throughout the year will be intermixed with oscillations resulting from unexpected weather conditions: solar systems are usually designed for the former and not the latter conditions. In general, patterns of air temperature and weather will be more useful to system design than averages because they take into account variations in conditions, which may cluster for several days.

Humidity. The humidity of the air can be described in two ways: *absolute* and *relative.* Absolute humidity is the amount of moisture actually present in a unit mass or unit volume of air. The amount of moisture the air can hold depends on its temperature. Relative humidity is the ratio of the actual amount of moisture in the air (absolute humidity) to the amount of moisture the air could hold at a given temperature. Relative humidity is the more useful of the two quantities because it gives a direct indication of evaporation potential. This information is crucial for maintaining indoor occupant comfort and for designing solar cooling systems that use evaporative cooling techniques.

Humidity per se has no meaning as an environmental index without knowledge of the accompanying dry-bulb (i.e., standard thermometer) air temperature. High humidity at low dry-bulb temperatures has a negligible effect on human comfort; air temperature is the governing factor for dwelling and solar-system design under these conditions. When relative humidities of 60 percent or more are accompanied by dry-bulb temperatures above 65°F (18°C), conditions unfavorable to human comfort are likely to occur, thereby requiring natural or mechanical relief or curtailment of activities.

The total heat content of air during conditions of high humidity and high temperature is substantially larger than during periods of low humidity and high temperature. Cooling is the most effective method of restoring comfort. This can be accomplished by natural ventilation or mechanican cooling. Under certain conditions, it is not desirable or feasible to restore comfort by natural air movements alone, and mechanical equipment will be needed. Solar cooling should be considered if conditions of high humidity and high temperatures are prevalent during a large percentage of the year. In general, humidity is only a problem in summer. Dwelling and solar-system designs for areas experiencing high temperature and high humidity will require special attention to air movement and perhaps the use of solar cooling systems.

Areas where cooling is the primary design condition and the humidity is low can make use of nocturnal cooling. Evaporation facilitated by the low humidity is the primary process of nocturnal cooling. Convection and radiation will also play a part if night temperatures are low, winds constant, and skies clear. The cooling medium—usually water—loses moisture during the evening, thereby cooling the water that is stored for daytime or continuous use.

Air Movement. Air movement is measured in terms of wind velocity and direction. The wind in a particular location may be useful for natural ventilation at certain times of the year and detrimental to the thermal performance of a building or solar collector at other times. Consequently, air movement in and around the building and site become an important consideration for dwelling and solar-system design.

Summer winds, if properly directed by the natural topography or site design and captured by the dwelling, can substantially reduce or eliminate the need for a mechanical cooling system. This will be particularly helpful in areas where the cooling requirement is small and the cost-effectiveness of solar cooling unlikely. Solar cooling at present is quite expensive, and by eliminating the need for it, significant benefits will be achieved in terms of cost, maintenance, and operation.

Winter winds, on the other hand, can be quite detrimental to the thermal behavior of the dwelling or solar system. Cold winds will increase the surface conductance of the dwelling's exterior walls, thereby increasing its heat loss. Heat loss can be reduced by the careful selection and combination of building materials, and attention to the shape and position of the dwelling on the site, and the location of landscaping around the building. A similar heat-loss condition will exist for solar collectors. Cold winds blowing across the transparent face of the collector will increase convective heat losses. A single layer of glass or plastic in some instances may not reduce heat losses to an acceptable level. When this occurs, two or more transparent surfaces may be desirable. Also, landscaping or other devices can be used to reduce wind striking the collector as long as solar radiation collection is not impaired.

Wind can become an issue in snowfall areas where drifting is likely or in high-wind areas where additional structural support of large collectors may be necessary. Solar dwellings should be designed to reduce snow drifting on or in front of the solar collector. Adequate wind support of the collector and cover plates should also be provided in high-wind areas. This is particularly true for large free-standing or sawtooth collector arrangements. Pressure differences caused by air movement around the building can also result in collector and cover-plate instability.

Solar Design Determinants

Climate. The climatic factors that influence the design of solar dwellings and systems include:

- Type of solar radiation reaching the site—direct, diffuse, or reflected—and percentage of each time of maximum energy demand.
- Geographic location of the site—tilt of the earth's axis alters the relationship of sun to site throughout the year. Therefore, dwelling, site, and solar system should be designed to accommodate variations in sun angle to assure proper exposure to solar radiation.
- Capture the full spectrum of solar radiation from ultraviolet to infrared—dependent on collector design and materials.

- Heating or cooling load of the dwelling—expressed in hourly and daily units—mean or average long-term loads expressed in weekly or monthly units.
- Numerous regional, local, and site climate anomalies, which could be potentially beneficial or detrimental to dwelling or solar-system performance. Dwelling and solar collector should be shielded from potentially detrimental winds (cold winds increase heat loss, lower collector efficiency) or meteorological events (hail, snowstorms, and tornadoes may necessitate special protective devices or precautions). Conversely, proper exposure of the dwelling and solar collector to sun and wind will temper indoor climate and reduce demand on or need for heating and cooling systems.
- Climate is a primary factor shaping the design of buildings and sites for the utilization of solar energy for heating and cooling. In addition, demand for heating and cooling can be reduced by properly exposing to and/or shielding the building from those climatic elements that may be beneficial or detrimental to the building's thermal performance.

Comfort: The Desirable Conditions. A building designer's major task is to create the best possible environment (indoor as well as outdoor) for the occupants' activities. In other words, the designer must provide total human comfort, which may be defined as the sensation of complete physical and mental well-being. Air temperature, humidity, radiation, and air movement, all of which affect human comfort, must be considered simultaneously if an acceptable residential indoor environment is to be provided. The designer must consider these factors whether the building is to be heated (or cooled) by solar energy or fossil fuels.

To effectively design dwellings for human comfort, it is necessary to understand the basic thermal processes of the human body. How the body generates and loses heat is crucial for identifying an occupant comfort zone and for designing heating, cooling, and humidity control systems. Two complementary approaches for the provision of human thermal comfort have developed. One seeks to maintain thermal conditions within an established comfort zone, while the other attempts to modify the comfort zone. Both approaches are used in solar-dwelling design. In a solar dwelling, the manner in which solar energy is collected, stored, and distributed can greatly affect the comfort of the dwelling's occupants.

The human body continuously produces heat. Everyday activities such as sleeping, walking, working, and playing are all heat-producing. The entire portion of the body's energy requirement is supplied by the consumption and digestion of food. The process of transforming food into usable energy is called *metabolism*. Of all the energy generated by the metabolic process, the body uses only 20 percent, and the remaining 80 percent must be lost to the environment.

Body temperature, in contrast to skin temperature, must be maintained at 98.6°F (37°C) for the body to adequately perform its functions. To maintain the constant temperature balance, all surplus heat must be dissipated to the environment. Heat gained from the environment—solar radiation, for example—must also be dissipated. The body loses approximately 80 percent of its heat to the environment by convection and radiation. The remaining 20 percent is lost by evaporation, with a very small percentage of heat lost by conduction.

The sum total of the body's heat gain and loss should at all times equal zero—a constant body temperature of 98.6°F. If the body's heat gain is more than its corresponding heat loss, an uncomfortable feeling will occur and sweating will begin. Likewise, if the heat loss is more than the heat gain, body temperature will drop and shivering will occur.

The body has numerous regulatory mechanisms for maintaining a constant temperature. Blood circulation may increase or decrease, sweat glands may open or close, and shivering may begin to raise the body's temperature. Also, continuous exposure to similar climatic conditions, called *acclimatization,* can cause a change in the basal metabolism process, an increased sweat rate, or a change in the quantity of blood. It is not surprising, therefore, to learn that Eskimos prefer cooler temperatures than equatorial Africans.

Heat Loss in Various Thermal Environments. Human comfort is shaped by four major factors: air temperature, humidity, air movement, and radiation. Accordingly, the heat-exchange process between the human body and its environment may be aided or impeded by these climatic variables. For example, convective heat loss is severely impeded by high air temperatures, and evaporative heat loss may be simultaneously restricted by high humidity. Different regions will have different dominant climatic features that will affect human comfort. Indoor comfort in hot-humid climates, for example, is very much dependent on internal airflow and temperature to control humid conditions. To effectively design comfortable indoor environments, the relationship of outdoor climatic conditions to indoor activities and to thermal controls must be properly balanced.

Subjective variables. A number of subjective or individual factors will also influence thermal preferences. These include clothing, age, sex, body shape, state of health, and skin color. In addition, there are psychological and sociological variables that will influence thermal comfort. Whether one is happy or sad, active or confined, alone or in a group will influence thermal preferences.

Thermal Comfort Scale. An assessment of the impact of local climate on the body's heat-dissipation process is important in providing a comfortable indoor environment. Four independent climatic variables (temperature, humidity, radiation, and air movement) must be assessed simultaneously. The difficulty of this task has led to the development of "thermal indices" or "comfort scales," which combine the effect of these four variables. A comfort scale is the composite of the interactions between climatic variables. Through

observation and measurement in the laboratory, the characteristics of human comfort (comfort zone) have been identified and may be compared to local climatic conditions, thereby defining the need for and the type of thermal controls.

Comfort Zone. The comfort zone is established by analyzing the relationship between air temperature and three climatic variables: mean radiant temperature (the temperature of the surrounding surfaces), humidity, and air velocity.

Such an analysis will establish the range of thermal conditions (comfort zone) over which the majority of adults feel comfortable. At best, the comfort zone is an approximation of human thermal comfort, due to human preferences, physiological and psychological characteristics, and the nature of the activity being performed. However, it does provide the designer and builder with an estimate by which outdoor climatic conditions of a locale may be evaluated so the appropriate methods of achieving a comfortable indoor climate are chosen.

Providing Human Comfort. There are two ways of looking at the provision of human comfort. The first involves establishing a comfort zone based upon the occupants' thermal preferences and proposed activities and then comparing this comfort zone to existing or anticipated climatic conditions. In this manner, the appropriate methods for returning the climatic conditions to within the comfort zone will be established. The second viewpoint accepts the existing or anticipated climatic conditions as given and identifies methods of alternating the comfort zone to be compatible with the climate.

A building functions in much the same manner. An occupant comfort zone can be established and mechanical equipment provided to assure that thermal conditions remain in the zone. On the other hand, the building can be designed to respond to the positive and negative effects of temperature, humidity, wind, and radiation so that the occupant comfort zone is altered as external climatic conditions change. Given the scarcity and rising cost of conventional fuels and the cost of mechanical climate control equipment, the most effective and inexpensive approach may be to design the building as a whole to respond to the external climatic conditions, thus altering the occupant comfort zone. Only when the natural thermal control processes have been exploited to a realistic maximum should mechanical energy systems be employed to make up the remaining comfort burden.

Human Comfort and Solar Energy. Solar systems are designed to maximize the collection and storage of thermal energy. The thermal energy, in turn, is used to heat spaces or domestic hot-water supply or to power heat-activated cooling systems. All of these processes necessitate the capture, transport, storage, and distribution of an incredibly large amount of energy. Properly designed and regulated, this energy will provide an indoor climate suitable for the occupant's activities. If, however, the energy systems are not carefully organized and arranged, human comfort and activities may be seriously impaired. All channels of thermal energy flow should be considered in terms of their potential impact on human comfort. For example, heat-transport piping or ductwork between the solar collector and storage will lose considerable heat to the occupied spaces if not properly insulated. Large concentrations or reservoirs of thermal energy must also be evaluated in terms of their effect on occupant comfort. This is particularly true of large solar collector arrays or storage areas that are in close proximity to a habitable space. The maintenance of the indoor climate within the comfort zone should be the goal of building and solar-system design.

Comfort. The comfort factors that influence the design of solar dwellings and systems include:

- Body's heat loss and heat gain—the degree of each is related to activity, climatic condition, and thermal preferences.
- Zone of human comfort and its relation to regional, local, and site climates. Will identify need for and design of type(s) of thermal control(s) necessary to return conditions to comfort zone.
- Relation of heat sources to occupied space. Proper control of solar collection, storage, and distribution is crucial for providing and maintaining indoor comfort.

The sole purpose for housing, solar systems, and other thermal controls is to provide a comfortable, suitable residential environment for the occupant. Designers and builders should be aware of this when designing buildings in general and solar buildings in particular. A solar dwelling that is designed to capture, store, and distribute the tremendous energy provided by the sun is especially vulnerable to situations unfavorable to human comfort. The individuals who will inhabit the dwelling should always be considered first.

Buildings

The Controlling Variable. The thermal characteristics of a building are extremely important for the design of solar dwellings and the provision of human comfort. It is in the building itself that the demands of climate and human comfort must be resolved.

Buildings are constructed to moderate the extremes of external climate to maintain the building interior within the narrow ranges of temperature and humidity that support occupant comfort. Building design can begin to accomplish this role, by working with instead of against climatic impacts. A building, like the human body, can be examined by its heat-exchange process with the outdoor environment. How buildings gain and lose heat can be examined, and methods of thermal control can be developed to assure satisfactory performance of a building's heat-exchange processes. Mechanical controls, such as heating and cooling systems, whether conventional or solar-powered, can be evaluated by their performance in compensating for a building's heat loss or gain; structural (nonmechanical) controls such as insulation, shading devices, or building shape can be evaluated by their performance in reducing the demand on or the need for mechanical systems by controlling a building's heat loss or gain.

In most cases, a mix of mechanical and structural controls will be necessary to assure the best possible indoor climate. Appropriate thermal controls for a particular building will depend on the local climatic and site conditions, cost, construction practice, and the architectural requirements of the design. The design of dwellings that incorporate energy-conserving techniques, particularly solar-heating and/ or cooling systems, will modify in varying degrees the construction practices and the architectural image of today's dwellings. In essence, a building is the controlling variable, which can be modified in numerous ways to reconcile the demands of climate and comfort.

Heat Exchange. A building's provision of thermal comfort may be examined by its heat-exchange processes with the outdoor environment. A building gains or loses heat by conduction, convection, evaporation, radiation, internal heat sources, and mechanical systems.

The flow of heat by conduction through walls, floors, and ceilings may occur in either direction. Generally, conductive heat losses will occur in winter, while conductive heat gains happen in the summer. The material composition of the walls, floors, and ceilings of buildings will determine the rate of conduction. With some solar systems, the material composition of the building is critically important to the satisfactory performance of the solar system. For instance, if south-facing windows are used as solar collectors, the thicknesses of the floor and wall material are extremely important for achieving proper absorption and reradiation of solar heat.

Heat exchange by convection can occur through building surfaces by the movement of air between areas of different temperatures. For example, movement of air between the outside and inside of a building in winter around doors and windows is considered a convective heat loss. The heat exchange between the interior of a building and the outdoor air may be unintentional air infiltration—"leakage"—or deliberate air regulation—"ventilation."

Radiation of heat through glass or other transparent surfaces can add considerable heat to a building. Conversely, thermal radiation from interior surfaces to cool exterior walls will influence to a small degree a building's heat loss. The amount of radiant or solar heat gain is influenced by window area, building orientation, and shading. Properly placed windows or other transparent surfaces that can be insulated when the sun is not shining can provide a substantial portion of a building's heat demand.

Internal heat sources such as human bodies, lamps, motors, and appliances can provide as much as 25 percent of a building's heat load. However, this source of heat is rarely considered in residential design.

Mechanical equipment may introduce or remove heat from a building by utilizing some form of outside energy such as natural gas, oil, electricity, or solar radiation. The amount of heat and cold produced by such systems is controlled by the designer and depends on the heat lost or gained by the other factors. Mechanical equipment, as the dependent variable, can therefore be adjusted according to the thermal balance of the other factors.

Finally, a building may lose heat by evaporation from its surfaces or from sources within the building. Evaporation produces a cooling effect as water vapor is removed.

The thermal balance of a building is maintained if the heat lost and gained from the preceding sources equals zero. If a building loses more heat than it gains, it will cool off. If it gains more heat than it loses, it will warm up. Regulating a building's heat-loss and heat-gain cycle within a level of occupant comfort through the day and year is a function of building design and mechanical system selection.

A dwelling designer and builder share the responsibility of selecting proper building materials, determining building size, volume, and orientation; and sizing and orientating windows, doors, overhangs, and other thermal controls to assure occupant comfort. Each of these considerations will influence a building's heat-exchange processes. The trade-offs between heat-change factors will most likely be based on climate, cost, and construction practices. For example, since mechanical equipment is expensive, it may be appropriate to reduce heat loss and gain by structural methods. By working with instead of against climatic impact, a building's need for mechanical equipment may be reduced. This, in turn, may reduce the amount of fossil fuel or the size of a solar system required to adequately heat and/or cool a dwelling.

Table 6-1. House heating factors.

In cold weather, a house is heated up by:
    Sun's rays, through radiation.
    People, lights, cooking, washing, drying, pilot lights, hot water.
    Solar, fossil fuel, electric, wood-space, and hot-water heating.

In cold weather, a house is cooled down by:
    Radiation to dark sky.
    House walls and glass, by conduction to cold outside air.
    Cold outside air, by convection (infiltration, ventilation, and combustion air).
    Humidification (air temperature drops as humidity rises unless external heat is applied).
    Ground, if basement is warmer than ground.
    Drains and flues—heat is lost down the drain, up the flue.

In hot weather, a house is heated up by:
    House walls and glass, by conduction when outside air is hot.
    Hot and/or humid outside air, by convection (infiltration, ventilation, and combustion air).
    People, lights, cooking, washing, drying, hot water, pilot lights.

In hot weather, a house is cooled down by:
    Radiation to dark sky.
    House walls and glass, by conduction when outside air is cool.
    Cool outside air, by convection (ventilation).
    Ground, by conduction from the basement or crawl space.
    Humidification, if air is dry (adiabatic cooling).

Thermal Control. Thermal controls are devices (e.g., furnace) or methods (e.g., dwelling orientation to capture summer winds) for moderating the extremes of outdoor climate to bring the interior within the narrow ranges of temperature and humidity that support human comfort. Thermal controls can be discussed by either the nature of the control—mechanical or structural—or the climatic variable regulated—wind, sun, or temperature. In either case, the objective of thermal controls can be briefly stated as follows:

When cold discomfort conditions prevail:

- Minimize heat loss
- Maximize use of heat gain from the sun and internal sources
- Compensate for any net heat loss by heating which uses some form of energy supply (preferably a renewable resource)

When hot discomfort conditions prevail:

- Minimize heat gain
- Maximize heat loss
- Remove any excess heat by cooling that uses some form of energy supply (preferably a renewable resource)

When conditions vary diurnally between hot and cold discomfort:

- Even out the variations
- Compensate for both excesses by a flexible heating/cooling system.

It has been said that the degree of sophistication (in environmental controls) is largely a socioeconomic question. In other words, the provision of any specified set of indoor conditions is possible, but preferences and sophistication of controls will be influenced by an individual's social status, standard of living, and financial means. Deciding what degree of comfort is to be achieved at what degree of cost involves a value judgment. This is the perplexing issue facing the United States today regarding energy in general and solar energy in particular.

All elements of climate can be moderated to maximize heat retention, solar heat gain, or internal airflow as needed in different combinations for various climatic conditions. It may be more useful, therefore, to discuss thermal controls by the climatic variable regulated. The thermal control strategies are organized by climatic variables—temperature, sun, and wind. Numerous other techniques that can be combined for climate control are available to the designer. The emphasis here is on those control strategies that do not require a conventional energy input for their operation.

Temperature Controls. Control strategies for temperature can be classified into two broad categories: thermal retention and thermal regulation. Thermal retention is simply the capture of heat produced within the building or from a surrounding heat source. Underground massing of the building to reduce heat loss, and the use of resistive insulation and fireplace design to maximize the radiation of heat into occupied space are examples of thermal retention strategies.

Thermal regulation, on the other hand, is the manipulation of building design to moderate the indoor climate. Thermal regulation includes: reducing the heated/cooled area and volume of the dwelling during the day or throughout periods of maximum energy demand by the internal zoning of spaces so that portions of the dwelling may be closed off; locating vegetation and landforms around the building to reduce climatic impact of temperature fluctuation in the building; and, in regions with low vapor pressure and a clear night sky, designing buildings to make use of nocturnal cooling.

Sun Controls. Strategies for sun control can be organized into two categories: solar exposure and light regulation. Solar-exposure strategies moderate the exposure of the building and adjacent site to solar radiation. Light regulation, on the other hand, regulates the amount of sunlight reaching the interior of the dwelling; however, one technique will serve both functions. For example, site vegetation—trees and high shrubbery—can be placed to block unwanted radiation from striking the building during certain periods of the day or year and can also be used to regulate the amount of sunlight entering the dwelling's openings.

Other solar-exposure strategies include capacity insulation to regulate (by thermal capacity) solar heat gain through the building's walls and roof (best for areas with large diurnal temperature swings and high solar-heat gain); ground cover to control reflected radiation gain and ground temperature; and solar collectors positioned to capture solar radiation for space-heating/cooling and domestic water heating.

Light-regulation strategies include window designs that regulate the sunlight entering the building, interior shading devices such as shutters (preferably insulated), draperies, louvers, and proper location and orientation of the dwelling's openings.

Wind Controls. Strategies for wind control can be classified into two general categories: wind regulation and internal airflow. Wind regulation primarily moderates the impact of wind on the building and adjacent site, while internal airflow strategies manipulate air movement within the building.

Wind-regulation techniques range from designing the building itself as a wind-controlling form through aerodynamic massing, to the placement of natural or synthetic elements (i.e., vegetation, landforms, or fences) on the site surrounding the building to regulate wind direction and force. Similarly, internal airflow strategies are quite varied and range from the use of roof monitors to create a "thermal chimney" for increased ventilation, to channeling wind-flow for use in evaporative cooling.

The designer must select a combination of thermal controls that delivers the level of occupant comfort at a justifiable cost in construction and energy consumption. From energy-cost escalation in the past few years, it is apparent that previous design assumptions are no longer valid, that is, energy-consuming mechanical equipment cannot make up in brute force for what building designs disregard in natural climatic impact.

Building Characteristics. The building characteristics that influence the design of solar dwellings and systems include:

- Heat loss or gain—will determine heating and cooling requirement.
- Total volume of heated/cooled area—will affect heating and cooling requirement.
- Total surface area exposed to the outdoor environment.
- Orientation—critical for controlling heat exchange with outdoor environment and solar collection.
- Thermal controls—means of tempering or utilizing outdoor climate to provide a comfortable indoor climate. May substantially reduce a building's heating and cooling requirement.
- Material composition—will determine rate of heat loss/gain and may be used to store solar energy.

Understanding local climatic conditions, human comfort requirements, and thermal characteristics of buildings will enable the designer to identify the most effective (and energy-conserving) methods for providing human comfort in building design—including the use of a solar system. Numerous thermal control strategies are available, and their use can substantially reduce the energy requirements of the building. As a result, a smaller solar energy system will be sufficient for the building's energy requirement and will thereby achieve a considerable cost saving.

Architectural Design Implications

Solar Energy Systems. As noted earlier, solar energy systems vary widely in design and function. They may utilize complex technology or require only common-sense design decisions. They may be used for space heating, cooling, and domestic hot-water heating. The components that make up a solar system—collector, storage, and distribution—may be arranged in numerous combinations depending on functions, component capability, climatic conditions, required performance, and architectural requirements.

A number of design implications are related to climate, comfort, thermal characteristics of buildings, and solar systems. Each presents a unique set of criteria, which must be adhered to within a prescribed range if the desired level of occupant comfort and system performance is to be achieved.

Since there are a variety of methods for collecting, storing, and distributing solar energy and an almost infinite number of ways that these concepts may be utilized in a building design, the following architectural implications of solar systems are necessarily simplified. However, some design issues are of a general nature and should be considered early in the design of dwellings incorporating solar systems. These range from the location, size, and orientation of a solar collector to the placement of room-air outlets for comfort heat distribution. To help the reader understand the architectural implications of solar systems, they are discussed according to type of system component. For example, the design implications associated with solar collectors are presented in one area followed by the design implications of solar storage in another and so on.

Solar Collectors

Location. Solar collectors are either detached from, attached to, or integrated with a building. The appropriateness of a collector location will depend, for example, on the type and size of collector, the climatic and site conditions, and whether the collector is for new or existing construction. These will all influence collector location and building design for any given project.

Orientation. Collector orientation is critical for optimum exposure to solar radiation. For stationary collectors, as opposed to "tracking" collectors which follow the sun across the sky, an orientation of true south is usually the best position. However, research indicates that a 20° variation, either side of true south, does not significantly alter the size or performance of most solar collectors. In fact, in some cases it may be necessary to orient the collector away from true south. For instance, if incoming radiation will be obstructed from striking the collector because of adjacent buildings and trees, or if morning fog or haze may interfere with collection, a deviation from true south may be absolutely necessary. When such a deviation does occur, a larger collector will generally be needed to provide a similar level of performance. Each collector area provides 50 percent of the dwelling's heating demand. As can be seen, the collector area varies significantly within and between climatic regions. Therefore, collector orientation can be crucial to a dwelling's design and should not be understated or overlooked.

Tilt. Collector tilt is an important consideration for dwelling and system design. Again, we are referring primarily to stationary collectors and not to tracking and concentrating collectors. The angle at which a collector is tilted toward the path of incoming radiation greatly affects its performance characteristics and the building design of which it is a part. The tilt of a collector, established at the time of installation, is largely determined by the geographic location of the dwelling (site latitude) and the functional requirement of the solar system (space heating, cooling, or domestic water heating).

Due to the tilt of the earth's axis, the angle at which the sun strikes every location on earth varies throughout the year. Therefore, a proper collector tilt for winter heating and summer cooling will be different for each geographic location. This variation in tilt fluctuates about 30° from heating optimum to cooling optimum. Consequently, most collectors are positioned within this range, generally with their location nearer the angle associated with the locality's primary energy requirement—heating or cooling.

The snowfall characteristics of an area may influence the appropriateness of an optimum collector tilt. Where snow builds up on or drifts in front of the collector, tilt angles of 40° or more are generally required to induce natural "avalanche" of the collector.

Numerous guidelines have been proposed for determining collector tilt based upon site latitude. The general rule of thumb is latitude plus 15° for heating and latitude plus 5° for heating and cooling. For example, a collector tilt for northern climates such as Minneapolis, Minnesota—where the primary energy requirement is for heating—would be Minneapolis' latitude, 42°, plus 15°, for a collector tilt of 57°. A variation of 10° to 15° either side of this optimum will not significantly alter collector performance. However, caution should be used. For instance, in the example just mentioned, 57° was indicated as an optimal collector angle without considering the reflective properties of snow usually associated with northern climates. Since the winter sun angle will be low for the Minneapolis area, the reflective radiation could be considerable, and the designer may wish to consider a vertical collector.

The angle at which a collector is positioned has a direct relationship to the amount of collector area required to provide equal levels of performance. This relationship is easily illustrated. Assuming a northern climate and a collector sized to provide 50 percent of the heating demand of a 1000-square-foot house, the change in collector area resulting from changes in tilt angle would be as follows:

- Using a collector size at an angle of 50° facing due south as the base for comparison, the collector size will increase 18 percent if the collector tilt is lowered to a 40° angle.
- Increasing the base angle 10° to 60° increases collector size 6 percent. However, increasing collector angle to 90° and assuming a reflective radiation gain from snow or another reflective surface in front of the collector decreases the base collector size 14 percent.

This simple example shows the importance of performing a general system analysis before proceeding into detail design. The impact of the collector angle upon the design of the dwelling that is to incorporate the solar system can be substantial, and careful consideration should be given to all factors before a decision is made.

Shadowing of Collector. Solar collectors should be located so that they are not in the shadow of adjacent buildings, landscaping, or another collector during hours when collection is desired. In some cases, however, shadowing of the collector may be desired. For instance, if the collector is to provide heating only, it may be desirable to shadow the collector and residence during hot summer months. This would be particularly true for windows designed as solar collectors. By properly calculating an overhang, the hot summer sun will be blocked before it can enter the dwelling.

Type and Size

The type and size of a solar collector will have a significant impact on the physical appearance of a solar dwelling. Each method of solar energy collection has a unique set of criteria for the effective capture and utilization of solar radiation, which will influence a dwelling's design. For example, the design requirements of a solar pond differ significantly from those of a liquid-cooled flat-plate collector. In addition, the performance characteristics of each collection method will vary with climate, building characteristics, and comfort requirement, thereby requiring different size collector areas to achieve an equal level of performance.

The influence of collection type and size on dwelling design is best illustrated by examining the relationship of the solar collector to the building. There are two basic relationships between a solar collector and a building. The first is where the entire building or various components of the building (e.g., walls, roof) acts as the solar collector—a passive collector. In this case, the collector and building are one and the same. The type and size of collector directly affect how the building will look. For example, if the method of collection is a thermosiphoning wall or a solar pond, the building design is strongly influenced by the size, orientation, and volume of the collection surfaces—for example, the David Wright, Atascadero, and Trombe/Michel houses.

In the second relationship, the collector and building are seen as independent elements that may be organized in many ways with varying degrees of impact on each other. The size and type of solar collector, in most situations, will indicate whether the collector should be attached, detached, or integrated with the dwelling. In some cases, it may not be possible to locate the collector on the building, and therefore the collector may be best located away from or adjacent to the dwelling. As a result, the site planning problem may become more significant than the architectural problems. If, however, the collector and building are to be attached, the type and size of collector are significant architectural considerations. For instance, when a large vertical collector area is required and south-facing windows, balconies, and patios are also desired, it becomes difficult to balance the need for collection area with the desire for architectural amenities.

Even with a rooftop collector, it is not just a simple matter of placing solar collectors on a dwelling's roof. The volume, roof slope, floor area, structure, and building orientation will all influence collector type, size, and location.

Solar Storage

Type and Size. The type and size of solar heat storage will have varying influence on a dwelling's design. Storage type and design will be closely related to the solar collector used, and the size of storage will depend on its performance characteristics and the dwelling's storage capacity requirement (determined by energy load and days without collection).

The influence of solar storage on the physical appearance of a dwelling may be unnoticeable with some systems and immediately apparent for others. For example, a storage tank located in the basement or buried adjacent to a dwelling will not affect its physical appearance. However, if the solar system combines solar collection and storage in the same element, the solar collector/storage component will be immediately visible and will substantially influence the architectural character of the dwelling. For most storage designs, however, there will be only minimal architectural modifications because of storage type or size.

Sufficient building or site area must be provided for the large volume associated with most storage components—especially with water or rock storage. Since the storage component is generally sized to provide for the dwelling's heating/cooling requirement over several days—usually 1 to 3 days—the size of storage can become quite large. It is not uncommon to have storage size range from 500 to 2000 gallons of water or 10 to 50 tons of rock for a single-family dwelling. This will require 65 to 270 cubic feet of space for water storage and 160 to 800 cubic feet for rock storage. A saving in storage size is possible for multifamily or other large buildings where a centralized storage component services all the living units. This method will generally be advantageous in terms of cost, efficiency, and operation over many individual storage elements.

Location. Location of the storage component can be critical to the efficient operation of the solar system. If it is located at a considerable distance from the collection and distribution components, heat transfer between components may be inefficient, and the total energy reaching the occupied space will be reduced.

There are three basic locations for the storage components: within or outside the building, or integrated with the building or other solar-system components. The location will be determined primarily by the collector type, storage performance characteristics, required thermal capacity (size), and the architectural requirements imposed by the building design.

Within the building, the storage component can be located wherever there is sufficient room. In most cases, however, it is located in the basement, if there is one, or buried beside the dwelling, if there is not. Other storage locations within the building include: an attic space, a storage closet (especially in the case of physical/chemical storage), or a specially built storage room or element.

Outside the building, the storage component is usually buried alongside the exterior wall closest to the mechanical equipment room. Where burying is not feasible (because of a low water table, for instance), the component can be placed in a well-insulated free-standing element adjacent to the building. Burying is recommended, however, because of the added insulation value of the earth. Also, because the storage component is outside the building, interior space does not have to be provided.

The solar storage component, which is integrated with the building or other system components, in certain instances, encompasses the best features of the previous two storage locations. Integrated storage elements serve several functions simultaneously. For example, the walls, floors, and roof of a building can be used to store solar energy as well as perform their intended function. The Pueblo Indian structures and the houses designed by George and Fred Keck are examples of this storage concept.

The storage component can also be integrated with the other components of a solar system. Thus, a solar system that combines several functions in the same element can also be used as an architectural element. For instance, the solar pond of the Atascadero house not only collects, stores, and distributes solar energy, it is also the roof of the house. Other solar systems can be used for exterior or interior walls—such as those found in the Trombe/Michel house.

Regardless of location, the component should be designed to control heat loss. For storage outside the building, heat loss should be reduced to a minimum by enclosing the element with insulation and burying it to a sufficient depth. Storage within the building should be regulated so that unwanted heat loss does not escape to the outside or adjacent interior spaces. Where the storage element is integrated with building or solar-system components, the heat loss from storage must be carefully controlled so that the occupied space served by storage does not overheat.

Solar Heat Distribution

Type and Size. The distribution method chosen for use in a dwelling may affect its architectural appearance significantly. For instance, distribution dependent on natural convection and radiation will require careful attention to room size and shape, opening placement, ceiling height, and building material selection. The same degree of attention would not be required if a forced warm-air or warm-water distribution were used. Selection of the distribution method should not be quickly decided without some analysis of the architectural implications of the choice.

How the solar radiation is collected and stored will usually determine the means of distribution. The number of possible combinations between solar-system components is limited. Consequently, for any specific collector-storage combination, there will be a limited number of compatible distribution methods. Once collection and storage components have been selected, a number of distribution methods may be identified and analyzed.

The size of the distribution component is dependent on the operating temperature of the solar collector and storage, the amount of area/volume to be heated/cooled, and the distance between collector/storage and the point of use. In some instances, size will not be associated with distribution.

For example, the solar pond of the Atascadero house is placed on top of metal ceiling panels above the serviced area. The panels, heated by solar radiation striking the pond, distribute collected energy to the interior of the house, not with conventional ducts and pipes, but instead by natural radiation and convection. However, solar systems that rely on mechanical devices for energy distribution generally require more attention to distribution size and location than conventional heating and cooling systems. Warm-air distribution, for instance, will usually require larger ducts and blower because solar-produced temperatures in storage are often in the low range. Consequently, the layout of distribution ducts can significantly influence a dwelling's spatial arrangement to accommodate efficient distribution and the added duct size.

Location. The location of solar-distribution components within a dwelling are much the same as conventional heating and cooling distribution components. Ducts and pipes can be located in a basement, attic, dropped ceiling space, or under the floor slab if the house is built on a grade. The use of larger ducts due to lower storage temperatures may significantly affect the location of duct runs through the dwelling. This in turn can influence the location and arrangement of rooms and the placement of outlet diffusers. Distribution methods dependent on natural radiation and convection, which in most cases do not rely on ducts or pipes, require special attention to energy source and target area. Since radiation and convection are dependent on the differential temperature of surfaces and air, respectively, the location of the energy source—collector or storage—relative to occupied spaces is important for efficient distribution. This in turn will suggest room sizes, shapes, and materials that are compatible with radiation and convection distribution.

Solar Design Determinants—Solar Systems. Factors that can influence the design of solar dwellings, systems, and sites include:

- Solar collector's location—whether the collector is detached, attached, or fully integrated with the building structure will significantly influence the design of the dwelling, site, and system.
- Collector type(s) and size(s)—dependent on climatic conditions, thermal characteristics of dwelling, efficiency of collector, and functional requirement (e.g., of space heating). Various collector types may be used simultaneously.
- Solar collector's orientation—generally $10°$ to $20°$ either side of true south.
- Tilt of solar collector—should be normal to the sun's rays during periods of collection. Annual optimum collector angle for space heating is generally considered latitude plus $15°$ and latitude plus $5°$ for space heating and cooling. Seasonal optima will vary from December to June within a total range of approximately $30°$.

- Energy loss at the back and side of collector—should be properly insulated to reduce unwanted heat loss from collector to external climate or internal spaces.
- Storage type(s) and design—must be compatible with collector and distribution systems.
- Storage size—should accommodate collector(s) performance characteristics, building's energy demand (generally for 1 to 3 days), and climatic variations.
- Minimized heat loss from storage to surrounding environment—storage should be well insulated to reduce heat loss on cool days or days when collection does not occur. However, in an integral collector/storage system such as a thermosiphoning wall, heat loss should be properly calculated to deliver heat at the rate desired.
- Efficient transfer of heat from collector to storage, storage to distribution—an important consideration for regulating the efficiency of solar-radiation collection and distribution.
- Reliability of storage medium—long life regeneration of storage medium.
- Placement of storage to maximize efficiency and minimize human discomfort—location of storage should be compatible with collector and distribution systems, but not in conflict with human comfort considerations. For example, storage may be located within or below an occupied space and may cause overheating of that space if not properly designed or insulated.
- Distribution type(s) and design—must be compatible with collector and storage systems.
- Distribution size—system may have to be larger than conventional distribution systems because temperatures are generally lower.
- Placement of air-distribution outlets—moving air that is only a few degrees above room temperature feels relatively cool to the occupants, even though it is actually warming the space. With the lower delivery temperatures common in solar warm-air heating systems, warm-air outlets should be positioned to avoid creating drafts.

Recognition of the many design implications is crucial for the efficient capture and utilization of solar energy for heating and cooling. In addition, the architectural appearance of the dwelling will be shaped by the design criteria associated with the various solar components included in the design. To be successful, the design must reconcile any conflict between these design issues and skillfully coordinate them into a dwelling design responsive to the climatic conditions and the occupant's needs.

SITE PLANNING FOR SOLAR ENERGY UTILIZATION

The building site is extremely important in solar design. Together with the solar design factors, the conditions and characteristics of the building site influence both dwelling and system design. Existing vegetation, geology, topography, and climate are primary site characteristics considered during site planning and design. These factors influence not only the design of dwellings incorporating solar heating and cooling systems but also the layout and organization of groups or neighborhoods of solar dwellings.

Every building site has a unique combination of site conditions. As a result, the same solar dwelling placed on different sites will require completely different site-planning and design decisions. Therefore, the site for a solar dwelling should be selected with care and modified as necessary to maximize the collection of solar energy and to minimize the dwelling's need for energy.

The Site-Planning Process

Site planning is concerned with applying an objective analysis and design process to specific site-related problems at increasingly smaller scales. While the building site and the dwelling design may vary significantly from one project to another, the process of site planning is replicable and easily adapted to the requirements of most projects. In the case of solar-dwelling design, the process is altered to include design criteria related specifically to the use of solar heating and cooling systems.

Site planning for the utilization of solar energy is concerned with (1) access to the sun and (2) location of the building on the site to reduce its energy requirement. The placement and integration of the solar dwelling on the site in response to these concerns entail numerous decisions made at a variety of scales. The process may commence at a regional climatic and geographic scale and terminate at a specific location on the building site. At every scale, decisions regarding site selection, building orientation and placement, and site planning and design are made.

Site Election

At times, a builder, developer, or designer may have the option of selecting a site or determining the precise location on a larger site for the placement of the solar dwelling or dwellings. The following factors should be carefully analyzed:

- *Geography of the surrounding area*—What are the daily and seasonal paths of the sun across the site? What are the daily and seasonal wind-flow patterns around or through the site? Are there earthforms, which may block the sun or wind? Are there low areas where cold air could settle?
- *Topography of site*—How steep is the slope? Can it be built upon economically? Are slopes beneficial or detrimental to energy conservation and solar energy utilization?
- *Orientation of slopes*—Are slopes south-facing slopes for maximum solar exposure? West-facing for maximum afternoon solar exposure? East-facing slopes for maximum morning solar exposure? North-facing slopes for minimum solar exposure?

- *Underlying geology*—What type of rock is on the site? How deep is it? Are there unsuitable building areas?
- *Existing soil potential and constraints*—Are soils able to support structures? To support vegetation?
- *Existing vegetation*—Would size, variety, and location of vegetation impair solar collection? Would building sites disturb existing vegetation to a minimum? Would size, variety, and location of vegetation assist in energy conservation?
- *Climatically protected areas*—Are areas protected at certain times of the day or year? By topography? By vegetation?
- *Climatically exposed locations on the site*—Are areas exposed to sun and wind? Are areas exposed primarily in winter? Summer? All seasons of the year?
- *Natural access routes*—Are there adjacent streets for vehicular access to the site? Adjacent walkways for pedestrian access?
- *Solar radiation patterns*—Daily and monthly? Seasonal? Impediments (e.g., vegetation that may cover the site or shadow buildable areas on the site)?
- *Wind patterns*—Daily and monthly? Seasonal? Impediments (e.g., thick vegetation or underbrush that may block air movement on or through the site)?
- *Precipitation patterns*—Fog movement, collection, or propensity patterns? Snow drift and collection patterns? Frost "pockets"?
- *Temperature patterns*—Daily and monthly? Seasonal? Warm area? Cold area?
- *Water or air drainage patterns*—Seasonal airflow or waterflow patterns? Daily airflow or waterflow patterns? Existing or natural impediments to airflow or waterflow patterns?

Tools for site analysis include air photos, topographic maps, climatic charts, and direct observations on the site. Site selection at whatever scale must take into account the distinctive characteristics of the major climatic regions of the United States mentioned earlier. Once the data are collected and organized, it can be used to evaluate, rate, and eventually select a specific location or site for the placement of the dwelling, solar system, and other site-related activities.

- *Altitude and Scope*—The topography is analyzed in both plan and cross section to locate buildable areas on upper and middle slopes.
- *Orientation and winds*—The site is assessed for areas oriented in a southerly direction for maximum solar exposure. Also, the prevailing and storm winds that move regularly or occasionally across the site are plotted.
- *Vegetation and moisture*—Existing vegetation and moisture patterns on the site are related to their potential for assistance in the creation of sun pockets and for providing wind protection. The density and type of vegetation are analyzed and graphically depicted in order to gain an understanding of the patterns of shade or protection and air or moisture flow.

- *Composite showing preferred sites*—From the preceding factors, a composite is prepared showing a ranking or a rating of the preferred sites for placement of a solar dwelling (*a* being best, *b* next best, and so on).

Siting and Orientation

Optimum solar energy utilization is achieved by proper placement and integration of the dwelling, solar collectors, and other site-related activities and elements on the building site.

In addition to the dwelling, the most common activity areas found on residential sites include:

- means of access (entrances to the site and to the dwelling)
- means of service (service and storage areas)
- areas for outdoor living (patios, terraces, etc.)
- areas for outdoor recreation (play areas, pools, courts, etc.)

On sites where the dwelling(s) will be heated or cooled by solar energy, additional factors must be considered for accommodating solar collection—either by dwelling or on-site collector.

Each of the four major climatic regions in the United States has different siting and orientation considerations.

In *cool regions,* maximum exposure of the dwelling and solar collector to the sun is the primary objective of site planning. Sites with south-facing slopes are advantageous because they provide maximum exposure to solar radiation. Outdoor living areas should be located on the south sides of buildings to take advantage of the sun's heat. Exterior walls and fences can be used to create sun pockets and to provide protection from chilling winter winds.

Locating the dwelling on the leeward side of a hill or in an area protected from prevailing cold northwest winter winds—known as a *window shadow*—will conserve energy. Evergreen vegetation, earth mounds (berms), and windowless insulated walls can also be used to protect the north and northwest exterior walls of buildings from cold winter winds. In addition, structures can be built into hillsides or partially covered with earth and planting for natural insulation.

In the *temperate region,* it is vital to assure maximum exposure of the solar collectors during the spring, fall, and winter. To do so, the collector should be located on the middle to upper portion of any slope and should be oriented within an arc 10° either side of south. The primary outdoor living areas should be on the southwest side of the dwelling for protection from north or northwest winds. Only vegetation that sheds its leaves in fall (i.e., deciduous) should be used on the south side of the dwelling since it provides summer shade and allows for the penetration of winter sun.

The cooling impact of winter winds can be reduced by using existing or added landforms or vegetation on the north or northwest sides of the dwelling. The structure itself can be designed with steeply pitched roofs on the windward side, thus deflecting the wind and reducing the roof area affected by the winds. Blank walls, garages, or storage areas can be placed on the north sides of the dwelling. To keep cold winter winds out of the dwelling, north entrances should be protected with earth mounds, evergreen vegetation, walls, or fences. Outdoor areas used during warm weather should be designed and oriented to take advantage of the prevailing southwest summer breezes.

In *hot-humid regions* where the heating requirement is small, solar collectors for heating-only systems require maximum exposure to solar radiation primarily during the winter. During the remainder of the year, shading and air movement in and through the site are the most important design considerations. However, for solar cooling or domestic water heating, year-round solar-collector exposure will be required. Collector orientation within an arc 10° either side of south is sufficient for efficient solar collection.

In *hot-arid regions* the objective of siting, orientation, and site planning is to maximize the duration of solar radiation exposure on the collector and to provide shade for outdoor areas used in late morning or afternoon. To accomplish these objectives, the collector should be oriented south-southwest and the outdoor living areas should be located to the southeast of the dwelling in order to take advantage of the early morning sun and afternoon shade.

Indoor and outdoor activity should take maximum advantage of cooling breezes by increasing the local humidity level and lowering the temperature. This may be done by locating the dwelling on the leeward side of a lake, stream, or other water body. Also, lower hillside sites will benefit from cooler natural air movement during early evening and warm air movement during early morning.

Excessive glare and radiation in the outdoor environment can be reduced by providing:

Small shaded parking areas or carports
Turf adjacent to the dwelling unit
Tree-shaded roadways and parking areas
Parking areas removed from the dwelling units
East-west orientation of narrow roadways

Exterior wall openings should face south but should be shaded either by roof overhangs or by deciduous trees in order to limit excessive solar radiation into the dwelling. The size of the windows on the east and west sides of the dwelling should be minimized in order to reduce radiation heat gain into the house in early mornings and late afternoons. Multiple buildings are best arranged in clusters for heat absorption, shading opportunities, and protection from east and west exposures.

Table 6-2. Site Orientation

| | OBJECTIVES | | | |
ADAPTATIONS	COOL	TEMPERATE	HOT-HUMID	HOT-ARID
	Maximize warming effects of solar radiation. Reduce impact of winter wind. Avoid local climatic cold pockets	Maximize warming effects of sun in winter. Maximize shade in summer. Reduce impact of winter wind but allow air circulation in summer.	Maximize shade and wind	Maximize shade late morning and all afternoon. Maximize humidity Maximize air movement in summer
Position on slope	Low for wind shelter	Middle-upper for solar radiation exposure	High for wind	Low for cool airflow
Orientation on slope	South to southeast	South to southeast	South	East-southeast for afternoon shade
Relation to water	Near large body of water	Close to water, but avoid coastal fog	Near any water	On left side of water
Preferred winds	Sheltered from north and west	Avoid continental cold winds	Sheltered from north	Exposed to prevailing winds
Clustering	Around sun pockets	Around a common, sunny terrace	Open to wind	Along east-west axis, for shade and wind
Building orientation[a]	Southeast	South to southeast	South, toward prevailing wind	South
Trees	Deciduous trees near building. Evergreens for windbreaks.	Deciduous trees nearby on west. No evergreens near on south.	High canopy trees. Use deciduous trees near building.	Trees overhanging roof if possible
Road orientation	Crosswise to winter wind	Crosswise to winter wind	Broad channel, east-west axis	Narrow: east-west axis
Material coloration	Medium to dark	Medium	Light, especially for roof	Light on exposed surfaces, dark to avoid reflection

[a]Must be evaluated in terms of impact on solar collector, size, efficiency, and tilt.

Each climatic region has its own distinctive characteristics and conditions that influence site-planning and dwelling design for solar energy utilization and for energy conservation. Table 6-2 suggests the general objectives of site planning and dwelling design for each climatic region as well as some methods for achieving these objectives. The chart reflects the seasonal tradeoffs made between climatic optima. In all cases, a detailed analysis should be undertaken to identify the site trade-offs between the optima for solar energy collection and those for solar conservation.

Integration of the Building and Site

Ideally, a building is designed for the specific site on which it is to be placed. Commonly, however, a building design may be replicated with only minor changes on different sites and in different climates.

Site-planning solutions are not as easy to replicate, because each site has a unique geography, geology, and ecology. The most appropriate way to integrate any building and its site is first to analyze the site very carefully, and then to place the building on the site with a minimum of disruption and the greatest recognition and acceptance of the site's distinctive features.

It is possible, however, to provide general techniques for integrating buildings with their sites. Historically, a number of such techniques have evolved, among which are indigenous architectural characteristics adapted to local site conditions; architectural extensions to the building such as walls and covered walks; the use of native materials found on the site; and techniques for preserving or enhancing the native ecology.

In each climatic region, guidelines can be determined to help apply the many techniques available for integrating a building and its site in ways appropriate to the particular region. These guidelines can be particularly helpful in maximizing energy conservation and increasing the opportunity for successful use of solar heating and cooling.

Detailed Site Design

The detailed design of a site for optimum solar energy utilization and energy conservation entails the use of a variety of types of vegetation, paving, fences, walls, overheat canopies, and other natural and synthetic elements. These elements are used to control the solar exposure, comfort, and energy efficiency of the site and the dwelling.

The materials used in site design have the ability to absorb, store, radiate, and deflect solar radiation as well as to channel warm or cool airflow. For instance, trees of all sizes and types block incoming and outgoing solar radiation, deflect and direct the wind, and moderate precipitation, humidity, and temperature in and around the site and dwelling. Shrubs deflect wind and influence site temperature and glare. Ground covers regulate absorption and radiation. Turf influences diurnal temperatures and is less reflective than most paving materials. Certain paving surfaces,

fences, walls, canopies, trellises, and other site elements may be located on the site to absorb or reflect solar radiation, channel or block winds, and expose or cover the dwelling or solar collector.

Onsite Solar Collectors

In some situations it may be desirable or necessary to place the solar collector at some distance from the dwelling. When this happens, the site designer has three alternatives available. The first is to screen the solar collector from view. The second approach is to integrate the collector with the site by the use of earthforms, vegetation, or architectural elements. A third approach is to emphasize the collector as a design feature.

Regardless of the approach taken, onsite solar collectors, together with any structural supports or additional equipment, may be unsightly, hazardous, and subject to vandalism. Earthforms, planting, and other site elements can be used efficiently to hide the collector and associated apparatus and to prevent easy access. In some cases, the space under the collector can be used as long as it does not interfere with its performance.

Summary

Site selection, planning, and design can significantly influence the effective use of solar energy for residential heating and cooling. The topography, geology, soils, vegetation, and local climate of a building site should be considered prior to site selection or building placement. Each climatic region has its own distinctive characteristics and conditions that influence site planning and design for solar energy utilization and for energy conservation. These should be recognized during the development of design objectives for each proposed site in terms of building-collector placement and orientation; relation of building collector to wind, water, and existing vegetation; and the merits of clustering, new vegetation, and material selection. In addition, the careful selection and location of all forms of planting, paving, fences, canopies, and earthforms can contribute to the effective and efficient use of solar energy. Dwelling design and site design for the utilization of solar energy are complementary and should be considered simultaneously throughout the design process.

IMPACT OF SOLAR ENERGY UTILIZATION ON TRADITIONAL DWELLING DESIGN

The use of solar energy imposes certain design requirements upon a dwelling design. Some relate to climatic issues such as achieving optimum capture of solar radiation by optimum tilt of the collector; others involve the solar-system components themselves, such as the area requirement for flat-plate collectors. Consequently, the integration of solar heating and cooling systems into traditional housing styles may, in some instances, require significant changes in the architectural appearances of these dwellings.

The majority of housing styles in the United States have developed from a historical tradition of dwelling designs that are responsive to local climate conditions. However, over the years, the relation of housing design to climate has diminished, due largely to the availability of mechanical heating and cooling systems that are not climate-dependent. Excessive cold, heat, or humidity could be overcome simply by the turn of a switch, which activated the mechancial equipment.

The United States is now faced with a situation where the fuel to power the mechanical equipment is in short supply. As a consequence, we are beginning to look to an infinite fuel source—the sun—to heat and cool our homes and to heat our domestic water. Because solar energy comes directly from the sun, a solar dwelling must necessarily respond to the climate. This, in turn, will influence the architectural appearance of the dwelling.

The more housing design responds to climatic factors—using the sun for its energy, the wind for cooling, the earth for insulation, and trees and vegetation for shading and protection—the less of an impact solar energy systems will have on dwelling design. In other words, the more energy-conserving the house design, the easier it will be to solar heat or cool.

The solar-dwelling concepts presented in this section illustrate what traditional housing styles may look like when modified for solar heating and cooling in different areas of the country. The design concepts were developed by the architectural firm of Massdesign Architects and Planners, Inc., located in Cambridge, Massachusetts. The traditional flavor of the housing styles has been retained whenever possible. The housing concepts are illustrative of the changes in solar dwelling and system design resulting from climate differences. The design concepts have been prepared for single-family housing, both attached and detached, and for low-rise multifamily housing. Only solar systems that are readily available today have been integrated into the designs.

A rough approximation of solar-system performance is presented for each design concept. This provides a general indication of the dwelling's heating and domestic hot-water load supplied by solar energy. The figures are rough estimates of expected solar-system performance for conceptual dwelling designs and for generalized climatic conditions. The solar designs are concepts only and are not presented as ready-to-build housing designs. Further assistance from design professionals will be necessary to develop a design and solar system that is responsive to a client's needs, climatic conditions, and site characteristics.

Cape Cod

The Cape Cod housing style was developed in response to the harsh climatic conditions of New England and the Atlantic seaboard area. Its compact floor plan and small exterior wall and roof area enabled the house to be easily heated by the centrally located fireplace. South-facing dormers allowed for ventilation, lighting, and views at the second level of the two-story house. It could be rapidly and easily constructed on a variety of sites and was relatively inexpensive. Although well suited to the natural resources and construction techniques of colonial America, the Cape Cod has become a favorite architectural style throughout the United States.

Very little alteration of the traditional Cape Cod style is necessary to incorporate the components of a solar heating system. One side of its double-pitched roof has the necessary area for solar-collector installation, sufficient to provide a major portion of the dwelling's annual heating demand in a cool climate. Also, the traditional steeply pitched roof is within acceptable angles of tilt for winter solar collection for most northern climates. Solar heat storage and distribution can be located in a basement, crawl, or attic space with little or no alteration of the traditional Cape Cod design.

In the solar design concept shown, in Figure 6-11, the only modification of the traditional Cape Cod has been the orientation of the dormers to the north, leaving the entire south-facing roof area free for solar collectors.

The Cape Cod is equally adaptable to either a warm-air or warm-water solar heating system. Natural ventilation is used in lieu of mechanical solar cooling due to the small cooling requirement of the New England climate. Massdesign has selected a warm-water solar heating system for the design concept shown in the figure. The system consists of a liquid-cooled flat-plate collector, with automatic drain-down to avoid freezing, forming the 45° roof, an insulated water-storage tank located in the basement, and associated piping, valves, and controls. Solar domestic water preheating is provided by running the supply line for the conventional water heater through a heat exchanger in the storage tank. The solar system is designed to supply approximately two-thirds of the annual heat and domestic-hot-water required annually by this 1900-square-foot house in a New England climate.

Saltbox

Like the Cape Cod, the saltbox design developed in response to the harsh climatic conditions of New England. The arrangement of spaces and use of materials are similar in the two styles. However, the saltbox, unlike the Cape Cod, has an extended north roof sloping to the first-floor ceiling level and full two-story exposure. These design changes increase the enclosed floor area, decrease the exposure of interior spaces to cold northwest winds, and increase the southern exposure of habitable spaces to the warm winter sun.

The saltbox is easily adapted for solar heating. Its large roof area can accommodate sufficient collector area at an acceptable tilt to provide the major portion of the dwelling's heating demand in the New England climate, provided that the house's traditional orientation is reversed, with the extending roof slope facing south instead of north. A consequence of this orientation is that the house should be

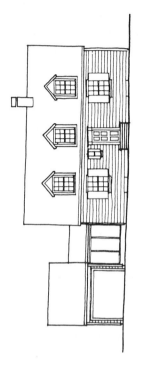

North Elevation

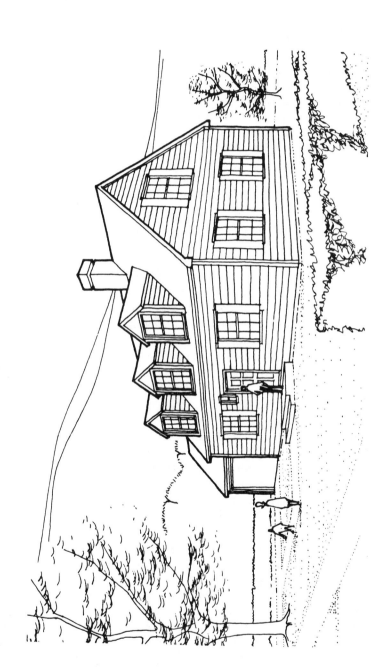

Figure 6-11. Cape Cod.

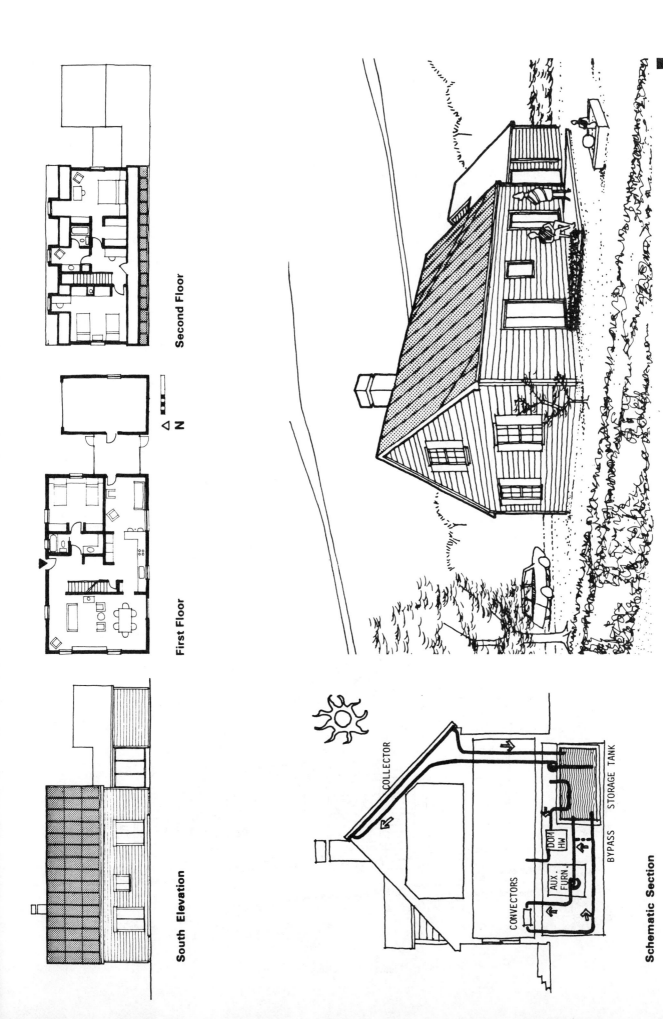

South Elevation

First Floor

Second Floor

N

Schematic Section

COLLECTOR

CONVECTORS

AUX. FURN.

DOM HW

BYPASS

STORAGE TANK

Figure 6-11. (continued)

placed on the site with the front entrance facing north. The storage component can be located in a basement, crawl space, or special storage room. Distribution ducts or pipes can easily be accommodated in floor, ceiling, and wall area of the saltbox.

The saltbox is equally adaptable to either a warm-air or a warm-water solar system. A warm-air system has

been incorporated in the design concept shown in Figure 6-12.

The solar heating and solar domestic-hot-water preheating are supplied by a system consisting of an air-cooled, flat-plate collector, forming a 45° roof; an insulated heat-storage bin filled with fist-sized rocks located in the basement; a hot-water preheat tank within the rock bin; and

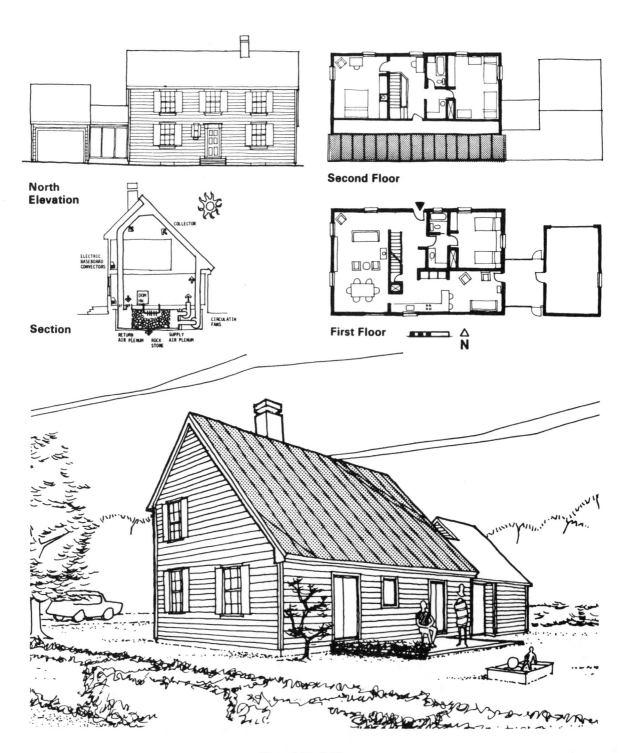

Figure 6-12. Saltbox.

associated fans, ductwork, dampers, controls, and insulation. This system is designed to provide approximately 50 to 75 percent of the heat and domestic hot water required annually by this 1900-square-foot house in a New England climate.

Farmhouse

The early housing styles of the Midwest and the Great Plains borrowed numerous design and construction techniques from the East. They evolved, however, into unique designs suitable to the new climatic and geographic conditions. The houses were easily heated, resistant to strong winds, and adaptable to variations in orientation and topography. Originally consisting of one-room long cabins or sod huts, housing styles gradually changed as the population grew and larger settlements developed.

A case in point is the traditional farmhouse. The house design consisted of several buildings organized around a central yard and sometimes connected. The buildings were usually one-story high with a sleeping loft. They were sited to block cold winter winds and to capture the cooling summer breezes. The farmhouse design has changed considerably over the years, yet it continues to find widespread application through out the United States.

Massdesign has taken the best of the farmhouse design features for adaptation to solar heating (see Figure 6-13). The roof above the living areas rises at a steep pitch to accommodate the collector, while the bedroom wing retains a conventional lower pitch for economy. Two different

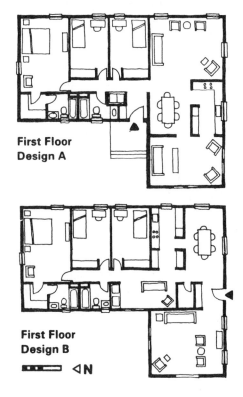

Figure 6-13. Farmhouse.

one-story L-shaped plans permit any orientation of the house on the site as shown in the site plan diagrams. Each plan for this three-bedroom house is accomodated economically in less than 1600 feet, including garage.

The collector area requirement is greater for cool regions than for other climatic regions. As a result, to achieve a substantial portion of the dwelling's seasonal heating requirement by solar energy, the farmhouse design maximizes roof area for solar collection and minimizes energy consumption by increased insulation, double-glazed windows, and insulated shutters. Also, solar heat storage is located beneath the house in a basement or crawl space so that any heat loss is to an occupied space.

The design concept is adaptable to both warm-air and warm-water solar heating systems. The warm-water system incorporated in the design in Figure 6-13 is sized to supply approximately 50 percent of the dwelling's space-heating and domestic-hot-water requirements for a cool climate. The solar heating system consists of a liquid-cooled, flat-plate collector; a buried water-storage tank; and associated piping, pumps, valves, and controls. Domestic hot water is preheated by running the supply line through a heat exchanger in the storage tank, and then a conventional water heater.

Southern Ranch

With their hot-humid climate, the southern Atlantic seaboard area, the gulf coast, and the southern portion of California are suited to housing styles that allow maximum heat loss and minimum heat gain during the summer. A ranch-type housing style has developed, which is responsive to both of these concerns. It incorporates large window and louver areas for natural ventilation and extended roof overhangs for protection from direct solar heat gain. The one-story arrangement containing 1750 square feet is open and spread out to capture air and channel its movement through the house. Although cooling is the primary design condition, heating is required 4 to 5 months out of the year. Therefore, the house is also properly insulated for easier heating during the winter. Walls and roof insulation, double-layered glass in all doors and windows, and large southern-exposed window area add substantially to the thermal performance of the house.

The energy-conserving design, along with the low heating requirement of the climate, makes the southern ranch a candidate for solar heating (see Figure 6-14). Since the heating load is relatively small, the roof area required for solar collectors is also small. The ranch design shown in the figure accommodates the required collector area and tilt by the use of a raised-roof, located over the living-dining room. The ceiling of a portion of the living-dining room follows the shape of the raised-roof, thus forming an interesting interior space. Solar heat storage and distribution are easily accommodated with only minor changes to the architectural style. The solar heating system consists of an air-cooled, flat-plate collector; a buried rock-storage bin and associated ductwork; fans, dampers, and controls. Solar

domestic hot-water heating is provided by a preheat tank located in the storage bin. Solar cooling is not included in lieu of natural ventilation. The collector can be mounted on either side of the raised roof. This, combined with rotation of the L-shaped plan, allows the house to fit most suburban lots, regardless of the direction of the street. The design has the additional advantage that the collector is far from the lot line, thus minimizing collector shading from neighboring trees. The collector and storage area is sufficient to provide approximately 80 percent of the dwelling's heat and domestic hot water in a hot-humid climate.

Adobe

The architectural style of the Southwest evolved from a combination of climatic and cultural influences. Adobe buildings are constructed with bricks made of local clay and straw and dried in the sun. Adobe construction, which is characteristic of the region, is well suited to the wide fluctuation in temperature experienced between day and night. The adobe walls act as a heat sink, absorbing the heat during the day and radiating it to the interior at night. The various Indian communities of the region had developed this unique construction technique prior to the arrival of the Spaniards in 1540. The Spaniards called the native inhabitants *Pueblo* Indians because of their communal environment. The Spaniards used the same materials of adobe and native rock to construct their houses.

The Massdesign concept (Figure 6-15) retains the architectural flavor of the region, while also incorporating solar heating. The traditional flat roof of the pueblo structures simplifies the integration of the solar collectors into the housing design. Two raised enclosures along the south edge of the roof house the two banks of flat-plate collectors. The collectors are positioned at a steep but acceptable angle for collecting maximum radiation during the heating season. The storage and distribution components of the solar system are easily integrated into the one-story structure. The storage bin and distribution ducts are both located under the concrete floor slab. Dampers and filters are reached by access panels located in the floor. This allows adjustment and cleaning without unnecessary disruption to the house.

Solar heating and solar domestic-hot-water preheating are supplied by a system consisting of an air-cooled, flat-plate collector; a rock-pile storage bin; a preheat tank located in storage; and associated ducts, dampers, controls, and insulation. The solar heating system is designed to supply approximately 85 percent of the heat and domestic hot water of this 2000-square-foot dwelling.

Brick Colonial

House styles based upon the beautiful brick Queen Anne and early Georgian homes of the Atlantic seaboard and Tidewater Virginia enjoy a continuing popularity south of New England. Typical of the brick colonial style are the decorative center entrance, spacious central hallway, flanking chimneys at each end, brick exterior walls, and frequently symmetrical wings. The brick colonial design shown in Figure 6-16 retains the architectural flavor of the traditional colonial style and at the same time incorporates solar heating. The 1900-square-foot design has a central entrance and generous porch area to capture east, south, and westerly breezes. The sloped south-facing roof has sufficient area for solar collectors, which can provide a considerable portion of the dwelling's heating requirement in a mid-Atlantic climate. The screened-porch roof could be coated with a reflective surface to increase incident radiation on the collector, thus improving solar-system performance. Ample space in the basement (reached from the garage) is provided for solar heat storage and distribution.

The solar heating system consists of a liquid-cooled, flat-plate collector, a water-storage tank, and associated piping, pumps, valves, and controls. Domestic hot water is preheated by running the supply line through the heat exchanger in the storage tank and then to the conventional water heater. The system is designed to supply approximately 85 percent of the annual space heating and domestic-hot-water requirement for a hot-humid or hot-arid climate. In a cooler climate such as in the northeastern United States, the system could supply about 40 percent space heating and 70 percent water heating.

Contemporary Split-Level

The split-level house is a relatively new housing style that has become popular throughout the United States during

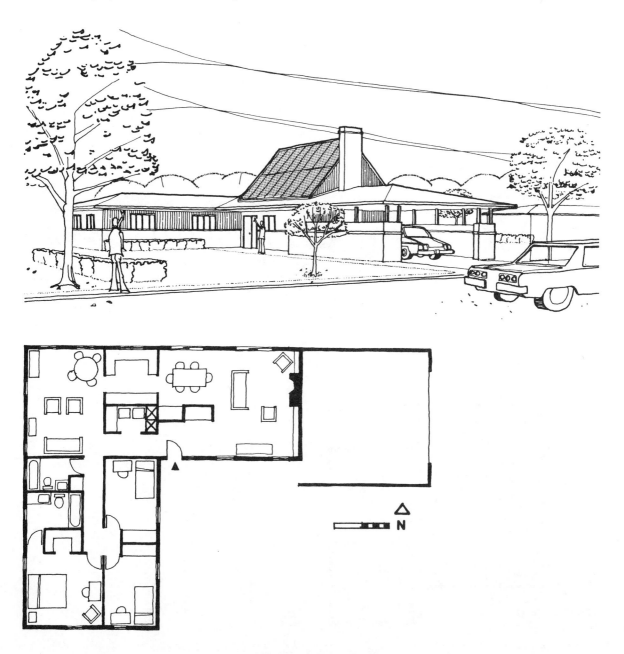

Figure 6-14. Southern ranch.

the past 15 years. The split-level design achieves housing economies by staying within a rectangular plan and by having its lower story built into a hillside. The design developed by Massdesign is typical of the more contemporary versions, with an overhanging upper floor, simple construction details, and a deck off the upper floor.

Heating the split-level with solar energy requires modifying the conventional roof pitches to accommodate the solar collectors and orienting the house so that it receives optimum solar radiation. From an architectural standpoint, the ideal orientation is to have the back of the house facing south with the main entrance facing north. By reversing the roof pitches, the reverse orientation is also possible; however, the results are less desirable, primarily because the living spaces would be oriented away from the sun. For

either orientation, accommodating the necessary collector tilt requires increasing the pitch of the traditional roofs. As shown in Figure 6-17, the roof pitch on the south side is steeper than on the north, and the roof design is thus asymmetrical.

The house plan, which contains about 1850 square feet, places one bedroom upstairs and two downstairs, with the primary living spaces along the south side of the house. As long as the curtains are not drawn on sunny winter days, the living spaces will overheat somewhat, and considerable solar heat can be captured through large areas of south glass. The use of insulating drapes or shutters at night will greatly reduce the heat loss through these large areas of glass, retaining the captured heat and making a useful contribution to the heating needs of the house. (The design

Figure 6-15. Adobe.

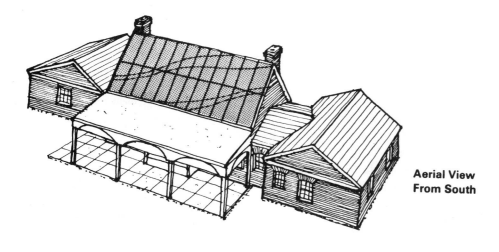

Aerial View
From South

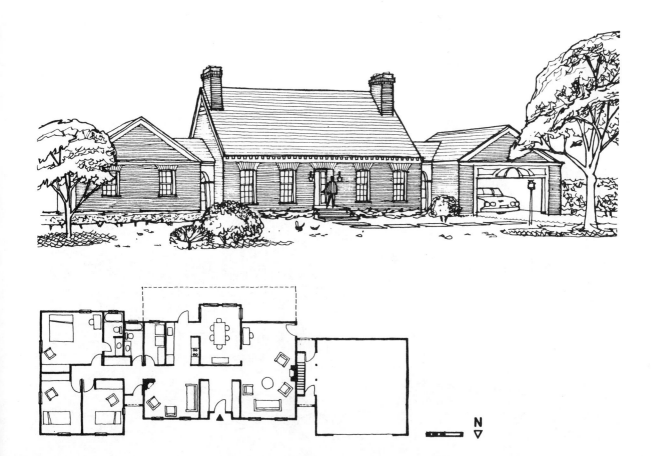

N
▽

Figure 6-16. Brick colonial.

concept shown in Figure 6-17 has only a moderate amount of south-facing glass area.)

A warm-water solar-heating system is incorporated into the design. A warm-air solar system could also have been selected, provided sufficient wall and floor area were set aside for ducts from collector to storage, and from storage to distribution throughout the house. The warm-water system consists of a liquid-cooled flat-plate collector; an insulated water storage tank located beneath the entry hall; and associated piping, pumps, valves, and controls. Distribution of heat throughout the house is by baseboard convectors, although a forced-air duct system could have been chosen. Domestic hot water is preheated by passing the supply line through a heat exchanger in the storage tank before feeding a conventional water heater.

The solar system will provide an estimated 50 to 60 percent of the dwelling's heat and domestic hot water in a cool climate and up to 70 or 80 percent in a temperate or warmer climate.

Northern Townhouse

The architectural form of the northern townhouse design (Figure 6-18) borrows many features from the Cape Cod house and adapts them to attached houses (houses built in rows with shared party walls separating adjacent units). The design shown is typical of many townhouses built in recent years in New England and other cold and temperate climates.

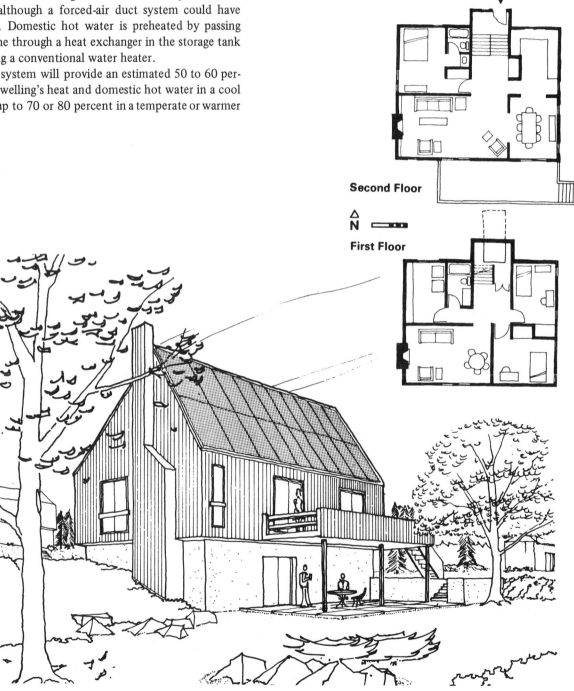

Second Floor

△
N

First Floor

Figure 6-17. Contemporary split entry.

Some of the traditional details of the Cape Cod, such as shutters, are eliminated in this design in favor of clean, simple moldings, but the traditional clapboard siding is retained.

Four dwelling units are included in the townhouse cluster; each has three bedrooms, two-and-a-half baths, a small family room, a kitchen, and a large living-dining room in a compact plan of 1690 square feet. A two-story space under the collector provides a dramatic high ceiling over the living room and stairway. The sloping south-facing roof, which houses the solar collector, is penetrated by a small protected terrace that provides light and air to the master bedroom.

The solar collectors located on each dwelling unit in one row feed a single, shared storage tank buried outside the townhouse cluster or located in a basement. The economic advantage of such an arrangement, combined with the intrinsically lower heat loss of compact attached dwellings, makes townhouses excellent candidates for solar heating. Piping, controls, and other components of the solar system are easily integrated within the townhouse structure, thus requiring little alteration to the building design.

A warm-water solar system is incorporated into the design concept shown in Figure 6-18. The system would provide approximately 80 percent of the total townhouse heating and domestic hot-water heating in a cool or temperate climate.

Northern Rowhouse

The rowhouse design concept by Massdesign is another modern adaptation of a traditonal architectural style (see

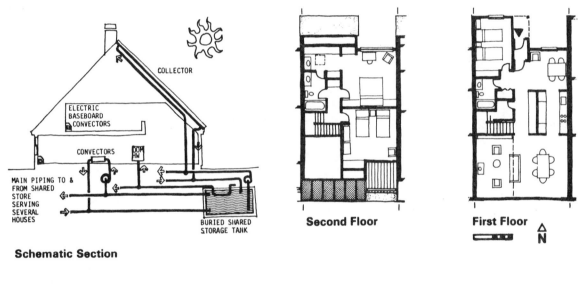

Schematic Section

Second Floor

First Floor

Figure 6-18. Northern townhouse.

Figure 6-19). Each two-story rowhouse has a compact, economical plan with three bedrooms, and two-and-a-half baths in less than 1600 square feet. The houses are oriented to show how solar collectors could be integrated with the roof structure of buildings facing in east and west directions (unlike the northern townhouse). Such an arrangement would be necessary if several rows of houses were organized around a central courtyard, where some rows would be oriented perpendicular to the others. This kind of flexibility can be very helpful in developing interesting site plans,

or adapting housing designs to difficult contours or restricted site boundaries.

Each bank of collectors ends in a traditional gable, with a near-flat roof below. In areas of heavy snow, it is advisable to raise the bottom of the collector higher above this intermediate roof than is shown in the figure to avoid an excessive buildup of snow at the base of the collector. Between each gable at the front is a sloping roof, which reaches down to the first story, helping to break up the scale of these small houses and to avoid a factorylike "sawtooth" roof. The

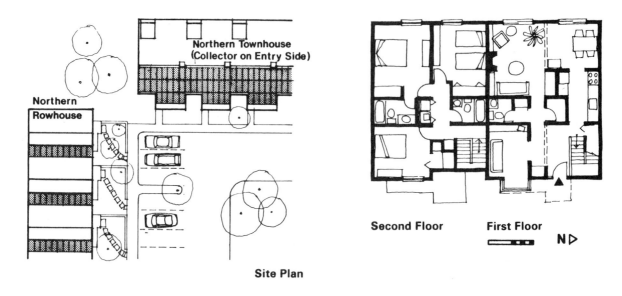

Site Plan

Figure 6-19. Northern rowhouse.

angle between the top of the collector and the base of the one adjacent must be calculated carefully to avoid excessive self-shading.

An individual or shared solar collection and storage system can be utilized in the rowhouse concept. However, as with the northern townhouse, there are significant economic and operational advantages with a shared collection and storage solar system. Either a warm-air or warm-water solar system can be integrated into the rowhouse design. Water is perhaps the more economical choice because of the transport distances involved and the fire-separation problems between houses caused by air-duct penetration through fire walls.

The solar system is designed to supply approximately 50 percent of the heating requirement of each rowhouse in a cold climate. The solar domestic hot-water system, on the other hand, should supply about 75 percent of the hot-water demand of each rowhouse.

Brick Townhouse

Solar heating and cooling can be as much a part of an urban situation as a suburban or rural setting. The brick townhouse design concept shown in Figure 6-20 illustrates the application of a simple roof-mounted solar-collector arrangement

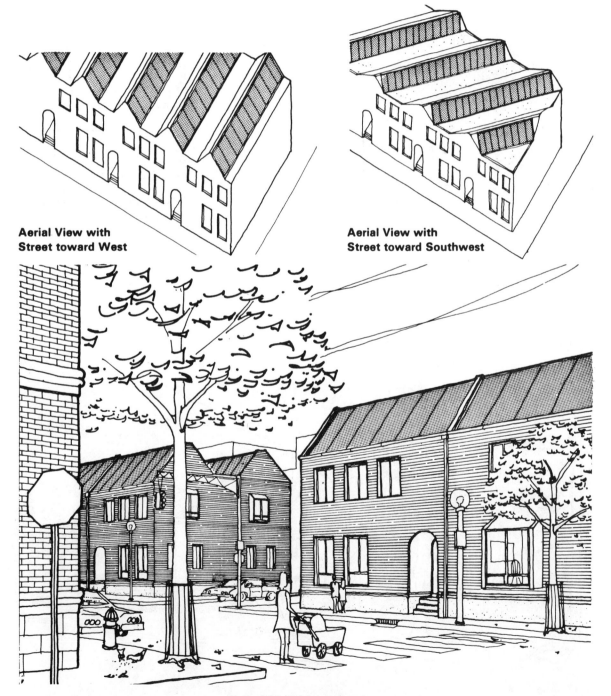

Aerial View with Street toward West

Aerial View with Street toward Southwest

Figure 6-20. Brick townhouse.

to a typical medium-density urban situation. The design has two stories; however, it is possible to use the same collector concept on a three- or four-story design.

The design concept has wide applicability for infill row housing in the built-up sections of a city. However, certain designs requirements must be met in order to assure the satisfactory performance of the solar system. For example, when the collectors run parallel to the street, the street must run roughly in an east-west direction for practical solar collection. A design with a gable end toward the street (similar to the northern rowhouse) is shown in Figure 6-20,

but the resulting architectural expression is visually less satisfactory in an urban context than the one shown, i.e., the collectors are integrated with the townhouse in the form of a glazed mansard roof. To solve every possible street orientation, a design with collectors running at an angle to the building would also be required, as shown.

For all the various design conditions, however, the ridge of each bank of collectors must be far enough from the next bank of collectors to the north, to avoid excessive shading during periods of solar collection. In addition, the buildings and trees around the houses should not impair

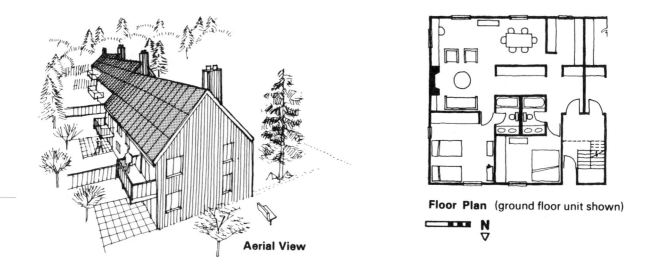

Aerial View

Floor Plan (ground floor unit shown)

N

Figure 6-21. Mid-Atlantic garden apartments.

solar collection to the extent that it becomes economically unfeasible. A neighborhood with continuous two-, three-, and four-story buildings and relatively small street and backyard trees would be ideal.

Although the solar-heat storage component can be placed anywhere within the building structure, it is generally located in a basement space. Piping or ducts run from the collector to storage are easily accommodated within mechanical chases generally associated with townhouse construction. With careful analysis and design integration, solar heating in the heart of the city can be a practical and attractive possibility.

Mid-Atlantic Garden Apartments

The garden apartment is a relatively new housing style in the United States. Whereas rowhouses and townhouses are most often associated with urban settings, garden apartments are generally associated with the suburbs. With more developable land areas available than in an urban setting, suburban garden-apartment designers can utilize the natural topography, vegetation, and conditions of the site to their best advantage.

The garden apartments shown in Figure 6-21 step up a gentle slope in blocks of four units. Each unit has two bedrooms, although various combinations could easily be developed. The living unit is entered through a gallery, which leads to the kitchen and living room. A fireplace, located in the living room, is included in each apartment unit. Since the flue of the fireplace would pass through the solar collectors mounted on the south roof if extended vertically, the chimney flue slopes within the attic space to emerge on the north side of the roof. Also, each apartment unit block is set back from its neighbor as they step up the hill, thus allowing the plane of the solar collectors to continue unbroken, while avoiding problems of collector shading caused by breaks in the roof plane.

The design incorporates a warm-water solar system. A central water-storage tank is buried in an accessible location adjacent to the building and is carefully insulated and protected from moisture. Transport lines from the collector to storage and from storage to apartment distribution are also heavily insulated to minimize heat loss. The large attic space created by the 45° roof pitch necessary for solar collection is used for apartment storage in the absence of a basement space in the design.

With the large area of solar collectors shown in Figure 6-21, the design should be capable of supplying 80 percent of the garden apartments' heating demand in a temperate climate and almost all of the domestic hot water.

In-town Apartments

The three-story brick-faced apartment building shown in Figure 6-22 is designed for a southern or mid-Atlantic urban or high-density suburban area. Although thoroughly modern in appearance, with a concrete frame, large glass areas, and precast concrete lintels, the building's curved arches, traditional brick, and black metal railings retain the architectural spirit of older buildings in the South. In many historic urban settings, this building could take its place without destroying the continuity and scale of the older buildings.

The metal railings serve two purposes. First, they help to unite the new buildings with the old settings and, second, the railing helps to subdue the large-scale tilted solar-collector banks on the roof behind a familiar visual element. Unlike a solid parapet, however, the open railings do not completely shade the ends and lower parts of the collectors during the hours of low sun—common during the heating season. The screening is accomplished without a noticeable loss in system performance.

The rows of rooftop collectors are mounted at the optimum tilt on rigid frames attached to the roof structure. The design employs a warm-water solar system with piping running between the collector banks to a large central water-storage tank in the basement. The heated water is pumped from storage to the individual apartments for distribution by fan-coil units. With such an arrangement, the architectural modifications required to integrate the system are minor, and, as long as the building is designed to be energy-conserving, the solar system should satisfy a major percentage of the building's heating needs. For climates with a large cooling requirement, solar-assisted cooling may be feasible. It may be necessary in these cases to design the collectors so that the angle of tilt may be adjusted from heating optimum to cooling optimum. The design shown has sufficient collector area and storage capacity to provide approximately 90 percent of the building's heat and domestic hot water with solar energy.

Balcony-Access Apartments

The balcony-access apartments shown in Figure 6-23 are organized around a central garden courtyard. The apartment design is typical of units that are built in climates where exterior access is acceptable. The design incorporates several solar collector concepts. Concentrating collectors appropriate for regions with clear sunny weather are mounted on the roof; liquid-cooled flat-plate collectors are mounted vertically on the railings of each floor.

The roof collector is a linear concentrator in which the long axis runs east-west. A tracking mechanism keeps the sun focused on the absorber by rotating the reflector and its glazed cover around the absorber to follow the sun's apparent up-and-down motion as it crosses the sky. The concentrated solar energy produces high temperatures in the working fluid, which allows for summer cooling as well as winter heating.

Supplementing the energy captured by the concentrating collectors on the roof are fixed vertical flat-plate collectors mounted on the railings of the balconies serving the upper floors. These collectors operate only during winter since in summer their vertical tilt angle allows them to

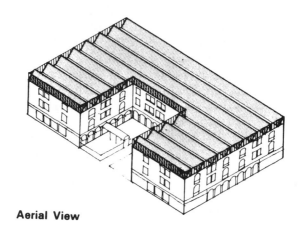

Aerial View

Figure 6-22. In-town apartments.

collect relatively little heat in southern latitudes. For the railing collectors to operate efficiently, the courtyard must be wide enough and the vegetation in them low enough to let the low winter sun strike the collectors.

Energy captured by both collector systems is removed by a working fluid and stored in a central compartmentalized heat-storage tank. During the heating season, the heated fluid is pumped from storage to fan-coil units located in each apartment. An energy boost is supplied by a central auxiliary boiler if storage temperatures are not sufficient for apartment heating. During the cooling season, the heated fluid is used to power a low-temperature absorption-cycle cooling system. The cooled fluid is again pumped to the fan-coil units located in each apartment for distribution. Domestic hot water is preheated by passing the supply line through a heat exchanger located within the central storage tank. The system should be able to supply most of a building's yearly heating and hot water load, and a good part of the cooling load, for the coldest climate in which balcony-access apartments are acceptable.

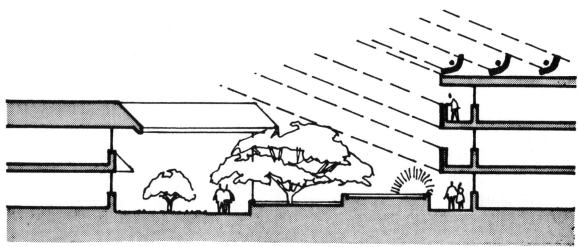

Site Section

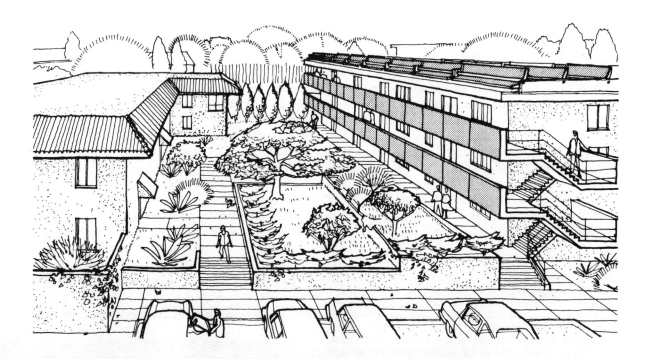

Figure 6-23. Balcony-access apartments.

7 Buying Solar

Solar energy* for your home is here. It may help you individually, and it certainly will help the United States collectively.

Whether it will help you individually in terms of producing real savings depends upon a number of factors, including where you live, the type of home you have or intend to build, the quality of insulation in your home, your present energy costs, and the type of solar system you intend to purchase.

The purpose of this chapter is to give you information on these five factors so that you, as an informed customer, can make decisions related to solar space heating and cooling, and domestic water heating that are in your best interests. By doing so, you will not only be doing a service to yourself, but to your fellow citizens.

Those who do use solar systems will help the nation save its precious fossil fuels. As you know from the headlines, our domestic supplies of oil and gas are very limited. There are just not enough domestic supplies available to meet all our future needs. And when Mother Nature says "that's all," that's all. The only way we can prevent or delay those fatalistic words is by conserving energy now, and using, where possible, sources of unlimited energy such as solar.

We won't have to worry about the sun running out of energy for another several billion years or so. Besides being an almost infinite energy source, solar has other advantages as well. When we use conventional sources to deliver energy to the home, we must deal with serious problems, ranging from the safe disposal of radioactive wastes to pollution from fossil-fueled generating plants. With solar, we will still need these conventional sources of energy, but we won't need as much. That means that we can reduce our environmental and safety problems.

*When *solar energy* is mentioned, the term is generally interpreted to involve a solar collector that will provide energy to the home. Actually, solar energy has a broader meaning. It can involve: photovoltaic energy, the direct conversion of the sun's energy into electricity; bioconversion, the utilization of agricultural or municipal wastes to provide fuel; ocean thermal, providing power by harnessing the temperature difference between the surface waters and the ocean depths; wind, harnessing wind energy to generate power; and solar thermal electric, concentrating the sun's rays to obtain high temperatures and thus generate electric power.

Furthermore, owners of solar homes have extra protection against energy inflation and energy shortages. When utility costs go up, and they will, owners of solar homes may not have to face altering their lifestyles just to pay the utility bill. Cutoffs or curtailments of conventional fuels won't affect solar-home owners as much as others.

With all these advantages, one might wonder why solar power is not more developed. The primary reason is economics. Until recently, it was just not economical for a homeowner to install a solar unit when there were cheap sources of conventional energy around. But those days are gone forever, and now solar is becoming increasingly competitive with electricity and oil. The present gas cost advantage over solar may change in the near future as a number of experts believe gas will triple in cost in the next few years.

Because solar energy has not been fully developed in the past, it poses some significant problems for the consumer in the present. For example, the consumer has not had any experience in buying solar equipment. What questions do potential buyers ask? How can they compare competing brands? How do they get a unit repaired if it goes on the blink? What do they do if their neighbor plants a tree that throws a long shadow on the collector? Can buyers trust a seller's claims for a particular solar unit? What happens to the unit when the owners go on vacation? What happens if a vandal throws a rock at the collector?

In some cases, such as solar rights (a citizen's guaranteed right to access to the sun's rays without encroachment), there are no complete answers today. In other areas, such as measuring the efficiency of a collector and evaluating the different components, there are a few answers. But only an expert—a mechanical or architectural engineer who has a background in solar—can give them to you. This is because what is a "smart" solar purchase for one consumer in one area may be foolish for a consumer in another area.

SOLAR SUBSYSTEMS—THE COLLECTOR

Several years ago, the pocket calculator was but a wink in an inventor's eye. Now, you can buy one in almost any retail outlet.

Similarly, when stereo systems first arrived in the marketplace, Americans heard a totally new language—woofers, tweeters, antistatic devices, distortion levels, etc. To choose properly, consumers had to learn what these terms meant, and then they had to learn how to compare competing products. Because so many Americans took the time and trouble to learn these essentials, stereo manufacturers by and large fought for the consumer's business by selling quality rather than imagery. Two important results occurred—superior products and satisfied customers.

Will solar evolve, as stereos did, into a widespread beneficial industry? To a great extent, that answer depends upon the American consumer. The more people who take the time to become educated buyers, the greater the chance that solar power will reach its full potential—providing safe, economical energy to millions of American families.

To be an educated buyer of solar systems does not mean you have to become a mechanical or architectural engineer. However, you should call on one of these experts for specific advice for your home. But before you consult an expert, you can do much to decide whether solar is for your home and, if so, what type of unit is best for your needs. Perhaps the best place to start is with a discussion of how solar works, and of the various main subsystems that make up a solar system—the collector, storage, distribution network, and controls.

The solar collector is the subsystem most people think about when solar energy is discussed. This is the component whose main function is to capture the sun's energy. There are many types of collectors available: high-performance collectors such as the focusing collector, which tracks the sun; a vacuum-sealed collector, which has very low heat loss; and the more conventional flat-plate collector. To understand the principle upon which a collector operates, let's take a look at the flat-plate collector, which has been used successfully for residential and commercial purposes.

Beyond the casing and insulation, the flat-plate collector has three main elements—the transparent cover or covers, the collector plate known to engineers as the *absorber,* and the channels in the collector plate.

The transparent cover can be made out of glass or plastic. It is hard to generalize about these materials because various products differ in quality. Glass holds its transparent characteristics well over the years. However, various quality characteristics such as transparency, strength, etc., vary from product to product. The same can be said of plastics. Some contain high-transparency characteristics for long periods of time, whereas others do not. Some turn yellow, reducing the capability to transmit solar radiation. Some glass and plastic covers are almost vandalproof, while others can be damaged very easily.

Some nonconductive material should separate the cover from the frame to decrease conduction losses.

In selecting collectors of different materials, the consumer should ask for test reports that show anticipated durability and transparency characteristics. Your engineering consultant can then evaluate these reports.

The transparent cover serves many purposes. It keeps outside air from carrying away the heat that has been trapped. It also keeps out the wind and the elements, protecting the inside components, and reducing energy loss by convection. In warm areas, one cover is usually all that is necessary, but, in colder climates, two transparent covers or insulated glass are generally considered necessary.

Here is how the typical flat-plate collector works. Solar radiation passes through the transparent cover (a small portion is absorbed or reflected off the cover's surface) and hits the absorber plate. Most of the radiation is absorbed by the plate and picked up by the fluid (air, water, or other liquids) passing through the channels in or against the plate. Some of the radiation is reflected off the plate back to the cover—how much depends upon the absorbing and reflecting characteristics of the coating on the collector plate. The better the absorbing quality, the more radiation captured and the less reflected back to the cover. Special coatings have been developed, which are highly absorptive with low radiation. Don't be turned off by a collector plate because it is black or a dark color; dark colors absorb radiation much better than light colors. Conversely, you can't always tell by the color whether the coating has the desired selectivity characteristics.

Some manufacturers are developing "selective surfaces" for the collector plate. These are not painted, but rather are specially coated metals that appear to be a technical improvement over flat black paint because reradiation losses are decreased. Selective surfaces cost more initially than flat black paint, and the extra cost must be weighed against the value of increased efficiency and the life expectancy of the coating. No matter what coating or metal is used, however, some portion of the incoming radiation will be radiated back, and, of that portion, the transparent cover will either allow some to pass through or will absorb some. The reason for two transparent covers in some collectors is to improve the insulation, just as storm windows on a house reduce the loss of heat through the windows. More than two covers are not necessary. Still other portions of the collected energy are lost from the collector plate through convection to the airspace above the plate, through conduction by the metal in the plate, and through the insulation on the back and side. How much is lost through conduction depends upon overall design of the collector. The net result of all these losses is that only a certain portion is absorbed by the passing fluid. The fewer the losses, the more efficient the collector.

To reduce losses through convection, some collectors have honeycomb material located between the transparent cover and the collector plate. This limits the formation of convection currents, which are responsible for some of the loss. The advantages of honeycomb in reducing convection losses may be offset by the greater expense and by the increasing conduction losses if the honeycomb conducts heat from the collector plate to the transparent cover. In addition, the honeycomb should help the collector take stagnation temperatures—high temperatures that can occur when the collector is not working. A properly designed honeycomb structure can add to the strength of a collector, increasing resistance to hail, and adding to the capability of the collector to replace the roof.

Depending upon local climatic conditions, an important design factor for a flat-plate collector is its position in relation to the sun. If a solar system is designed for heating, the collector surface in most instances should be nearly perpendicular to the sun's rays in the middle of January, generally the coldest days of the year. This is usually an effective position for capturing the sun's rays. This perpendicular position is the sum of the latitude of the location of your home plus 15°.

If a solar system is for heating and cooling, the optimum angle is near the latitude. If a system is for cooling only, the collector should be mounted at an angle about 10 to 15 degrees less than the latitude.

If a collector is within 10° to 15° of the optimum, it will not make much difference in the energy recovered by the collector. Thus, if you live in Phoenix and you decide to put a flat-plate collector on your roof, and the roof has a 45° tilt, there's no need to go to the added expense of building in the other 3° on your collector. Similarly, if your hire an architect to design a solar house, don't get upset if he or she shades a few degrees off here or there to obtain some other desired end result. Furthermore, some solar systems have a mirror or reflective surface next to the collector to augment the radiation reaching the absorber by reflecting extra sunlight into it. This is often done with cooling systems where higher temperatures are needed.

The ideal position of a collector for the greatest annual and seasonal energy recovery is facing due south. However, a variation of as much as 15° in either direction will not significantly reduce the efficiency of the collector.

Concentrating collectors can only collect direct solar radiation, so they aren't much good on cloudy days. Flat-plate collectors do not have this limitation and can pick up diffused radiation. The extra cost of the reflector must be weighed against the Btu's on sunny days only.

Beyond the main elements within the collector, it is important to keep in mind that soundness of construction is an important characteristic in obtaining an efficient collector. It must keep out wind, rain, and dust, so good sealants are necessary.

Another important feature of the collector is the insulation under the absorber plate to reduce downward heat losses. Some insulation does not hold up well over a long period of time under the very hot conditions present under the absorber. Fiberglass, 3 to 4 inches thick, holds up well. Without efficient insulation, much of the heat collected will be lost. If the insulation does not hold up, the owner could be faced with periodic replacement and higher repair bills.

The consumer is likely to hear a lot about the *efficiency* of a collector–that is, the ability of the collector to capture the sun's power. For instance, if a collector has a 50-percent efficiency, it means that, under certain conditions, the collector captures 50 percent of the total energy hitting the surface of the collector.

There are many conditions that will affect the efficiency of a collector. Of primary importance is the amount of radiation you can expect in your area for the particular period in which you plan to use a solar system. For instance, if you plan to install a space-heating system, you should know the efficiency of the collector from December through February, when you will need the system the most. How much radiation you can expect will depend upon the amount of sunshine and the atmosphere where you live.

In some limited areas, there are so many cloudy and partially cloudy days that the amount of radiation reaching the collector will usually be found to be so small as to make it unlikely that it would be advantageous for installation of a solar system. Furthermore, the efficiency decreases as the temperature difference between the outdoors and the absorber plate increases. To put it another way, the colder it gets on the outside, the more inefficient the collector will be. This is because the difference in temperatures causes an increase in energy loss through conduction and reradiation. Remember the definitions of convection and radiation:

- Heat will always flow from a hot surface to a cold one when there is a medium such as air to transport the heat.
- When this occurs, it is called *convection.* In the collector, the airspace serves as the medium, and the net result is convection currents. In colder weather, convection increases.
- When an object is warmer than its surroundings, it radiates invisible heat waves to the colder areas. This is *radiation.* An object that has been warmed often radiates heat back.

To show how efficiency varies among competing collectors in terms of temperature differences, the National Bureau of Standards has developed standards by which efficiency can be measured. For instance, an increase in the flow of the fluid through the collector can help to reduce losses. If there is no movement at all, the fluid will become very hot and lose a great deal of heat. As the fluid starts moving, the temperature is reduced, and heat losses are reduced. Another factor affecting efficiency is the evenness of the flow of the fluid through the channels and the arrangement of the channel system; the more even the flow, the more efficient the system.

The temperature at which heat must be stored also affects a system's efficiency. In some systems, heat must be stored at high temperatures (100° to 175°F) to keep the home warm so that collector must produce hotter water or air. The hotter the system operates, the lower the efficiency of the solar system. In other systems, heat is stored at lower temperatures (85° to 125°F), and so the solar collector system operates at higher efficiency. Collector efficiency is only one of several considerations in system performance and quality, so the preceding comparison does not necessarily indicate that the lower temperature system should be the choice.

Now, why all the explanations about collector efficiency? First, to point out that there are a number of complex engineering factors that have a bearing on how efficient a given collector is. Second, the efficiency of a collector is a major factor in determining a whole system's performance. Third, to give the consumer some ideas for evaluating collectors.

Let's say that an engineer has evaluated two competing collectors. Both are estimated to have a 20-year life span and require little maintenance and few repairs. Collector A cost $10 per square foot installed, while Collector B costs $6 per square foot. On a given day, both receive 200 Btu per square foot per hour. Collector A has a 60-percent efficiency and will deliver 120 Btu per hour per square foot. Collector B has a 50-percent efficiency and will deliver 100 Btu per hour per square foot. Which collector does the consumer choose? If the two collectors are the same in all other aspects, the consumer chooses Collector B because it delivers 20 Btu per dollar whereas Collector A delivers only 10 Btu per dollar.

But it must be stressed that there are other considerations besides first cost and efficiency. The wise consumer will want to know whether the collector is covered by a warranty, what the warranty covers, and for how long.

A warranty is either "full" or "limited." If it is limited, it will only cover certain features, not the whole product. The following are major considerations, some of which can only be determined by an expert:

- Durability of the product. How long is it expected to last? Is it weatherproof? Does it shed water?
- Ease of repair. If something goes wrong, who will fix it? How long will it take to fix it? Are repair parts easily obtainable? And what will various repairs cost?
- Susceptibility to either freezing or overheating. There are a number of adequate solutions to both problems, but you should be sure your collector has built-in protection for both extremes. You should also be sure that if the collector is used for hot-water purposes, and if antifreeze fluid is used in the collector, it cannot mix with water for domestic use.

Systems using water or "corrosion-inhibited" water solutions can be built so that the collector and exposed piping will drain when temperatures drop close to freezing or when the collector is not being used. Alternatively, a nonfreezing fluid can be used in place of water in the collector. Water-antifreeze (that is, ethylene-glycol–type antifreeze formulations such as those used in automotive cooling systems) are not recommended for use in consumer solar applications. Exposure of ethylene-glycol antifreeze mixtures to high temperatures such as those encountered in most systems in the summertime accelerates the rate at which the ethylene gylcol will degrade, forming acids which are corrosive to most metals, including copper and aluminum.

The "nonfreezing" fluid selected should be one that will not degrade at maximum temperatures expected, which can reach 400°F.

Electric warmers can be used to prevent freezing, and they are acceptable in southern areas of the country where freezing temperatures occur infrequently. In colder areas, however, the cost of operating an electric warmer might be prohibitive, and power failures can occur just when severe freezing weather might be encountered. Obviously, any such use of electric energy tends to defeat the purpose of a solar energy system.

Overheating could have a detrimental effect on a solar system, if there are no built-in protective devices in active systems. So ask the seller what will happen to a collector if you go on vacation for several weeks during the middle of summer.

Corrosion and leakage problems must be considered. Metal corrosion can cause irreparable damage to a solar system and shorten a system's life span. More importantly, corrosion can cause serious health problems if the water is used directly by the user.

Suitable protection against corrosion depends upon the system and the metals used. *Potable water* is the water one uses for drinking and washing. In *inhibited water,* chemicals have been added to prevent corrosion. Nonfreezing, nonaqueous heat-transfer fluids are various liquids used in a solar system to prevent freezing. Most of these fluids are noncorrosive to the three metals commonly used in solar systems, but are not drinkable.

These three metals are copper, steel, and aluminum. Copper is more expensive, but it can be used with most potable waters with inhibitors. Steel and aluminum are more economical, but inhibitors must be added to prevent corrosion in most systems. Copper will hold up over a long time without inhibitors, whereas steel and aluminum will fail quickly unless they have special coatings. In inhibited water, all three will last indefinitely as long as the inhibition is maintained. All three will hold up indefinitely in suitable nonaqueous, nonfreezing, heat-transfer fluids.

Steel and aluminum must be electrically isolated from each other and from copper to avoid galvanic corrosion. Isolation is obtained by using insulated rubber or plastic connections to join dissimilar metals.

For leakage, another rule of thumb is that the greater the number of joints, the greater the possibility of leakage. You should know what to expect if a leak occurs, how it will be fixed, who will fix it, and how much it will cost. Some liquid collectors, with unique channel systems, have reduced the need for soldered joints considerably without hurting the efficiency of the collector.

Solar collectors that supply hot air are generally free from problems associated with freezing, corrosion, and boiling. Air leakage in piping and joints between collector sections, although not potentially as damaging as in liquid collectors, can reduce efficiency. The purchaser should therefore have assurance that the entire collector array will perform as well as individual sections or modules.

If possible, try to obtain the names of other buyers of the collector, and learn whether they are satisfied with its performance and whether the collector has lived up to the claims of the seller.

You should also ask the seller whether the collector meets interim standards of the National Bureau of Standards. If the manufacturer claims that such standards are met, you should get that in writing. These standards do provide a degree of assurance for the consumer, and if a seller claims such standards are met, he is legally accountable for that claim.

The seller's reputation is important too. If the seller has a history of backing up and servicing the products he or she sells, then your chances of getting satisfactory service are improved.

A final point on collectors. To the buyer, the most important points are total costs (including installation), the expected performance during the months the system is needed most, and how long the system is expected to last before it needs replacement.

THE STORAGE SUBSYSTEM

After the collector, storage facility is the next important subsystem for most solar systems. Basically, the collector captures heat from the sun and transfers it to storage for later use. There are three main methods of storage—water, rocks, and change-of-state storage. Each has its advantages and disadvantages.

Water

Water has low cost and a high heat capacity (that is, the ability to contain heat effectively within a limited amount of space). Water-storage tanks must be protected from freezing if located outside. In addition, tanks must be protected from corrosion, either through use of a corrosion inhibitor in the water or by constructing them with corrosion-resistant material. Storage tanks must be heavily insulated to prevent the stored heat from being lost.

To insure that steel has a long life, rust and corrosion inhibitors may be added to the materials used for water storage. However, because inhibitors are toxic, much care must be taken to insure that the water in the system does not mix with the potable water.

Plastic tanks do not have corrosion problems, but the type of plastic needed to store large quantities of water at high temperatures is rather expensive when compared with steel. Concrete is safe, durable, and economical, but if a leak develops, it is hard to repair. Great care must be taken to make certain that the tank is properly sealed when it is first built and installed. Some experts recommend bladders or diaphragms lining the tank to insure water tightness.

Two thousand gallons of water weigh about 9 tons. Therefore, no matter where the storage system is located—in the basement or the floor—adequate structural support must be provided for, and this is an area where you will need an engineering expert to give specific advice for your home.

As discussed earlier, to prevent damage to the collector under freezing conditions, either a nonfreezing fluid must circulate through the collector or the collector must be built so that it drains when temperatures drop close to freezing. Although the amount of fluid needed in the collector itself is small and its cost is not an important factor, it is generally not practical to fill the large storage tank with the nonfreezing fluid. Therefore, a heat exchanger must be used to transfer heat from the fluid in the collector system to the water in storage if the collector fluid is expensive. A draining system eliminates this heat exchanger and its cost, as well as temperature losses occurring from the exchange. However, a draining system using water or inhibited water as the fluid depends in most instances upon the reliability of the draining mechanism to prevent costly freeze-up. Building reliability into such systems is generally expensive. Despite its limitations, water is at present the most practical method for storing heat for liquid-type solar collectors.

The heat-storage system must be well-insulated. Sprayed polyurethane foam, mineral wool insulation, fiberglass, and vermiculite are all excellent insulation materials for the outside of the tank.

Rocks

For air-type solar collectors, rocks or pebbles are the most cost-effective way to store the collected heat. Rocks are easily obtainable, economical, and are not subject to problems stemming from corrosion, freezing, or leaking. Furthermore, there is no need for an expensive container. However, rocks and pebbles do have two main disadvantages: they take up a lot of space and are much heavier than water. The space needed for storing heat with rocks is roughly three times the amount needed for water. This can be a major problem for some homes.

Rocks can also be used to store heat at temperatures above 212°F, while water containers designed to carry temperatures over that point need to be able to withstand pressure and can be costly unless a fluid that has a high boiling point is used. The ability of the rocks to reach higher temperatures is important because such heat will not be wasted if the collector delivers it.

For best results, the rocks or pebbles should be roughly spherical, of uniform size, and up to 2 inches in diameter. These provide minimum resistance to the air passing through. Other solids can also be used, but they are usually much more expensive.

Change of State

Change-of-state storage uses materials or substances that change composition from solid to liquid when heated. The changes permit the storage of more heat per pound than if the material did not change composition. When a material such as Glauber's salts is cooled and goes from a liquid to solid state, it gives off this extra heat. This method, however, is still in the experimental stage.

The advantage of change-of-state storage is that it can contain a great amount of heat in a limited space and at limited weight. Let's say that the objective of a storage system is to hold 200,000 Btu at 100° to 160°F. Water systems would need 53 cubic feet at 3300 pounds and rocks would need 175 cubic feet at 17,500 pounds, whereas Glauber's salts would require only 19 cubic feet at 1740 pounds. In addition, the cost of such salts is relatively reasonable.

Salt storage, however, has some major limitations. The salts can only go through so many cycles before their natural state is altered and they lose their capacity to successfully store heat. There are some chemical additions which act to prolong the number of cycles considerably, and there are some interesting experiments, which, if successful, may dramatically prolong the life span.

Second, although salts are economical, present costs for containerization and transportation are fairly expensive. Mass production could bring these costs down dramatically, but there are at present no marketable systems.

In choosing a heat-storage system, you should determine whether and what type of warranties are available. Depending upon the system, you will need to know answers to such specific questions as:

- (For water) What steps have been taken to provide protection against corrosion and freezing? How long is this protection expected to last? Is this lifespan estimate given in a guarantee?
- What types of maintenance are necessary? How much? Who will do it (owner or service personnel)? Is there easy access to the storage area?
- What is the heat-storage capability of the system measured as Btu stored within a specified temperature range? (Ask for the engineering calculations upon which this estimate was based. An engineering consultant should be able to easily determine the accuracy.) Remember, the storage should be large enough to absorb the full energy output of the collectors on on an average sunny day.
- Are the insulation and structural design sufficient?
- Is the warranty transferable?

THE SOLAR ENERGY SYSTEM

Before you comprehend how a solar energy system works or decide whether to invest in one, you must first understand the basic operations of a collector and a storage system.

Residential solar energy systems are used mainly for heating water for domestic use; heating swimming pools; heating the home (space heating); and heating and cooling the home (space heating and cooling). Of course, there are various combinations of these uses. A system could be used to heat a house during winter and a swimming pool in late spring or early fall when house heating is not needed. Or, a unit could be designed to both heat a home and supply hot water.

There are two basic types of systems. An active system uses mechanical means, such as pumps and valves, for operating. A passive system uses natural forces such as gravity, convection, and nocturnal radiation to accomplish the same objective. Both types of systems have their advantages and disadvantages.

Passive systems are more economical to operate. In some cases in certain climatic regions, heating and cooling by solar energy is economically competitive with conventional forms of energy. However, these systems are limited by seasonal conditions and architectural restraints. They work well in hot, dry areas, but in cold temperatures or humid areas they are subject to such problems as freezing or excessive humidity. They may not be able to provide uninterrupted comfort levels. In most cases, the requirements of a passive system will dictate the overall design of the structure more so than with active systems. For instance, one type of passive cooling system requires a water pond on the roof with movable shutters, which cool off the home by evaporation and nocturnal radiation. The shutters are closed during the daytime to prevent reheating. This system is most effective in areas of low humidity and clear night skies.

Active systems, properly designed, can work in almost any area. Initial costs for an active system are generally higher than those for a passive system, but the costs may be more economical in terms of Btu delivered per dollar invested. Active systems are not as architecturally restrictive as passive systems.

Some solar manufacturers utilize both passive and active techniques, combining some of the economic values of the passive system with some of the versatilities of the active method. These are called *hybrid systems*.

A conventional furnace (oil, gas, electricity) serves as a backup energy source for most space heating and cooling, hot water, and some swimming-pool solar systems. For cost-effective reasons, most active solar systems are not designated to handle all the anticipated demand. If, for instance, a solar system were designed to handle 100 percent of the heating load during the middle of the winter, the collector and storage system would have to be large enough to provide energy for extended periods when the sun didn't shine. As these extended units would only be used during limited periods, the extra cost normally isn't justified by the savings in fuel cost.

Beyond providing heat when the sun doesn't shine, conventional heat sources are often used directly in conjunction with solar heat. For instance, solar energy can be used to heat water up to a temperature of 100°F, and conventional energy used to boost that temperature another 40° for a dishwasher.

Some designers have used alternate energy systems such as wind power and wood stoves as a heat and power backup, thus eliminating the need for utility service. However, the user would have to be prepared to accept inconvenience and occasional discomfort, especially in the case of wind power—for example, when the wind doesn't blow and it has been cloudy for three days.

The best known solar method is hot-water heating. Hot-water heaters have been in use in Florida since the early 1930s, and they have an overall record of acceptable performance when properly constructed and maintained. Total costs depend upon the quality and size of the unit, the warranty offered, installation costs, and other variables.

Because of fuel savings, they can return the original costs or pay for themselves within a reasonable period of time if replacing electricity or high-price fuel. The payback period depends upon the cost of conventional heating, the amount of water used by your household, the insulation rate (i.e., the R factor of the house), and the efficiency of the unit. Generally, these units save 50 to 80 percent of water-heating costs, again depending upon the unit and locale.

In addition to asking the questions about the collector, a consumer considering the purchase of a hot-water unit should ascertain that the system meets local toxicity and potable water standards. These standards are established to protect people against unsafe drinking water.

For hot-water systems, the size of the collector should be about 1 square foot for every gallon of water required every day. Not counting dishwashers and washing machines, the average American family uses about 15 gallons of hot water per day per person, and so, for a family of four, a typical hot-water installation could be about 60 square feet. This would take care of 50 to 80 percent of a family's total hot water needs, depending upon climate and with a conventional system supplying the remainder. Of course, a system could be designed to handle nearly all hot-water needs.

Some solar heaters for swimming pools operate on the same basic principle as a hot-water system, except that the storage area is not the hot-water tank but the swimming pool. A number of experts suggest that solar swimming-pool systems be designed so that they could be used to heat domestic hot water. If you should decide to have a dual-purpose system, make sure that you select the collector carefully.

Some solar swimming-pool heaters use unglazed collectors because the temperature to which it is desired to heat the pool is only a relatively few degrees above the outside air temperature. Unglazed collectors are very inefficient when operated at high temperatures, however, and may not be suitable for dual purposes.

One of the most effective and economical ways to heat a swimming pool is by using the pool itself as the solar collector. To do so, an easily removable durable transparent plastic membrane is placed on the water surface when the pool is not being used. Water temperatures can thus be kept above 80° for several months without fuel supplements.

As with hot water systems, it is important that the swimming-pool water meet health and safety requirements.

If a separate collector is used, it should either be of the draining type or otherwise able to withstand freezing. Even though the night temperature of the air may be 40°F or more, the collector could go below freezing due to nocturnal radiation. Nocturnal radiation can also be used to cool the pool in summer months if the water becomes uncomfortably warm.

Structural stability, the capability of the solar unit to withstand high winds, and the compatibility of solar with conventional heating systems (if the two are supposed to work together) are other important considerations. You should obtain professional guidance on these matters.

Costs for swimming-pool solar systems vary according to the temperature desired, how long the pool will be operated in the spring and fall months, and the local insolation rate.

Pros and Cons

Solar space-heating systems promise to dramatically reduce monthly energy costs for the consumer. Earlier, advantages and disadvantages of various components within a system were discussed, which, of course, carry over to the system itself. Let's add up some of the pros and cons of air and water systems.

Air

Advantages

- Moderate cost.
- No freezing problems.
- Minor leaks of little consequence.
- As air is used directly to heat the house, there are no temperature losses due to heat exchangers (devices that transfer heat from one fluid to another) when the system is used for space heat.
- No boiling or pressure problems.

Disadvantages

- Can only be used to heat homes; cannot presently be economically adapted to cooling.
- Large air ducts needed.
- Larger storage space needed for rocks.
- Heat exchangers needed if system is to be used to heat water.

Water or Liquid

Advantages

- Holds and transfers heat well.
- Water can be used as storage.
- Can be used to both heat and cool homes.
- Compact storage and small conduits.

Disadvantages

- Leaking, freezing, and corrosion can occur.
- Corrosion inhibitors are needed with water when using steel or aluminum. There are liquids that are noncorrosive and nonelectrolytic; however, they are toxic and some of them are flammable.
- A separate collector loop using a nonfreezing fluid and a heat exchanger or, alternatively, a draining water system, are required to prevent freezing. In warm regions, where freezing is infrequent, electric warmers can be used.

COMBINING SOLAR SYSTEMS WITH HEAT PUMPS

Some manufacturers combine solar systems with heat pumps for the purpose of reducing auxiliary electricity costs. A heat pump is a device that has the capacity to extract heat from either cool air or cool liquid and increase the temperature to a useful level for heating. When used for cooling, the pump extracts heat from the house and then raises the temperature high enough to be dumped outside.

A refrigerator is a type of heat pump. It simply extracts heat from inside the refrigerator and pumps it out, leaving the inside of the refrigerator cold. It does this through the circulation of a refrigerant, usually Freon. Liquid Freon expands and evaporates in the coils inside the refrigerator. Upon evaporating, the Freon—now a gas—becomes much colder than the inside of the box; thus, it is actually heated by the inside of the refrigerator. A pump compresses this "warmed" gas, causing it to heat up to temperatures higher than the room the refrigerator is in. The compressed gas then delivers its heat to the room through condenser coils where it "coils" and condenses back to a liquid, and the cycle begins again.

So, imagine that liquid Freon is moving through a tube that passes through water in storage. The Freon extracts the heat from the water by evaporation of this liquid. Then, as the Freon vapor is compressed, its temperature is raised to a range of $100°$ to $150°F$. This hot Freon vapor then flows through a heat exchanger, which uses the heat in the Freon to heat water or air. The Freon is cooled and condensed to liquid in the exchanger; after the liquid leaves the exchanger, it is allowed to expand to a low pressure. At the lower pressure, it evaporates at a low temperature, (colder than storage temperature), so it then can pick up more heat from the stored water to begin the cycle all over again.

Needless to say, the warmer the water in storage, the less work the heat pump needs to do to deliver useful heat. However, the colder the stored water is, the more efficiently the solar collector will function.

A heat pump can also be used to cool a house in summer. Here the cycle is reversed by operating valves, and then the process is identical to the description of the refrigerator where the inside of the house is the same as the inside of the refrigerator.

The heat that a heat pump delivers a house is the sum of the heat extracted from storage plus the electric energy delivered to the compressor of the heat pump. The total Btu delivered by the heat pump to the house, divided by the Btu supplied by electricity, is what engineers call the *coefficient of performance* (COP). Thus, if 3 Btu were delivered to the house for every Btu equivalent of electricity, the COP would be 3.

A problem with heat pumps, however, is that the colder the source used to evaporate the Freon, the lower the COP. This is where solar energy comes in. It can be used to heat the air or water, thus reducing the work that the heat pump must do and thereby increasing the COP. As you can see, the higher the COP, the less electrical energy required for a given amount of useful heat delivered to the house.

One key advantage to the use of solar energy with heat pumps is that solar can be allowed to operate more effectively. As stated earlier, the efficiency of a solar energy system increases as the temperature of the fluid pumped through the collector decreases. Therefore, a solar collector can be operated at a lower temperature when used with a heat-pump system and thus will collect more of the sun's energy for use than will a nonheat pump system. A properly designed heat-pump solar energy system will have its collector working at peak efficiency and its heat pump supplying only the additional heat required of the house. In other words, the higher the average COP, the better.

In addition, a common air-to-air heat pump can be used in a solar heating system as a furnace to increase the temperature of air being supplied from the solar system to the distribution ducts. Outdoor air is used as the heat source for the heat pump, and electricity provides the necessary temperature increase in the Freon cycle. Although this system does not have the higher collector efficiency of the design employing solar storage as the heat-pump source, it does not suffer from midwinter depletion of storage to temperatures consistently too low for direct solar heating without electricity supplement. Annual electricity usage in the two systems is not greatly different.

The heat pump replaces the furnace in a house at a considerably higher cost. On the other hand, if a house is cooled in the summer anyway, a heat pump costs only a little more than a conventional Freon air conditioner. Systems using heat pumps for summer cooling may be able to use the storage system to store "coolness" during off-peak hours and thus alleviate the summer peaks that the electric utilities experience on hot summer afternoons. If your utility system adopts off-peak pricing (costs are higher per kilowatt during periods of high use, lower during periods of low use), this solar-heat pump combination would work even further to your advantage. In the Washington, D.C., area, a house that requires a 4-ton air-conditioning system will find that it actually operates only about 25 percent of the time. If water storage were used, a 1-ton unit could be used to handle the load by operating 100 percent of the time or the 4-ton unit could be used only 6 hours a day at offpeak times.

SOLAR ENERGY AND THE HOME

There are three basic questions consumers must ask to determine whether solar energy can be used in their homes:

Is it suitable for my house?
Is it a worthwhile investment?
Can I afford it?

This section and the next will discuss the first two questions. Only you can answer the third.

To begin with, solar energy is not suitable for every house. For example, can your house or lot accomodate a large collector and storage system? (The collector will have to face a southern direction and receive adequate unobstructed sunlight.) For your specific area, how much usable radiation will fall during the time you intend to use your solar system? As will be explained, the amount of usable radiation, along with the system's efficiency and the size of the collector, will allow you to determine how many Btu you will receive.

Here are a few approximate space needs for various systems:

- For hot-water systems, the general rule of thumb is 1 square foot of collector area for every gallon of water required per day. In other words, for a family using 80 gallons a day, an 80-square-foot collector will be needed for the solar heater to furnish about three-fourths of the hot water in a sunny climate.
- For swimming pools, the rough estimate is 1 square foot of collector for every 2 square feet in the pool. If a pool is 600 square feet, you will need about a 300-square-foot collector. A transparent pool cover-heater is the same area as the pool.

There are many variables to consider in determining the approximate space needs for a solar heating system. As a rough calculation, estimate 1 square foot of collector for every 2.5 to 4 square feet of house. So for a well-built, 1500 square-foot house in a cold, sunny climate, about 500 square feet of collector can supply from two-thirds to three-fourths of the annual heat requirements. As the efficiencies of collectors increase in the future, the size of the collector required will decrease.

For rock systems the approximate space needed for storage is about 0.5 cubic foot of rock storage for every square foot of collector; for water, about 1 to 2 gallons (0.13 to 0.25 cubic feet) of storage for every square foot of collector; for salt (when commercialized), about 1 cubic foot of storage for every 9 square feet of collector.

A number of factors will determine the actual size of the system. For example, in winter, Maine does not receive as much sun as Virginia. Consequently, those who live in Maine will need a larger collector than those in Virginia in order to accomplish the same end results. (Yet the economics of solar heating are much better in Maine than in Miami simply because Maine has a longer heating season.) Similarly, one may decide for any number of reasons to limit the use of a collector, having it provide only 50 percent of a hot-water load, for instance, rather than 80 percent.

In considering whether a solar space-heating system will be suitable for your home, an important question concerns insulation. If you have any holes in your walls, you had better patch them up with sufficient insulation before using a solar system.

Be sure the money to buy solar equipment wouldn't be better spent on reducing heat losses. Some utility companies will do a free cost analysis on where the greatest losses are and how much it will cost to reduce them.

With improved insulation in a house and with some attention to heat recovery, the collector and storage space needs could be reduced dramatically. It cannot be stressed often enough that the key to conserving conventional fuel and reducing utility bills is to insulate rather than insolate. The reason for this is simply sound economics. You will be buying solar energy in order to save money on your heating bill. Thus, if you are losing heat because of poor insulation, it is more economical to save heat through better

insulation than by adding heat through a solar system. Once you have a soundly insulated house, then you will be able to determine whether the purchase of a solar energy system is worth the investment. You will have made every Btu count, and you will not be using a solar energy system to heat up the outside.

Let's look at some of the ways an engineer calculates heat losses and at why insulation is so important in factoring solar. Heat losses are expressed through what is known as a "U value." The U value stands for the amount of Btu lost per hour per square foot for every degree of temperature difference between the inside and the outside of the house.

For instance, let us say that a single pane of window glass has a U value of 1.5. The glass would therefore lose 1.5 Btu per hour per square foot for every degree difference between the inside and outside of the house.

If we wanted to find out how many Btu were being lost a day through the windows, we would simply take the square footage of the windows and multiply the temperature difference times 1.5 times 24 (for the hours in the day). As an example, let's say that the temperature inside the house was 65°F and that the outside was 0°F, and there were 100 square feet of windows in the house. We would be losing $1.5 \times 100 \times 65 \times 24 = 234,000$ Btu through the windows every day! Now let's take a look at how U values can come down considerably with proper insulation:

Material	U Value	Number of Btu lost per day per ft² with a 50°F temperature difference
Single pane of glass	1.5	1800
Double pane	0.55	660
Double pane with insulated shutter	0.10	120
Insulated wall	0.07	84
Super insulated wall (etc.)	0.04	48

You can see that these losses can really add up by looking at the comparison in Table 7-1 of two similar homes, one insulated, the other not. The total savings per day in the insulated house is 453,345 Btu! In other words, the uninsulated house would cost a lot more to heat in the winter.

Although you won't be able to completely eliminate heat losses, you can reduce them substantially. And once you have, you and your engineering consultant can better determine whether solar power is a sound investment for you.

If you plan to build a new home, you are in an excellent position to utilize a solar energy system and insulation. It is important to bring in an architect as soon as possible—even before the site is selected. Of course, be sure that the architect is experienced in designing solar energy systems. Solar designing is a speciality; it requires knowledge of how the mechanical and architectural systems work together.

Table 7-1. Heat losses in an insulated and an uninsulated house.[a,b]

	AREA (ft²)	°T	HR/D	UNINSULATED		INSULATED		SAVINGS (BTU/D)
				U	LOSS (BTU/D)	U	LOSS (BTU/D)	
Walls	900	25°	24	0.25	135,000	0.10	54,000	81,000
Ceiling	625	25°	24	0.65	243,750	0.075	28,125	215,625
Floor	625	25°	24	0.38	142,500	0.06	22,500	120,000
Windows	84	25°	24	1.13	56,952	0.53	26,712	30,240
Doors	40	25°	24	0.64	15,360	0.37	8,800	6,480

[a]Simple 25-by-25-foot houses, frame construction, crawl space below, attic above, seven windows, two doors, 65°F inside.
[b]Booklets on insulation are available from U.S. Department of Commerce and the U.S. Department of Housing and Urban Development (see Bibliography).

A good solar architect can put it all together. He or she can make the collector part of the roof, thus saving you money. The architect can also make sure that the site is suitable for solar and reduce the possibility of future arguments about legal solar rights. In addition, the architect can check out all the building codes, tax provisions, and other applicable local laws, and can make maximum use of energy-conservation techniques, such as proper placement of windows and natural convection.

If you are planning to retrofit your existing house, it is still wise to bring an engineering consultant on board. But before doing so, take a good look at your property to determine whether you could possibly become involved in a future dispute over solar rights. Briefly, solar rights is an unresolved legal issue concerning access to the sun's rays. If your neighbor plants a tree that blocks sunlight from reaching the collector, can you do anything about it? Or, instead of a tree, suppose your neighbor erects a tall building. What legal rights do you have?

Right now, there are few answers to these questions. Many states have passed laws preventing the shading of a neighbor's solar collector, and others may follow. Meanwhile, you could have a lawyer draw up a convenant with your neighbor, which would protect you against blockage. Other than that, you will have to use common sense. If your area is safely zoned, single-family residential and if you have enough space for a solar system and are situated so that your neighbors won't encroach on your sun space, then you are in a good position to plan. But if you are close to a developing commercial area, then you are taking risks.

Whether you plan on building a new home or retrofitting an old one, you should check with your local building and zoning authorities to insure that your solar system doesn't run counter to present legal provisions. And don't forget the tax assessor. You may find to your surprise that a solar system will cost you more in taxes. So even though you are doing your country a favor by saving energy, you still may be taxed for it. (On the other hand, some states already give a property-tax exemption.*) If this is the case in your area, let your local politician know about it. It will make good campaign material.

You will need to talk to your insurance agent about insuring the system against vandalism and possible damage from leaking, corrosion, hail, etc. You shouldn't have much difficulty getting proper insurance coverage, but before you sign on the dotted line, ask whether your present homeowners policy covers vandalism and whether a collector would be included in the coverage. You may not need extra protection.

As discussed earlier, you may need a full-size furnace for those periods when no solar heat is in storage and the weather is very cold. Such a backup is ordinarily an inexpensive furnace identical with those used in coventional systems.

SOLAR AS AN INVESTMENT

Is solar a good investment for you? To find out, one might compare buying solar with buying a house as opposed to renting one. When you buy a house, you slowly but surely gain ownership. When you finally make the last mortgage payment, the house is all yours. No more payments. Even if you have to sell before you make the last installment, you certainly won't lose much from your investment; in fact, you probably will receive more for your house than you paid for it, while also saving on tax deductions. But to realize a profit on your investment, you will have to undertake regular maintenance and repair. If you rent, you don't have to worry about upkeep, but when it comes time to move, all those rental dollars are lost forever.

Buying solar is somewhat the same, but with a few important differences. When you buy a solar system, you are concerned about initial and lifetime costs. Some new-home builders claim that costs for their residences with hybrid or passive solar systems are close to prices for comparable homes with conventional heating and cooling systems. If these systems function properly, then the consumer can expect a very short payback period—that is, the time needed for the savings from using solar energy to pay for the extra cost of the system.

If you are going to buy an active system, however, the initial cost will be much larger than that for a conventional system, and the payback period will be much longer. In a

*States that have passed property-tax incentives include Indiana, Arizona, Colorado, Illinois, Maryland, Montana, New Hampshire, North Dakota, Oregon, and South Dakota. Arizona and New Mexico have passed legislation providing income-tax incentives to buyers of solar systems. Texas has exempted the sale, lease, or rental of solar equipment from taxation.

sense, you will be paying for several years' supply of energy all at once. As with passive systems, you will have to be concerned about durability. Maintenance and repair costs will add to the time needed to pay back on the system. However, when the solar system is paid for, you will have free energy coming into your home. If you relied on a conventional system, it would be like renting—no hope of ever regaining the dollars spent for the operation of a conventional system.

However, as with renting a home, renting or leasing a solar system may provide you with some advantages. If, for instance, you rented a solar system for $25 a month, repairs and maintenance included, with an estimate that it would save you $40 a month in utility costs, then you would come out ahead by $15 a month or $180 a year if the projections were correct. You would never own the system, but you would obtain some savings with no repair or maintenance problems to worry about.

There are times when a solar system is not a smart purchase and a conventional system is a better buy. You may find that initial costs are prohibitive, or that the time period to pay back your investment is too long for your needs. Or, some of the unknowns, such as solar rights and property appreciation or depreciation, may seem too risky.

You should know how the engineers make the appraisals, so that at least you will know how to evaluate sales literature. What the engineers do (and what many businesses do) is to undertake a life-cycle cost analysis. Simply put, this means an analysis of all the money spent on a system versus all the savings. On the debit side for a solar system, there are initial costs, anticipated annual maintenance costs, interest and principal payments on a loan (which may be pivotal), extra insurance, possible tax increases, and possible depreciation. On the plus side, there are annual savings (as opposed to operating costs for a conventional system), possible extra appreciation for the house due to the solar system, possible tax breaks from the federal or state government, tax deductions for interest payments, and energy inflation (which may also be pivotal). Some of these factors, such as taxes, are sometimes difficult to analyze, and you or your engineering consultant will have to determine from local, state, and federal authorities what the tax breaks are when it comes time to buy. Furthermore, because some solar systems are relatively new, annual maintenance costs may be hard to calculate. But again, a good engineering consultant will give you some estimates as to how durable a system may be.

How long will it take to recover your investment? If the payback period is, for example, 10 years, and you have a sound, durable product, you can be fairly assured that the system will hold up for that period before any major extensive repairs or replacements are needed. If the payback period is longer, then your risks will be appreciably higher.

Any life-cycle analysis of a solar system begins with an evaluation of the Btu requirements for the specific end use of your solar system. For hot water systems, you need to know how much hot water you and your family use per month. You can compute that roughly from your present bills (from 15 to 20 percent of the total bill in cases where hot water and heat come from the same system), or you can base usage on the national average. For swimming pools, determine the Btu needed by the size of the pool, the temperature desired, and the period of extended use.

Determining the heating and cooling needs of a house is a complex process. Basically, you will need the design heat (or cooling) loss for the house, the degree days per month, or the cooling days per month. Next, calculate approximately how many Btu of solar heat will be delivered to your house. This calculation is determined by the average amount of insolation falling during a particular period in your area per square foot, the size of the collector, the efficiency of the system, and the portion of the collected solar heat that can be used. (Heat has to be thrown away during some periods.)

The payback period estimation prepared by the system manufacturer is based on the competing costs of conventional fuel, the amount of solar insolation, and the yearly efficiency of the system. There is also an adjusted payback period based on the premise that the solar home will appreciate considerably. In looking at the price and payback period, remember that these are the estimates of just one manufacturer. Other costs and savings estimates may differ considerably.

Note that the payback period did not take into consideration possible maintenance costs, loan interest, and possible tax disadvantages. On the other hand, neither did the calculations take into account energy inflation, which probably will be considerable, and possible tax advantages. Furthermore, the gas and oil estimates are conservative and do not reflect inefficiencies in gas- and oil-heating systems. In other words, although a gallon of oil is equal to 140,000 Btu about 55,000 Btu are lost when the oil is burned in a heater. The estimates did not take these losses into consideration. Another assumption is that the house is well insulated, with no major preventable heat losses.

In addition, remember that such overall calculations apply to a general area; they may not apply to a specific location. In other words, you may live in an area where, because of both high costs for conventional fuel and a high insolation rate, solar systems appear to be particularly attractive. Yet the particular site of your home—in the shadow of a mountain, for instance—may negate these advantages.

Table 7-2 can be helpful in determining the basic advantages of solar in your general area. By finding out the costs of available competing fuels and the insolation rate you can quickly determine the competitiveness of solar in your area.

The basic ingredients for a life-cycle analysis are the total installed costs of a solar system versus the anticipated benefits, which basically are the number of Btu you can properly anticipate to meet your needs. The first part of this basic equation, total installed costs, should be easy to obtain. The amount of Btu you can anticipate for various solar systems for your specific home, however, is not that easy to determine.

Table 7-2. Normal total heating degree-days (base 65°F).

STATE AND STATION	JULY	AUG.	SEP.	OCT.	NOV.	DEC.	JAN.	FEB.	MAR.	APR.	MAY	JUNE	ANNUAL
ALA. BIRMINGHAM	0	0	6	93	363	555	592	462	363	108	9	0	2551
HUNTSVILLE	0	0	12	127	426	663	694	557	434	138	19	0	3070
MOBILE	0	0	0	22	213	357	415	300	211	42	0	0	1560
MONTGOMERY	0	0	0	68	330	527	543	417	316	90	0	0	2291
ALASKA ANCHORAGE	245	291	516	930	1284	1572	1631	1316	1293	879	592	315	10864
ANNETTE	242	208	327	567	738	899	949	837	843	648	490	321	7069
BARROW	803	840	1035	1500	1971	2362	2517	2332	2468	1944	1445	957	20174
BARTER IS.	735	775	987	1482	1944	2337	2536	2369	2477	1923	1373	924	19862
BETHEL	319	394	612	1042	1434	1866	1903	1590	1655	1173	806	402	13196
COLD BAY	474	425	525	772	918	1122	1153	1036	1122	951	791	591	9880
CORDOVA	366	391	522	781	1017	1221	1299	1086	1113	864	660	444	9764
FAIRBANKS	171	332	642	1203	1833	2254	2359	1901	1739	1068	555	222	14279
JUNEAU	301	338	483	725	921	1135	1237	1070	1073	810	601	381	9075
KING SALMON	313	322	513	908	1290	1606	1600	1333	1411	966	673	408	11343
KOTZEBUE	381	446	723	1249	1728	2127	2192	1932	2080	1554	1057	636	16105
MCGRATH	208	338	633	1184	1791	2232	2294	1817	1758	1122	648	258	14283
NOME	481	496	693	1094	1455	1820	1879	1666	1770	1314	930	573	14171
SAINT PAUL	605	539	612	862	963	1197	1228	1168	1265	1098	936	726	11199
SHEMYA	577	475	501	784	876	1042	1045	958	1011	885	837	696	9687
YAKUTAT	338	347	474	716	936	1144	1169	1019	1042	840	632	435	9092
ARIZ. FLAGSTAFF	46	68	201	558	867	1073	1169	991	911	651	437	180	7152
PHOENIX	0	0	0	22	234	415	474	328	217	75	0	0	1765
PRESCOTT	0	0	27	245	579	797	865	711	605	360	158	15	4362
TUCSON	0	0	0	25	231	406	471	344	242	75	6	0	1800
WINSLOW	0	0	6	245	711	1008	1054	770	601	291	96	0	4782
YUMA	0	0	0	148	319	363	228	130	29	0	0	1217
ARK. FORT SMITH	0	0	12	127	450	704	781	596	456	144	22	0	3292
LITTLE ROCK	0	0	9	127	465	716	756	577	434	126	9	0	3219
TEXARKANA	0	0	0	78	345	561	626	468	350	105	0	0	2533
CALIF. BAKERSFIELD	0	0	0	37	282	502	546	364	267	105	19	0	2122
BISHOP	0	0	42	248	576	797	874	666	539	306	143	36	4227
BLUE CANYON	34	50	120	347	576	766	865	781	791	582	397	195	5507
BURBANK	0	0	6	43	177	301	366	277	239	138	81	18	1646
EUREKA	270	257	258	329	414	499	546	470	505	438	372	285	4643
FRESNO	0	0	0	78	339	558	586	406	319	150	56	0	2492
LONG BEACH	0	0	12	40	156	288	375	297	267	168	90	18	1711
LOS ANGELES	28	22	42	78	180	291	372	302	288	219	158	81	2061
MT. SHASTA	25	34	123	406	696	902	983	784	738	525	347	159	5722
OAKLAND	53	50	45	127	309	481	527	400	353	255	180	90	2870
POINT ARGUELLO	202	186	162	205	291	400	474	392	403	339	298	243	3595
RED BLUFF	0	0	0	53	318	555	605	428	341	168	47	0	2515
SACRAMENTO	0	0	12	81	363	577	614	442	360	216	102	6	2773
SANDBERG	0	0	30	202	480	691	778	661	620	426	264	57	4209
SAN DIEGO	6	0	15	37	123	251	313	249	202	123	84	36	1439
SAN FRANCISCO	81	78	60	143	306	462	508	395	363	279	214	126	3015
SANTA CATALINA	16	0	9	50	165	279	353	308	326	249	192	105	2052
SANTA MARIA	99	93	96	146	270	391	459	370	363	282	233	165	2967
COLO. ALAMOSA	65	99	279	639	1065	1420	1476	1162	1020	696	440	168	8529
COLORADO SPRINGS	9	25	132	456	825	1032	1128	938	893	582	319	84	6423
DENVER	6	9	117	428	819	1035	1132	938	887	558	288	66	6283
GRAND JUNCTION	0	0	30	313	786	1113	1209	907	729	387	146	21	5641
PUEBLO	0	0	54	326	750	986	1085	871	772	429	174	15	5462
CONN. BRIDGEPORT	0	0	66	307	615	986	1079	966	853	510	208	27	5617
HARDFORT	0	6	99	372	711	1119	1209	1061	899	495	177	24	6172
NEW HAVEN	0	12	87	347	648	1011	1097	991	871	543	245	45	5897
DEL. WILMINGTON	0	0	51	270	588	927	980	874	735	387	112	6	4930
FLA. APALACHICOLA	0	0	0	16	153	319	347	260	180	33	0	0	1308
DAYTONA BEACH	0	0	0	0	75	211	248	190	140	15	0	0	879
FORT MYERS	0	0	0	0	24	109	146	101	62	0	0	0	442
JACKSONVILLE	0	0	0	12	144	310	332	246	174	21	0	0	1239
KEY WEST	0	0	0	0	0	28	40	31	9	0	0	0	108
LAKELAND	0	0	0	0	57	164	195	146	99	0	0	0	661
MIAMI BEACH	0	0	0	0	0	40	56	36	9	0	0	0	141
ORLANDO	0	0	0	0	72	198	220	165	105	6	0	0	766
PENSACOLA	0	0	0	19	195	353	400	277	183	36	0	0	1463
TALLAHASSEE	0	0	0	28	198	360	375	286	202	36	0	0	1485
TAMPA	0	0	0	0	60	171	202	148	102	0	0	0	683
WEST PALM BEACH	0	0	0	0	6	65	87	64	31	0	0	0	253

Table 7-2. Normal total heating degree-days (base 65°F). (continued)

STATE AND STATION	JULY	AUG.	SEP.	OCT.	NOV.	DEC.	JAN.	FEB.	MAR.	APR.	MAY	JUNE	ANNUAL
GA. ATHENS	0	0	12	115	405	632	642	529	431	141	22	0	2929
ATLANTA	0	0	18	127	414	626	639	529	437	168	25	0	2983
AUGUSTA	0	0	0	78	333	552	549	445	350	90	0	0	2397
COLUMBUS	0	0	0	87	333	543	552	434	338	96	0	0	2383
MACON	0	0	0	71	297	502	505	403	295	63	0	0	2136
ROME	0	0	24	161	474	701	710	577	468	177	34	0	3326
SAVANNAH	0	0	0	47	246	437	437	353	254	45	0	0	1819
THOMASVILLE	0	0	0	25	198	366	394	305	208	33	0	0	1529
IDAHO BOISE	0	0	132	415	792	1017	1113	854	722	438	245	81	5809
IDAHO FALLS 46W	16	34	270	623	1056	1370	1538	1249	1085	651	391	192	8475
IDAHO FALLS 42NW	16	40	282	648	1107	1432	1600	1291	1107	657	388	192	8760
LEWISTON	0	0	123	403	756	933	1063	815	694	426	239	90	5542
POCATELLO	0	0	172	493	900	1166	1324	1058	905	555	319	141	7033
ILL. CAIRO	0	0	36	164	513	791	856	680	539	195	47	0	3821
CHICAGO	0	0	81	326	753	1113	1209	1044	890	480	211	48	6155
MOLINE	0	9	99	335	774	1181	1314	1100	918	450	189	39	6408
PEORIA	0	6	87	326	759	1113	1218	1025	849	426	183	33	6025
ROCKFORD	6	9	114	400	837	1221	1333	1137	961	516	236	60	6830
SPRINGFIELD	0	0	72	291	696	1023	1135	935	769	354	136	18	5429
IND. EVANSVILLE	0	0	66	220	606	896	955	767	620	237	68	0	4435
FORT WAYNE	0	9	105	378	783	1135	1178	1028	890	471	189	39	6205
INDIANAPOLIS	0	0	90	316	723	1051	1113	949	809	432	177	39	5699
SOUTH BEND	0	6	111	372	777	1125	1221	1070	933	525	239	60	6439
IOWA Burlington	0	0	93	322	768	1135	1259	1042	859	426	177	33	6114
DES MOINES	0	9	99	363	837	1231	1398	1163	967	489	211	39	6808
DUBUQUE	12	31	156	450	906	1287	1420	1204	1026	546	260	78	7376
SIOUX CITY	0	9	108	369	867	1240	1435	1198	989	483	214	39	6951
WATERLOO	12	19	138	428	1296	1296	1460	1221	1023	531	229	54	7320
KANS. CONCORDIA	0	0	57	276	705	1023	1163	935	781	372	149	18	5479
DODGE CITY	0	0	33	251	666	939	1051	840	719	354	124	9	4986
GOODLAND	0	6	81	381	810	1073	1166	955	884	507	236	42	6141
TOPEKA	0	0	57	270	672	980	1122	893	722	330	124	12	5182
WICHITA	0	0	33	229	618	905	1023	804	645	270	87	6	4620
KY. COVINGTON	0	0	75	291	669	983	1035	893	756	390	149	24	5265
LEXINGTON	0	0	54	239	609	902	946	818	685	325	105	0	4683
LOUISVILLE	0	0	54	248	609	890	930	818	682	315	105	9	4660
LA. ALEXANDRIA	0	0	0	56	273	431	471	361	260	69	0	0	1921
BATON ROUGE	0	0	0	31	216	369	409	294	208	33	0	0	1560
BURRWOOD	0	0	0	0	96	214	298	218	171	27	0	0	1024
LAKE CHARLES	0	0	0	19	210	341	381	274	195	39	0	0	1459
NEW ORLEANS	0	0	0	19	192	322	363	258	192	39	0	0	1385
SHREVEPORT	0	0	0	47	297	477	552	426	304	81	0	0	2184
MAINE CARIBOU	78	115	336	682	1044	1535	1690	1470	1308	858	468	183	9767
PORTLAND	12	53	195	508	807	1215	1339	1182	1042	675	372	111	7511
MD. BALTIMORE	0	0	48	264	585	905	936	820	679	327	90	0	4654
FREDERICK	0	0	66	307	624	955	995	876	741	384	127	12	5087
MASS. BLUE HILL OBSY	0	22	108	381	690	1085	1178	1053	936	579	267	69	6368
BOSTON	0	9	60	316	603	983	1088	972	846	513	208	36	5634
NANTUCKET	12	22	93	332	573	896	992	941	896	621	384	129	5891
PITTSFIELD	25	59	219	524	831	1231	1339	1196	1063	660	326	105	7578
WORCESTER	6	34	147	450	774	1172	1271	1123	998	612	304	78	6969
MICH. ALPENA	68	105	273	580	912	1268	1404	1299	1218	777	446	156	8506
DETROIT (CITY)	0	0	87	360	738	1088	1181	1058	936	522	220	42	6232
ESCANABA	59	87	243	539	924	1293	1445	1296	1203	777	456	159	8481
FLINT	16	40	159	465	843	1212	1330	1198	1066	639	319	90	7377
GRAND RAPIDS	9	28	135	434	804	1147	1259	1134	1011	579	279	75	6894
LANSING	6	22	138	431	813	1163	1262	1142	1011	579	273	69	6909
MARQUETTE	59	81	240	527	936	1268	1411	1268	1187	771	468	177	8393
MUSKEGON	12	28	120	400	762	1088	1209	1100	995	594	310	78	6696
SAULT STE. MARIE	96	105	279	580	951	1367	1525	1380	1277	810	477	201	9048
MINN. DULUTH	71	109	330	632	1131	1581	1745	1518	1355	840	490	198	10000
INT FALLS	71	112	363	701	1236	1724	1919	1621	1414	828	443	174	10606
MINNEAPOLIS	22	31	189	505	1014	1454	1631	1380	1166	621	288	81	8382
ROCHESTER	25	34	186	474	1005	1438	1593	1366	1150	630	301	93	8295
SAINT CLOUD	28	47	225	549	1065	1500	1702	1445	1221	666	326	105	8879

Table 7-2. Normal total heating degree-days (base 65°F). (continued)

STATE AND STATION	JULY	AUG.	SEP.	OCT.	NOV.	DEC.	JAN.	FEB.	MAR.	APR.	MAY	JUNE	ANNUAL
MISS. JACKSON	0	0	0	65	315	502	546	414	310	87	0	0	2239
MERIDIAN	0	0	0	81	339	518	543	417	310	81	0	0	2289
VICKSBURG	0	0	0	53	279	462	512	384	282	69	0	0	2041
MO. COLUMBIA	0	0	54	251	651	967	1076	874	716	324	121	12	5046
KANSAS	0	0	39	220	612	905	1032	818	682	294	109	0	4711
ST. JOSEPH	0	6	60	285	708	1039	1172	949	769	348	133	15	5484
ST. LOUIS	0	0	60	251	627	936	1026	848	704	312	121	15	4900
SPRINGFIELD	0	0	45	223	600	877	973	781	660	291	105	6	4561
MONT. BILLINGS	6	15	186	487	897	1135	1296	1100	970	570	285	102	7049
GLASGOW	31	47	270	608	1104	1466	1711	1439	1187	648	335	150	8996
GREAT FALLS	28	53	258	543	921	1169	1349	1154	1063	642	384	186	7750
HAVRE	28	53	306	595	1065	1367	1584	1364	1181	657	338	162	8700
HELENA	31	59	294	601	1002	1265	1438	1170	1042	651	381	195	8129
KALISPELL	50	99	321	654	1020	1240	1401	1134	1029	639	397	207	8191
MILES CITY	6	6	174	502	972	1296	1504	1252	1057	579	276	99	7723
MISSOULA	34	74	303	651	1035	1287	1420	1120	970	621	391	219	8125
NEBR. GRAND ISLAND	0	6	108	381	834	1172	1314	1089	908	462	211	45	6530
LINCOLN	0	6	75	301	726	1066	1237	1016	834	402	171	30	5864
NORFOLK	9	0	111	397	873	1234	1414	1179	983	498	233	48	6979
NORTH PLATTE	0	6	123	440	885	1166	1271	1039	930	519	248	57	6684
OMAHA	0	12	105	357	828	1175	1355	1126	939	465	208	42	6612
SCOTTSBLUFF	0	0	138	459	876	1128	1231	1008	921	552	285	75	6673
VALENTINE	9	12	165	493	942	1237	1395	1176	1045	579	288	84	7425
NEV. ELKO	9	34	225	561	924	1197	1314	1036	911	621	409	192	7433
ELY	28	43	234	592	939	1184	1308	1075	977	672	456	225	7733
LAS VEGAS	0	0	0	78	387	617	688	487	335	111	6	0	2709
RENO	43	87	204	490	801	1026	1073	823	729	510	357	189	6332
WINNEMUCCA	0	34	210	536	876	1091	1172	916	837	573	363	153	6761
N. H. CONCORD	6	50	177	505	822	1240	1358	1184	1032	636	298	75	7383
MT. WASH. OBSY.	493	536	720	1057	1341	1742	1820	1663	1652	1260	930	603	13817
N. J. ATLANTIC CITY	0	0	39	251	549	880	936	848	741	420	133	15	4812
NEWARK	0	0	30	248	573	921	983	876	729	381	118	0	4859
TRENTON	0	0	57	264	576	924	989	885	753	399	121	12	4980
N. MEX. ALBUQUERQUE	0	0	12	229	642	868	930	703	595	288	81	0	4348
CLAYTON	0	6	66	310	699	899	986	812	747	429	183	21	5158
RATON	9	28	126	431	825	1048	1116	904	834	543	301	63	6228
ROSWELL	0	0	18	202	573	806	840	641	481	201	31	0	3793
SILVER CITY	0	0	6	183	525	729	791	605	518	261	87	0	3705
N. Y. ALBANY	0	19	138	440	777	1194	1311	1156	992	564	239	45	6875
BINGHAMTON (AP)	22	65	201	471	810	1184	1277	1154	1045	645	313	99	7286
BINGHAMTON (PO)	0	28	141	406	732	1107	1190	1081	949	543	229	45	6451
BUFFALO	19	37	141	440	777	1156	1256	1145	1039	645	329	78	7062
CENTRAL PARK	0	0	30	233	540	902	986	885	760	408	118	9	4871
J. F. KENNEDY INTL	0	0	36	248	564	933	1029	935	815	480	167	12	5219
LAGUARDIA	0	0	27	223	528	887	973	879	750	414	124	6	4811
ROCHESTER	9	31	126	415	747	1125	1234	1123	1014	597	279	48	6748
SCHENECTADY	0	22	123	422	756	1159	1283	1131	970	543	211	30	6650
SYRACUSE	6	28	132	415	744	1153	1271	1140	1004	570	248	45	6756
N.C. ASHEVILLE	0	0	48	245	555	775	784	683	592	273	87	0	4042
CAPE HATTERAS	0	0	0	78	273	521	580	518	440	177	25	0	2612
CHARLOTTE	0	0	6	124	438	691	691	582	481	156	22	0	3191
GREENSBORO	0	0	33	192	513	778	784	672	552	234	47	0	3805
RALEIGH	0	0	21	164	450	716	725	616	487	180	34	0	3393
WILMINGTON	0	0	0	74	291	521	546	462	357	96	0	0	2347
WINSTON SALEM	0	0	21	171	483	747	753	652	524	207	37	0	3595
N. DAK. BISMARCK	34	28	222	577	1083	1463	1708	1442	1203	645	329	117	8851
DEVILS LAKE	40	53	273	642	1191	1634	1872	1579	1345	753	381	138	9901
FARGO	28	37	219	574	1107	1569	1789	1520	1262	690	332	99	9226
WILLISTON	31	43	261	601	1122	1513	1758	1473	1262	681	357	141	9243
OHIO AKRON	0	9	96	381	726	1070	1138	1016	871	489	202	39	6037
CINCINNATI	0	0	54	248	612	921	970	837	701	336	118	9	4806
CLEVELAND	9	25	105	384	738	1088	1159	1047	918	552	260	66	6351
COLUMBUS	0	6	84	347	714	1039	1088	949	809	426	171	27	5660
DAYTON	0	6	78	310	696	1045	1097	955	809	429	167	30	5622
MANSFIELD	9	22	114	397	768	1110	1169	1042	924	543	245	60	6403
SANDUSKY	0	6	66	313	684	1032	1107	991	868	495	198	36	5796
TOLEDO	0	16	117	406	792	1138	1200	1056	924	543	242	60	6494
YOUNGSTOWN	6	19	120	412	771	1104	1169	1047	921	540	248	60	6417

Table 7-2. Normal total heating degree-days (base 65°F). (continued)

STATE AND STATION	JULY	AUG.	SEP.	OCT.	NOV.	DEC.	JAN.	FEB.	MAR.	APR.	MAY	JUNE	ANNUAL
OKLA. OKLAHOMA CITY	0	0	15	164	498	766	868	664	527	189	34	0	3725
TULSA	0	0	18	158	522	787	893	683	539	213	47	0	3860
OREG. ASTORIA	146	130	210	375	561	679	753	622	636	480	363	231	5186
BURNS	12	37	210	515	867	1113	1246	988	856	570	366	177	6957
EUGENE	34	34	129	366	585	719	803	627	589	426	279	135	4726
MEACHAM	84	124	288	580	918	1091	1209	1005	983	726	527	339	7874
MEDFORD	0	0	78	372	678	871	918	697	642	432	242	78	5008
PENDLETON	0	0	111	350	711	884	1017	773	617	396	205	63	5127
PORTLAND	25	28	114	335	597	735	825	644	586	396	245	105	4635
ROSEBURG	22	16	105	329	567	713	766	608	570	405	267	123	4491
SALEM	37	31	111	338	594	729	822	647	611	417	273	144	4754
SEXTON SUMMIT	81	81	171	443	666	874	958	809	818	609	465	279	6254
PA. ALLENTOWN	0	0	90	353	693	1045	1116	1002	849	471	167	24	5810
ERIE	0	25	102	391	714	1063	1169	1081	973	585	288	60	6451
HARRISBURG	0	0	63	298	648	992	1045	907	766	396	124	12	5251
PHILADELPHIA	0	0	60	291	621	964	1014	890	744	390	115	12	5101
PITTSBURGH	0	9	105	375	726	1063	1119	1002	874	480	195	39	5987
READING	0	0	54	257	597	939	1001	885	735	372	105	0	4945
SCRANTON	0	19	132	434	762	1104	1156	1028	893	498	195	33	6254
WILLIAMSPORT	0	9	111	375	717	1073	1122	1002	856	468	177	24	5934
R. I. BLOCK IS.	0	16	78	307	594	902	1020	955	877	612	344	99	5804
PROVIDENCE	0	16	96	372	660	1023	1110	988	868	534	236	51	5954
S. C. CHARLESTON	0	0	0	59	282	471	487	389	291	54	0	0	2033
COLUMBIA	0	0	0	84	345	577	570	470	357	81	0	0	2484
FLORENCE	0	0	0	78	315	552	552	459	347	84	0	0	2387
GREENVILLE	0	0	0	112	387	636	648	535	434	120	12	0	2884
SPARTANBURG	0	0	15	130	417	667	663	560	453	144	25	0	3074
S. DAK. HURON	9	12	165	508	1014	1432	1628	1355	1125	600	288	87	8223
RAPID CITY	22	12	165	481	897	1172	1333	1145	1051	615	326	126	7345
SIOUX FALLS	19	25	168	462	972	1361	1544	1285	1082	573	270	78	7839
TENN. BRISTOL	0	0	51	236	573	828	828	700	598	261	68	0	4143
CHATTANOOGA	0	0	18	143	468	698	722	577	453	150	25	0	3254
KNOXVILLE	0	0	30	171	489	725	732	613	493	198	43	0	3494
MEMPHIS	0	0	18	130	447	698	729	585	456	147	22	0	3232
NASHVILLE	0	0	30	158	495	732	778	644	512	189	40	0	3578
OAK RIDGE (CO)	0	0	39	192	531	772	778	669	552	228	56	0	3817
TEX. ABILENE	0	0	0	99	366	586	642	470	347	114	0	0	2624
AMARILLO	0	0	18	205	570	797	877	664	546	252	56	0	3985
AUSTIN	0	0	0	31	225	388	468	325	223	51	0	0	1711
BROWNSVILLE	0	0	0	0	66	149	205	106	74	0	0	0	600
CORPUS CHRISTI	0	0	0	0	120	220	291	174	109	0	0	0	914
DALLAS	0	0	0	62	321	524	601	440	319	90	6	0	2363
EL PASO	0	0	0	84	414	648	685	445	319	105	0	0	2700
FORT WORTH	0	0	0	65	324	536	614	448	319	99	0	0	2405
GALVESTON	0	0	0	0	138	270	350	258	189	30	0	0	1235
HOUSTON	0	0	0	6	183	307	384	288	192	36	0	0	1396
LAREDO	0	0	0	0	105	217	267	134	74	0	0	0	797
LUBBOCK	0	0	18	174	513	744	800	613	484	201	31	0	3578
MIDLAND	0	0	0	87	381	592	651	468	322	90	0	0	2591
PORT ARTHUR	0	0	0	22	207	329	384	274	192	39	0	0	1447
SAN ANGELO	0	0	0	68	318	536	567	412	288	66	0	0	2255
SAN ANTONIO	0	0	0	31	207	363	428	286	195	39	0	0	1549
VICTORIA	0	0	0	6	150	270	344	230	152	21	0	0	1173
WACO	0	0	0	43	270	456	536	389	270	66	0	0	2030
WICHITA FALLS	0	0	0	99	381	632	698	518	378	120	6	0	2832
UTAH MILFORD	0	0	99	443	867	1141	1252	988	822	519	279	87	6497
SALT LAKE CITY	0	0	81	419	849	1082	1172	910	763	459	233	84	6052
WENDOVER	0	0	48	372	822	1091	1178	902	729	408	177	51	5778
VT. BURLINGTON	28	65	207	539	891	1349	1513	1333	1187	714	353	90	8269
VA. CAPE HENRY	0	0	0	112	360	645	694	633	536	246	53	0	3279
LYNCHBURG	0	0	51	223	540	822	849	731	605	267	78	0	4166
NORFOLK	0	0	0	136	408	698	738	655	533	216	37	0	3421
RICHMOND	0	0	36	214	495	784	815	703	546	219	53	0	3865
ROANOKE	0	0	51	229	549	825	834	722	614	261	65	0	4150
WASH. NAT'L. AP.	0	0	33	217	519	834	871	762	626	288	74	0	4224

One of the most practical of weather statistics is the "heating degree day." First devised some 50 years ago, the degree day system has been in quite general use by the heating industry for more than 30 years.

Heating degree days are the number of degrees the daily average tempertaure is below 65°. Normally heating is not required in a building when the outdoor average daily temperature is 65°. Heating degree days are determined by substracting the average daily temperatures below 65° from the base 65°. A day with an average temperature of 50° has 15 heating degree days (65 − 50 = 15) while one with an average temperature of 65° or higher has none.

Several characteristics make the degree day figures especially useful. They are cumulative so that the degree day sum for a period of days represents the total heating load for that period. The relationship between degree days and fuel consumption is linear, i.e., doubling the degree days usually doubles the fuel consumption. Comparing normal seasonal degree days in different locations

Table 7-2. Normal total heating degree-days (base 65°F). (continued)

STATE AND STATION	JULY	AUG.	SEP.	OCT.	NOV.	DEC.	JAN.	FEB.	MAR.	APR.	MAY	JUNE	ANNUAL
WASH. OLYMPIA	68	71	198	422	636	753	834	675	645	450	307	177	5236
SEATTLE	50	47	129	329	543	657	738	599	577	396	242	117	4424
SEATTLE BOEING	34	40	147	384	624	763	831	655	608	411	242	99	4838
SEATTLE TACOMA	56	62	162	391	633	750	828	678	657	474	295	159	5145
SPOKANE	9	25	168	493	879	1082	1231	980	834	531	288	135	6655
STAMPEDE PASS	273	291	393	701	1008	1178	1287	1075	1085	855	654	483	9283
TATOOSH IS.	295	279	306	406	534	639	713	613	645	525	431	333	5719
WALLA WALLA	0	0	87	310	681	843	986	745	589	342	177	45	4805
YAKIMA	0	12	144	450	828	1039	1163	868	713	435	220	69	5941
W. VA. CHARLESTON	0	0	63	254	591	865	880	770	648	300	96	9	4476
ELKINS	9	25	135	400	729	992	1008	896	791	444	198	48	5675
HUNTINGTON	0	0	63	257	585	856	880	764	636	294	99	12	4446
PARKERSBURG	0	0	60	264	606	905	942	826	691	339	115	6	4754
WIS. GREEN BAY	28	50	174	484	924	1333	1494	1313	1141	654	335	99	8029
LA CROSSE	12	19	153	437	924	1339	1504	1277	1070	540	245	69	7589
MADISON	25	40	174	474	930	1330	1473	1274	1113	618	310	102	7863
MILWAUKEE	43	47	174	471	876	1252	1376	1193	1054	642	372	135	7635
WYO. CASPER	6	16	192	524	942	1169	1290	1084	1020	657	381	129	7410
CHEYENNE	19	31	210	543	924	1101	1228	1056	1011	672	381	102	7278
LANDER	6	19	204	555	1020	1299	1417	1145	1017	654	381	153	7870
SHERIAN	25	31	219	539	948	1200	1355	1154	1054	642	366	150	7683

gives a rough estimate of seasonal fuel consumption. For example, it would require roughly 4½ times as much fuel to heat a building in Chicago, Ill., where the mean annual total heating degree days are about 6,200 than to heat a similar building in New Orleans, La., where the annual total heating degree days are around 1,400. Using degree days has the advantage that the consumption ratios are fairly constant, i.e., the fuel consumed per 100 degree days is about the same whether the 100 degree days occur in only 3 or 4 days or are spread over 7 or 8 days.

The rapid adoption of the degree day system paralleled the spread of automatic fuel systems in the 1930's. Since oil and gas are more costly to store than solid fuels, this places a premium on the scheduling of deliveries and the precise evaluation of use rates and peak demands.

For all solar systems—hot water, swimming pool, and space heating and cooling—you will need the following weather data to thoroughly calculate anticipated Btu return:

Monthly solar insolation data
Monthly average ambient temperature
Monthly average wind velocities
Heating degree-day data and the base for calculating degree-days
Cooling degree-day data and the base used in calculating degree-days
Unusual weather conditions, overcast, rain, hail, dust conditions

Heating degree-days is a unit used to determine the heating requirements of a house. It is determined by subtracting the average ambient temperature over a 24-hour period from a base of 65°F. In other words, if the average temperature for a given day is 45°F, then the degree-days for that day would be 20°F. For degree-days in a month or in a season, simply add up all the degree-days. The greater the number of degree-days, the greater your heating needs.

Additional data are needed in order to properly evaluate the anticipated benefits of the solar system. The data are also needed to determine the design specifications for the system.

For *hot-water systems,* the following information is needed:

Type and capacity of water heater now used?
Number of hot-water outlets?
Which appliances use hot water?
Number of persons in the household?
Fuel consumption of water heater?
Desired water temperature?

At what time of day is the most hot water used?
Temperature of cold-water supply?

Swimming-pool systems require:

Size of pool (width, length, average depth)?
Present pool heater?
Type of fuel used in present system?
Is pool covered when not used?
Length of swimming season without heat?
Desired length of use?
Temperature and source of water supply?
Is there a wind barrier?

For *space heating and cooling,* the following information is needed:

Orientation of the house?
Floor area of the house? Number of stories?
Type of construction used? Type and extent of insulation?
Mean U value or heat loss of the home, including infiltration and ventilation?
Previous bills and rates?
Desired temperature levels?

Beyond the weather and specific-use data, topographical information, latitude, altitude, and description of the specific site—hills, trees, valleys, etc.—will also be needed. If you buy a solar system without taking all of these factors into account, then, frankly, you are gambling.

Actually determining the amount of Btu delivered is a complicated process. So, if a manufacturer claims, for example, that its solar system will handle up to 90 percent of your heating needs, be wary. Such a claim may only apply to homes that have extravagant insulation. A more modest and more accurate claim is, "On a good sunny day, this solar system will handle from 40 to 60 percent of your heating needs." Knowing the difference among warranties can make a big difference in your pocketbook. Compare the following:

Full Warranty	Limited Warranty
This product is guaranteed against all defects in construction and against corrosion for a period of 5 years. Manufacturer will pay for all labor and parts costs to correct problems.	This product is guaranteed to be one of the finest solar systems ever manufactured. Manufacturer will pay for costs of parts to correct any problems.

PROTECTING YOUR POCKETBOOK

If there is one characteristic consumers have in common, it is the desire to get their money's worth on a commercial transaction. In the solar field, there are three main obstacles to satisfying this all-important objective:

1. The consumer's own lack of knowledge and inexperience in this field.
2. Manufacturers who build shoddy products or who are overenthusiastic about their products.
3. Deliberate fraud and misrepresentation.

The best weapon against these obstacles is for consumers to recognize their own limitations and to rely upon competent engineering counsel.

In addition to the guidelines already presented, there are some other steps you can take to insure that you get your money's worth in a solar system:

- *Ask for proof that the product will perform as advertised.* Such proof could come from an independent laboratory or a university. You should see the actual report, rather than accept what the manufacturer says is in the report. Have your engineering consultant go over the report.
- *Examine the warranty carefully.* Remember that according to the law, the manufacturer must state whether a warranty is full or limited. If it is limited, know what the limitations are. How long does the warranty last? Are parts, service, and labor covered? Who will provide the service? Does the equipment have to be sent back to the manufacturer for repairs? Make sure you understand the terms of the warranty before you buy. Ask the seller what financial arrangements, such as an escrow account, have been made to honor the warranties. Be sure your engineering counsel not only looks over the warranty, but the design itself to determine whether there are any important omissions.
- *Solar components are like stereo components—some work well together, others don't.* If the system you are purchasing is not sold as a single package by one manufacturer, then you should obtain assurance that the seller has had the professional experience of choosing properly.

- *Ask the person who owns one.* Ask the seller for a list of previous purchasers and their addresses. Contact the people, and inquire about their experiences with the system.
- *Beware of sellers who use post office box numbers.* Even though many legitimate businesses use such outlets as a convenient way to receive bills and orders, a common tactic of the fly-by-night artist is to use a post office box number, operate a territory until the law starts closing in, and then move on to take a new name in a new territory. Find out from the seller where the place of business is, how long it has been there, and ask for financial references.
- *Be sure you know specifically who will service the solar system if something goes wrong.* Don't settle for the response that any plumber or handyman will do.
- *Don't try a do-it-yourself kit, unless you are truly handy.* One or two mistakes could make a system inoperable, and you will have no one to blame but yourself.
- *Remember that the amount of Btu delivered for the final end use of the system is most important.* This amount can fluctuate widely. A good winter with much sunshine can produce performance levels beyond the manufacturer's projections. Conversely, an unusually bad winter with heavy cloud cover could cause the projections to drop dramatically. The seller will be working from historical averages. A good guide to performance is whether the season is typical or atypical. If it is typical, and your energy-use patterns haven't changed, then the savings projections may have been inaccurate.
- *Don't change your use habits simply because you are getting plenty of free energy.* Conservation of energy still counts if you want to bring your monthly bills down. Don't blame the seller of a solar heating system if you keep your doors open during the middle of winter.
- *Don't forget your local consumer protection office or the Better Business Bureau.* Both may be able to help you determine whether a seller is reputable. Find out if there is a local citizens' solar organization. If so, it can probably give you plenty of good advice.
- *If a seller makes verbal claims not reflected in the literature handed out, ask the seller to write those claims down and to sign it.* Save that statement.
- *If you have a legitimate complaint, notify the local district attorney's office immediately, the Better Business Bureau, and the local consumer protection agency.* Be as specific in your complaint as possible, and document it as well as you can.

QUESTIONS YOU ALWAYS WANTED TO ASK

Q: Suppose I sell the house in a year or two. Can I count on appreciation?

A: It depends on two basic factors: if it is aesthetically appealing, and if it saves sufficient energy purchases. In cases of retrofit, you should be able to prove whether the solar energy system works simply by saving your energy bills and comparing conventional fuel usage with past bill statements. In new homes, comparisons of operating costs for solar versus conventional homes can be helpful, but not conclusive because of large differences in heat use by different families even in identical homes. Year-to-year differences may also be large, so a call to your energy supplier (gas or electric company, or oil supplier) can help you establish what the relative energy use should have been.

As to aesthetics, if it looks good to you and your neighbors, the odds for appreciation are in your favor. But remember, it's the buyer's eye that counts the most.

Q: If I plan on a heating system, should I allot extra space for storage and collector so that I can later adapt to a cooling unit?

A: Ask your engineering consultant what the extra cost will amount to for your particular design. If you are building a new house and plan to live in it for some time, it will probably be easier to plan now for a cooling system, rather than retrofit later.

Q: Should I buy now or later when solar technology is sure to be improved?

A: There is no doubt that later solar systems will have improvements over present models, and there will not be as much risk for the buyer as there is today. However, costs for tomorrow's solar system may go higher, and you will lose all the money you could have saved by not using expensive conventional fuels. There's something to be said too about being a pioneer.

Q: Will I need a humidifier with my solar system?

A: Fundamentally, a solar system replaces "conventional" energy with solar energy. It does not necessarily have any effect upon the need for a humidifier. Many solar systems use a hot-water tank for energy storage, and the tank may be located within the structure. Generally, if this is the case, the tank should be covered and sealed so that water vapor (humidity) does not escape from the tank into the house in the summer when it is not wanted. An uncovered tank would help humidify the air in the house in the winter but the excessive humidity in the summer is generally too great.

Q: How can I contact a reliable solar engineer?

A: Call a local university and ask one of the engineering professors to recommend an adviser who is knowledgeable in solar energy. Or, contact one of the local engineering societies (such as the American Institute of Architects, the Society of Mechanical Engineers, or the American Society of Heating, Refrigeration, and Air Conditioning Engineers) and ask for a list of engineers who are knowledgeable about solar energy. When you contact those on the list, ask for references regarding their previous work in the field.

RETURN ON INVESTMENT

What "return on investment" may the consumer expect from the purchase of solar equipment in 5 years? In 10 years? To answer this, we must consider a number of factors:

Future inflation. Most economists expect that the inflation rate will continue to rise during the next decade. As a result, the value of property acquired now will appreciate substantially.

Reduction in the cost of solar equipment. Production of solar equipment is still on a modest scale. Many manufacturers expect that as volume increases and as further product development takes place, the cost of the equipment without considering inflation may decrease.

These two factors offset one another, and, to be conservative, we might assume that the price of a solar installation will not increase in the future; therefore, the value of an installation made today will not "appreciate" because of inflation. (If the producers were not able to "hold the line" on cost, equipment bought today would appreciate in value, and the purchaser's return on the investment would increase correspondingly.)

Depreciation. Good solar equipment should have a 25-year life. Thus, at the end of 5 years, 20 percent of the useful life of the equipment should be used up, and 40 percent should be used up at the end of 10 years.

Effect on resale value of residence. Since "average" homeowners move frequently, and so must sell their home, it is important to consider the effect of previously installed solar equipment on the resale value of a house. Most economists agree that the cost of "conventional" energy will rise more rapidly than the general rate of inflation. That is, the rate at which solar equipment saves money for the user will become greater as time goes on and so the "pay out" to the user 5 or 10 years from now will be materially better than it is today. Thus, a residence with a solar system already installed should be attractive to the next buyer. The next buyer would compare the price of a solar-equipped house with the total potential cost of an unequipped house (including the cost of retrofitting it with solar equipment). Of course, it will always cost more to retrofit a solar system to an existing house than to incorporate a solar system in a new house. The extra cost of retrofitting tends to become an element of "appreciation" in the value of the previously equipped house. In the examples that follow, there are two sets of calculations. One is based upon ignoring this element of appreciation. The other is based upon the assumption that retrofitting an existing house would cost 20 percent more than the cost of an installation made at the time the house was built.

Case 1. A 75-square-foot solar water heater is installed in a house on the East Coast as an alternative to the use of electricity for water heating. The original cost is estimated at $1200. Without appreciation, its value after 5 years is $960; after 10 years, it is $720. With appreciation, its value is $1152 after 5 years, and $864 after 10 years. In 5 years, $643 would have been saved if the current cost of electricity was 3.5¢ per kilowatt-hour and if this increased at 7.5 percent per year. Without considering appreciation, the original value would have depreciated by $240. Subtracting this figure from the 5-year savings of $643, we find that the Return on Investment is $403, which is 33.6 percent of the original investment. If we consider appreciation, the reduction in value is only $48, and the return on investment is $595 in the first 5 years, which is 49.6 percent.

Savings in 10 years are substantially larger—$1561. Again, ignoring appreciation, the depreciation in 10 years is $480; substracting this from the total savings, the return on investment in 10 years is $1081 or 90 percent of the original cost. If we include appreciation, we only deduct $336 from the 10-year savings, leaving a total return on investment of $1225 to the user, which is a little over 100 percent of the first cost.

If the next buyer pays the appreciation value for the solar installation when buying the solar-equipped house, the pay-out time becomes quite rapid. If the buyer pays $1152 after 5 years, then, due to the continuously increasing cost of electricity, the cumulative savings will equal the $1152 price in only 4.7 years. The pay-out for the second owner becomes even better after 10 years. If the price paid is the appreciated value of $864, the savings will equal the sales price in only 3.8 years. Obviously, resale of the solar-equipped house should represent a real bargain to the second buyer.

Case 2. A 50-square-foot solar water heater installed in Los Angeles as an alternative to electricity, with the current price of electricity being 3¢ per kilowatt-hour:

Original Cost	$900
Value after 5 years	
without appreciation	$720
with appreciation	$778
Value after 10 years	
without appreciation	$540
with appreciation	$583

Savings		
in 5 years	$569	
in 10 years	$1382	
Savings less loss in value		
without appreciation		
after 5 years	$389	(43.2%)
in 10 years	$922	(102.4%)
Savings less loss in value		
with appreciation		
after 5 years	$447	(49.7%)
in 10 years	$1065	(118.3%)

Case 3. A 50-square-foot solar water heater installed in Dallas, Texas, as an alternative to natural gas currently available at 17.5¢ per therm.

Original Cost	$900
Value after 5 years	
without appreciation	$720
with appreciation	$778
Value after 10 years	
without appreciation	$540
with appreciation	$583
Savings	
in 5 years	$199
in 10 years	$519

Without considering appreciation, savings in 5 years are $19 more than depreciation, and return on investment is 2 percent. With appreciation, return is $82 or 9.1 percent. This is not a very good return. However, after 10 years, even without appreciation, return is $159 or 17.7 percent of the original investment; with appreciation, it is $202 (22.4 percent).

Similar calculations may be made for other alternatives. In general, they show that return to the original purchaser in 5 or 10 years is: good to excellent when electric heat or electric hot water is the alternative; reasonably good when oil is the alternative; and only fair where low-cost natural gas is still available and remains available in the future.

8 Possible Effects of Solar Energy

ABSTRACT

The use of fossil, nuclear, and, to a lesser extent, geothermal energy sources throughout the world could cause an increase in average global temperature by the year 2100. The size and effects of such a temperature increase have not yet been calculated with any degree of certainty. The use of solar instead of "stored" energy sources would avoid the problems of waste-heat release and carbon dioxide emissions and thus avoid the possibility of such a global temperature increase. However, the large areas needed for solar energy collectors could pose a problem in land use and could affect climate through changes in solar absorption and energy distribution patterns. More extensive measurements and calculations of global climate changes are needed in order to understand the causes of such changes, particularly from human activities, and to predict their impact on the earth and on human society.

CONCLUSIONS AND RECOMMENDATIONS

The earth underwent major changes in average temperature—as much as 20°C—during prehistoric times. The average global temperature increased about 0.5°C from 1870 to around 1940 and then decreased about 0.25°C to the present.[1] The reasons for these temperature fluctuations, both prehistoric and recent, are not understood. Some authorities attribute the recent fluctuations to natural cycles of solar flux and earth reflectivity, and they anticipate a further cooling of the earth.[1A] Others attribute the increase since 1870 to the increasing carbon dioxide concentration in the atmosphere[25] and the decrease since 1940 to increasing particulate and aerosol concentrations in the atmosphere.[26-28] Many of these authorities predict a renewed warming of the earth.

The continued use of fossil fuels could further increase the concentration of carbon dioxide in the atmosphere and, depending on the degree of control of particulate emissions, could result in renewed increase in temperature. Also, the use of "stored" energy resources such as fossil fuels, nuclear fission and fusion, and, to a somewhat lesser degree, geothermal energy, could result in an additional increase in global temperature due to the ultimate release of energy as thermal waste, both at the point of consumer use and at electric generating stations.[3]

The current use of "stored" energy throughout the world is approximately 2×10^{17} Btu per year or 0.01 of 1 percent of the solar energy absorbed by the earth's surface (2×10^{21} Btu per year). The use of "stored" energy at a rate approximately 1 percent of the solar energy absorbed by the earth's surface (2×10^{19} Btu per year) is likely to produce a 2° to 3°F rise in average global temperature and a 10°F or more rise at the poles.[30,31] A world population of about 20 billion people, each using an average of 1000 million Btu per year (roughly three times the current United States average per capita use rate), would release heat at such a rate. The impact of this energy-release rate on the global climate, the span of time taken for the climate to change, and the impact of any climate change on human welfare are difficult to predict. However, any climate change is likely to have extremely serious consequences, at least in the short run, and could force major readjustments in land-use and living patterns throughout the world.

Solar energy could conceivably supply the heat, fuel, and electricity needed by 20 billion or more people, each using more energy than is used in the United States today, without causing global climate changes from waste-heat release or carbon dioxide and particulate emissions. On the other hand, local, mesoscale, or global climate changes might occur due to changes in solar energy absorption by the large collector surfaces needed and to redistribution of the collected energy.

Realistically, the energy supplies for the future will come from a mix of sources: fossil, nuclear, geothermal, and solar. Also, the world population may level off well below 20 billion and per capita energy use may never reach or exceed the current United States rate. Even so, it is recommended that the initiation of significant solar energy research and technology programs be undertaken now since historically the introduction of a new energy source has required 50 years.

It should be noted that all of the preceding conclusions are based on very preliminary and simplified calculations. We are just beginning to be able to calculate global climate changes using the largest computers. The global climate must be carefully monitored to maintain a careful watch on the impacts resulting from energy use and other activities of human society. It is further recommended that strong support be given to computer simulations of global climate changes due to natural and artificial causes in order to begin to understand the complex forces that may produce extreme consequences for humanity.

BACKGROUND

The average temperature of the earth has varied naturally over the past million years, by over 20°C. During the greater part of the earth's history, it has been considerably warmer than it is now; 50 million years ago, it was 10°C warmer. In fact, polar ice caps have existed only about 10 percent of the time since the seas and continents were formed.[1]

Glacial periods, during which the average temperature of the earth was 10° cooler than now, have occurred during at least three periods in the distant past, each lasting several million years: one or more glacial ages about 600 million years ago; the "Permo-Carboniferous" glacial age about 300 million years ago; and the most recent ice age, which started about 5 million years ago. The Antarctic ice sheet, which contains a mass of ice equivalent to about 59 meters of sea level, attained essentially its present size and volume about 4 million years ago. Continental ice sheets first appeared in the Northern Hemisphere during this ice age about 3 million years ago. During the last million years, the Arctic Ocean ice cover was never less than it is today, and the earth's climate has been characterized by an alternation of glacial and interglacial periods marked in the Northern Hemisphere by the waxing and waning of continental ice sheets. The total sea-level change from maximum to zero glaciation would be about 210 meters. Melting of all of the currently existing glaciers (which cover about 3 percent of the earth's surface) would raise the sea level about 70 meters, of which the Greenland ice cap represents about 10 meters.[1A]

There appears to be a rough climate cycle of about 100,000 years. To find an ancient period similar to the present interglacial, it is necessary to look back about 125,000 years. The most recent glacial maximum occured 18,000 years ago and ended rather abruptly about 10,000 years ago. The glaciers retreated at about 1 kilometer per year, reaching their present state roughly about 7000 years ago. The retreat was not steady. On one occasion, the glacier surged forward, covering a large part of eastern Europe within 100 years. There have been significant temperature fluctuations of ±1°C or so since then, including the "little ice age" from 1430 to 1850 A.D.[1A] During the last hundred years, the average temperature increased about 0.5°C until around 1940 when it began to decrease. Since 1940, it has dropped about 0.25°C worldwide and over 1°C above the Arctic Circle.[1]

There is some speculation that this cooling trend since 1940 represents the end of a 10,000-year interglacial period and that the cooling will continue. There is a finite probability, based on a variety of sources of cyclic paleoclimatic data, that a serious worldwide cooling could befall the earth within the next hundred years. In the past, the end of an interglacial period has been marked by major changes in vegetation, deciduous trees giving way to sparse shrubs, and a large increase in dust. Agricultural productivity would have been marginal.[1A] The reasons for these fluctuations are unknown, although changes in the sun's energy output, continental drift, and volcanic activity have been suggested as major primary causes in preindustrial eras. Certainly, a rough equilibrium exists among the sun-land-ocean-atmosphere-glacier-system, which, to date, has prevented the earth from either becoming completely iced over or continuously ice-free.

Also, the impact of these temperature fluctuations on people are not known. Certainly the human species has survived through the climate extremes of the past million years. However, their existence during periods of glaciation, flood, or drought could hardly have been easy. Now, a further question may be asked: can people so affect the global climate by their own activities as to bring about similar severe conditions leading to widespread disaster or even the demise of human beings?

Human influences on the global heat balance have arisen to date from five basic sources:[2]

- Changing the reflectivity or absorptivity of the earth (changing its *albedo*). Many human activities such as cutting forests, plowing grasslands, building cities, etc., that affect the albedo result in a net heating of the earth. (Albedo is the amount of electromagnetic radiation reflected by the earth's surface.)
- Irrigation. Although the local effect of irrigation may cause a cooling, its ultimate effect results in a net heating since evaporation of the irrigation water absorbs sunlight.
- Particulate emissions. Fires and industrial activities emit soot and ash into the atmosphere (although wind-blown dust and volcanic eruptions also cause major, if not much more significant effects), which probably produce a net cooling by reflecting or scattering sunlight into space.
- Carbon dioxide emissions from combustion of fossil fuels. Atmospheric carbon dioxide absorbs radiation and produces a net heating of the earth.
- Direct release of heat. All use of stored energy, whether fossil, nuclear, or (to a lesser extent) geothermal, results in a net heating of the earth's surface.

Human activities have undoubtedly affected the local, if not the global, climate of the earth since primitive peoples started clearing forests for fuel and agriculture. It is unclear, however, whether the effects of even the industrial age up to the present have been more significant than, for example, volcanic activity in causing the observed average temperature changes of the earth over the last few decades. The total atmospheric content of fine particles is currently estimated to be about 4×10^7 tons, of which only about one-quarter has come from human activities.

On the other hand, it is clear that in the foreseeable future, human activities can swamp natural forces and produce significant global effects. In particular, as H.A. Wilcox[3] has pointed out, a continued increase in the use of stored fuels at the rate of 4 percent per year indefinitely (the worldwide rate of increase was 5.7 percent per year until the recent energy crisis) would result in heat being released to the air and water at 0.1, 1, and 10 percent of the sun's input rate by the years 2050, 2110, and 2170, respectively. Wilcox calculates from a very simple model that the average temperature of the earth would have increased 5 to 8°C by about 2170, and that flooding from the melting of polar ice caps would be producing widespread damage to the coastal cities of the earth.

An additional point needs to be considered. If natural forces can change the global climate in the future and cause widespread human distress, perhaps humans should intentionally intervene and attempt to maintain a more stable climate, or even attempt to improve the climate in certain harsh regions. In particular, if a rapid global cooling is already occurring naturally and if human activities generally cause a heating of the earth, there may be every reason to increase those activities. Unfortunately, just as we cannot calculate with any certainty the effects of natural changes, we also cannot estimate either the direction or magnitude of any effects arising from purposefully manipulating one or more of the human activities influencing climate in the future.

PROJECTIONS OF WORLD POPULATION AND ENERGY USE

Clearly, the population of the world cannot increase indefinitely. The world population increased 44 percent between 1950 and 1970 (from 2.5 to 3.6 billion) at a rate of roughly 2.1 percent per year.[4] Continued at this rate, there would be 54 billion people on earth by the year 2100 and one person on each square foot of land and sea—5.5 million billion people—within less than 700 years. R. Revelle[5] has indicated that the earth can provide food for a population of 40 to 50 billion. A number of studies have suggested 15 to 20 billion people as a likely "stable" world population to be reached in 100 to 150 years.[6,7] D. Meadows and others[8] have raised the specter of a world population that uses up its resources and "crashes." D.J. Bogue[9] is unique in predicting a leveling off at only 6 billion.

H. Kahn[10] has developed a set of scenarios resulting in world populations ranging from 7.5 to 30 billion and has described their economic development such that, on the average, everyone on earth is very "well off." He bases his estimates of the leveling-off of the population on the "historical experience that as societies grow richer . . . population growth rates decline." Supporting his view is the fact that the population growth rates in 56 countries out of 66 surveyed by the World Bank are decreasing.[11] He warns

that no population forecast made so far in this century has proven to be correct and that all have been on the high side of the actual number. P. Demeny[12] notes that "the population of the underdeveloped countries account for nearly three-fourths of the human species . . . (and) will continue their rapid growth for the rest of the century. Control will eventually come through development or catastrophe."

Similarly, energy use in the world cannot increase indefinitely. The world use of energy has been increasing recently at a rate of roughly 5.7 percent per year, reaching 2×10^{17} Btu per year in 1970.[4] The average per capita use of energy throughout the world increased 57 percent from 1950 to 1970, reaching 53 million Btu per capita per year in 1970.[4] However, this use was highly skewed, with the United States using 35 percent of the total, about 68×10^{15} Btu per year at an average of roughly 340 million Btu per capita per year, and the other industrialized nations using another 35 percent at an average of about 150 million Btu per capita per year.[13] The underdeveloped and developing nations used roughly 55×10^{15} Btu per year at an average of perhaps 20 million Btu per capita per year in 1970, but the greatest increase in per capita use occurred between 1950 and 1970[4]: 73 percent in Africa, 122 percent in Latin America, and 197 percent in Asia.

Wilcox pointed out the dire results that would occur if such increases continued for only another 200 years or so.[3] By approximately the year 2170, human activities would be generating heat at a rate equal to 10 percent of the sun's input, and the polar ice caps would most likely be melting. By the year 2230, the oceans would be close to the boiling point!

Yet, as Wilcox has gone on to note,[14] there is a reason to believe that energy use per capita will "saturate" and that the total energy use in the world will level off perhaps in 100 to 125 years. Kahn has suggested that a "postindustrial" society is emerging in which most human occupations will consist of services and recreation.[15] Energy use per capita might be no greater or even less in such a society than exists now for upper-middle-class persons in the United States. He notes that at the 5 percent economic growth rate attained by the world during the 1960s, almost every country should be well off in 100 years, in terms of gross national product per capita, as the United States is today. Hence, in 100 to 125 years, the world could be approaching an era of stable population, with a relatively flat income and energy-use distribution, and thus rich enough not to want to increase either.

Some degree of saturation in energy use in the United States is already apparent. Between 1960 and 1970, the average per capita use of energy by families in the upper quarter of income distribution increased their energy use by only 2 percent as compared with 3 percent for the national average.[16] In a recent NASA-supported study,[17] it was noted that the "normal peacetime" increase in per capita energy use over the past 70 years was only about 2 percent per year. During World Wars I and II and the Korean and Vietnam conflicts, substantial increases in energy use occurred, which are reflected in the overall average 3 percent growth curve. O.A. Zraket[18] has suggested that a 2 percent annual increase in total energy use in the United States is achievable within 20 to 30 years and that, in the long run, major reductions in energy use can be attained by more efficient energy generation and end-use systems without limiting economic growth.

Since significant inequities exist in the United States in terms of both income and energy use, it is possible to envision an increase by the poorer segments of the economy without a proportionate increase in the overall energy use. Thus, a value of 500 million Btu per capita per year can be regarded as a value representing a very healthy economy, even for the year 2100. However, recognizing the potential desire of everyone in most developed and developing countries to obtain and use the goods and services available to the upper middle class, a figure twice as large, 1 billion Btu per person per year (a rate roughly three times greater than the present United States use), can be suggested as the maximum energy required to fuel a very affluent world society for the indefinite future beyond the year 2100. In this regard, the mainland Chinese represent a very large population with an apparent satisfaction with a relatively low per capita use of energy. If they become "energy conscious," the energy use in the world could increase rapidly.

The purpose of this chapter is not to forecast the world of the year 2100, but to illustrate the consequences of some basic choices in providing energy to fuel human activities until then and from that time on. Let us assume at this point that a stable world population can be achieved without catastrophe at 6, 10, or 20 billion, and that the average per capita energy use levels off at 100 million Btu per year (about twice the current world average rate), 300 million (nearly the current United States average), 500 million, or 1 billion Btu per year. Table 8-1 is a list of the total energy used by the world under this range of conditions. The range is from six-tenths of a "Q" Btu per year (Q = 10^{18}) to 200 Btu per year. For comparison, the current rate of world energy use is about two-tenths of a "Q" Btu per year, and the rate of energy being absorbed at the earth's surface from the sun is 2400Q Btu per year. In other words, at the maximum condition postulated here, human society would be using energy at a rate of a little less than 1 percent of that abosrbed from the sun.

A. Weinberg and R.D. Hammond[1] arrived at a figure of 12Q Btu per year for the future world energy-use rate, and

J.P. Holdren and P.R. Ehrlich[6] quote G.M. Woodwell[19] and E.D. Sellers[10] at a figure of 4.5Q Btu per year (15 billion people at 10 thermal kilowatts each). Kellogg[21] arrives at a figure of 24Q Btu per year, estimating 20 billion people each using 40 thermal kilowatts (based on Kahn's estimate of an average income of $20,000). Wilcox[14] suggests 200Q to 300Q Btu per year as the maximum tolerable rate of consumption of the "thermally polluting" energy on a global basis.

SOURCES OF ENERGY

Fossil

A significant fraction of energy utilized in the twenty-first century could be from fossil fuels. Coal- and oil-fired power plants with useful lives of 40 to 50 years are still being constructed. No replacement for natural hydrocarbons for use in the internal combustion engine is yet in development. Potentially available resources of fossil fuels total perhaps 2×10^{20} Btu according to the *UN Statistical Yearbook, 1971,*[4] enough for 100 to 300 years at the lowest per capita use rate in Table 8-1, but only about 10 years at the highest total use rate. On the other hand, A.B. Cambel[22] has estimated total fossil resources to be over 10^{22} Btu. We will adopt the lower figure on the basis that economically recoverable reserves would be lower than total resources.

The use of fossil fuels puts both particulate and carbon dioxide into the atmosphere. As noted previously, they possibly have opposite effects on the earth's average temperature. There is considerable discussion of the nature and magnitude of these effects both for the recent past and in the future. Combustion of fossil fuels has caused a 9 percent increase in atmospheric carbon dioxide concentration since 1870. A 30 percent increase is projected by 2000. G.N. Plass[23] has estimated that the increase in atmospheric carbon dioxide concentration over the past 100 years could have been responsible for the observed 0.5°C rise in average global temperature. This leaves unexplained the 0.25°C drop since 1940. R.A. Bryson[24,25] and M.J. Budyko[26] believe the decrease is due to the particulate emissions. M.M. Mitchell, Jr.[27] and W.W. Kellogg and S.H.Schneider[28] doubt this since aerosols in the atmosphere over land, where the particulates are generally emitted, should lead to a net heating in contrast to aerosols distributing over the oceans, which should lead to a cooling. Kellogg[29] suggests that the cooling trend may be unrelated to humanity's activities.

Looking to the future, if particulate emissions are indeed the cause of the earth's cooling, and if efficient electrostatic precipitators are used to prevent particulate emissions, the carbon dioxide emitted would be the dominant effect. The global radiation models of S. Manabe and R.T. Wetherald[30] and of S.I. Rasool and S.H. Schneider[31] indicate that the average surface temperature of the earth might increase 1°C or slightly more by the middle of the twenty-first century from atmospheric carbon dioxide radiation absorption.

Table 8-1. Range of total world energy use, Btu per year.

World Population	AVERAGE PER CAPITA ENERGY USE (MILLION BTU/YEAR)			
	100	300	500	1000
6 billion	0.6Q[a]	1.8Q	3Q	6Q
10 billion	1Q	3Q	5Q	10Q
20 billion	2Q	6Q	10Q	20Q

[a]Q = 10^{18}.

The temperature at the poles could rise somewhat higher, perhaps 2°C or more.

In the longer term, if fossil fuels were used until they were exhausted, approximately seven times as much carbon dioxide would be added to the atmosphere as is currently contained therein. We do not know at present the rate at which some or a large fraction of this atmospheric carbon dioxide would be absorbed in green plants and in the oceans. (Currently, the oceans contain 50 times as much carbon dioxide as the atmosphere, and much greater amounts of carbon dioxide are contained in carbonate rocks.) Even so, a significant warming effect on global climate is a very likely consequence, with an even greater impact at the poles than elsewhere.

In the much longer term, if fossil fuels were no longer used, the level of carbon dioxide in the atmosphere might decrease somewhat due to incorporation in larger amounts of green plants (because of the need for increased vegetation to feed the possibly very large population) and due to slow reaction leading to precipitation of additional carbonates from the oceans. Also, if methanol or some other "synthetic" hydrocarbon were made from hydrogen produced by decomposition of water by nuclear or solar energy and from atmospheric carbon dioxide, the storage of large quantities of such products could decrease the carbon dioxide concentration in the atmosphere slightly.

Nuclear Fission

Sufficient uranium is available at today's prices to fuel the nuclear reactors projected to be constructed in the next decade for only about 30 years.[32] However, at prices 5 to 10 times higher, perhaps 10 times as much uranium would be available.[33] In fact, E. Teller[34] believes that there is adequate uranium for the foreseeable future even utilizing non-breeder reactors. If the breeder reactor is developed in the near future, roughly 100 times more uranium would be available and possibly the very large world resources of thorium could be "burned." In this case, there is no question that sufficient nuclear fuels exist to supply the world for many centuries at even the maximum rate projected for 2100.

One basic question must be answered: can the radioactive wastes be collected safely and stored to prevent man-made radiation emitted to the biosphere from reaching levels that are close to the natural radiation background? For the purpose of this chapter, we will assume that the answer to this question is affirmative.

Given this explanation, nuclear fission would affect the global climate primarily through the heat released by its conversion to useful forms of energy. As will be described, all of the heat content of the nuclear fuel is ultimately released into the biosphere. At some rate of use, the global temperature is certain to increase significantly.

Nuclear Fusion

Scientific feasibility of the production of useful energy from a nuclear fusion process has yet to be demonstrated.

However, such demonstration may occur late in this century, and the commercial production of electricity or other forms of energy could be initiated early in the twenty-first century. By the end of that century, it is conceivable that all energy for human society could be generated by the fusion process. There is sufficient fuel in the oceans and the earth's crust to furnish energy at the rate of 2×10^{19} Btu per year for the indefinite future.[35] The potential release of radiation from projected fusion power plants is estimated to be perhaps 10,000 times less than for a nuclear fission plant. Also there will be little or no long-lived fission products to be disposed of. Hence, the radioactive waste problem may again be ignored for the purposes of this chapter.

As with nuclear fission, however, what cannot be ignored is the heat released by the fuel. Again, as will be described, too great a rate of use will certainly affect global climate.

Geothermal

Heat continually flows from the interior of the earth to the surface at an average rate of 0.02 Btu per hour per square foot or about 0.06 thermal watts per square meter. Gustavson estimates that of this total useful heat flow, only about 10^{11} thermal watts[36] could be utilized worldwide. Other estimates are much larger.[37] An energy system that would utilize all the heat flowing out from 1 square mile would have a continual supply of 140 thermal kilowatts (roughly 40 electrical kilowatts assuming a conversion efficiency of 30 percent). These systems, which do not add to the heat flow to the surface, would not cause an increase in the average temperature of the earth.

However, to get additional, more concentrated energy, most geothermal systems are designed to extract more energy per square foot than is provided by the natural heat flow. Hence, in general, geothermal systems must be considered as "stored" energy systems, which at sufficiently high rates of use would ultimately heat up the world. In addition, there might be a pollution problem from hydrogen sulfide and other noxious gases or salts in the geothermal steam.

Geothermal systems are currently in use in Italy, Japan, Iceland, New Zealand, and the United States as well as in other countries. Approximately 300 megawatts of electrical generating capacity exist in the United States today, and many thousands of acres of likely geothermal sources under public lands were recently leased to public utilities and others by the United States Department of the Interior. The National Science Foundation has obtained an estimate that at least 10^5 megawatts of generating capacity could be developed in the United States by the year 2000. Geothermal energy will undoubtedly be an important source of heat and electricity in many countries for the indefinite future. However, it is unlikely to be the major source of energy for the world at the rates listed in Table 8-1, primarily because of the limited availability of near-surface hot rock or dry steam sites.

Other Sources

M.R. Gustavson has given estimates of the total amounts of hydroelectric and tidal energy that could be developed throughout the world: 9×10^{16} Btu per year for the total tidal-dissipation energy, of which perhaps 1.9×10^{15} Btu per year is usable, and 2.7×10^{17} Btu per year for the total hydrologic runoff, of which potentially 8.6×10^{16} Btu per year is usable.[38] Both of these sources are nonthermally polluting, although there may be serious ecological disturbances caused by their construction and operation. Again, these sources will undoubtedly be very important in certain localities, but will not be able to supply the amounts of power listed in Table 8-1.

Solar Energy

The solar input to the earth is 130 watts per square foot for a total of 1.78×10^{17} watts or 5.3×10^{21} Btu per year,[39] of which about 3.6×10^{21} Btu per year are retained within the earth's atmosphere. Roughly 2.4×10^{21} Btu per year are absorbed by the surface. In terms of supplying the world's energy needs, roughly 450 thermal megawatts of solar energy fall on a square mile at latitudes and under weather conditions similar to those of the United States. Thus, for a collector of 50 percent efficiency (typical of thermal collectors) and 50 percent spacing to avoid self-shielding, about 250,000 square miles are needed to produce 1Q Btu per year. For a collector of 20 percent efficiency (which is possible with advanced photovoltaic devices) and 50 percent spacing, about 10,000 square miles would be needed to produce 440,000 electrical megawatts (0.44 trillion watts), the current electrical generating capacity of the United States.

The great advantage of using solar energy is that no additional heat is released at the earth's surface (unless of course a collector in space is used). The great disadvantages of using solar energy are (1) the tremendous areas involved in collecting it, and (2) its diurnal nature requiring some form of energy storage. The largest figure above, 250,000 square miles, is roughly the area of Arizona and New Mexico. Hence, innovative dual use of the land may have to be arranged. Also, the ocean may have to be used as a collector area.

Ocean-based systems for producing electricity and fuels may eventually be feasible. The current state-of-the-art is the ocean thermal difference technique with an efficiency of only 1 to 3 percent in converting thermal differences in the ocean into electricity. Future devices may be based on photovoltaic or photogalvanic effects, perhaps with higher efficiency. If 1 percent is taken as the overall efficiency of converting sunlight to electricity, roughly 7 million square miles of ocean (about 5 percent of the total ocean area) would be serving as a collector to produce 1Q Btu per year or roughly 30 trillion watts. Of course, one major advantage of the ocean thermal technique is that the ocean serves as an energy-storage system so that the power

plants can run day and night, not just when the sun is shining.

Many other combinations of solar energy techniques can be used to decrease the total area devoted to collectors. First, most systems permit dual use of the area such as putting collectors on top of buildings and on roads. By increasing the spacing between collector elements, the land can be farmed underneath and crop wastes turned into fuel. The total amount of energy available from the conversion of organic materials to fuels could be substantial. Gustavson estimates that the solar input to photosynthesis is about 1.2×10^{18} Btu per year.[39] Naturally, only a fraction of this could be collected and processed into fuel. Another estimate gives 1.7×10^{18} Btu per year as a possible amount of energy available from organic material producible in the ocean.[40] Thus, the use of organic materials for fuels may contribute appreciably to the large worldwide energy requirements considered here (note that the carbon dioxide produced in consuming the organic material would have come from the atmosphere in the first place). Another large solar-derived source of energy is the wind. An estimate of the energy that could be extracted from winds over the United States has been calculated to be about 1.5×10^9 megawatt hours of electricity produced per year,[41] or a rate of roughly 0.2 trillion watts. Worldwide, this might amount to 2 trillion watts or more. This may be an important energy source but not the major contribution to the world's energy supply in the year 2100.

It should be noted that large-area solar energy collectors could increase the heat absorbed from the sun by changing the effective albedo. An efficient collector spread over desert area (highly reflective) could result in a net additional heat load to the earth.

In summary, solar energy can probably supply the total world's energy supply at a rate of 20×10^{18} Btu per year or more (20 billion people using 1 billion Btu per year on the average) for the indefinite future. Several techniques for energy conversion would need to be developed, including thermal collectors, photovoltaic or photogalvanic devices, wind-driven generators, ocean thermal power plants, organic material processors, and additional hydroelectric plants. The total area required for collectors would be large, ranging from 150,000 (0.6Q) to 5 million square miles of land (20Q) and up to 7 million square miles of ocean (1Q). However, these areas could serve dual purposes. On the other hand, the ecological impact of large land-based arrays and ocean power plants has yet to be determined. Although the use of solar energy does not add pollutants and thermal waste to the biosphere, the very large systems described here would redistribute the received solar energy significantly, and there might well be a major local climatic effect.

IMPACT OF USE OF ENERGY ON CLIMATE
Thermal and Carbon Dioxide Pollution

As Wilcox[3] and others have pointed out, any increase in the release of energy on earth must result in an increase in temperature to radiate the energy into space. It does not matter significantly whether the energy source is fossil or nuclear, or even upon the efficiency of utilization of the energy. Essentially, the entire heat content of each of these forms of "stored" energy results ultimately in additional thermal release. Geothermal energy may yield slightly less net thermal waste than fossil or nuclear sources, but only tidal or solar energy utilization results in no additional release of heat and ensuing temperature rise. Furthermore, the utilization of solar energy must occur in such a way that no net additional energy is absorbed by the earth—i.e., the solar collectors must not change the albedo of the earth significantly.

There may be other significant impacts on climate, both local and global, caused by human activities. For the purpose of this discussion, however, these other impacts will be assumed to be negligible in comparison to thermal and carbon dioxide pollution.[42]

SCENARIOS

The following three basic scenarios represent (I) maximum, (II) medium, and (III) minimum use of energy by a stable world society in the year 2100 and beyond. For each scenario, two primary options are described: the use of (A) stored fuels and (B) nonstored fuels.

Scenario I: World energy use reaches 20Q Btu per year by the year 2100 (population of 20 billion at 1 billion Btu per capita per year).

Option A: All fossil-fuel reserves are exhausted during the twenty-first century (2×10^{20} Btu generated). Nuclear energy becomes the major energy source late in the twenty-first century and the dominant source by the year 2100.

Option B: The use of fossil fuels is curtailed by the year 2050 so as not to exceed 420 parts per million of carbon dioxide in the atmosphere. Nuclear energy becomes the major source for an interim period in the middle part of the twenty-first century, but then solar energy takes over by the year 2100.

Scenario II: World energy use reaches 3Q Btu per year by the year 2100 (population of 10 billion at 300 million Btu per capita per year).

Option A: Use of fossil fuels is curtailed by the year 2050 so as not to exceed 420 parts per million of carbon dioxide in the atmosphere. Nuclear energy becomes the major energy source before the middle of the twenty-first century and the dominant source by the year 2050.

Option B: Similar to Option A except solar energy takes over from nuclear energy by about the year 2075.

Scenario III: World energy use reaches 0.6Q Btu per year by the year 2100 (population of 6 billion at 100 million Btu per capita per year).

Option A: Use of fossil fuels is curtailed by the year 2000 so as not to exceed 400 parts per million of carbon dioxide in the atmosphere. Nuclear energy becomes the major energy source early in the twenty-first century and the dominant source by the year 2025.

Option B: Similar to Option A except solar energy takes over from nuclear energy by about the year 2050.

These scenarios and options are intended to span the range of likely conditions to be met in the next 100 to 125 years and to highlight the primary choices that can be made in the use of fuels to limit human impacts on the global climate.

Impact of the Scenarios

Scenario I

Option A. The impact of using stored fuels at a rate of 20Q Btu per year has been described by W.W. Kellogg[43] based on independent models of W.D. Sellers[44] and M.I.Budyko.[45] Basically, for a 1 percent increase in the release of heat at the earth's surface (20Q Btu per year is roughly 1 percent of the solar energy absorbed at the earth's surface), the average global temperature would increase 2° to 3°C, and the average temperature at the poles would increase perhaps 10°C or more.

W.W. Kellogg believes the Arctic Ocean ice pack might indeed melt, more or less permanently (without, by the way, increasing the level of the oceans since the Arctic Ocean ice is floating), although the ice has persevered for the last few hundred thousand years. He feels the Greenland and Antarctic ice caps would melt over a period of centuries, if at all, and that no models currently available can predict this result. A.S. Monin believes that the polar ice is more stable than many others predict.[46]

W.M. Washington has calculated global heat balance, using a general circulation model (GCM) of the atmosphere, on the large computer at the National Center for Atmospheric Research, Boulder, Colorado. His first calculation assumed that additional heat load, equivalent to 50 percent of the incoming solar radiation, was released uniformly over the land areas of the world—about 75Q Btu per year. Temperatures increased 8° to 10°C over the northern parts of Asia and all of North America.[47]

In a second calculation, Washington assumed a total additional heat load of 9Q Btu per year distributed proportionally to existing populations around the world. The results were indeterminate relative to a random error introduced into the model and load of 9Q Btu per year.[48] He concludes that for a heat load of 9Q Btu per year, there is a relatively small modification of the model earth-atmosphere heat balance, but that the calculations have the serious shortcoming of assuming fixed ocean-surface temperatures. He adds that the calculations should be repeated with a coupled atmosphere-ocean model.

The use of fossil fuels to exhaustion within 125 years or so may be difficult to accomplish from the sheer magnitude of mining the tremendous quantities of coal and heavy oil deposits. However, it may be necessary to do so to achieve the eventual 20Q Btu per year rate. On the other hand, the net result very likely would be to cause an increase of at least 2° to 3°C in the average global temperature since the carbon dioxide emitted would be about seven times the amount already in the atomosphere.[49] A large fraction of the carbon dioxide would be absorbed in the oceans (which exchange carbon dioxide with the atmosphere at a rate of about 10 percent per year); some would be absorbed by green plants, but enough carbon dioxide would probably be left in the atmosphere to cause a substantial temperature increase. In fact, this temperature increase might reduce the capacity of the oceans to hold carbon dioxide and cause a larger-than-expected effect.[1] Furthermore, the impact of the heat load of 20Q Btu per year would be in addition to the carbon dioxide effect. Clearly, Scenario I, Option A leads to highly uncertain but possibly severe impacts on the global climate.

A major question still to be answered concerning the first scenario is the distribution of the release of the waste heat. Global effects could be much less or much greater than described above, depending on the location and manner in which the heat is released.

The utilization of energy at rates approaching 20Q Btu per year will undoubtedly be distributed unevenly over the continental areas of the earth. During the next 125 years, it is unlikely that civilization will spread out more over the land, much less establish ocean-based colonies. In fact, the opposite is more likely—urban areas will tend to grow. Although it is conceivable that most large electric generating plants could be constructed at the seashore or at sea to utilize the ocean for cooling, and the electricity transmitted long distances via super-conducting cables to the cities, most energy use would still occur in the urban areas. For instance, if one-half of all energy use goes to make electricity at an average efficiency of 50 percent, then 25 percent of the global use of energy would be released as waste heat at the generating site and 75 percent at the point of ultimate use.

R.B. Panero[51] has introduced the concept that three types of civilizations still occur throughout much of the world, namely:

A Country—twentieth century, urban
B Country—seventeenth century, modernized rural
C Country—the frontier, unexplored, unexploited

The United States and Europe still have plentiful A and B Country, while Canada, South America, Africa, Australia, Southeast Asia, China, etc. have pockets of A Country isolated by much B and C Country. It is doubtful that in 125 years the world will be homogenized. There may still be plentiful C Country and much B Country left. On this premise, Hudson Institute[52] has estimated the actual population distribution for 20 billion people living in the year 2100. By and large, currently existing metropolitan areas grow and merge. These areas, then, will be the locale for the release of most of the 20Q Btu per year.

For this case, the power plants are assumed to be located in or adjacent to the cities they serve. The nonuniform distribution of heat release is clear. It might be noted that the energy used along the northeastern shore of South America, 7 trillion watts or 2×10^{17} Btu per year, is just equal to

the current use of energy in the world. The impact of these large "heat islands" has not yet been estimated. Some significant effect on local climate seems assured. For instance, the heat rising from nonequatorial latitudes might interfere with the natural poleward flow of heat via upper atmospheric flow of air heated in the tropics.

To sum up the impact of Scenario I, Option A, the risk of truly severe climate changes, even if the use of fossil fuels is curtailed sharply at the beginning of the twenty-first century, makes the use of stored fuels at such high rates extremely risky without greatly increased knowledge of heat balance effects on climate long before such rates are achieved.

Option B. The shift to solar energy early enough to avoid serious thermal problems from using nuclear fuels at too high a rate does not avoid the problems of carbon dioxide pollution from using up the fossil fuels. Therefore, the use of fossil fuels must be curtailed early enough to limit the total amount of carbon dioxide emitted such that the likely temperature increase is small enough to be tolerated. Kellogg has estimated that fossil fuels could be burned until the middle of the twenty-first century with an increase in average global surface temperature 1°C, or perhaps a bit more.[53] To hold the likely temperature increase to 0.5°C or less, fossil fuel use would have to be curtailed sharply by the year 2000.

Thus, the impact of this option appears at first to avoid the entire problem of waste heat and resulting global temperature rise and to suffer only about 1°C increase in temperature due to limited use of fossil fuels. However, two other factors enter: (1) the distribution of the use and release of the collected solar energy, and (2) the change in land use with a resulting change in albedo of the earth.

The area of collectors to supply the energy needed by each megalopolis can be estimated by multiplying number of trillion watts (3×10^{16} Btu per year) given in Figure 8-1 by (a) 8200 square miles for thermal collectors having 50 percent efficiency and 50 percent spacing, and (b) 23,000 square miles for photovoltaic generators having 20 percent efficiency and 50 percent spacing. For instance, the northeast corner of South America is listed as 7 trillion watts. If half of this is provided as thermal energy ($3.5 \times 8200 = 28,700$ square miles) and half as electric energy ($3.5 \times 23,000 = 80,500$ square miles), the total area needed would be about 110,000 square miles. Actually dual-purpose collectors are likely to be developed so that perhaps the area could be cut to 75,000 to 85,000 square miles. Now, this area can be both distributed on rooftops, parking lots, roads, etc., and centralized in huge single facilities. Even the latter can be raised and spread out to accommodate other activities underneath such as farming. Brazil has an area of 3,286,473 square miles of which the Amazon Basin makes up 1,465,637 square miles. The megalopolis expected to grow up around Recife and Natal would have perhaps 325 million people at a density of roughly 5000 per square mile covering an area of about 65,000 square miles.[52] Therefore, in this case the collection and utilization of the energy could occupy

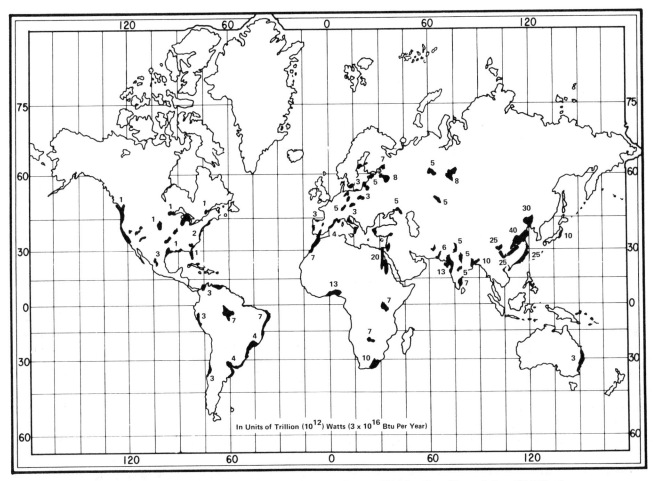

Figure 8-1. Estimated thermal waste from urban areas in year 2100 (total world population—20 billion).

essentially the same area. On the other hand, there may be large urban areas that would be difficult to cover with collectors, such as a downtown urban center, with skyscrapers, for instance.

In these cases, offshore ocean-thermal power plants may be desirable. However, to provide the 3.5 trillion watts of electric generating capacity being considered in the Recife-Natal area, 3,500 power plants of 1000-megawatt capacity each would have to be constructed. At 3 percent overall efficiency,* they would utilize the sunlight falling on about 210,000 square miles of ocean—for example, a strip extending 1200 miles along the coast and 200 miles wide (the power plants would not physically occupy this space—they might be the size of oil-drilling rigs spaced 5 to 10 miles apart). Hence, from this initial, rough analysis, it may very likely be possible to install very large solar energy systems close to or above the urban areas where most of the energy would be utilized.

In other areas of the world, there may indeed be difficulty in collocating the solar energy system with urban areas. In this case, it may be possible to construct a very large system in a remote area and to manufacture hydrogen or a

synthetic hydrocarbon for piping to the urban centers. In effect, for the Paris megalopolis for instance, 5 trillion thermal solar watts might be collected in the Sahara (a collector covering 41,000 square miles), converted to hydrogen by electrolysis of seawater piped in from the Mediterranean Sea, and dissipated in Paris. Conceivably, this might not only change the character of the Sahara, but also might affect the climate in Europe by the "heat island" effect, as described above for Option A. Unfortunately, without more rigorous calculations of these cases in the manner of Washington,[47] we can only speculate on the magnitude of these effects.

Scenario II

Option A. The impact of using stored fuels at a rate of 3Q Btu per year borders on having a significant effect on the global climate. If we assume a roughly linear effect, 3Q Btu per year should produce a little more than 10 percent of the effect of 20Q Btu per year, namely, an 0.2 to 0.3°C increase in average global temperature and 0.5° to 1°C increase at the poles. In addition, there will be the effect of carbon dioxide from the use of fossil fuels. We are arbitrarily limiting this to about 1°C increase in average global temperature by constraining fossil fuel burning so that the carbon dioxide

*One percent efficiency is the current state-of-the-art. An improvement to 3 percent in 100 years or so is considered very likely.

concentration in the atmosphere is limited to 420 parts per million (ppm). Thus, the total anticipated impact would be an average global temperature increase of 1.2° to 1.3°C and an average polar temperature increase of 2.5° to 3°C.

The net result of this temperature increase is hard to predict. The assumption of linearity of effect is very dubious, and 3Q Btu per year may have much less impact than 20Q Btu per year. Fossil-fuel burning could be curtailed to give a carbon dioxide concentration less than 420 ppm, and thus a temperature increase less than 2° to 3°C at the poles.

The Arctic and Antarctic may well retain their ice at temperatures only 2° to 3°C higher than now. A.S. Monin[46] believes that the Viking chronicles from the eighth through tenth centuries, which told of little glaciation in the Arctic, do not support the hypothesis that if the Arctic Ocean ice melts, it will never refreeze. On balance, we would have to say that the impact of this option would most likely be noticeable in the polar regions at least, but would be unlikely to have disastrous consequences.

Option B. The impact of replacing nuclear energy with solar energy by about 2075 and reaching the same total "steady-state" energy use of 3Q Btu per year by the year 2100, primarily avoids the 0.2° to 0.3°C temperature increase caused by the use of stored energy, but does not avoid the 1°C increase caused by carbon dioxide even though the use of fossil fuels would be curtailed and eliminated when the carbon dioxide concentration in the atmosphere reached 420 ppm. On the other hand, this option would largely avoid the "heat island" effects, although some redistribution of energy from large central collectors such as in the southwestern United States or the Sahara to urban centers might be necessary. The total area needed for solar collectors would be about 750,000 square miles or a little more than 1 percent of the total land area of the world. As noted before, the area needed for collectors may not be available around some urban centers, and remote sites may have to be utilized.

Scenario III

Option A. The impact of using stored fuels at a rate of 0.6Q Btu per year, while limiting fossil fuel use such that the atmospheric carbon dioxide concentration does not exceed 400 ppm, appears to be minimal. Temperature increases on a global scale may be measurable but, from the calculations made to date, would appear to have little impact even at the poles. However, "heat island" effects may be noticeable on a micro- and even mesoscale.

The major question with Scenario III is not whether the energy use will have a deleterious impact on the global or polar climate, but whether a stable human society can be achieved at such a low population and per capita energy-use rate (only twice the current world average) without a major

Table 8-2. Summary of impacts on climate.

	AVERAGE TEMPERATURE INCREASE (°C)	
	GLOBAL	AT POLES
Scenario I		
Option A	2°–3°	10° or more
Option B	1°[a]	2°–3°[a]
Scenario II		
Option A	1.2°–1.3°[a]	2.5°–3°[a]
Option B	1°[a]	2°–5°[a]
Scenario III		
Option A	0.5°[a]	1°[a]
Option B	0.5°[a]	1°[a]

[a]The predominant cause of these temperature increases is the CO_2 concentration in the atmosphere due to combustion of fossil fuels.

catastrophe such as nuclear war. It may be that the development of much more efficient energy generation and utilization equipment and techniques will be accomplished. Also, the projections of Bogue[9] may be realized by rapid dissemination of population control techniques. In this case, nuclear energy could well be used from a global heat balance standpoint.

Option B. The replacement of nuclear energy with solar energy at a total energy-use rate of 0.6Q Btu per year may have to be justified on the basis of limiting radioactive emissions or waste-storage problems rather than on the basis of avoiding global climate problems. Weinberg and Hammond[7] and others claim that the emissions can be contained and the wastes stored at a total energy-use rate of 10^{14} watts (3Q Btu per year). On the other hand, these steps may prove to be expensive and difficult. Certainly, storing nuclear wastes for thousands of years safely poses a novel problem to human society from the standpoint of long-range planning. However, this concern is beyond the scope of this report.

Summary of Impacts

The impacts on climate are summarized in Table 8-2. Clearly, the use of fossil fuels should be curtailed so that atmospheric carbon dioxide concentration does not exceed 400 to 420 ppm, and the use of "stored" fuels should be limited to a total rate of 3Q Btu per year or less to avoid a significant temperature rise at the poles. The use of solar energy permits total energy use to become 20Q, assuming fossil fuel is curtailed, without a significant global or polar temperature increase.

All of these conclusions must be considered extremely tentative. Hence, the primary action that needs to be taken should be to institute better long-range monitoring and modeling activities in order to make much better estimates before the year 2000.

REFERENCES

1. W.W. Kellogg, *Climate Change and the Influence of Man's Activities on the Global Environment*, M72-166, The MITRE Corporation, September 1972, pp 2-5, 10-13.
   See also:
   (A) U.S. Committee for the Global Atmospheric Research Program, "Understanding Climatic Change: A Program for Action," DRAFT, National Research Council, National Academy of Science, Washington, D.C., September 1974
   (B) A.S. Monin "Weather Forecasting as a Problem in Physics," (translated by P. Superak) (Cambridge: M.I.T. Press, 1972), p. 13.
   (C) H. Flohn, *Climate and Weather* (New York: McGraw-Hill, 1969).
   (D) J.M. Mitchell, Jr., *Causes of Climate Change*, Meteorological Monographs, 8, No. 30, Lancaster Press, 1968.
   (E) E.D. Sellers, *Physical Climatology* (Chicago: University of Chicago Press, 1965).
   (F) M.I. Budyko, "Comments on A Global Climate Model Based on the Energy Balance of the Earth-Atmosphere System," *J. Appl. Met*, 9, 1970.
   (G) C.L. Wilson, Director, *Man's Impact on the Global Environment* (Cambridge: M.I.T. Press, 1970), pp 44-45.
   (H) W.H. Mathews et al, eds, *Man's Impact on the Climate* (Cambridge: M.I.T. Press, 1971), p. 3.
   (I) W.W. Kellogg, "Long Range Influences of Mankind on the Climate" (Presented at the World Conference on "Toward A Plan of Action for Mankind," Institute de la Vie, Paris, France, September 9, 1974).
   (J) C.E.P. Brooks, *Climate Through the Ages* 2nd rev. ed. (New York: Dover Publications, Inc., 1970).
2. W.W. Kellogg, op. cit., pp 13-14.
3. H.A. Wilcox, "The Energy Growth—Present Trends and Future Prospects for the World and the USA" (A paper presented to the Marine Technology Society, Washington, D.C., June 12, 1973), p. 36.
4. *Statistical Yearbook*, 1971, (New York: United Nations Statistical Office, 1972).
5. R. Revelle, "Food and Population," *Scientific American* 231, no. 3 (September 1974), p. 160.
6. J.P. Holdren and P.R. Ehrlich, "Human Population and the Global Environment," *American Scientist* 62, 1974, p. 288.
7. A. Weinberg and R.D. Hammond, "Limits to the Use of Energy," *American Scientist* 58, 1970, p. 412.
8. D. Meadows et al., "The Limits to Growth," Potomac Associates, 1972.
9. D.J. Bogue, University of Chicago, mentioned by H. Kahn, *Symposium on Energy, Resources and the Environment*, M72-154, The MITRE Corporation, May 1972, p. 178.
10. H. Kahn, "Prospects for Mankind," Vol. II of the *1973 Synoptic Context on the Corporate Environment: 1975–1985*, Draft, Hudson Institute, Croton-on-Hudson, N.Y., 1973, Card II-4.
11. "A Little More Time," *The Economist*, 29 June, 1974, p. 16.
12. P. Demeny, "The Population of the Underdeveloped Countries," *Scientific American*, 231, No. 3, September 1974, p. 149.
13. E. Cook, "The Flow of Energy in an Industrial Society," *Scientific American* 224, No. 3, 1971, p. 134.
14. H.A. Wilcox, op. cit., p. 39.
15. H. Kahn, op. cit., Card II-5.
16. J.H. Bridges, Director of National Energy Programs, Center for Strategic and International Studies, Georgetown University, Washington, D.C. in a talk to The MITRE Corporation, June 26, 1974. See also D. Chapmann et al, *Science*, 178, No. 4062, 17 November 1972; and A.L. Levinson, University of California, Berkeley, "Reducing Residential Demand for Electric Power" (Presented at the Sierra Club Meeting, January, 1972).
17. Terrastar, NASA/ASEE Systems Design Summer Faculty Program Final Report, NASA CR-129012, Marshall Space Flight Center, Ala., September, 1973, pp 1-13.
18. C.A. Zraket, "Growth and the Conservation of Energy," MTP-389, The MITRE Corporation, February 1974, p. 11.
19. G.M. Woodwell, "The Energy of the Biosphere," *Scientific American* 223 (3), September 1970, p. 64.
20. W.D. Sellers, *Physical Climatology* (Chicago: University of Chicago Press, 1965).
21. W.W. Kellogg, "Mankind as a Factor in Climate Change," *The Energy Question*, F. Erickson ed., (Toronto: University of Toronto Press, 1974). (in press).
22. A.B. Cambel, "Energy R&D and National Progress," U.S. Office of Science and Technology, 1964.
23. G.N. Plass, "Carbon Dioxide and Climate" *Scientific American* 201, No. 1, 1959.
24. R.A. Bryson, "Climatic Modification by Air Pollution," *The Environmental Future*, N. Polunin, ed. (New York: MacMillan, 1972), pp. 133-154.
25. R.A. Bryson, "Climatic Modification by Air Pollution," II. *The Sahalian Effect*, Rept. 9, Madison Institute for Environmental Studies, U. of Wisconsin, 1973, and "A Perspective on Climatic Change," *Science* 184, 1974, pp. 753-760.
26. M.J. Budyko, "The Future Climate," *EOS* 53, No. 2, 1972, pp. 868-874.
27. J.M. Mitchell, Jr., "A Reassessment of Atmospheric Pollution As a Cause of Long-Term Changes of Global Temperatures," *Global Effects of Environmental Pollution*, S.F. Singer, ed. (Dordrecht, Holland: Reidel Publ. Co., 1974).
28. W.W. Kellogg and S.H. Schneider, "Climate Stabilization: For Better or for Worse?" *Science* 186, 1974, pp. 1163-72.
29. W.W. Kellogg—Reference 1 (I) p. 6.
30. S. Manabe and R.T. Wetherald, "Thermal Equilibrium of the Atmosphere with a Given Distribution of Relative Humidity," *J. Atmos. Sci* 24, pp. 241-59.
31. S.I. Rasool and S.H. Schneider, "Atmospheric Carbon Dioxide and Aerosols: Effects of Large Increase on Global Climate," *Science* 173, pp. 138-141.
32. National Petroleum Council, U.S. Energy Outlook: "An Initial Appraisal 1971-1985," 1, July 1971.

33. M.R. Gustavson, *Dimensions of World Energy*, M71-71, The MITRE Corporation, November 1971, p. 13.

34. E. Teller, *Symposium on Energy, Resources and the Environment*, M72-154, The MITRE Corporation, May 1972, p. 95.

35. M.R. Gustavson, op. cit., p. 14.

36. M.R. Gustavson, op. cit., p. 5.

37. See for example J. Barnea, *Technology Assessment*, M74-30, The MITRE Corporation, March 1974, p. 18; and Geothermal World Directory (New York: Geothermal Energy Institute, 1972).

38. M.R. Gustavson, op. cit., pp. 6-7.

39. M.R. Gustavson, op. cit., p. 4.

40. National Academy of Sciences, *Resources and Man* (San Francisco: W.H. Freeman and Co., 1969), p. 107.

41. F.R. Eldridge, "Wind Energy Conversion Systems," Draft Report to the National Science Foundation, June 1974, p. 17.

42. For a complete discussion see:
    (A) *Man's Impact on the Global Environment*, Report of the Study of Critical Environmental Problems (Cambridge: MIT Press, 1970).
    (B) *Man's Impact on the Climate*, edited by W.H. Mathews, W.K. Kellogg, and G.D. Robinson (Cambridge: MIT Press, 1971).

43. W.W. Kellogg, op. cit, (Ref. 17), P. 109.

44. W.D. Sellers, "A Global Climatic Model Based on the Energy Balance of the Earth-Atmosphere System," *J. Appl. Met.* 12, 1973, pp. 241-254.

45. M.I. Budyko, "The Effect of Solar Radiation Variations on the Climate of the Earth," *Tellus* 21, 1969, pp. 611-619.

46. A.S. Monin, op. cit. (Ref. 1A), p. 13.

47. W.M. Washington, "On the Possible Uses of Global Atmospheric Models for the Study of Air and Thermal Pollution," Chapt. 18 in *Man's Impact on the Climate*, W.H. Mathews, et al, eds. (Cambridge: MIT Press, 1971).

48. W.M. Washington, "Numerical Climate-Change Experiments: The Effect of Man's Production of Thermal Energy," *J. Appl. Met.* 11, No. 5, August 1972, pp. 768-772.

49. M.R. Gustavson, op. cit., p. 12.

50. M.R. Gustavson, op. cit., p. 7, 8.

51. R.B. Panero, *Symposium on Energy, Resources and the Environment*, M72-23, The MITRE Corporation, January 1972, pp. 187-191.

52. Hudson Institute Report, "Population Distributions for the Year 2100," Croton-on-Hudson, N.Y. 1973.

53. W.W. Kellogg, op. cit. (Ref. 21), p. 101.

9 State Solar Legislation

Increasing public interest in solar energy has prompted many states to pass laws encouraging or facilitating the adoption of solar heating and cooling systems. Legislative support for solar energy utilization is particularly important on the state level. Local tax structures can encourage the choice of solar systems as alternatives to more conventional energy sources. Guaranteeing the future viability of solar systems may require altering zoning and building ordinances. Accordingly, a number of states have taken steps (1) to provide tax incentives for users or vendors of solar systems or equipment and (2) to safeguard the solar energy system user's access to sunlight.

Tax-incentive legislation aims at reducing the financial burden of installing solar heating and cooling systems. The addition of a solar unit may increase the market value of a building, and therefore some states exempt a building's owner from the increased property taxes resulting from the addition. Other states allow solar system owners to claim a certain percentage of the cost of a solar system as a tax credit against their personal state income tax. Others exempt vendors of solar energy equipment from sales tax on their receipts.

Three states have passed laws to safeguard a solar system user's access to sunlight, either by providing procedures for creating voluntary solar easements or by requiring that solar energy considerations be taken into account in zoning requirements and land-use planning. In addition, some states have provided funds for programs to promote research into and development of solar energy. Some states also require that feasibility studies on alternative energy systems be conducted before construction or substantial renovation contracts in public buildings are executed.

The following state legislation concerning solar tax incentives and access to sunlight is currently in effect.

ARIZONA

Bill # CH. 129 Laws of 1976

Provides for amortization of the cost of solar energy devices over 36 months in computing net income for state income-tax purposes. Applies to all types of buildings.

Contact: Neal Trasente, Director, Dept. of Revenue
Capitol Building, West Wing
Phoenix, AZ 85017
(602) 271-3393

Bill # CH. 165 Laws of 1974

Provides exemption from property-tax increases that may result from addition of solar system to new or existing housing.

Contact: Same as above

Bill # CH. 81 Laws of 1977

Provides income-tax credit of 35 percent of cost of residential solar system up to $1000. Credit percent declines 5 percent per year until program expires in 1984.

Contact: Same as above

Bill # CH. 42 Laws of 1977

Exempts devices used for production of nonfossil energy from Transaction Privilege Sales Tax.

Contact: Same as above

ARKANSAS

Bill # CH. 535 Laws of 1977

Allows individual homeowner/taxpayer to deduct entire cost of energy-saving equipment, including solar equipment, from gross income for year of installation.

Contact: Dept. of Revenue, Income Tax Section
Seventh and Wolfe Sts.
Little Rock, AR 72201
(501) 371-2193

CALIFORNIA

Bill # CH. 168 Laws of 1976

Provides for state income-tax credit (up to $1000) of 10 percent of the cost of installing a solar system.

Contact: Franchise Tax Board
Sacramento, CA 95814
(916) 355-0370

Bill # CH. 1 Laws of 1978

Authorizes the Department of Housing and Community Development to conduct an interest-free loan program ($2000 per dwelling) for residential installations of solar water- and space-heating systems. Only those owners of dwellings damaged or destroyed by disaster in areas where the governor has declared a state of emergency on or after July 1, 1977, are eligible. The installation must be made in conjunction with the repair or replacement of the home on the same site and must be of a type approved by the Energy Resources Conservation and Development Commission. Law expires on December 31, 1980.

Contact: Department of Housing & Community Development
Division of Research & Policy Development
921 10th Street
Sacramento, CA 95814
(916) 445-4725

Bill # CH. 670 Laws of 1976

Provides that any city or county may require new buildings subject to state housing law to be constructed so as to permit the future installation of solar devices. Specific building features mentioned include roof pitches and alignments.

Contact: Division of Codes & Standards
Department of Housing & Community Development
921 10th Street
Sacramento, CA 95814
(916) 445-9471

COLORADO

Bill # CH. 344 Laws of 1975

Provides that all solar systems be assessed at 5 percent of their original value for property-tax purposes.

Contact: Raymond E. Carper
Property Tax Administrator
Col. Division of Property Taxation
614 Capitol Annex
Denver, CO 80203
(303) 892-2371

Bill # CH. 326 Laws of 1975

Provides procedures for recording voluntary solar easements.

Contact: Relevant county clerk and county recorder

Bill # HB. 1519 Passed 1977

Allows personal and corporate income-tax deduction equal to the cost of installation, construction, reconstruction, remodeling, or acquisition of a solar energy device in a building.

Contact: Department of Revenue
Income Tax Division
1375 Sherman St.
Denver, CO 80261
(303) 839-3781

CONNECTICUT

Bill # Public Act 409 (1976)

Authorizes local taxing authorities to exempt property with solar system from increased assessment due to system.

Contact: Relevant local assessor or board of assessors

Bill # Public Act 457 (1977)

Provides sales and use tax exemptions for solar collectors.

Contact: State Tax Department
Audit Division
92 Farmington
Hartford, CT 06115
(203) 566-2501

Bill # Public Act 409 (1976)

Provides that commissioner of planning and energy policy establish standards for solar energy systems sold in state.

Contact: Department of Planning and Energy Policy
20 Grant Street
Hartford, CT 06115
(203) 566-2800

FLORIDA

Bill # CH. 246 Laws of 1976

Directs Florida Solar Energy Center to develop standards for solar systems sold in state and testing procedures to evaluate them. All solar systems manufactured or sold in the state may optionally display results of approved performance tests.

Contact: Florida Solar Energy Center
300 State Rd. 401
Cape Canaveral, FL 32920
(305) 783-0300

Bill # CH. 361 Laws of 1974

Provides that no single-family residence shall be constructed in the state unless the plumbing is designed to facilitate future installation of solar water-heating equipment.

Contact: Bureau of Codes and Standards
2571 Executive Center Circle E.
Tallahassee, FL 32301
(904) 488-3581

GEORGIA

Bill # Act No. 1030 (1976)

Provides that purchasers of equipment for solar systems will receive refund of sales and use tax paid on such equipment.

Contact: Dept. of Revenue, Sales Tax Unit
Room 310, Trinity-
Washington Bldg.
Atlanta, GA 30334
(404) 656-4071

Bill # Resolution No. 167 (1976)

Enabled constitutional amendment authorizing governing authority of any county or municipality to exempt value of solar system from ad valorem property tax.

Contact: Relevant city council or county board of supervisors

Bill # Act 1446 Laws of 1978

Provides procedures for the creation of voluntary solar easements and makes them subject to the same conveyancing and instrument recording requirements as other easements.

Contact: Relevant clerk or recorder

HAWAII

Bill # Act 189 (1976)

Provides income-tax credit for up to 10 percent of cost of solar system for year of installation. Also, provides exemption from any property-tax increase resulting from addition of solar system.

Contact: Tax Department
Hale Auhdo Bldg.
425 Queen St.
Honolulu, HI 96813
(808) 548-2211

IDAHO

Bill # CH. 212 Laws of 1976

Provides that cost of residential solar systems may be deducted from taxable income over a period of 4 years. Deduction shall not exceed $5000 in any one taxable year.

Contact: Income Tax Division
State Tax Commission
Box 36
Boise, ID 83722
(208) 384-3290

Bill # H.B. 333 (Passed 1978)

Provides procedures for creation of voluntarily negotiated solar easements. Declares that such an easement shall be presumed to be attached to the real property and shall pass with the property when title is transferred to another owner.

Contact: Relevant clerk or recorder

ILLINOIS

Bill # Public Act 79-943 (1975)

Provides that when a solar system has been installed on real property, owner of property may claim improvement value of a conventional system if that value is less than value of solar system.

Contact: Frank A. Kirk, Director
Dept. of Local Government Affairs
303 E. Monroe St.
Springfield, IL 62706
(217) 782-6436

For information on the standards being developed, contact:

Division of Energy
Illinois Dept. of Business & Economic Development
222 S. College
Springfield, IL 62706
(217) 782-7500

INDIANA

Bill # Public Law 15 (1974)

Allows owner of real property with solar system an annual deduction from the assessed value of the property equal to the lesser of: (1) the assessed value of the property with solar system minus the assessed value without the system, or (2) $2000.

Contact: County auditor

KANSAS

Bill # CH. 434 Laws of 1976

Allows individuals and businesses to deduct 25 percent of cost of solar system (up to $1000 for individuals and $3000 for businesses) from state income tax.

Contact: Kent Kalb
Secretary of Revenue
Second Floor, State Office Building
Topeka, KS 66612
(913) 296-3041

Bill # CH. 346 Laws of 1977

Allows business taxpayer to deduct amortized (60 months) cost of a solar system installed in a business or commercial building from taxable income. Can be applied in addition to the income-tax credit provided by CH. 434, Laws of 1976.

Contact: Same as above

Bill # CH. 345 Laws of 1977

Provides for reimbursement of 35 percent of total property tax paid for the entire building or building addition if solar provides 70 percent of energy needed for heating or cooling of that building on an average annual basis.

Contact: Same as above

Bill # CH. 227 Laws of 1977

Provides procedures for creation of voluntarily negotiated solar easements between property owners.

Contact: County clerk or recorder

MAINE

Bill # CH. 685 Laws of 1978

Appropriates $16,000 for a solar water-heater demonstration program. Grants of $400 are to be made to home builders, homeowners, schools, and hospitals that demonstrate intent to install solar water-heating systems and meet any other qualifications deemed necessary by the Office of Energy Resources. No firm, institution, or individual is eligible for more than one $400 grant.

Contact: Office of Energy Resources
55 Capitol St.
Augusta, ME 04330
(207) 289-2196

Bill # CH. 542 Laws of 1977

Exempts solar-heating systems from property taxation for 5 years from the date of installation. Exemption must be applied for. Also provides refund of sales or use tax paid on solar equipment certified as such by the Office of Energy Resources. Both tax provisions expire on January 1, 1983.

Contact:
For Property Tax: Local assessor or board of assessors
For Sales Tax: Office of Energy Resources
55 Capitol St.
Augusta, ME 04330
(207) 289-2196

MARYLAND

Bill # CH. 740 Laws of 1976

Authorizes Baltimore City and city within a county or any county to provide a credit against local real-property taxes for those residential or nonresidential buildings using solar heating and cooling units. Amounts and definitions are at the discretion of local jurisdictions.

Contact: Relevant city or county department of revenue

Bill # CH. 509 Laws of 1975

Provides that solar system shall not cause property tax assessment of new or existing building to be greater than it would be with conventional system.

Contact: Dept. of Assessment and Taxation
302 W. Preston St.
Baltimore, MD 21201
(301) 383-2526

Bill # CH. 934 Laws of 1977

Provides that negotiated restrictions on the use of land or water for the purpose of protecting solar access shall be enforceable in law and equity. Restrictions can be created by voluntarily negotiated easements, convenants, restrictions, or conditions between property owners.

Contact: Relevant clerk or recorder

MASSACHUSETTS

Bill # CH. 28 Laws of 1977

Authorizes banks and credit unions to make loans with extended payback periods and increased maximum amounts for financing alternative energy systems including solar. Payback period is extended to 10 years. Maximum amount is increased to $7000 for banks and $9500 for credit unions.

Contact: Local bank or credit union

Bill # CH. 433 Laws of 1976

Provides that a corporation may deduct cost of a solar system in computing taxable income for year of installation. Also provides that solar system will not be subject to tangible property tax.

Contact: Harvey Michaels
Asst. to Director for Solar Energy
Energy Policy Office
John McCormack State Office Building
One Ashburton Pl.
Boston, MA 02108
(617) 727-4732

Bill # CH. 734 Laws of 1975

Provides for real-estate tax exemption for solar system.

Contact: Same as above

Bill # CH. 487 Laws of 1976

Provides that a corporation may deduct the cost of a solar system from a taxable income for year of installation. Also provides that the system will be subject to tangible property tax.

Contact: Commissioner of Corporation & Taxation
100 Cambridge St., Rm. 806
Boston, MA 02204
(617) 727-4201

Bill # CH. 989 Laws of 1977

Exempts solar energy systems used in an individual's principal residence from retail sales taxes.

Contact: Department of Corporation & Taxation
Sales Tax Division
100 Cambridge St., Rm 605
Boston, MA 02204
(617) 727-4620

MICHIGAN

Bill # Public Act 132 (1976)

Provides that receipts from sale of tangible property to be used in solar system shall not be used to compute tax liability for business-activities tax.

Contact: Edward Kane, Director
Dept. of Treasury
State Tax Commission, State Capitol Bldg.
Lansing, MI 48922
(517) 373-2910

Bill # Public Act 133 (1976)

Provides that tangible personal property used for solar devices shall be exempt from excise tax on personal property.

Contact: Same as above

Bill # Public Act 135 (1976)

Provides that solar system shall not be considered in assessing value of real property.

Contact: Same as above

MINNESOTA

Bill # CH. 333 Laws of 1976

Directs building-code division of Department of Administration to develop performance criteria for solar systems made or sold in the state. These standards must be in reasonable conformance with federal interim performance criteria for residential and commercial solar systems. Manufacturers and retailers are required to disclose the extent to which each system meets these standards.

Contact: Building Code Division
Dept. of Administration
Seventh & Roberts Sts.
St. Paul, MN 55101
(612) 296-4626

Bill # CH. 786 Laws of 1978

Excludes the market value of a solar system installed in a building prior to January 1, 1984, from the market value of the building.

Contact: Minnesota Department of Revenue
Property Equalization Division
Centennial Office Bldg.
St. Paul, MN 55145
(612) 296-5131

Bill # CH. 786 Laws of 1978

Provides procedures for the creation of voluntarily negotiated solar easements. Allows solar energy considerations to be included in comprehensive planning, local zoning, and subdivision regulations.

Contact: Relevant clerk or recorder (for easements); relevant zoning body (for zoning)

MONTANA

Bill # CH. 548 Laws of 1975

Allows property-tax exemptions on capital investments for energy conservation or alternative energy systems. Also allows public utility to lend capital for energy conservation/alternative energy systems at rate not exceeding 7 percent per year. Provides utilities with license-tax credit for difference between 7 percent and prevailing rate of interest.

Contact: William Groff, Director
Dept. of Revenue
Mitchell Building
Helena, MT 59601
(406) 449-2460

Bill # CH. 574 Laws of 1977

Provides solar income-tax credit for residential solar energy system installed prior to December 31, 1982. Credit is for 10 percent of the first $1000 spent and 5 percent of the next $3000. If the federal government provides a similar tax credit, the amount of credit is halved. If credit exceeds tax liability, it can be carried forward for up to 4 years.

Contact: Department of Revenue
Income Tax Audit Section
Mitchell Building
Helena, MT 59601
(406) 449-2837

Bill # CH. 576 Laws of 1977

Allows personal and corporate income-tax deduction for a capital investment in a building for a solar system. Maximum deduction for a residence is $1800 and $3600 for a nonresidential building according to the following schedules:

Residential	Nonresidential
100 percent of the first $1000 spent	100 percent of the first $2000 spent
50 percent of the next $1000 spent	50 percent of the next $2000 spent
20 percent of the next $1000 spent	20 percent of the next $2000 spent
10 percent of the next $1000 spent	10 percent of the next $2000 spent

This provision replaces the property-tax exemption on capital investment for a solar system (CH 548 Laws of 1975).

Contact: Director
Department of Revenue
Mitchell Building
Helena, MT 59601
(406) 449-2460

NEVADA

Bill # CH. 345 Laws of 1977

Provides property-tax allowance for amount of assessed value of property with solar systems minus assessed value without. Allowance may not exceed total value of property tax accrued or $2000, whichever is less.

Contact: Local county assessor

NEW HAMPSHIRE

Bill # CH. 391 Laws of 1975

Allows each city and town to adopt (by local referendum) property-tax exemptions for solar systems.

Contact: Local or municipal tax assessor

NEW JERSEY

Bill # CH. 256 Laws of 1977

Allows owners of real property with solar system an annual deduction from property taxes equal to the remainder of assessed value of property with system minus assessed value without. Deduction must be applied for, and it expires on December 31, 1982. System must meet standards to be established by State Energy Office.

Contact: Director
Division of Taxation
W. State and Willow Sts.
Trenton, NJ 08625
(609) 292-5185

NEW MEXICO

Bill # CH. 12 (Special Session of 1975)

Provides income-tax credit for 25 percent of cost of residential solar system up to $1000 credit maximum.

Contact: Marshall Morton
Bureau of Revenue
Manuel Lujan, SR, Bldg.
St. Francis Drive at Alta Vista
Sante Fe, NM 87503
(505) 827-3221

Bill # CH. 114 Laws of 1977

Allows income-tax credit for a solar system used for irrigation pumping. Credit may be claimed for 3 years and aggregate credit cannot exceed $25,000. System must be certified by the Energy Resources Board.

Contact: Same as above

Bill # CH. 169 Laws of 1977

Declares that access to sunlight is a transferable property right. Provides that disputes over solar access shall be resolved by prior appropriation rule modified by court decisions. Effective July 2, 1978.

Contact: City or county zoning authority

NEW YORK

Bill # CH. 322 Laws of 1977

Provides property-tax exemption for solar and wind systems in amount of assessed value of property with system minus assessed value without. All exempted systems must be approved by state energy office. Exemption extends for 15 years after approval. Systems must be installed prior to July 1, 1988.

Contact: Relevant assessor or board of assessors

NORTH CAROLINA

Bill # CH. 792 Laws of 1977

Allows personal and corporate income-tax credit equal to 25 percent of the cost of the solar system installed in a building (limit $1000). System must meet performance criteria prescribed pursuant to the Solar Demonstration Act of 1974.

Contact: North Carolina Dept. of Revenue
Individual Income Tax Div.
Box 2500
Raleigh, NC 27640
(919) 733-4682

Bill # CH. 965 Laws of 1977

Provides that buildings equipped with solar energy systems shall be assessed for property-tax purposes as if they were equipped with a conventional system only. Effective January 1, 1978, and expires December 1, 1985.

Contact: Local assessor or board of assessors

NORTH DAKOTA

Bill # H.B. 1069 Passed, 1977

Provides procedures for the creation of voluntary solar easements.

Contact: Relevant clerk or recorder

Bill # CH. 508 Laws of 1975

Provides that solar heating and cooling systems utilized in new or existing buildings will be exempt from property taxes for 5 years following installation.

Contact: John Hulteen
Supervisor of Assessments
State Tax Department, State Capitol
Bismarck, ND 58505
(701) 224-3461

Bill # CH. 537 Laws of 1977

Provides income-tax credit for installation of solar or wind system equal to 5 percent a year for 2 years.

Contact: Same as above.

Bill # CH. 425 Laws of 1977

Provides procedures for the creation of voluntary solar easements.

Contact: Relevant clerk or recorder

OKLAHOMA

Bill # H.B. 1322 (1977)

Provides income-tax credit of 25 percent of the cost of a solar system installed in a private residence. Credit limitation is $2000. If allowance credit exceeds tax liability, the credit may be carried over for five succeeding taxable years. Credit expires on December 31, 1987.

Contact: Assistant Director
Income Tax Division
N.C. Conners Bldg.
2501 Lincoln Blvd.
Oklahoma City, OK 73194
(405) 521-3121

OREGON

Bill # CH. 460 Laws of 1975

Provides exemption from ad valorem taxation for any increased value of property resulting from installation and use of a solar system.

Contact: Donald M. Fisher, Administrator
Assessment and Appraisal Division
506 State Office Bldg.
Salem, OR 97310
(503) 378-3378

Bill # CH. 153 Laws of 1975

Adds solar energy considerations to comprehensive planning. Allows city and county planning commissioners to recommend ordinance governing building height for solar purposes.

Contact: Lon Topaz, Director
Dept. of Energy
528 Cottage St. NE
Salem, OR 97310
(503) 378-4128

Bill # CH. 196 Laws of 1977

Allows personal income-tax credit of 25 percent ($1000 limit) of cost of installing alternative energy device (including solar) in home. System must provide at least 10 percent of home's energy requirements and must meet performance criteria adopted by Department of Energy. Credit must be applied for and can be claimed for a device installed between January 1, 1978 and January 1, 1985.

For information on income-tax credit:
Oregon Dept. of Revenue
Box 14555
Salem, OR 97310
(503) 378-8303

For information on standards being developed:
Director
Dept. of Energy
528 Cottage St., NE
Salem, OR 97310
(503) 378-4128

Bill # CH. 315 Laws of 1977

Permits veterans to obtain subsequent loan ($3000 limit) above the maximum amount allowed from Oregon War Veterans' Fund for the purpose of installing alternative energy device (including solar) in a home. System must meet performance criteria established by Director of Veterans' Affairs and Department of Energy, and must provide at least 10 percent of the home's energy requirements.

Contact: Assistant Director
Department of Veterans' Affairs
1225 Ferry St., SE
Salem, OR 97310
(503) 378-6851

RHODE ISLAND

Bill # CH. 202 Laws of 1977

Provides that a solar heating/cooling system installed in a residential or nonresidential building shall be assessed at no more than the value of a conventional heating/cooling system necessary to serve the building. Law expires on April 1, 1997.

Contact: Local assessor or board of assessors

SOUTH DAKOTA

Bill # CH. 111 Laws of 1975

Allows owner of residential real property an annual deduction from assessed value for installation of solar device. Deduction may be equal to the lesser of: (1) difference between assessed value with system and without, or (2) $2000.

Contact: County auditors/assessors and
Lyle Wendall, Secretary
Dept. of Revenue, State Capitol
Pierre, SD 57501
(605) 224-3311

Bill # H.B. 1354 (Passed 1978)

Allows property-tax assessment credit for installation of a renewable resource energy system (including solar) as a part of an improvement to either residential or commercial real property. Residential credit is equal to the greater of (1) the difference between assessed valuation of real property with system and assessed value without, or (2) the actual installed cost of the system. Commercial credit is equal to 50 percent of the actual installed cost of the system. Residential credit may be claimed for 5 years and the commercial credit for 3 years. The credit will be adjusted to include any federal income-tax credit that may become law.

Contact: County Auditors/Assessors and/or Secretary
Department of Revenue
State Capitol
Pierre, SD 57501
(605) 224-3311

TEXAS

Bill # CH. 719 Laws of 1975

Provides exemption from sales tax on receipts from sale, lease, and rental of solar devices. Business tax exemption granted for corporations that exclusively manufacture, install, and sell solar devices. Corporation may deduct amortized (60 months) cost of solar system from taxable capital.

Contact: Bob Bullock, Comptroller
Director of Public Accounts
LBJ Bldg.
17th and Congress Sts.
Austin, TX 78711
(512) 475-6001

TENNESSEE

Bill # CH. 837 Laws of 1978

Exempts solar energy systems from property tax. Requires the Tennessee Energy Office to develop standards and specifications for solar systems.

Contact: Local assessor or board of assessors

Also provides a loan program to low- and moderate-income persons and families for installation of solar hot-water heaters in residential housing.

Contact: Tennessee Housing Development Authority
Hamilton Bank Bldg.
Nashville, TN 37219
(615) 741-2400

VERMONT

Bill # Public Law 226 (1976)

Allows towns to enact real- and personal-property tax exemptions for solar systems.

Contact: State Energy Office
State Office Bldg.
Montpelier, VT 05602
(802) 828-2768

Bill # Act 210 Laws of 1978

Allows personal and business income-tax credits for the installation of solar energy systems prior to July 1, 1983. Personal credit is equal to the lesser of (1) 25 percent of the cost of purchasing and installing a system in the taxpayer's dwelling, or (2) $1000. Business credit is equal to the lesser of (1) 25 percent of the cost of purchasing and installing a system on real property used in a trade or business or held for the production of income, or (2) $3000. If credit exceeds tax liability, excess credit may be carried forward for up to four succeeding tax years.

Contact: Vermont Tax Department
Income Tax Division
State St.
Montpelier, VT 05602
(802) 828-2517

VIRGINIA

Bill # CH. 782 Laws of 1976

Enabled constitutional amendment authorizing legislature to permit localities to enact tax exemptions on property used for solar systems.

Contact: Individual state senators or delegates

Bill # CH. 561 Laws of 1977

Allows county, city, or local governing body to exempt solar equipment from property taxation.

Contact: Relevant tax governing body

Bill # CH. 323 Laws of 1978

Provides procedures for the creation of voluntary solar easements and makes them subject to the same conveyancing and instrument recording requirements.

Contact: Relevant clerk or recorder

WASHINGTON

Bill # CH. 364 Laws of 1977

Exempts solar energy systems from property taxation. Exemption must be applied for, is valid for 7 years, and can only be applied to equipment meeting HUD's minimum property standards. Opportunity to apply for exemption extends to December 31, 1981.

Contact: Director
Property Tax Division
Dept. of Revenue
General Administration Bldg.
Olympia, WA 98504
(206) 753-2057

WISCONSIN

Bill # CH. 313 Laws of 1977

Allows businesses to deduct (in the year paid), depreciate, or amortize (over 5 years) the cost of designing, constructing, and installing a solar energy system. Applies to expenses incurred between April 20, 1977 and December 31, 1984.

Allows individual income-tax credit for the cost of design, construction, equipment, and installation of a solar energy system on the individual's property. If credit exceeds tax liability, refund is paid. Credit applies to expenditures exceeding $500 but not more than $10,000 incurred between April 20, 1977 and December 31, 1984. Amount of credit varies according to the following schedules:

- If the structure on which the system is installed appeared on the local tax roll prior to or on April 20, 1977:

For expenses in 1977 and 1978, 30 percent
For expenses in 1979 and 1980, 24 percent

For expenses in 1981 and 1982, 18 percent
For expenses in 1983 and 1984, 12 percent

- If the structure on which the system is installed appeared on the local tax roll after April 20, 1977:

For expenses in 1977 and 1978, 20 percent
For expenses in 1979 and 1980, 16 percent
For expenses in 1981 and 1982, 12 percent
For expenses in 1983 and 1984, 8 percent

Systems for the business and personal income-tax provisions must meet performance standards promulgated by the Department of Industry, Labor and Human Relations.

Contact: Wisconsin Department of Revenue
Income Tax Division
3648 University Ave.
Madison, WI 53708
(608) 266-2772

10 How a Nuclear Power Plant Works

THE NATURE OF NUCLEAR POWER

People tell pollsters that their greatest worry about nuclear power is that nuclear reactors will explode. A homemaker asks whether the electricity generated by a nuclear plant is radioactive. A congressional representative refers to the "nuclear explosions" that produce energy at these generating stations. Actually, nuclear plants are not as different from other power plants as some people think. Understanding what they are and how they work is important to forming attitudes about nuclear power.

Electricity is produced at all power plants by spinning the shaft of a huge generator, in which coils of wire and magnetic fields interact to create electricity. In most plants—thermal or steam-electric stations—this spinning is done by high-pressure steam "blowing" the propeller-like blades of a turbine connected to the generator shaft.

Heat needed to boil water into steam at these plants is produced in one of two ways: (1) by burning coal, oil, or gas—the fossil fuels—in a furnace or (2) by splitting certain atoms of uranium in a nuclear reactor. Nothing is burned or exploded in a nuclear reactor. The fuel consists of many tons of ceramic pellets made from an oxide of uranium or other fissionable metal. The cylindrical pellets, each about the size of the end of your little finger, are carefully arranged in long, vertical tubes within the reactor.

Inserted throughout bundles of these fuel tubes are many "control rods." These rods regulate a process that results in atoms invisibly flying apart, or fissioning. As the atomic pieces plow through the fuel pellets, they generate heat by a kind of friction—something like the heat generated when you rub your hands together.

The atoms involved are those of the uranium metal within the ceramic fuel. As the nucleus of each atom fissions, it shoots out particles called *neutrons,* which cause more fissions when they hit the nuclei of other uranium atoms. This sequence of one fission triggering others, and those triggering still more, is called a *chain reaction.*

The control rods contain a material that absorbs neutrons and prevents them from hitting fissionable atoms, enabling the rods to speed up or slow down the chain reaction. Operation of the reactor is controlled by varying the number of rods withdrawn and the amount of their withdrawal.

REACTOR TYPES AND MATERIALS

Currently, two types of commercial reactors are widely used in the United States: the *boiling water reactor (BWR)* and the *pressurized water reactor (PWR).* Both are called *light-water reactors (LWR)* because their coolant (or medium to transfer heat) is ordinary water, incorporating the light isotope of hydrogen.

BWRs and PWRs use essentially the same fuel and operate on basically the same principles. The fuel is an oxide containing two kinds, or isotopes, of uranium atoms, designated by their differing weights. Uranium-235 is the fissionable isotope, which constitutes less than 1 percent of natural uranium. Uranium-238 is practically nonfissionable but makes up all the rest of the element as it is found in nature.

The uranium in the fuel has passed through an enriching process that increases the concentration of uranium-235 to 3 or 4 percent. This allows the reactor to be smaller than it would be if fueled with natural uranium; it also allows ordinary water to be used for cooling instead of a more expensive and less convenient fluid. Nevertheless, the concentration of Uranium-235 is still so low that a bomblike nuclear explosion is impossible.

BUILT-IN SAFETY

Prefired as a ceramic, the fuel resists the effects of high temperature and corrosion during reactor operation. And the dilute mixture of Uranium-235 with other, nonfissionable, materials in the pellet is one of the reactor's natural safety features because it tends to slow down the chain reaction as it gets hotter.

The pellets are stacked end-to-end in 12-foot-long tubes made of zirconium alloy. These fuel rods are then precisely arranged as bundles within the reactor, with spaces between them for control rods. Water flowing up through the bundles removes the heat and puts it to work.

The water serves as a "moderator," slowing down the neutrons to increase the probability of fission. Rather than "clobbering" the nucleus, the neutron must combine with it momentarily. Thus, the neutron must go fast enough to combine with the nucleus but not so fast as to skip over it—like a ball putted toward the hole in golf.

FUEL REPLACED, RECYCLED

Certain changes take place in the fuel pellets during their time in the reactor. Rather than causing fissions in Uranium-235, some neutrons hit atoms of Uranium-238 and turn them into plutonium, another fissionable element. Some of this plutonium is subsequently fissioned in place by other neutrons, contributing to the production of heat. Because it can breed these fissionable atoms, Uranium-238 is known as a "fertile" isotope.

Most of the fragments of fission are radioactive; during the life of the fuel enough fragments collect within the pellets to contribute substantially to the heat through radioactive decay—the spontaneous emission of particle and gamma rays.

The fuel remains in a reactor for 3 or 4 years before these trapped fission products reduce the efficiency of the chain reaction—like ashes smothering a fire. All the fuel is not removed at once; from one-fourth to one-third of the rods are discharged and replaced each year.

Enough unfissioned uranium and newly generated plutonium remain to make recovery worthwhile, so this spent fuel is saved for reprocessing and eventual recycling of the fissionable materials.

HARNESSING THE ENERGY

The cooling system is a main link in the chain that converts fission energy to electrical energy.

Following is the chief difference between BWRs and PWRs.

In BWRs, made by the General Electric Co., the water boils to steam directly in the reactor vessel. In PWRs, made by the Babcock & Wilcox Co., Combustion Engineering, and Westinghouse Electric Corp., the reactor water is pressurized so as not to boil. Instead, it is pumped through a heat exchanger, where a separate supply of water is heated to produce steam.

At both types of plant, this steam is then used to make electricity in exactly the same way as at a plant burning fossil fuel—by spinning the turbine, which drives an electric generator. This equipment is essentially the same at all thermal power plants. The product, electricity, is identical.

A nuclear power plant, then, is nothing more than a steam-electric generating station in which a nuclear reactor takes the place of a furnace, and the heat comes from the fissioning of uranium fuel rather than from the burning of fossil fuel.

11 Shipping Nuclear Fuel

SAFETY IN MOTION

No nuclear power plant is an island. As with electric generating stations burning coal, oil, or gas, its fuel must be brought in and its waste taken away. And, once removed from the multiple safeguards of a reactor for shipment over rail or highway, this nuclear material is subject to the same transportation hazards as any other freight or cargo. Conditions of this environment cannot be controlled. However, they can be predicted. Accordingly, multiple layers of protection surround the fuel to make certain that significant amounts of radioactivity will not be released, even during the most severe accident. The same principle of defense-in-depth applied to every power reactor ensures also the safety of the public while nuclear fuel is in transit.

NEED FOR TRANSPORTATION

In today's commercial reactor, the fuel consists of uranium dioxide pellets stacked into 14-foot-long metal tubes. A square array of these sealed tubes makes up an assembly, from 5 to 8 inches on a side, weighing from one-quarter to three-quarters of a ton. Several hundred assemblies in turn form the fuel core. During reactor operation, Uranium-235 fissions to produce heat for the steam generation of electricity. And the fragments of fission accumulating within the pellets and their cladding gradually reduce the efficiency of the chain reaction.

Therefore, after approximately three years of producing power, the assemblies are replaced with fresh fuel. From one-fourth to one-third of the core—several dozen assemblies—is discharged each year. This spent fuel, however, also contains substantial quantities of unfissioned Uranium-235 and of newly formed plutonium. Both can be recovered and eventually returned to the reactor. Delivering fresh fuel to the generating station and carrying spent fuel away for reprocessing create the primary transportation link in the nuclear fuel cycle.

TRAFFIC FLOW MINOR

Shipments to meet the needs of a 1-million-kilowatt power reactor are much fewer than material transfers for many light industrial or even commercial facilities. Assemblies of fresh fuel are forwarded by truck: some 20 trips for initial loading, about 6 for annual refueling. Assemblies of highly radioactive spent fuel are retrieved in massive, heavily shielded casks; with those of recent design, about 25 truckloads or 10 railcars will be needed each year.

Over the 40-year operating lifetime of the 100 reactors expected to be on-line by 1980, the total fuel transportation requirements would be:

- *Fresh fuel:* for initial loading, about 2000 truckloads; for refueling, some 600 annually.
- *Spent fuel:* with projected truck and train service, some 1400 trips by highway and 400 by railway per year.

Compared to the 100 thousand shipments of other hazardous materials—flammable, toxic, explosive, corrosive—that take place in the United States each year, these few thousand movements annually are quite small.

ACCIDENTS HIGHLY UNLIKELY

The risk to public health and property from nuclear fuel on the move is small. First of all, the probability is extremely low that a vehicle conveying this type of freight will meet with a traffic mishap. Why? From the millions of shipments of all types of radioactive material made so far and the much more extensive experience with transferring other hazardous substances, the frequency of such incidents can be estimated.

Over the lifetime of 100 reactors, there may occur 35 truck accidents with fresh fuel and 66 with spent fuel; 14 train accidents with spent fuel may occur. Just because a shipping vehicle is involved in a crash, however, does not mean that radioactive materials will be released. In fact, less than 1 percent of these projected mishaps are expected to produce conditions severe enough to potentially damage the cargo. These figures, however, tell only half the story: the extremely remote likelihood that a shipment of nuclear fuel may experience a serious accident. The other half of the story lies in packaging.

PACKAGING REDUCES RISK

Although every precaution is taken to avoid an accident, the primary safety factor in nuclear fuel transportation is the shipping container itself: even under severe stress it must ensure that there will be no loss of radioactive contents and no significant increase in external radiation. This protection minimizes the consequences of potential accidents.

Cold or unirradiated fuel on its way to a plant poses little radiological hazard: uranium dioxide fuel pellets may be held safely in the hand. Clamped and cushioned inside steel boxes or cylinders, the pellets are packaged to protect them from external damage. On the other hand, the full spectrum of defense-in-depth—successive layers of protection—is provided for containment and shielding of the irradiated assemblies discharged from a reactor. First of all, the fuel pellets within their metal rods are designed to retain more than 99.9 percent of the radioactive fission products during high-temperature, high-pressure reactor operation. Similar driving forces would not be found in a transportation environment. Then the assemblies are stored onsite in water filled concrete pools for several months. When ready for shipment, about 95 percent of their radioactivity at the time of power generation has dissipated. Nevertheless, spent fuel is transported in the most carefully designed, fabricated, and tested shipping containers available.

In one cask design, the assemblies are sealed into a water-filled stainless-steel cylinder with walls 0.5 inch thick, clad with 4 inches of heavy metal shielding enclosed by a shell of 1.5-inch-thick steel plate, surrounded by 5 inches of water, and encircled by a corrugated stainless-steel outer jacket. The overall package measures 5 by 17 feet and weighs 70 tons. This multiple barrier containment ensures that the public is fully protected from radiation during routine transport of spent fuel.

And to make certain that the cask preserves this integrity even during abnormal circumstances, its design must withstand a series of "torture tests":

1. a 30-foot fall on a flat, unyielding surface, followed by
2. a 40-inch drop onto a vertical steel rod, followed by
3. exposure to a 1475°F fire for 30 minutes, followed by
4. submersion under 3 feet of water for 8 hours

The container must undergo these destructive forces, in sequence, with no breach of containment and with no significant reduction in shielding.

PERFECT RECORD OF SAFETY

During more than 30 years of transporting nuclear material of all kinds, no one has ever been injured in the United States by the radioactive nature of the cargo. However, accidents have occurred. Most have been minor, with only minimal damage to the vehicle. In one incident, however, a truck carrying spent fuel overturned, killing the driver. The cask remained intact.

Although every effort is taken to prevent such mishaps involving a shipment of nuclear fuel, some 115 transportation incidents are nevertheless expected to take place over the operating lifetime of 100 reactors. However, defense-in-depth makes certain that the packages will survive the most severe abuse foreseeable in such situations—in none of these forecast events is either significant radiation damage to the environment or exposure of the public expected.

Conservative engineering design and careful operating practices enforced by strict regulatory standards ensure that the shipping of fuel to and from nuclear power plants will continue safely and efficiently as this means of generating electricity expands to meet our increasing energy needs.

12 Safeguarding Nuclear Material

Since the dawn of the atomic age in the early 1940s, the importance of protecting nuclear materials against theft or diversion and the necessity of protecting nuclear facilities against sabotage has been recognized and precautions have been taken. Certain nuclear materials, such as plutonium and highly enriched uranium, can be used to make nuclear weapons. And since plutonium is potentially hazardous if sufficient quantities enter the body, it is extremely important that our protective systems be highly effective and that they be continually evaluated to assure their effectiveness. Sabotaging of nuclear facilities could create a serious public-safety problem and in some cases adversely affect the national security. Therefore, an intense safeguards effort has been under way for a number of years, both in the United States and other countries.

SAFEGUARDS DEVELOPMENT IN THE UNITED STATES

Prior to the passage of the Atomic Energy Act of 1954, all the strategic nuclear materials and facilities in the United States were under the control of the United States government because they were government property. The act permitted commercial organizations to possess strategic* nuclear materials for peaceful uses of nuclear energy. Because relatively small amounts of nuclear materials were in use at the time, the government depended upon the high value of the materials and the severe criminal penalties for unauthorized use to protect them against misuse. However, in the mid-1960s, it became apparent that strategic nuclear materials would be in the possession of a large number of commercial organizations involved in the expanding nuclear power industry. At that time, the government began a program for the development and implementation of methods for safeguarding nuclear materials and facilities.

Two government agencies have safeguards responsibilities. The Energy Research and Development Administration (ERDA) is responsible for implementing safeguards in government-owned facilities and conducting research and development on domestic and international safeguards policies and requirements. The Nuclear Regulatory Commission (NRC) establishes safeguards regulations for domestic commercial activities and inspects such activities for compliance with regulatory requirements. It is the policy of both organizations to assure that safeguards are equally effective for both government and private industry. An extensive amount of research and development on safeguards standards and regulations and on nuclear material measurement techniques, physical protection methods, and effective safeguards systems required for their implementation has already been carried out.

SAFEGUARDS SYSTEMS TODAY

The safeguards systems for protecting facilities and materials consist of three basic components:

- a strong physical protection system
- a system, which includes a personnel reliability program, to carefully control strategic nuclear materials so that theft or diversion is difficult and can be detected promptly
- a strong materials management and accounting program to provide timely information as to material location and custodianship

A physical protection system for a plant that processes or uses strategic nuclear materials is designed to provide multiple obstacles between the materials and a would-be thief or saboteur. It includes the use of trained armed guards, multiple physical barriers that are equipped with alarm devices, strict access control to the facility and critical operations, televised surveillance of critical areas, multiple communication systems with local law-enforcement authorities, and security-control facilities that are fortified against armed attack.

Strategic nuclear material is shipped under armed guard, and the shipment is traced from a control center that is in frequent radio or telephone communication with the transporting vehicle.

*Strategic quantities of nuclear materials is a term that describes amounts greater than 2 kilograms of plutonium and 5 kilograms of uranium enriched to more than 20 percent.

The nuclear-material control system for a processing plant involves dividing the plant into a series of separate control areas that are physically separated from each other and carefully measuring and recording the amounts of nuclear materials introduced to and removed from each area. The inventory of nuclear materials in each area is measured periodically and compared to the recorded inventory. The significance of the differences is determined mathematically to detect the possibility of unauthorized removal of nuclear materials from a control area or from the plant.

INTERNATIONAL SAFEGUARDS

Ways have been devised for countries without nuclear technology to participate in the development and use of the peaceful atom, while assuring that the required nuclear materials were properly safeguarded against diversion to military uses. At first, this assurance was provided by written agreements under which the country supplying the nuclear materials and technology had the right to inspect the foreign country's activities to ensure that the materials or technology were not misused.

In 1957, the International Atomic Energy Agency (IAEA) was organized, under the framework of the United Nations, to foster the peaceful uses of nuclear energy. In 1961, the IAEA established a system for nuclear material safeguards whereby nonnuclear weapons states agree to inspections by specifically trained IAEA staff. The IAEA system of safeguards embodies the same basic principles as those imposed by the United States government on domestic nuclear activities.

The Treaty on Non-Proliferation of Nuclear Weapons was drawn up at the Geneva Disarmament Conference in 1966 and endorsed by the United Nations in 1968. The treaty came into force in 1970, and by May 1975 a total of 108 countries had signed it. Signers of this treaty are prohibited from transferring nuclear weapons to any country that does not have a nuclear weapons capability. Countries not having such a capability are prohibited from manufacturing nuclear weapons or acquiring them. They are also obligated to adopt IAEA safeguards and to accept IAEA inspection of all their nuclear facilities to ensure that there is no diversion of nuclear materials to military applications.

FUTURE PROSPECTS

As amounts of nuclear materials and the numbers of such facilities increase with the growth in use of nuclear power, safeguards will be constantly improved and expanded. Research and development work by the government, private industry, and other countries is also continuing and is directed toward meeting the safeguards needs of the future.

13 Transportation

The transportation sector is a prime conservation target. It accounts for one quarter of the United States' annual energy consumption, it uses energy inefficiently and it primarily uses oil.

The internal combustion engines, which predominate in transportation, are not very efficient—only about 15 percent of the fuel energy is actually used to move vehicles. It is not, of course, possible to build 100 percent efficient engines, but there is much room for improvement.

Transportation figured importantly in the conservation efforts during the 1973-1974 oil embargo. The national speed limit was lowered to 55 mph, and efforts were made to reduce the number of miles traveled. Both these efforts succeeded in part. The average speed of automobiles on interstate highways declined by 7.5 mph in 1974, and total miles traveled during the first 8 months of 1974 were 4 percent lower than the similar period in 1973. Gasoline consumption was reduced by 3.4 percent. An appreciated side effect of lowered speed was 17 percent reduction (9500 fewer deaths) in highway fatalities.

While the savings from these energy measures were important and continue to have an effect (the 55 mph limit is still federal law), there are many other ways to reduce consumption.

RESOURCES AND STRATEGIES

Urban and intercity passenger travel has doubled since 1950. Intercity bus transport has dropped from 5 percent of the total passenger miles in 1950 to 3 percent in 1970; rail transport has dropped from 7 percent to 2 percent. Air travel, on the other hand, has increased. In urban travel, bus transport has dropped from 20 percent in 1950 to 2 percent in 1970. It is clear then why many of the conservation strategies focus on the automobile.

Moving People

Transportation has two purposes: moving people and moving things. The United States spends the largest fraction of transportation energy on people. We have already discussed engine efficiency, but a more pointed measure is the energy per passenger mile. The energy price we pay for speed and convenience is evident.

There are three strategies that can bring about more efficient passenger transport. "Load factors," the percentage of capacity carried in a car, plane, or bus, can be increased; engine efficiencies can be improved; and we can try to switch passengers from one mode to another (from the urban auto to mass transit, for instance).

Car-pooling is an attempt to improve urban load factors from the present 28 percent (an average of 1.4 passengers in a five-passenger car). The convenience of driving alone to work and the subsidized parking provided by employers, however, works strongly against car-pooling, and little progress has been made.

The overall efficiency of a car as measured in miles per gallon (mpg) depends more on car weight and add-on systems such as air conditioning, automatic transmission, and emission control than on the engine itself. The national average has dropped over the years and was down to 13.9 mpg in 1974 as compared to 15.5 mpg in 1967. It improved slightly to 15.6 mpg in 1975 and rather dramatically to 17.6 mpg in 1976. While lighter smaller cars contributed to this improvement, a significant share of credit belongs to the improvement in emission control due to the catalytic converter.

The improvements are continuing, and the federally mandated "sales weighted average" of 20 mpg by 1980 and 27.5 mpg by 1985 should be achieved. Much of the improvement will be brought about by the reduction in weight; each 400-pound reduction in a medium-sized car produces an improvement of about 1 mpg. (Considerations of safety and ridability will ultimately limit this approach.) We will eventually see new engine designs, and we have already seen increased use of more efficient diesel engines.

Changes in travel modes could have significant effects but will come slowly. In fact, mass transit has such a small share of the market that even doubling its share will not make a big dent in energy consumption. New transit systems and improvement in older systems will likely have to be combined with penalties for automobile use before a shift to transit becomes an important part of the conservation picture.

Table 13-1 lists FEA estimates of possible savings (in barrels of oil per day) of three of the measures discussed herein. The total potential savings are greater than our present domestic oil production.

In intercity travel, the automobile is used a bit more efficiently, averaging 2.4 passengers. Shifting to trains or buses would still save considerable energy. Airplane travel is both the most energy-intensive and the most rapidly increasing form of intercity travel. The reduction of airline routes and other measures introduced in the past few years have resulted in some savings; a 13 percent reduction in fuel use in 1974 over 1973 was reported. In the long run, we may have to approach energy conservation by marking out certain distances for different types of transportation— for example, emphasizing autos for 100 miles or so, trains and buses for 100 to 500 miles, and airplanes for the longer distances.

Moving Things

The energy efficiency of various freight modes is shown below:

Mode	Energy Cost (Btu needed per ton-mile)
Waterway	680
Pipeline (oil)	1,850
Railroad	700
Truck	2,800
Airplane (domestic)	62,000

The data show the energy penalty we pay for door-to-door delivery by truck and the speed of air freight. These are our most rapidly growing modes of freight transport. We are somewhat limited in conservation strategies for freight transport. Load factors are already high in most cases. Trucks do get better gas mileage at 55 mpg (despite all the controversy), but they move less tons per mile per day this way. Some aerodynamic improvements can be made in truck design (reduction of air drag), which could save 10 percent or so of their fuel. Larger gains, however, would come from shifts between modes, from trucks and planes to trains, for instance.

NEW TECHNOLOGIES

Despite persistent rumors, there are no ultraefficient new engines hidden in the automobile companies' parent files. What we will see, in the near future, are many small improvements such as the electronic ignitions and stratified charge engines that are already being marketed. There is also considerable ERDA support for longer-range improvements. A new, continuously variable transmission that could improve fuel economy by as much as 26 percent is under development. Two new engines, the gas turbine and the Stirling engine, are receiving much attention. In the gas turbine, high-velocity, hot gases from fuel combustion turn

Table 13-1. Energy impacts of transportation conservation measures.

MEASURE	ESTIMATED FUEL SAVINGS IN 1980 (BARREL/DAY)
Increased new-car fuel economy by 40% between 1974 and 1980 (20 mph in 1980)	690,000
Increased rush hour auto load factor from 1.2 to 1.6 (car-pooling)	440,000
Double fraction of urban travel carried by transit from 2.5% in 1973 to 5% in 1980	52,000
	1,182,000

a turbine. The Stirling engine is an external combustion engine. The fuel is burned outside the cylinder, and the hot, high pressure is then used to drive the piston. Both these engines have important advantages. They can burn any kind of fuel including nonpetroleum fuels such as alcohol. They will be more efficient and much cleaner than the gasoline or diesel engines and will not require emission controls.

Electric Cars

In many ways, the ideal engine for transportation would be the electric car. It is efficient, quiet, and does not emit pollutants or use oil-based fuels. Although the electric vehicle has a long history and is marketed today, it does not really compete with gasoline-powered vehicles.

The major unsolved problem of the electric automobile is energy storage. The common lead-acid automobile storage battery is too heavy. It does not store enough energy per pound, cannot be discharged and recharged enough times, and is too expensive. Lead batteries account for 40 to 50 percent of the weight of existing electric vehicles, and the set, which costs about $600, needs replacing every 3 years or so.

Work is also ongoing on such refinements as "regenerative braking"—slowing the vehicle by running the motor as a generator to recharge the battery. With its higher efficiency and the fact that it could largely draw on the nighttime capacity of large power plants, the use of electric autos in urban situations could save considerable energy. It has been estimated that for each 1 percent of urban passenger car miles taken over by electric cars, 4 billion kilowatt-hours of electricity would be consumed and 12.6 million barrels of gasoline saved. If it is further assumed that the electricity is produced in a relatively pollution-free way, the potential improvement becomes most attractive.

Other New Technologies

By the next century, an even more interesting substitution of electricity in the transportation sector may take place— video communication instead of business flights. It appears that in many of the knowledge or information industries (insurance, finance, education, government, etc.), face-to-face telecommunicators may eventually replace physical meetings. The energy savings potential is great. It is estimated, for instance, that a gallon of gasoline used in an automobile is equivalent in energy to that required for 80 hours of telephone conversation. This may turn out to be the most practical way to replace gasoline by electricity.

SUMMARY

The transportation sector uses one quarter of our total energy directly (at an efficiency of about 15 percent), and as much as 40 percent of the total if indirect energy consumption is added in. It uses a premium and scarce fuel.

Many of the strategies we have described will require changes in attitudes and life-styles in the private sector, but even those that seemingly call for federal intervention face a dispersal of responsibility across many governmental departments and agencies.

14 Tomorrow's Cars

One of the United States' energy problems is that we are using too much oil. Cars and trucks run on gasoline or diesel fuel; trains run on diesel fuel; aircraft run on high-octane gas or kerosene; and ships run mainly on diesel fuel. The country's entire transportation system depends on oil. In addition, 17 percent of all of our electric power comes from oil-burning power stations.

If we continue to use oil lavishly, we will exhaust our supplies. The United States is already importing about 40 percent of its oil, and projections show that by 1985 this will increase to 50 percent. This country can neither control the price of imported oil nor be assured of a steady supply. We are in a situation that is both economically and strategically bad.

The following measures should be looked into:

1. Passenger cars consume nearly 30 percent of all oil, a figure that must be reduced by all possible means. This means smaller and lighter cars with more efficient engines and transmissions, reduced rolling friction, and lower aerodynamic drag. The gasoline fuel saved in this way will give us more time to solve the problems involved with advanced power systems and their environmental impact.
2. The entire transportation sector should be removed from its petroleum fuel base, and new, clean, and efficient power plants with multifuel capability should be developed.
3. New technology should be developed for long and short-term energy storage. Such research efforts will spur progress in ways to store electricity, hydrogen, and synthetic liquid fuels.

Improving the Internal Combustion Engine

An immediate changeover to advanced power units is impossible because the auto industry is a mass production one, and the only type of engine it is equipped to produce is the piston-type internal combustion engine. Also, it takes about 5 years to get a new engine into production, and an engine line is usually used for 20 years. Consequently, a big effort must be undertaken to make the conventional engine more efficient and more economical. But the increasing concern over pollution limits the fuel economy that can be achieved with that engine.

The catalytic converters in today's car engines do not in themselves improve fuel economy; they simply allow the engine to run "dirtier" because of their ability to clean up the exhaust gas. At best, converters may be only a short-term answer.

Diesel engines are more economical to use, but create noise, an unpleasant odor, smoke, and cost problems. Diesel engines serve half of the heavy truck market, and automakers in Europe and Japan produce diesel cars. United States auto manufacturers are gradually switching to diesel engines by making them available in light trucks and cars.

Alternative Electric and Hybrid Systems

Looking away from heat engines to on-board energy storage, we find that the electric-battery car has a practical potential for replacing cars and vans in local use. New batteries with higher power density are in the experimental stage and promise to more than double the speed and range of current electric vehicles. In the long term, electric-battery drive is attractive because batteries can be charged from energy sources that otherwise cannot be used in transportation, such as nuclear power, solar energy, or hydroelectric power.

Hybrid electric systems are another attractive approach. A very small engine can charge a battery pack, which then provides power to drive an electric motor. The engines can run at constant speed where they run most efficiently. Such systems are complex, however, with a certain duplication of subsystems, because of the multistage energy-conversion process.

Advanced Heat Engine Systems

Alternative power systems with multifuel capability are being developed by a number of companies. Stirling-cycle engines, which are clean, quiet, vibrationless, and efficient, are external combustion engines that can be built in all the sizes needed by the automotive industry.

Steam-power systems are under study for a number of reasons. The operating principles and hardware are simple; the engine runs with low emission levels and can use a number of fuels. The system involves many separate units, however, and various operational problems have not been overcome. Current experimental units show very disappointing fuel economy, but researchers believe that this can be substantially improved.

The automotive gas-turbine engine is very different from the jet engines on an airplane and shares only the basic operational principles. The automotive gas turbine must work in a more demanding environment and needs heat exchangers, for instance, to assure part-load fuel economy. In return, it requires no cooling system. It is clean, vibrationless, efficient, and has a very long service-life potential. Its current problems are high initial cost, complexity, high speed of rotation, and low part-load fuel economy.

Future Fuels

Our principal fossil-fuel resource is coal. At the current rate of consumption, the United States has enough coal to last for about 300 years. Coal can be liquefied into gasoline or methanol, or gasified.

Methanol is compatible with all heat engines and burns in a clean fashion. But it has lower energy density than gasoline, which means that the fuel tank must be 2.5 times larger to give the same range. Also, minor operational problems must be overcome.

Hydrogen is an attractive fuel for the long-term future. It packs more power per ounce than any known fuel, is abundant in nature, could be a by-product of nuclear power stations, is compatible with all manner of heat engines, has low-emission potential, and its sources are potentially inexhaustible. However, the supply of hydrogen in quantity and at a reasonable price is tied to abundant nuclear energy, which is needed to separate the hydrogen from water. There are storage and distribution problems—liquid hydrogen can only be stored at low temperatures, which is possible on ships and other heavy equipment but not practical for automobiles. Various types of chemical storage have been proposed, and a metal hydride storage system looks promising.

Finally, there are fuel cells, which convert chemical energy directly into electricity. In automobiles, fuel cells would be part of a hybrid electric-propulsion system. The basic type of fuel cells work by reacting hydrogen with oxygen, but other fuel cells using lower-cost fuels, such as hydrazine, methanol, or carbon, are under development.

15   Experimental Electric Cars

Three experimental electric vehicles were used to transport Energy Research and Development Administration (ERDA) staff between downtown federal government offices and to Capitol Hill on official business. These vehicles were part of a program to test and evaluate the available electric cars under Washington, D.C., traffic conditions.

Records were maintained with each vehicle regarding range in stop-and-go-city driving, efficiency, performance comparability with conventional automobiles, drivability, maintenance requirements, and overall reliability. These tests and maintenance records enable ERDA to assess more accurately the role of electric vehicles in energy-conservation programs and also to assist in the guidance of research and development programs to advance electric vehicle technology.

The electric vehicles used battery-electric energy to run a motor that propelled it. Theoretically, the motor could be fully engaged as soon as the switch was turned on. In practice, however, a multispeed transmission provided the flexibility of operation that best fit the modern traffic pattern and helped to obtain improved vehicle performance. The goal, of course, was to make electric cars more practical, economical, and, therefore, more available to the public.

Some of the more frequently asked questions about these three electric vehicles are answered in the following pages.

- *Who conducted the test?* The ERDA Division of Transportation in the ERDA Office of Conservation conducted the test.
- *Who manufactured these cars?* The three cars were manufactured by Electric Vehicle Associates, Inc., of East Liberty, Ohio.
- *How were the cars selected?* The selection of the cars from Electric Vehicle Associates was the result of an extensive survey and brief test program. Twenty-six potential manufacturers were contacted to determine who could provide immediate delivery of three four-passenger vehicles manufactured from components approved for highway operation. At the time the survey was conducted, Electric Vehicle Associates was the only manufacturer that could meet these requirements.

- *What are the basic characteristics of these cars?* Each car can carry four passengers, has a top speed of 55 mph, and can go about 25 miles in city stop-and-go traffic.
- *How much do these cars cost?* At the present time, each costs more than $10,000. These are not in large-scale production; and therefore, the cost per vehicle does not reflect quantity production-line costs.
- *What kind of battery load do they carry?* Each car carries about 1100 pounds of lead-acid batteries.
- *Why aren't electric cars in wide use now?* The major impediments facing electric vehicles today are life cost, performance, and range. Each of these factors is related to the low energy storage and power capacity per unit weight for existing batteries. The result is that a high percentage of the total vehicle weight is made up by the energy-storage system (batteries). Due to the limited life of current batteries, the battery pack must be changed several times throughout the life of the vehicles. Thus, the overall life cost of today's electric vehicles is high.
- *Aren't there clean-air advantages to the electric car?* True, battery-electric drive is largely pollution-free; however, it should be noted that without proper safeguards, increased production of electricity can also result in adverse environmental impact.
- *What are the chances of a breakthrough in battery technology?* Manufacturers of lead-acid batteries are working hard toward improving the weight versus energy-storage ratio of batteries, and an improvement of about 10 percent may be possible. However, other types of batteries are under development, which have the potential for much higher energy-and-power-storage capacity per unit weight. The successful development of these new systems will significantly reduce vehicle weight and increase battery life. This will have a major effect on lowering the vehicle life-cycle cost, and also on improving performance and range.
- *Do electric cars save energy?* In applications involving a large amount of engine idling time, the electric car can save energy since it does not consume energy

at standstill (the gasoline-fueled car would consume a cup of gas for every 6 minutes idling). However, the primary energy benefits result from the fact that the electricity for battery recharging could be generated by power plants using nonpetroleum energy sources such as coal, nuclear, hydroelectric, etc. Additional petroleum savings, although minor, are also realized since electric cars use very little lubricating oil.

- *What kind of new batteries are in the future?* In time, new batteries with 5 to 10 times the energy versus weight ratio may be developed. This would more than double the range and speed capacity for batteries of similar weight and size as compared with the lead-acid battery. At least six different types of batteries are in various stages of development at the present time. These include: nickel/zinc, iron/nickel, zinc/chlorine, lithium/metal-sulfide, and sodium/sulfur.

- *What is the future of the electric car?* For the near future, there is the obvious use as a town car—one not competitive with the gasoline-fueled automobile because of the limited range. But, in the long run, the electric car may compete with the familiar internal combustion auto.

A key advantage of the electric car is the fact that it can reduce United States reliance on scarce petroleum-based fuels, shifting the burden to coal, nuclear, and eventually to renewable energy sources such as solar for the production of electricity.

Yet another advantage of the electric car is the potential it offers to allow more efficient use of the nation's electric generating capacity. By recharging auto batteries during off-peak hours, generating capacity that would otherwise be idle is earning revenues for utilities, which could lower the per-kilowatt-hour cost of electricity.

- *To date, what kind of results have been obtained on the electric experimental vehicles?* Tests of the four-passenger Electric Vehicle Associates car at the Transportation Research Center, Cleveland, Ohio, resulted in the following performance figures:

Range *At Steady Speeds of*	*Test Results* *(3750 pounds gross weight)*
25 mph	56 miles
45 mph	32 miles
53 mph	28 miles
Acceleration	
0–30 mph	12–13 seconds
0–45 mph	38–39 seconds
Braking	
30–0 mph	47 feet
50–0 mph	104 feet

16   New Electric Light Bulb

Lighting for homes and businesses uses approximately 20 percent of the electrical energy generated in the United States today, or about 5 to 6 percent of the total national energy consumed. Inefficient incandescent bulbs, which use 220 billion kilowatt-hours annually, are the leading consumer of energy among lighting devices.

Energy needed for lighting could be reduced by 70 percent by replacing incandescent with fluorescent lighting, which produces three times the light with the same amount of electricity. In many instances, however, the cost of replacing incandescent fixtures with new ones to hold the currently bulky fluorescent tube inhibits the switch to more energy-efficient lighting.

A new electrodeless fluorescent bulb, an energy-efficient light that fits a standard incandescent fixture, is being refined and readied for market by the Lighting Technology Corporation in a research-and-development project supported by ERDA. With this new LITEK bulb, which looks like a conventional bulb, incandescent users will be able to relamp for greater energy efficiency without replacing existing fixtures. Since the new bulb contains no electrodes or filaments to burn out, it may provide efficient and inexpensive lighting for several years.

HOW DIFFERENT BULBS WORK

An *incandescent* bulb, most commonly used in the home, glows when its filament is heated to a high temperature by the electricity flowing through it. Incandescent lighting is very inefficient—only about 10 percent of the energy input becomes visible light, and the rest is wasted heat.

In a *fluorescent* light, electrical energy is used much more efficiently. Electrons emitted by an electrode in the conventional fluorescent light excite atoms, producing ultraviolet light. This ultraviolet light causes the phosphor layer, which coats the inside of the fluorescent tube, to fluoresce, emitting visible light but very little waste heat.

In the new *electrodeless fluorescent* light, a small electronics package in the bulb base produces a radio-frequency signal, which sets up a magnetic field. This magnetic field, like the electrons emitted in the conventional fluoresent light, excites atoms. The bulb will use only one-third of the energy needed for incandescent lighting and will waste less energy as heat. Energy and cost savings, along with longer bulb life and compatibility with standard fixtures, give this new light source attractive advantages over conventional bulbs.

ENERGY CONSERVATION

Generating the 220 billion kilowatt-hours of electricity used annually for incandescent lighting requires the equivalent of over 1 million barrels of oil per day. Nationwide use of the new electrodeless fluorescent bulb could reduce the fuel required for lighting by the equivalent of up to 500,000 barrels per day.

Mainly because of the electronics component, the new bulb will cost about $7.50, although volume production is expected to bring this initial cost down. However, reduced operating costs and fewer bulb replacements will mean substantial savings with the new electrodeless fluorescent bulb. Since it will produce three times as much light per unit of electricity as an incandescent bulb, the new bulb will soon repay its user with lower operating costs.

In a room now lighted 8 hours a day with a 100-watt incandescent bulb, the LITEK will pay for itself within a year (assuming electricity costs 3.5 ¢ per kilowatt-hour) and thereafter provide more cost savings. The projected operating cost of the LITEK bulb over its 20,000-hour operating lifetime is $30.80, compared to $83 to operate the twenty-six 100-watt incandescent bulbs it could replace. (Conventional bulb life is only about 750 hours.)

17 Basic Energy Conservation Code

The private sector has responded to the need to conserve energy by developing a standard for energy conservation in new building design, that is, American Society of Heating, Refrigerating, and Air Conditioning Engineers (ASHRAE) Standard 90-75. Although it is not the final answer to the problem of energy conservation in buildings, the standard represents the current state of the art and has been shown to effectively reduce energy consumption.

The United States government has also responded to this need as can be witnessed by two public laws currently in effect. The first is the Energy Policy and Conservation Act (PL 94-163, December 22, 1975). This act requires each state to develop and implement a plan for a reduction in energy consumption of 5% by 1980 and requires, among other things, mandatory thermal efficiency standards and mandatory lighting efficiency standards. The second law is the Energy Conservation and Production Act (PL 94-385, August 14, 1976), which directs the U.S. Department of Housing and Urban Development to establish thermal performance standards for buildings. It is expected that the provisions of ASHRAE Standard 90-75 will be acceptable criteria to satisfy the requirements of these laws. In any case, they provide a means of implementing effective energy-conservation measures immediately.

This chapter presents all applicable energy-conservation requirements and applicable reference standards presently found in the Building Officials & Code Administrators (BOCA) Basic Building, Mechanical, and Plumbing Codes (as amended 1976). The energy-conservation requirements of the BOCA Basic Codes are based on the provisions of ASHRAE Standard 90-75.

The BOCA Basic Codes are kept up-to-date through the review of changes proposed by code enforcement officials, industry and design professionals, and other interested persons and organizations. Proposed changes are discussed in a public hearing, carefully reviewed by committees, and acted upon by code-enforcement officials in an open meeting of the organization. Those changes approved are published annually in supplements to the codes in convenient form for adoption by local governments. New editions are then prepared every 3 years and contain all approved changes since the previous editions. When any changes are approved that affect energy-conservation requirements, this document will be updated to reflect such changes.

For more information, write:

Building Officials & Code Administrators International
1313 E. 60th St.
Chicago, IL 60637
(312) 947-2580

Following is the text of the BOCA Basic Codes:

ARTICLE 1 SCOPE AND APPLICATION

Section 100.0 General

100.1 Scope: These provisions regulate the design and construction of the exterior envelopes and selection of HVAC, service water heating, electrical distribution systems and equipment required for the purpose of effective use of energy and shall govern all buildings and structures, or portions thereof, hereafter erected that provide facilities or shelter for human occupancy.

Exceptions:

1. Buildings and structures, or portions thereof, which are neither heated nor cooled.
2. Buildings and structures, or portions thereof, whose peak design rate of energy usage is less than one (1) watt per square foot or three and four-tenths (3.4) Btu per square foot of floor area for all purposes.

Section 101.0 Plans and Specifications

101.0 General: Plans, specifications and necessary computations shall be submitted to indicate conformance with this section and other applicable sections of this code.

101.2 Details: The plans and specifications shall show in sufficient detail all pertinent data and features of the building and the equipment and systems as herein governed, including but not limited to: exterior envelope component materials, U values of elements, R values of insulating materials, size and type of apparatus and equipment, equipment and system controls and other pertinent data to indicate conformance with the requirements herein.

ARTICLE 2 DEFINITIONS

Section 200.0 General

200.1 Scope: The provisions of this article contain the definitions of terms relating to energy conservation requirements of all buildings and structures.

200.2 Application of terms: The terms herein defined shall be used to interpret all the applicable provisions contained herein.

Section 201.0 General Definitions

201.1 Meaning: Unless otherwise expressly stated, the following terms shall, for the purpose of this code, have the meaning indicated in this section.

201.2 Tense, gender and number: Words used in the present tense include the future; words used in the masculine gender include the feminine and neuter; the singular number includes the plural and the plural the singular.

201.3 Terms not defined: Where terms are not defined, they shall have their ordinarily accepted meanings or such as the context may imply.

Coefficient of performance (COP)–Cooling: The ratio of the rate of net heat removal to the rate of total energy input, expressed in consistent units and under designated rating conditions. For definition of specific net heat removal and inputs, see Sections 402.1.1 and 402.1.2.

Coefficient of performance (COP)–Heat pump, heating: The ratio of the rate of net heat output to the rate of total energy input, expressed in consistent units and under designated rating conditions.

The rate of net heat output shall be defined as the change in the total heat content of the air entering and leaving the equipment not including supplementary heat.

Total energy input shall be determined by combining the energy inputs to all elements, except supplementary heaters, of the heat pump, including, but not limited to, compressor(s), pump(s), supply-air fan(s), return-air fan(s), outdoor-air fan(s), cooling-tower fan(s), and the heating, ventilating and air conditioning system equipment control circuit.

Degree day, heating: A unit, based upon temperature difference and time, used in estimating fuel consumption and specifying nominal heating load of a building in winter. For any one (1) day, when the mean temperature is less than sixty-five (65) degrees F, there exist as many degree days as there are mean temperature for the day and sixty-five (65) degrees F.

Energy efficiency ratio (EER): The ratio of net cooling capacity in Btu to total rate of electric input in watts under designated operating condition.

Exterior envelope: The elements of a building which enclose conditioned spaces through which thermal energy may be transferred to or from the exterior.

Floor area, gross: Gross floor area shall be the floor area within the perimeter of the outside walls of the building under consideration, without deduction for hallways, stairs, closets, thickness of walls, columns, or other features.

Packaged terminal air-conditioner: A factory-selected combination of heating and cooling components, assemblies or sections, intended to serve a room or zone. (For the complete technical definition see Air Conditioning and Refrigeration Institute Standard 310-70.)

Reheat: The application of sensible heat to supply air that has been previously cooled below the temperature of the conditioned space by either mechanical refrigeration or the introduction of outdoor air to provide cooling.

Residential buildings: All buildings and structures or parts thereof shall be classified in the residential (R) use group in which families or households live, or in which sleeping accommodations are provided for individuals with or without dining facilities, excluding those that are classified as institutional buildings.

Use group R-1 structure: This group shall include all hotel and motel buildings, lodging houses, boarding houses and dormitory buildings arranged for the shelter and sleeping accommodation of more than twenty (20) individuals.

Use group R-2 structures: This use group shall include all multiple-family dwellings having more than two (2) dwelling units; and shall also include all dormitories, boarding and lodging houses arranged for shelter and sleeping accommodations by more than five (5) and not more than twenty (20) individuals.

Use group R-3 structures: This use group shall include all buildings arranged for the use of one- or two-family dwelling units including not more than five (5) lodgers or boarders per family.

Resistance, thermal (R): A measure of the ability to retard the flow of heat. The R value is the reciprocal of a heat transfer coefficient, as expressed by U. $R = 1/U$.

Thermal transmittance (U): Overall coefficient of heat transmission or thermal transmittance (air to air) expressed in units of Btu per hour per square foot per degree F. It is the time rate of heat flow. The U value applies to combinations of different materials used in series along the heat flow path and also to single materials that comprise a building section, and include cavity air spaces and surface air films on both sides.

Thermal transmittance (Uo): Overall (average) heat transmission or thermal transmittance of a gross area of the exterior building envelope, expressed in units of Btu per hour per square foot per degree F.

The Uo value applies to the combined effect of the time rate of heat flow through the various parallel paths, such as windows, doors, and opaque construction areas, comprising the gross area of one or more exterior building components, such as walls, floor, or roof/ceiling.

Thermostat: An instrument which measures changes in temperature and controls device(s) for maintaining a desired temperature.

Zone: A space or group of spaces within a building with heating or cooling requirements sufficiently similar so that comfort conditions can be maintained throughout by a single controlling device.

ARTICLE 3 BUILDING ENVELOPE

Section 300.0 General

300.1 Scope: The provisions of this article regulate the thermal performance and air leakage of the exterior envelope of buildings and structures.

Section 301.0 Exterior Envelope Requirements

301.1 General: The intent of this section is to provide minimum requirements for exterior envelope construction in the interest of energy conservation. Calculation and measurement procedures and information contained in the ASHRAE Standard 90 . . . shall be used, except where otherwise noted, to determine conformance with the requirements herein, in accordance with recognized standards.

In addition to the criteria set forth in this article, the proposed design may take into consideration the thermal mass of the building in considering energy conservation.

301.1.1 Thermal performance: All buildings and structures that are heated or mechanically cooled shall be constructed so as to provide the required thermal performance of the various components.

The required thermal transmittance value (Uo) of any one component, such as roof/ceiling, wall or floor, may be increased and the Uo value for other components decreased provided that the overall heat gain or loss for the entire building envelope does not exceed the total resulting from conformance to the required Uo values.

301.1.2 Different requirements: A building that is designed to be both heated and cooled shall meet the more stringent of the heating or cooling requirements of the exterior envelope as provided in this section when requirements differ.

301.1.3 Exterior walls: For the purpose of this article the gross area of exterior walls consists of all opaque wall areas, including foundation walls above grade, peripheral edges of floors, window areas including sash, and door areas, where such surfaces are exposed to outdoor air and enclose a heated or mechanically cooled space.

301.1.4 Roof assembly: For the purpose of this article a roof assembly shall be considered as all components of the roof/ceiling envelope through which heat flows, thereby creating a building transmission heat loss or gain, where such assembly is exposed to outdoor air and encloses a heated or mechanically cooled space.

The gross area of a roof assembly consists of the total interior surface of such assembly, including skylights, exposed to the heated or mechanically cooled space.

Where air ceiling plenums are employed, the roof/ceiling assembly shall:

- For thermal transmittance purposes not include the ceiling proper nor the plenum space as part of the assembly, and
- For gross area purposes be based upon the interior face of the upper plenum surface

Table 301.2.1a. Maximum allowable "Uo" values for gross exterior wall assemblies.

ANNUAL HEATING DEGREE DAYS*	DETACHED ONE- & TWO-FAMILY	ALL OTHER RESIDENTIAL
500	0.30	0.38
1000	0.29	0.37
2000	0.28	0.35
3000	0.26	0.33
4000	0.25	0.31
5000	0.23	0.29
6000	0.22	0.27
7000	0.20	0.26
8000	0.19	0.24
9000	0.17	0.22
10,000 or more	0.16	0.20

*As specified in Chapter 43 ASHRAE Handbook-Systems .

301.2 Criteria for residential buildings: The requirements herein shall apply to all buildings and structures or portions thereof of use groups R-1, R-2 and R-3 (residential, hotels, multi-family and one- and two-family) that are heated or mechanically cooled when not more than three (3) stories or forty (40) feet in height.

301.2.1 Walls: The gross area of exterior walls above grade, including foundation walls, should have a combined thermal transmittance value (Uo) not exceeding those specified in the following Table 301.2.1a.

Exceptions:

1. In location with less than five hundred (500) Heating Degree Days there shall not be a maximum Uo requirement if only heating is provided and the Uo shall be 0.30 maximum if the building is mechanically cooled.

2. The opaque exterior wall areas may be constructed having thermal transmittance (U) values in conjunction with glazed opening areas in accordance with Table 301.2.1b.

301.2.2 Roof/ceiling: The roof-ceiling assemblies shall have a combined thermal transmittance value (Uo) or shall be provided with thermal insulation having an "R" value as specified in the following Table 301.2.2.

Exception: Roof/ceiling assemblies in which the finished interior surface is essentially the underside of the roof deck, such as a wooden cathedral ceiling, may have a "Uo" value not to exceed 0.08 Btu per hour per square foot per degree F. for any Heating Degree Day area.

301.2.2 Floors over unheated spaces: The floor of a heated or mechanically cooled space located over an unheated space shall have a combined thermal transmittance value (Uo) or shall be provided with thermal insulation having an "R" value as specified in the following Table 301.2.3.

301.2.4 Slab-on grade floors: For slab-on grade floors, the perimeter of the floor shall be insulated with a material

Table 301.2.1b. Maximum allowable "U" values for above-grade exterior wall sections and corresponding maximum allowable glazed opening areas.

REQUIRED "U" OPAQUE WALLS BTUH PER SQUARE
FOOT PER DEGREE F. (3 STORIES OR LESS)

YEARLY DEGREE DAYS	GLAZED OPENINGS	USE GROUP R 3 PER CENT GLAZED OPENINGS				ALL OTHER RESIDENTIAL PER CENT GLAZED OPENINGS			
		10	15	20	25	15	20	25	30
2500 or less	Single	.21	.15	.09	.03	.25	.19	.13	.07
	Double	.26	.24	.21	.18	.33	.31	.29	.27
2501 to 4500	Single	.17	.12	.06	.02	.20	.14	.08	.03
	Double	.23	.20	.18	.14	.29	.26	.24	.21
4501 to 6000	Single	.14	.08	.02	NP	.15	.09	.03	NP
	Double	.19	.17	.14	.10	.24	.21	.18	.15
6001 to 8000	Single	.12	.06	.01	NP	.13	.07	.01	NP
	Double	.17	.14	.11	.08	.21	.19	.16	.13
8001 to 10,000	Single	.09	.02	NP	NP	.08	.02	NP	NP
	Double	.14	.11	.08	.04	.17	.14	.10	.06
10,001 or more	Single	.05	NP	NP	NP	.04	NP	NP	NP
	Double	.11	.07	.04	NP	.12	.09	.05	NP

Note 1. NP = not permitted.

Note 2. For glazed opening percentages other than those specified above, linear interpolation may be utilized.

Note 3. For combinations of single and double glazing, the "U" values above may be interpolated in proportion to the single and double glazed areas utilized.

Note 4. To obtain credit for triple glazing or superior quality sash, or to utilize combinations of single and double glazing not permitted by this table use Table 2102.2.1a [in the ASHRAE Handbook].

Note 5. Interpolation between given "U" values and between degree days is not permitted.

Table 301.2.2. Maximum allowable "Uo" values and alternative minimum allowable "R" values of added insulation for roof/ceiling assemblies.

ANNUAL HEATING DEGREE DAYS	MAXIMUM "Uo"	MINIMUM "R"
8000 or Less	0.05	19
More than 8000	0.04	22

Note 1. These values presume no significant thermal transmission through framing members, skylights or other interruptions in the roof envelope. If such interruptions occur, calculations must be made showing conformance to the required "Uo" values.

Table 301.2.3. Maximum allowable "Uo" values and alternative minimum allowable "R" values of added insulation for floors over unheated spaces.

ANNUAL HEATING DEGREE DAYS	MAXIMUM "Uo"	MINIMUM "R"
500*	0.36	-
1000	0.32	-
2000	0.25	4
3000	0.18	6
4000	0.11	9
4500 or more	0.08	11

*Table values may be interpolated.

having a thermal resistance value (R) not less than those specified in the following Table 301.2.4.

The insulation shall extend downward from the top of the slab for a minimum distance of twenty-four (24) inches or downward to the bottom of the slab then horizontally beneath the slab for a minimum total distance of twenty-four (24) inches.

301.3 Other Buildings: The requirements herein shall govern all buildings and structures or portions thereof other than defined by Section 301.2.

Table 301.2.4. Minimum allowable "R" values of perimeter insulation for slab-on-grade floors.

ANNUAL HEATING DEGREE DAYS	HEATED SLAB	UNHEATED SLAB
500*	2.9	-
1000	3.3	-
2000	4.0	-
3000	4.8	2.8
4000	5.5	3.5
5000	6.3	4.2
6000	7.0	4.9
7000	7.8	5.5
8000	8.5	6.2
9000	9.3	6.8
10,000 or more	10.0	7.5

*Table values may be interpolated.

Table 301.3.1. Maximum allowable "Uo" values for gross exterior wall assemblies.

ANNUAL HEATING DEGREE DAYS	3 STORIES OR 40 FT. OR LESS	MORE THAN 3 STORIES OR 40 FT.
500	0.38	0.47
1000	0.37	0.46
2000	0.35	0.43
3000	0.33	0.41
4000	0.31	0.38
5000	0.29	0.36
6000	0.27	0.33
7000	0.26	0.31
8000	0.24	0.28
9000	0.22	0.28
10,000 or more	0.20	0.28

Table 301.3.2. Maximum allowable "Uo" values for roof/ceiling assemblies.

ANNUAL HEATING DEGREE DAYS	MAXIMUM Uo
3000 and less*	0.10
4000	0.092
5000	0.084
6000	0.076
7000	0.068
8000 and more	0.06

*Table values may be interpolated.

Table 301.3.3. Maximum allowable "Uo" values for floor assemblies over unheated spaces.

ANNUAL HEATING DEGREE DAYS	MAXIMUM Uo
500*	0.36
1000	0.32
2000	0.25
3000	0.18
4000	0.11
4500 or more	0.08

*Table values may be interpolated

Table 301.3.4. Minimum allowable "R" values of perimeter insulation for slab-on-grade floors.

ANNUAL HEATING DEGREE DAYS	HEATED SLAB	UNHEATED SLAB
500*	2.9	-
1000	3.3	-
2000	4.0	-
3000	4.8	2.8
4000	5.5	3.5
5000	6.3	4.2
6000	7.0	4.9
7000	7.8	5.5
8000	8.5	6.2
9000	9.3	6.8
10,000 or more	10.0	7.5

*Table values may be interpolated

301.3.1 Heating criteria for walls: All buildings and structures that are heated shall have a combined thermal transmittance value (Uo) for the gross area of exterior walls not exceeding those specified in . . . Table 301.3.1

301.3.2 Heating criteria for roof/ceiling: All buildings and structures that are heated shall have a combined thermal transmittance value (Uo) for roof/ceiling assemblies not exceeding those specified in . . . Table 301.3.2.

301.3.3 Heating criteria for floors over unheated spaces: The floor of a heated space located over an unheated space shall have a thermal transmittance value (Uo) not exceeding those specified in . . . Table 301.3.3.

301.3.4. Heating criteria for slab-on grade floors: For slab-on grade floors the perimeter of the floor shall be insulated with a material having a thermal resistance value (R) not less than those specified in . . . Table 301.3.4.

The insulation shall extend downward from the top of the slab for a minimum distance of twenty-four (24) inches or downward to the bottom of the slab then horizontally beneath the slab for a minimum total distance of twenty-four (24) inches.

301.3.5 Cooling criteria for walls: All buildings and structures that are mechanically cooled shall have an overall thermal transfer value for the gross area of exterior walls not exceeding those specified in . . . Table 301.3.5.

301.3.6 Cooling criteria for roof/ceilings: All buildings and structures that are mechanically cooled shall have a combined thermal transmittance value (Uo) for roof/ceiling assemblies the same as specified in Table 301.3.2 for heating.

301.4 Air leakage: The requirements of this section shall apply to all buildings and structures and apply only to those locations separating outdoor ambient conditions from interior spaces that are heated or mechanically cooled and are not applicable to separation of interior spaces from each other.

301.4.1 Standard: Compliance with the criteria for air-leakage shall be determined by ASTM E-283, Standard Method of Test for Rate of Air Leakage through Exterior Windows, Curtain Walls and Doors, listed . . . at a pressure differential of one and five-hundred-sixty-seven thousandths (1.567) lb/ft^2 which is equivalent to the effect of a twenty-five (25) mph wind.

Table 301.3.5. Maximum overall thermal transfer values for gross exterior walls.

DEGREES NORTH LATITUDE	MAXIMUM OVERALL THERMAL TRANSFER VALUE BTU PER SQUARE FOOT
24	29.0
32	31.3
40	33.5
48	35.7
56	38.0

301.4.2 Acceptance criteria: The air infiltration rate for windows shall not exceed five-tenths (0.5) cfm per foot of sash crack.

The air infiltration rate of sliding glass doors in residential buildings shall not exceed five-tenths (0.5) cfm per square foot of door area.

The air infiltration rate of swinging doors in residential buildings shall not exceed one and twenty-five hundredths (1.25) cfm per square foot of door area.

The air infiltration rate for swinging, revolving or sliding doors in other than residential buildings shall not exceed eleven (11) cfm per lineal foot of door crack.

301.4.3 Caulking and sealants: Exterior joints around windows and door frames, between wall cavities and window or door frames, between wall and foundation, between wall and roof, between wall panels, at penetrations or utility services through walls, floor and roofs, and all other openings in the exterior envelope shall be caulked, gasketed, weatherstripped or otherwise sealed.

ARTICLE 4 WARM AIR HEATING, VENTILATING AND AIR CONDITIONING SYSTEMS AND EQUIPMENT

Section 400.0 General

400.1 Scope: This article applies to air duct systems employing mechanical means for movement of air used for warm air heating, ventilation, air conditioning systems, exhaust systems and combination heating and air conditioning systems, except that this article shall not apply to systems for the removal of flammable vapors or residues or to systems for conveying dust, stock or refuse by means of air currents.

Heating, ventilating and air conditioning systems of all buildings and structures or portions thereof shall be designed and installed for efficient use of energy as herein provided. Calculations of heating and cooling loads shall be in accordance with the ASHRAE Standard 90 . . . .

For special applications such as hospitals, laboratories, thermally sensitive equipment, computer rooms, and manufacturing processes, the design concepts and parameters shall conform to the requirements of the application at minimum energy levels.

Section 401.0 Design Requirements

401.1 Design Conditions: For calcualtions under this Article, the following design temperature shall apply:

a. Outdoor design temperature shall be selected for listed locations in Chapter 33 of the ASHRAE Handbook of Fundamentals, . . . from columns of ninety-seven and one-half (97½) percent values for heating and two and one-half (2½) percent values for cooling.

b. Indoor design temperature shall be seventy (70) degrees F. for heating and seventy-eight (78) degrees F. for cooling.

c. Indoor design relative humidity for heating shall not exceed thirty (30) percent. For cooling, the actual design relative humidity within the comfort envelope as defined in ASHRAE Standards 55-74 "Thermal Environmental Condition for Human Occupancy" shall be selected for the minimum total heating, ventilating and air conditioning system energy use.

401.2 Cooling with outdoor air: Each fan system shall be designed to use up to and including one hundred (100) percent of the fan system capacity for cooling with outdoor air automatically whenever its use will result in lower usage of energy than would be required under its normal operation.

Exceptions: Cooling without outdoor air is not required under any one (1) or more of the following conditions.

a. Fan system capacity less than five thousand (5,000) cfm or one hundred thirty-four thousand (134,000) Btu/hr total cooling capacity.

b. The quality of the outdoor air is so poor as to require extensive treatment of the air.

c. The need for humidification or dehumidification requires the use of more energy than is conserved by outdoor air cooling.

d. The use of outdoor air cooling may affect the operation of other systems (such as return or exhaust air fans or supermarket refrigeration) so as to increase the overall energy consumption of the building.

e. Internal/external zone heat recovery or other energy recovery is used.

f. Annual heating degree days are less than twenty-five hundred (2500).

g. When all space cooling is accomplished by a circulating liquid which transfers space heat directly or indirectly to a heat rejection device such as a cooling tower without the use of a refrigeration system.

401.3 Mechanical ventilation: Each mechanical ventilation system shall be equipped with a readily accessible means for either shut-off or volume reduction and shut-off when ventilation is not required.

401.4 Simultaneous heating and cooling: Systems that employ both heating and cooling simultaneously in order to achieve comfort conditions within a space shall be limited to those situations where more efficient methods of heating and air conditioning cannot be effectively utilized to meet system objectives. Simultaneous heating and cooling by reheating or recooling supply air or by concurrent operation of independent heating and cooling systems serving a common zone shall be restricted as specified herein.

401.4.1 Recovered energy, provided the new energy expended in the recovery process is less than the amount recovered, may be used for control of temperature and humidity. New energy is defined as energy, other than recovered, utilized for the purpose of heating and cooling.

401.4.2 New energy may be used, when necessary, to prevent relative humidity from rising above sixty (60) percent for comfort control or to prevent condensation on terminal units or outlets.

401.4.3 New energy may be used for control temperature if minimized as specified in Sections 401.4.4 through 401.4.8.

401.4.4 Reheat systems: Systems employing reheat and serving multiple zones, other than those employing variable air volume for temperature control, shall be provided with control that will automatically reset the system cold air supply to the highest temperature level that will satisfy the zone requiring the coolest air. Single zone reheat systems shall be controlled to sequence reheat and cooling.

401.4.5 Dual duct and multizone systems: These systems shall be provided with control that will automatically reset the cold deck air supply to the highest temperature that will satisfy the zone requiring the coolest air and the hot deck air supply to the lowest temperature that will satisfy the zone requiring the warmest air.

401.4.6 Recooling systems: Systems in which heated air is recooled, directly or indirectly, to maintain space temperature, shall be provided with control that will automatically reset the temperature to which the supply air is heated to the lowest level that will satisfy the zone requiring the warmest heat.

401.4.7 For systems with multiple zones, one or more zones may be chosen to represent a number of zones with similar heating/cooling characteristics. A multiple zone heating, ventilating and air conditioning system that employs reheating or cooling for control of not more than five thousand (5000) cfm or twenty (20) percent of the total supply air of the system, whichever is less, shall be exempt from the supply air temperature reset requirements of Sections 401.4.4 and 401.4.6.

401.4.8 Concurrent operation of independent heating and cooling systems serving common spaces and requiring the use of new energy for heating or cooling shall be minimized by one or both of the following:

a. By providing sequential temperature control of both heating and cooling capacity in each zone.
b. By limiting the heating energy input, through automatic reset control of the heating medium temperature (or energy input rate), to only that necessary to offset heat loss due to transmission and infiltration and, where applicable, to heat the ventilation air supply to the space.

Section 402.0 Equipment Performance Requirements

402.1 General: The requirements of this section apply to equipment and component performance for heating, ventilating and air conditioning systems. Where equipment efficiency levels are specified, data furnished by the equipment supplier or certified under a nationally recognized certification program or rating procedure shall be used to satisfy these requirements.

402.1.1 System equipment: Heating, ventilating and air conditioning system equipment whose energy input in the cooling mode is entirely electric shall show a coefficient of

Table 402.1.1a. Minimum EER and COP for electric heating, ventilating and air conditioning system equipment.

STANDARD RATING CAPACITY	EER	COP
Under 65,000 Btu/hr (19,050 watts)	6.1	1.8
65,000 Btu/hr (19,050 watts) and over	6.8	2.0

performance (COP) and energy efficiency ratio (EER) not less than the values specified in . . . Table 402.1.1.1a.

These requirements apply to, but are not limited to, unitary cooling equipment (air and water source); packaged air conditioners; and room air conditioners. These requirements do not apply to equipment used in areas having open refrigerated food display cases. For determining coefficient of performance (COP), the rate of net heat removal shall be defined as the change in the total heat contents of the air entering and leaving the equipment (without reheat). Total energy input shall be determined by combining the energy inputs to all elements of the equipment, including but not limited to, compressor(s), pump(s), supply-air fan(s), cooling-tower fan(s), and the system equipment control circuit.

Heat operated cooling equipment shall show a coefficient of performance (COP) in the cooling mode not less than the values specified in the following Table 402.1.1b. These requirements apply to, but are not limited to, absorption, engine-driven and turbine-driven equipment. The coefficient of performance (COP) is determined excluding the electrical auxiliary inputs.

402.1.2 System components: Heating, ventilating and air conditioning system components whose energy input in the cooling mode is entirely electric shall show a coefficient of performance (COP) and energy efficiency rate (EER) not less than the values specified in the following Table 402.1.2.

For determining coefficient of performance (COP), the rate of heat removal is defined as the difference in total heat contents of the water or refrigerant entering or leaving the component. Total energy input shall be determined by combining the energy inputs to all elements and accessories of the component, including but not limited to, compressor(s), internal circulating pump(s), condenser-air fan(s), evaporative-condenser cooling water pump(s), purge, and the component control circuit.

Table 402.1.1b. Minimum COP for heating, ventilating and air conditioning system heat operated cooling equipment.

HEAT SOURCE	MINIMUM COP
Direct fired (gas, oil)	0.40
Indirect fired (steam, hot water)	0.65

Table 402.1.2. Minimum COP for electrically driven heating, ventilating and air conditioning system components.

COMPONENT	CONDENSING MEANS	AIR		WATER		EVAPORATION	
		EER	COP	EER	COP	EER	COP
Self-contained water chillers	Centrifugal	7.5	2.2	12.9	3.8	–	–
	Positive Displacement	7.2	2.1	10.9	3.2	–	–
Condenserless water chillers	Positive Displacement	8.9	2.6	10.0	3.2	–	–
Compresser and condenser units 65,000 Btu/hr. (19,050 watts) and over	Positive Displacement	7.8	2.3	11.3	3.3	11.3	3.3

Table 402.1.3. Minimum COP for heat pumps, heating mode.

SOURCE AND OUTDOOR TEMPERATURE (°F.)	MINIMUM COP
Air source–47 DB/43 WB	2.2
Air source–17 DB/15 WB	1.2
Water source–60 entering	2.2

[Note: DB = dry-bulb temperature.
WB = wet-bulb temperature.]

402.1.3 Heat pumps: Heat pumps whose energy input is entirely electric shall show a coefficient of performance (COP), heating, not less than the values specified in ... Table 402.1.3.

402.1.4 Supplementary heaters: The heat pump shall be installed with a control to prevent supplementary heater operation when the heating load can be met by the heat pump alone.

Supplementary heater operation is permitted during transient periods, such as start-ups, following room thermostat setpoint advance, and during defrost.

A two-stage room thermostat, which controls the supplementary heat on its second stage, shall be accepted as meeting this requirement. The cut-on temperature for the compression heating shall be higher than the cut-on temperature for the supplementary heat, and the cut-off temperature for the compression heating shall be higher than the cut-off temperature for the supplementary heat. Supplementary heat may be derived from any source of electric resistance heating or combustion heating.

402.1.5 Combustion heating equipment: All gas- and oil-fired comfort heating equipment shall show a minimum combustion efficiency of seventy-five (75) percent at maximum rated output. Combustion efficiency shall be determined in accordance with the ASHRAE Standard 90. ...

Section 403.0 Duct Insulation

403.1 Insulation: All duct systems, or portions thereof, exposed to nonconditioned spaces shall be insulated to provide a thermal resistance, excluding film resistance, of

$$R = \frac{t_i - t_o}{15} \text{ (hr) (sq.ft.) (F)/BTU}$$

where $t_i - t_o$ is the design temperature differential (absolute value) between the air in the duct and the surrounding air.

Exceptions: Duct insulation, except when needed to prevent condensation, is not required in any of the following cases:

a. Where $t_i - t_o$ is twenty-five (25) degrees F or less.
b. When the heat gain or loss of the ducts, without insulation, will not increase the energy requirements of the building.
c. Exhaust air ducts.
d. Supply or return air ducts installed in crawl spaces with insulated walls, basements or cellars in one- and two-family dwellings.

Where required to prevent condensation, insulation with vapor barriers shall be installed in addition to insulation required above.

Section 404.0 Controls

404.1 System controls: All heating, ventilating and air conditioning systems shall provide controls as specified herein.

404.2 Temperature: Each heating, ventilating and air conditioning system shall be provided with at least one (1) thermostat for the regulation of temperature. Each thermostat shall be capable of being set from fifty-five (55) degrees F to seventy-five (75) degrees F, where used to control heating only and from seventy (70) degrees F to eighty-five (85) degrees F where used to control cooling only. Where used to control both heating and cooling it shall be capable of being set from fifty-five (55) degrees F to eighty-five (85) degrees F, and shall be capable of operating the system heating and cooling in sequence. It shall be adjustable to provide a temperature range of up to ten (10) degrees F. between full heating and full cooling, except as allowed in 401.4.8.

404.3 Humidity: If a heating, ventilating and air conditioning system is equipped with a means for adding moisture to maintain specific selected relative humidities in spaces or zones, a humidistat shall be provided. This device shall be capable of being set to prevent new energy from being used to produce space relative humidity above thirty (30) percent RH. Where a humidistat is used in a heating, ventilating and air conditioning system for controlling moisture removal to maintain specific selected relative humidities in spaces or zones, it shall be capable of being set to prevent new energy from being used to produce a space relative humidity below sixty (60) percent.

404.4 Temperature zoning: In all building and structures of use Group R-3 (residential, one- and two-family), at least one (1) thermostat for regulation of space temperature shall be provided for each separate heating, ventilating and air conditioning system. In addition, a readily accessible manual or automatic means shall be provided to partially restrict or shut off the heating or cooling input to each zone or floor, excluding unheated or uncooled basements and garages.

In all buildings and structures of use Group R-2 (residential, multi-family) each individual dwelling unit shall be considered separately and shall meet the requirements for one- and two-family dwellings above.

In all buildings and structures other than use Group R-3 (residential, one- and two-family dwellings) and in spaces other than dwelling units in use Group R-2 (residential, multi-family), at least one (1) thermostat for regulation of space temperature shall be provided for each separate heating, ventilating and air conditioning system and for each floor of the building.

404.5 Set back and shut-off: In all buildings and structures, or portions thereof of use Group R-3 (residential, one- and two-family), the thermostat, or an alternative means such as a switch or clock, shall provide a readily accessible, manual or automatic means for reducing the energy required for heating and cooling during periods of non-use or reduced need.

In all other buildings and structures, or portions thereof, each heating, ventilating and air conditioning system shall be equipped with a readily accessible means of reducing the energy used for heating, ventilating and air conditioning during periods of non-use or alternate uses of the building spaces or zones served by the system, such as with manually adjustable automatic timing devices, manual devices for use by operating personnel, or automatic control systems.

Lowering thermostat set points to reduce energy consumption of heating system shall not cause energy to be expanded to reach the reduced setting.

Section 405.0 Steam and Hot Water Heating Piping

405.1 Piping insulation: All piping serving as a part of a heating or cooling system installed to serve buildings and within buildings shall be thermally insulated as shown in the following Table 405.1.

Insulation thickness is based on insulation having thermal resistance in the range of four (4.0) to four and six tenths (4.6) per inch of thickness on a flat surface at a mean temperature of seventy-five (75) degrees F. Minimum insulation thickness shall be increased for materials having R value less than four (4.0) or may be reduced for materials having R values greater than four and six-tenths (4.6) per inch of thickness as follows:

405.1.1 For materials with thermal resistance greater than R = 4.5, the minimum insulation thickness may be reduced as follows:

$$\frac{4.6 \times \text{Table 405.1 Thickness}}{\text{Actual R}} = \text{New Minimum Thickness}$$

405.1.2 For materials with thermal resistance less than R = 4.0, the minimum insulation thickness shall be increased as follows:

$$\frac{4.0 \times \text{Table 405.1 Thickness}}{\text{Actual R}} = \text{New Minimum Thickness}$$

Table 405.1. Minimum pipe insulation.

| PIPE SYSTEM TYPES | FLUID TEMPERATURE RANGE, F. | INSULATION THICKNESS IN INCHES FOR PIPE SIZES | | | | | |
		RUNOUTS UP TO 2	1 & LESS	1½–2	2½–4	5 & 6	8 & LARGER
Heating systems							
Steam & Hot water							
High pressure/temp	306–450	1½	1½	2	2½	3½	3½
Medium pressure/temp	251–305	1½	1½	2	2½	3	3
Low pressure/temp	201–250	1	1	1½	1½	2	2
Low temperature	120–200	½	¾	1	1	1	1½
Steam condensate (for feed water)	any	1	1	1	1½	1½	2
Cooling systems							
Chilled water,	40–55	½	½	¾	1	1	1
Refrigerant, or brine	Below 40	1	1	1½	1½	1½	1½

Exceptions: Piping insulation, except when needed to prevent condensation, is not required in any of the following cases:

   a. Piping installed within heating, ventilating and air conditioning equipment.
   b. Piping at temperatures between fifty-five (55) degrees F. and one hundred twenty (120) degrees F.
   c. When the heat loss or heat gain of the piping, without insulation, does not increase the energy requirements of the building.
   d. Piping installed in basements or cellars in one- and two-family dwellings.

Where required to prevent condensation, insulation with vapor barriers shall be installed in addition to insulation required above.

ARTICLE 5 PLUMBING SYSTEMS

Section 500.0 General

500.1 Scope: This article sets forth provisions for design and equipment selection for energy conservation in service water heating systems.

Section 501.0 Fixtures

501.1 Lavatories: Lavatories in restrooms of public facilities shall be equipped with self-closing outlet devices which limit the flow of hot water to a maximum of five-tenths (0.5) gpm, devices which limit the outlet temperature to a maximum of one hundred ten (110) degrees F. and self-closing valves which limit the quantity of hot water to a maximum of twenty-five hundredths (0.25) gallons.

501.2 Showers: Showers used for other than safety reasons shall be equipped with flow control devices to limit total flow to a maximum of three (3) gpm per shower head.

Section 502.0 Insulation

502.1 Piping insulation: Piping in required return circulation systems shall be insulated so that heat loss is limited to a maximum of twenty-five (25) Btu per square foot of external pipe surface for above ground piping and a maximum of thirty-five (35) Btu per square foot of external pipe surface for underground piping. Maximum heat loss shall be determined at a temperature differential equal to the maximum water temperature minus a design ambient temperature no higher than sixty-five (65) degrees F.

Exceptions: Conformance with Table 405.1 for "low temperature piping systems" shall be deemed as complying with this section.

502.2 Tanks: Unfired hot water storage tanks shall be insulated so that heat loss is limited to a maximum of fifteen (15) Btu per square foot of external tank surface area. For purposes of determining this heat loss, the design ambient temperature shall be no higher than sixty-five (65) degrees F.

Section 503.0 Equipment

503.1 Pump Operation: Circulating hot water systems shall be arranged so that the circulating pump can be conveniently turned off either automatically or manually when the hot water system is not in operation.

503.2 Performance efficiency

503.2.1 Electric water heaters: All automatic electric storage water heaters shall have a stand-by loss not exceeding four (4) watts per square foot of tank surface area. The method test of stand-by loss shall be as directed in Section 4.3.1 of ANSI C72.1 Household Automatic Electrical Storage-Type Water Heaters.

503.2.2 Gas- and oil-fired water heaters: All gas- and oil-fired, automatic storage heaters shall have a recovery efficiency, E, not less than seventy-five (75) percent and a stand-by loss percentage, S, not exceeding $S = 2.3 + 67/V$, where V = rated volume in gallons. The method of test of E and S shall be as described in Section 2.7 of ANSI Z21.103 Circulating Tank, Instantaneous and Large Automatic Storage Type Water Heaters, Approval Requirements for Gas Water Heaters.

Section 504.0 Controls

504.1 Temperature controls: All hot water supply systems shall be equipped with automatic temperature controls capable of adjustments from the lowest to the highest acceptable temperature settings for the intended use.

504.2 Shut-down: A separate switch shall be provided to terminate the energy supplies to electric hot water supply systems. A separate valve shall be provided to turn off the energy supplied to the main burner of all other types of hot water supply systems.

ARTICLE 6 ELECTRICAL SYSTEMS

Section 600.0 General

600.1 Scope: This article sets forth provisions for efficient distribution of electrical energy in the building distribution systems.

Section 601.0 System Requirements

601.1 Power factor: The power factor of the overall electrical distribution system in a building shall be not less than ninety (90) percent under rated design installed load of the building, either by utilization equipment design or by the use of power factor corrective devices. The power factor corrective devices may be installed on individual equipment, rated greater than one thousand (1000) watts and switched therewith, regionally grouped, located at the service equipment or power force correction achieved by other equivalent means. The choice among these corrective methods should be made based upon engineering evaluation of each distribution system.

601.2 Service voltage: Where a choice of service voltages is available, the voltage resulting in the least energy loss shall be used.

601.3 Voltage drop: In any building, the maximum total voltage drop shall not exceed three (3) percent in branch circuits or feeders, for a total of five (5) percent to the farthest outlet based on steady state design load conditions.

601.4 Lighting switching: Switching shall be provided for each lighting circuit, or for portions of each circuit, so that the partial lighting required for custodial or for effective complementary use with natural lighting may be operated selectively.

601.5 Separate metering: In all multi-family dwellings (use Group R-2) provisions shall be made to determine the electrical energy consumed by each tenant.

ARTICLE 7 ALTERNATIVE SYSTEMS

Section 700.0 General

700.1 Performance alternative: Alternative building systems and equipment design may be approved by the building official when they can be shown to have energy consumption not greater than that of a similar building with similar forms of energy requirements, designed in accordance with the provisions of this code.

700.1.1 Non-depletable sources: When such alternative systems utilize solar, geothermal, wind or other non-depletable energy sources for all or part of its energy sources, such non-depletable energy supplied to the building shall be excluded from the total energy chargeable to the proposed alternative design.

700.2 Documentation: Proposed alternative designs, submitted as requests for exception to the standard design criteria, must be accompanied by an energy analysis prepared in accordance with the ASHRAE Standard 90. . . .

18   U.S. Department of Energy

The U.S. Department of Energy (DOE) was established by the Department of Energy Organization Act, approved August 4, 1977, and effective October 1, 1977, pursuant to Executive Order 12009 of September 13, 1977. The act consolidated the major federal energy functions into one cabinet-level department, transferring to DOE all the responsibilities of the Energy Research and Development Administration; the Federal Energy Administration; the Federal Power Commission; and the Alaska, Bonneville, Southeastern, and Southwestern Power Administrations, formerly components of the U.S. Department of the Interior (DOI), as well as the power-marketing functions of the DOI Bureau of Reclamation. Also transferred to DOE were certain functions of the Interstate Commerce Commission and of the U.S. Departments of Commerce, the Navy, the Interior, and Housing and Urban Development. Figure 18-1 illustrates the DOE's organization.

The secretary, deputy secretary, and under secretary are the three principal officers of the department. They are responsible for the overall planning, direction, and control of departmental activities. Supporting DOE in departmental matters are crosscutting staff offices, including those of the assistant secretaries for policy and evaluation, intergovernmental and institutional relations, and international affairs; the general counsel; the controller; the inspector general; the director of administration; and the director of procurement and contracts management.

Programs requiring large budget outlays are the responsibility of the director of energy research and the five program assistant secretaries (energy technology, resource applications, conservation and solar applications, environment, and defense programs).

The DOE's organization also includes the Economic Regulatory Administration, the Energy Information Administration, and the Federal Energy Regulatory Commission, which is an independent regulatory organization within the department. In addition to headquarters' offices, the department has an extensive field structure, which plays an integral part in the implementation and management of departmental projects and programs.

OFFICE OF THE SECRETARY

The *secretary of energy* directs and supervises the administration of the DOE, decides major energy policy issues, and acts as the principal energy adviser to the President. The secretary is the principal spokesperson for the department. The *deputy secretary* assists the secretary in representing the DOE before Congress and the public, and supervises administrative and general support functions of the department. The deputy secretary also has primary responsibility for overseeing the Economic Regulatory Administration and the Energy Information Administration. The *under secretary* assists the secretary in overall departmental program management and specifically oversees DOE programs requiring major budget outlays, including conservation and solar applications, energy research, energy technology, resource applications, environment, and defense programs. The under secretary also assists the deputy secretary in DOE administrative matters. The *special assistant,* an aide to the secretary, deputy secretary, and under secretary, acts as the liaison officer with the White House and performs special projects as assigned. The *director of the office of the executive secretariat* ensures the necessary coordination and follow-up of the secretary's decisions and also provides current, accurate, and complete communication of secretarial decisions.

STAFF OFFICES

General Counsel

The Office of the General Counsel provides legal counsel to Components of the department, except the Federal Energy Regulatory Commission. These services include counsel on legislative activities, litigation, advice on international cooperation agreements and projects, management of the department's patent program, and assistance as required in general administrative matters.

Policy and Evaluation

The Office of the Assistant Secretary for Policy and Evaluation formulates and recommends the overall national energy

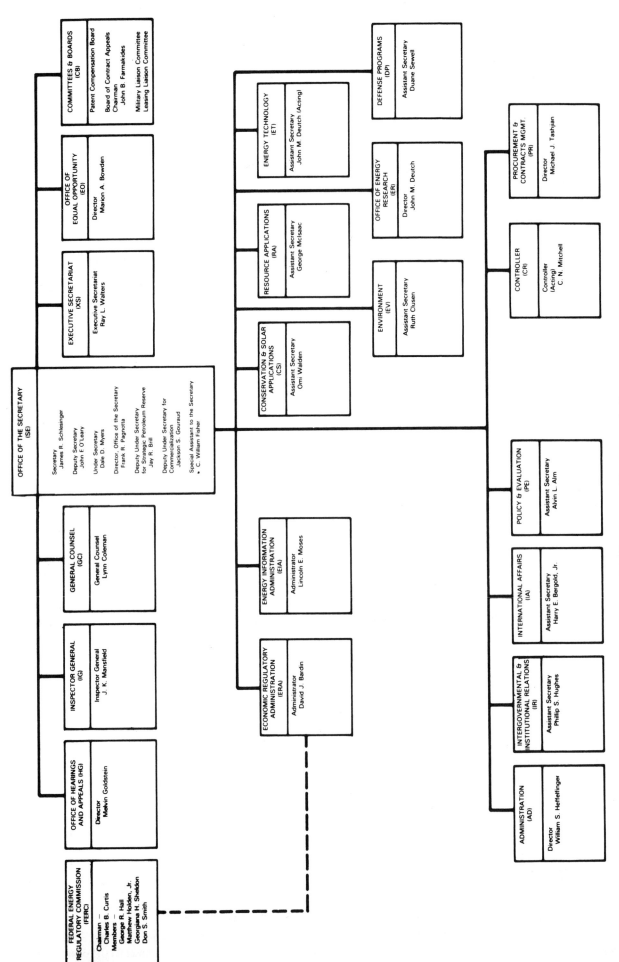

Figure 18-1. Department of Energy organization.

policy, coordinates the analysis and evaluation of policies and programs, and conducts a continual assessment of the nation's energy situation. It also ensures that all DOE policies and programs promote competition in the energy industry and that consumer impact is considered in decision-making. This office, in conjunction with the Office of the Controller, provides staff support to the department's formal system for policy and program planning. The office develops legislative proposals to support policy objectives and reviews program and management plans and budgets to ensure integration of new policy proposals.

Intergovernmental and Institutional Relations

The major outreach and consumer affairs coordination functions of DOE are the responsibility of the Office of the Assistant Secretary for Intergovernmental and Institutional Relations. These functions are designed to ensure that individuals and groups outside the department have a central point of contact with the department and that views and interests are communicated to policymaking and program decisionmaking personnel within the department. Specifically, the office oversees and maintains DOE relations with Congress, the news media, the states, regional and local agencies, private institutions, and consumer interests.

International Affairs

The Office of the Assistant Secretary for International Affairs is responsible for developing, managing, and directing programs and activities related to the international aspects of overall energy policy. The office ensures that United States international energy policies and programs conform with national goals, legislation, and treaty obligations and assists the secretary in providing the president with independent technical advice on international energy negotiations. The office assesses world price and supply trends and technological developments and also studies the effects of international actions on the United States energy supply. In addition to supporting national policies on international nuclear nonproliferation and the international fuel cycle, the office also coordinates cooperative international energy programs and maintains relationships with foreign governments and international organizations.

Controller

Charged with management of the department's financial resources, the Office of the Controller also participates in the development of program and management information systems, administers financing studies for loan guarantees, and conducts independent cost estimates for procurement and construction programs and for pricing of DOE services and products. The office also coordinates the department's overall authorization/appropriation process.

Procurement and Contracts Management

As the office responsible for policy development and advice on procurement, contracts, and other business agreements, the Office of Procurement and Contracts Management negotiates and administers all contracts and grants, cooperative agreements, and loan guarantees at headquarters and oversees them in the field. The office, headed by the Director of Procurement and Contracts Management, is also responsible for small-business and minority-business procurements and the management of federal personal property held by DOE contractors.

Administration

The director of administration is responsible for organization and management studies, directives, and records and reports management; personnel management and administration programs; federal manpower resource allocation; reporting and control; construction support to program officials; contractor-labor relations, wage and benefit administration, and manpower development; automatic-data-processing planning, policy, management, and operations; and telecommunications policy, management, and operations. Headed by the director of administration, the directorate also develops adminstrative services policy and provides printing, transportation, travel, space, personal property, and library services to headquarters. The directorate also processes Freedom of Information and Privacy Act requests .

Inspector General

The Office of the Inspector General coordinates, supervises, and conducts investigations and audits of all DOE internal activities, including those of the Federal Energy Regulatory Commission, to ensure honesty and efficiency. When necessary, these activities are coordinated with other federal, state, and local agencies and with nongovernment entities. The office also recommends corrective actions and identifies and arranges for prosecution of participants in fraud and abuse cases.

PROGRAMS

Energy Research

The Office of Energy Research advises the secretary on the physical and energy research and development programs of the department; the use of multipurpose laboratories, education, and training for basic and applied research; and financial assistance and budgetary priorities for these activities. The office also manages the basic energy sciences program, administers DOE programs supporting university researchers, and administers a financial support program for research and development projects not funded elsewhere in the department. In addition, the office monitors DOE research, development, and demonstration programs for any gaps or duplication of effort and, in conjunction with the assistant secretary for international affairs, monitors the international exchange of scientific and technical personnel.

Energy Technology

The assistant secretary for energy technology is responsible for research, development, and technology demonstration in all energy areas, including solar, geothermal, fossil, and nuclear energy. Activities coordinated under the assistant secretary for energy technology focus primarily on making new energy technologies available for public or private commercial application as early as possible. Research and development projects involving energy storage, electric energy systems, and improved energy conversion efficiency are assigned to the energy technology office since they are primarily developmental and long-term and support other technology development programs. The energy technology officials applied research and development programs support specialized projects managed by other DOE program organizations, such as resource applications or conservation and solar applications.

Resource Applications

The assistant secretary for resource applications is responsible for developing, managing, and directing policies and programs to increase domestic supplies of petroleum, natural gas, coal, and uranium; to reduce regulatory and financial constraints to resource development and utilization; and to demonstrate and encourage the commercial use of developed energy supply technologies. The assistant secretary manages the national uranium enrichment activities and the departmental energy resource leasing program. The assistant secretary is also responsible for the supervision and administration of the Alaska Power Administration, Bonneville Power Administration, Southeastern Power Administration, Southwestern Power Administration, Western Power Administration, the Naval Petroleum Reserves, and Oil Shale Reserves, and the Strategic Petroleum Reserves.

Conservation and Solar Applications

The assistant secretary for conservation and solar applications is responsible for formulating and directing the department's conservation and solar commercialization programs. These programs include those designed to improve energy efficiency and system utilization and reduce energy consumption in the transportation industry, public and private buildings, and agricultural and industrial process heating; and preparation of a solar commercialization plan. Within the Office of Conservation and Solar Applications is the Office of Small Scale Technology, which seeks to fully use the services of individual inventors and small business firms.

Environment

The assistant secretary for environment is responsible for ensuring that the implementation of all departmental programs is consistent with environmental and safety laws, regulations, and policies. The assistant secretary provides policy guidance for the secretary to ensure compliance with environmental protection laws and is responsible for reviewing and coordinating all environment impact statements prepared with the department. In addition, the assistant secretary monitors DOE programs with respect to the health and safety of both workers and the general public. The assistant secretary also conducts environmental and health-related research and development programs, such as studies of energy-related pollutants and their effect on biological systems. In addition, the office coordinates the department's responsibilities under the National Environmental Policy Act of 1969.

Defense Programs

The assistant secretary for defense programs directs the nation's nuclear-weapons research, development, testing, production, and surveillance program. In addition, the assistant secretary coordinates a safeguards and security program to provide accountability and physical protection of special nuclear materials, including research and development for improvements, testing, evaluation, and implementation of safeguards systems. Additional responsibilities include management of the laser fusion development program, classification and declassification of sensitive nuclear weapons information, analysis and coordination of international activities related to nuclear technology and materials, and support of the assistant secretary for international affairs with respect to nonproliferation controls.

Economic Regulatory Administration

The Economic Regulatory Administration (ERA) administers the department's regulatory programs, other than those assigned to the Federal Energy Regulatory Commission. These functions include the oil pricing, allocation, and import programs designed to ensure price stability and equitable supplies of crude oil, petroleum products, and natural gas liquids among a wide range of domestic users. The ERA ensures compliance with existing regulations and carries out new regulatory programs as assigned. ERA also administers other regulatory programs, including conversion of oil- and gas-fired utility and industrial facilities to coal, natural gas import/export controls, natural gas curtailment priorities and emergency allocations, regional coordination of electric power system planning and reliability of bulk power supply, and emergency and contingency planning. On behalf of the secretary, ERA organizes and manages an active intervention program before the Federal Energy Regulatory Commission and other federal and state regulatory agencies in support of departmental policy objectives.

Energy Information Administration

The Energy Information Administration (EIA) is responsible for the timely and accurate collection, processing, and publication of data on energy reserves, the financial status of energy-producing companies, production, demand, consumption, and other areas. Analyses of data to assist government and nongovernment users in understanding energy trends are also performed. Specifically, analyses are prepared on complex, long-term energy trends and the microeconomic and macroeconomic impacts of energy trends on regional and industrial sectors. Special-purpose analyses are prepared involving competition within the energy industries, the capital/financial structure of energy companies, and interfuel substitution. To ensure the validity of regulatory and other energy data, extensive field audits are conducted to determine the accuracy of data contained within the departmental system.

The EIA provides data publication and distribution services within DPE, throughout the government, and for the public. A clearinghouse for general information on energy, it coordinates its activities with the department's Technical Information Center.

Federal Energy Regulatory Commission

An independent, five-member commission within the DOE, the Federal Energy Regulatory Commission has retained many of the functions of the Federal Power Commission, such as the setting of rates and charges for transportation and sale of natural gas and for the transmission and sale of electricity and the licensing of hydroelectric power projects. In addition, the authority to establish rates or charges for the transportation of oil by pipeline, as well as the valuation of such pipelines, has been assigned to the commission from the Interstate Commerce Commission.

FIELD STRUCTURE

Regional Representatives

The secretary is represented in each of the 10 standard federal regions by regional representatives, whose duties include speaking for the secretary in all regional DOE activities; working with the state governors to establish Regional Energy Advisory Boards; ensuring the effectiveness of regional outreach programs, including DOE interaction with business, labor, and consumer groups and their appropriate involvement in DOE decisionmaking; providing feedback on the regional impact of DOE policies and programs; and performing assigned nonregulatory activities in regional planning, conservation, energy resource development, and energy data collection. The regional representatives report to the secretary, are supervised by the deputy secretary, and coordinate their day-to-day activities with the assistant secretary for intergovernmental institutional relations.

Regional Regulatory Programs

The DOE regional allocation and compliance program is conducted by directors of regional compliance, under the supervision and direction of the administrator of the Economic Regulatory Administration and with the legal support of the general counsel's field staff. These programs are separate from the outreach and other activities of the regional representatives.

The field activities of the Federal Energy Regulatory Commission will be separate from other department field activities but will have offices co-located with regional compliance or other departmental field offices in order to share common administrative support.

Operations Offices and Contractor-Operated Field Installations

The Operations Offices provide a formal link between department headquarters and the field laboratories and other operating facilities. The primary function of these offices is to administer contracts and manage programs and projects as assigned from headquarters. Operations Offices, except those primarily concerned with the weapons program and naval reactors, report to the under secretary. Day-to-day specific program direction for the Operations Offices is provided by the assistant secretaries and the director of energy research.

The DOE's contractor-run laboratories and other field installations, except those involved in the weapons program, are controlled by the director of energy research or the assistant secretary who is most concerned with each installation's functions. However, because of their multidisciplinary or emerging roles, the Argonne National Laboratory, the Oak Ridge National Laboratory, the Pacific Northwest Laboratory, and the Solar Energy Research Institute initially are controlled directly by the under secretary.

A field and Laboratory Coordination Council assists the under secretary in overseeing the assignment of program tasks to field activities.

DOE Headquarters
Organizational Listing

THE SECRETARY

The Secretary, ...	252-6210
Executive Assistant to The Secretary,	
Evelyn C. Irons	252-6210
Secretary, Lois E. Madden	252-6210
Secrétary, Margaret S. Sellers	252-6210
Receptionist, Andrea L. Catapano	252-6210

THE DEPUTY SECRETARY

The Deputy Secretary, John F. O'Leary	252-5500
Staff Assistant to The Deputy Secretary,	
Margaret (Peggy) Burris	252-5500
Executive Assistant to The Deputy	
Secretary, L. Andrew Zausner	252-5500
Executive Assistant to The Deputy	
Secretary, Robert I. Hanfling	252-5700
Confidential Assistant for Legislation	
to The Deputy Secretary,	
Kathleen M. Linehan	252-5600

THE UNDER SECRETARY

The Under Secretary, Dale Myers	252-5704
Confidential Assistant/Secretary to The Under Secretary, Phyllis M. Byrne ...	252-5704
Deputy Under Secretary for Commercialization, Jackson S. Gouraud............	252-5266
Deputy Under Secretary, Strategic Petroleum Reserve, Jay R. Brill............	634-5510

THE SPECIAL ASSISTANT

The Special Assistant to The Secretary, Deputy Secretary and	
Under Secretary, C. William Fischer	252-6230
Confidential Assistant/Secretary to The Special Assistant,	
Patricia Unkle	252-6230
Assistant to The Secretary, Jeffrey R. Cooper	252-5110
Assistant to The Secretary, Jeffrey L. Stanfield	252-5110
White House Fellow, Stephen R. Hill	252-5127

DIRECTOR, OFFICE OF THE SECRETARY

Director, Office of The Secretary, Frank R. Pagnotta	252-5777
Executive Assistant to the Director, Carole J. Gorry	252-5777
Confidential Assistant/Secretary to the Director, Eldyne (Dee) Bordner	252-5777
Assistant to the Director for Correspondence and Files,	
Donald C. Legates	252-5079
Assistant to the Director for Administrative Services, Frank A. Townsend	252-5090
Assistant to the Director for Special Projects, E. J. (Jim) Vajda	252-5090
Staff Assistant, Policy and Analysis Section, Patricia A. Gravatt	252-5911

NOTE: All telephone numbers given are in area code 202, unless otherwise noted.

EXECUTIVE SECRETARIAT

	Extension
Director, Executive Secretariat, Raymond Walters	252-5230
Confidential Assistant to the Director, Helen J. Ammen	252-5230
Confidential Secretary to the Director, Jo W. Muehleib	252-5230
Assistant Director (Analysis), Document Coordination and Analysis Center, Gordon M. Grant	252-5106
Chief, Document Control Center, Frances E. Hooks, Acting	252-5131
Assistant Director (ACTS Manager), Kenneth M. Pusateri	252-5115
Chief Historian, Richard G. Hewlett	252-5237

ASSISTANT SECRETARY
CONSERVATION AND SOLAR APPLICATION

Assistant Secretary, Omi G. Walden	376-4934
Deputy Assistant Secretary, Maxine Savitz	376-4943
Deputy Assistant Secretary, Kelly C. Sandy (Acting)	376-4940
Special Assistant, A. Holly Miller	376-4943
Executive Director, Kelly C. Sandy	376-4940
Director, Budget and Administrative Support, J. Keith Davy (Acting)	376-1632
Assistant Director, Budget and Program Support, James McKeown, Jr.	376-1629
Chief, Budget Branch, Fred Glatstein	376-4830
Assistant Director, Management Services, Jeremy A. Black	376-1623
Director, Policy, Planning, and Evaluation, J. Michael Power	376-4602
Chief, Data and Analysis Division, Gurmukh Gill	376-4604

Office of Commercialization

Director, Joseph Barrow (Acting)	376-1865

Office of Buildings and Community Systems

Director, Maxine Savitz (Acting)	376-4943
Assistant Director, Communications and Building Energy Systems, Gerald Leighton	376-4714
Assistant Director, Systems Analysis and Technology Transfer, Melvin H. Chiogioji	376-4711

Office of Industrial Programs

Director, Douglas, G. Harvey	376-4113
Assistant Director, Waste Energy Reduction Development, William B. Williams (Acting)	376-9056
Assistant Director, Industrial Process Energy Efficiency Development, Alan J. Streb (Acting)	376-1671

Office of Solar Applications

Director, Ronald D. Scott (Acting)	376-9610
Program Manager, Russell A. O'Connell	376-9610
Chief, Demonstration Program Branch, William L. Corcoran	376-9604
Chief, Research and Development Branch, Frederick H. Morse	376-9630
Chief, Agriculture and Industrial Process Heat Branch, Jamie Dollard	376-9633
Chief, Technology Transfer Branch, Lawnie Taylor	376-9114
Chief, Barriers and Incentives Branch, Roger Bezdek	376-9616
Assistant Director, Solar Commercialization, John Schuler (Acting)	376-4455

NOTE: All telephone numbers given are in area code 202, unless otherwise noted.

Office of State and Local Programs

	Extension
Director, Frank M. Stewart (Acting)	376-4344
Director, State Grant Programs, Sandy Delaney (Acting)	376-1797
Director, State-Specific Program, Michael Willingham (Acting)	376-5843
Dirctor, Weatherization Assistance Programs, Mary M. Bell (Acting)	376-1801
Director, Energy Extension Service, Judy Liersch (Acting)	376-4615

Office of Transportation Programs

Director, Vincent J. Esposito (Acting)	376-4524
Assistant Director, Highway Systems, John J. Brogan	376-4610
Chief, Heat Engines Branch, George M. Thur	376-4675
Chief, Alternative Fuels Branch, E. Eugene Ecklund	376-4892
Assistant Director, Electric and Hybrid Vehicle Systems, Paul J. Brown	376-4681

Office of Small Scale Technology

Director, Jerry Duane (Acting)	376-4711

ASSISTANT SECRETARY DEFENSE PROGRAMS

Assistant Secretary, Duane C. Sewell	376-4078
	353-5166
Deputy Assistant Secretary (Vacant)	376-4080
Deputy Assistant Secretary (Research), (Vacant)	376-4080
Executive Assistant, Robert L. Wainwright	376-4083
Executive Director, Joseph P. Flynn, Jr.	376-4086
Director, Policy Analysis and Operations, Joseph P. Flynn, Jr.	376-4086
Director, Resource Management, Samuel Rousso	353-3276
Director, Institutional Liaison and Communications, A. Bryan Siebert, Jr.	353-4227
Administrative Officer, Mary E. Whitley	376-9744

Office of Military Application

Director, Maj. Gen. Joseph K. Bratton, USA	353-4221
Deputy Director, James W. Culpepper (Acting)	353-2967
Associate Director, James W. Culpepper	353-2967
Technical Operations Officer, James M. McCulloch, CAPT., USN	353-5523
Director, Division of Financial Management, Lewis M. Groover	353-3374
Deputy Director, Ronald Bartell	353-3374
Chief, Capital Branch, Carl R. Forsberg	353-3374
Chief, Production Branch, Ralph J. Ross	353-3374
Chief, Research, Development and Testing Branch, Ronald Bartell	353-3374
Director, Division of Operations, Ralph E. Caudle	353-3441
Deputy Director, Randall E. Jaycox, Jr., CAPT., USN	353-3487
Chief, Production Operations Branch, Herbert R. Schmidt, COL., USA	353-3441
Chief, Systems Development Branch, Randall E. Jaycox, Jr., CAPT., USN	353-3487
Director, Division of Planning, Wayne L. Beech, CAPT., USN	353-4007
Deputy Director, Daniel P. Cannon, COL., USAF	353-4007
Chief, Analysis and Materials Branch, Charles A. Sommer	353-4007
Chief, Stockpile Planning Branch, Daniel P. Cannon, COL., USAF	353-4007
Director, Division of Program Support, John E. Rudolph	353-3618
Deputy Director, J. Donald McBride	353-4580
Chief, Management Support Branch, Mary E. Heitzman	353-5551
Chief, Policy and Analysis Branch, J. Donald McBride	353-4580

NOTE: All telephone numbers given are in area code 202, unless otherwise noted.

Extension

Chief, Weapons Information Branch, Roy C. Boger, Jr. 353-4376
Senior DOE Respresentative, Joint Atomic Information
   Exchange Group, Robert E. O'Brien 325-7579
Director, Division of Research, Development, and Testing, Alan R. Cole,
   COL., USAF 353-5342
Deputy Director, John H. Carlson 353-5342
Chief, Development Branch, Irvin L. Williams 353-3986
Chief, Research Branch, Paul O. Matthews 353-5492
Director, Division of Safety, Environment, and Emergency Actions,
   Roy E. Lounsbury, COL., USA 353-5277
Deputy Director, Gordon C. Facer 353-3011
Chief, Nuclear Operations Safety Branch, Ralph G. Shull 353-5306
Chief, Systems Safety and Emergency Actions Branch, Gene A. Strommen,
   CDR., USN 353-3544

Office of Safeguards and Security

Director, George Weisz... 353-5106
   Deputy Director, Leonard M. Brenner.................................... 353-5108
   Chief, Budget and Administration Branch, James E. Eason (Acting)....... 353-4440
   Chief, Mail and Records Section, Mary V. Ferguson...................... 353-5114
Assistant Director for Plans and Policy, Thomas S. D'Agostino (Acting)..... 353-5186
   Chief, Assessments Branch, Thomas S. D'Agostino....................... 353-5104
   Chief Coordinator, Domestic & International Liaison & Demonstration
      Section, William C. Bartels.. 353-5217
   Chief, Plans and Analysis Branch, Frank L. Martin (Acting)............ 353-4163
   Chief, Programs and Policy Branch, William E. Gilbert................. 353-5690
Assistant Director for Research and Development, Samuel C. T. McDowell...... 353-5067
   Chief, Material Control and Development Branch, Glenn A. Hammond....... 353-3649
   Chief, Physical Protection and Development Branch, Burton Newmark...... 353-5672
Assistant Director for Security Affairs, Martin J. Dowd.................... 353-3652
   Chief, Operations Security Branch, Barry Dalinsky (Acting)............ 353-4176
   Chief, Internal Security Branch 353-3601
   Chief, Personnel Security Branch, Barry Dalinsky (Acting)............. 353-5485
   Chief, Central Personnel Indices, Joan S. Wiley...................... 353-3654
   Chief, Headquarters Security Branch, ... 353-4243
   Chief, Central Document Control, Joyce Forsyth....................... 353-3767
   Sgt., DOE Guard Group, James D. Fitez............................... 353-4337
   Chief, Downtown Security Office, Elmer Heinlein...................... 252-5762
Assistant Director for Information Support (Vacant)....................... 353-4254
   Chief, Information Systems Operations Branch, Nicholas Ovuka (Acting)..... 353-5394
   Chief, Information Systems Development Branch (Vacant)................. 353-5433

Office of Laser Fusion

Director, Richard L. Schriever (Acting)................................... 353-3462
   Deputy Director, Richard L. Schriever................................ 353-3463
   Associate Director, L. E. Killion.................................... 353-3345
   Assistant Director for LF, Ernie D. Braunschweig (Acting)............ 353-3397
   Assistant Director for PPB, Sheldon Kahalas (Acting)................. 353-3398
   Assistant Director for Management, Robert S. Robenseifner............ 353-3346
   Budget Officer, John P. Murphy...................................... 353-3138
   Administrative Officer, Joyce E. Lewis.............................. 353-3772

Office of International Security Affairs

Director, Ray E. Chapman.. 376-1740
   Deputy Director, Julio L. Torres.................................... 376-1741
Director, Division of Systems and Technology (Vacant).................... 376-4540
   Chief, Resource Management Branch, John L. Meinhardt................ 376-4127

NOTE: All telephone numbers given are in area code 202, unless otherwise noted.

<u>Extension</u>

Chief, Technical Program Branch, John V. Walker.................................... 376-4130
Director, Division of Defense Intelligence, Col. Robert A. O'Brien, USA
   (Acting)... 376-4131
  Chief, Operations Branch, Wilmer K. Benson.. 376-1749
  Chief, Assessments Branch, Robert E. Upchurch.................................... 376-1748
Director, Division of Politico-Military Security Affairs, Vance H. Hudgins...... 376-5885
  Chief, Operations Branch, Jeremiah F. Krauz...................................... 376-4134
  Chief, Policy and Plans Branch, Samuel Thompson.................................. 376-5882
Director, Division of Arms Control, Robert T. Duff.............................. 376-4540
  Administrative Officer, Barbara A. Moyers.. 376-4126

Office of Classification

Director, J. A. Griffin... 353-3521
  Deputy Director, M. L. Nash (Dual Capacity).................................... 353-3526
Director, Division of Systems & Technology, M. L. Nash (Dual Capacity)......... 353-3526
  Program Manager for Weapons, W. A. Strauser...................................... 353-4869
  Program Manager for Isotope Separation, J. R. Patton............................ 353-4859
  Program Manager for Reactor & Safeguards, D. C. Little (Acting)............... 353-3689
Director, Division of Policy, Operations & Support, E. L. Ellman
   (Dual Capacity)... 353-3636
  Program Manager for Operations, I. L. Cucchiara................................. 353-4863
  Program Manager for Policy, E. L. Ellman (Dual Capacity)...................... 353-3636
  Program Manager for Education & Administration, H. B. Hawthorne............... 353-4856

Office of Nuclear Materials Production

Director, F. Charles Gilbert 353-3777
  Deputy Director, George B. Pleat 353-3757
Director, Production Operations, Richard D. Hahn 353-3782
Director, Materials Processing, Roger K. Heusser 353-5496
Director, Fiscal and Program Support, R. J. Galbo 353-3982

ADMINISTRATOR ENERGY INFORMATION ADMINISTRATION

Administrator, Lincoln E. Moses 633-9085
  Deputy Administrator (Vacant).. 633-8477
  Executive Assistant, Lawrence R. Klur 633-8474

Office of Project Accountability and Control

Director, Robert Brown 633-8480

Office of Management Services

Director, R. Eugene Odom 633-8198
Director, Financial Services Division, Robert D. Hull 633-8184
Director, Manpower and Management Services Division, Ann Daniels 633-8188

Office of Planning and Evaluation

Director, Christina L. Rathkopf 633-8696

NOTE: All telephone numbers given are in area code 202, unless otherwise noted.

Energy Data

	Extension
Assistant Administrator, Albert H. Linden, Jr.	633-9021
Director, Office of Energy Data and Interpretation, Jimmie L. Petersen	254-5147
Director, Coal and Electric Power Statistics Division, Charles Heath	252-6860
Director, Oil and Gas Statistics Division, William Park	634-1044
Director, Interfuel, Nuclear and Other Energy Sources Div., Frank Lalley	633-8500
Director, Data Collection and Operations Division, Adolph J. Barsanti	254-8450
Director, Dallas Field Office, Thomas Garland	(214) 729-2200
Director, Denver Field Office, William Henkes	(303) 234-5717
Director, Pittsburgh Field Office, Frank Doyle	(412) 729-2200
Director, Office of Energy Data Development, Joseph DiMarino	254-5077
Director, Energy Use Systems Development Division, Kenneth Vagts	634-5642
Director, Energy Source Systems Development Division, William Molloy	254-3364
Director, Regulatory Systems Development Division, John Yienger	252-6599
Director, Integrated Energy Systems Development Division, William Skinner	254-7303
Director, Systems Development Support Division, Martha Chin	252-6559
Director, Office of Energy Data Standards and Statistical Design, John Gross	252-5214
Director, Survey and Statistical Design Division, Irene Montie	252-5118
Director, Energy Data Standards Division, James Brown	252-5216
Director, Office of ADP Services, Vincent J. Iannuzzi	254-7210
Director, Computer Services Division, Larry Clinton	254-8720
Director, Systems Support Division, William Coakley	254-8720
Director, Office of the Energy Emergency Management Information System Project, Barry Yaffee	634-2079

Energy Information Validation

Assistant Administrator, Charles S. Smith	633-8800
Director, Office of Validation Resources, Ed Cresswell	633-8116
Director, Office of Validation Analysis, John Shewmaker	633-8800
Director, Validation Research Division, Richard W. Kline	633-8104
Director, Energy Model Validation Division, Vacant	
Director, Validation Methodology Division, W. Richard Johnsen	633-8770
Director, Office of Systems Validation, Wallace O. Keene (Acting)	633-8117
Director, Design Review Division, Craig Cranston	633-9461
Director, Validation Operations Division, Charles Shirkey (Acting)	633-8113

Applied Analysis

Assistant Administrator, C. Roger Glassey	633-8544
Deputy Assistant Administrator, Elizabeth C. MacRae	633-8544
Director, Office of Analysis Oversight and Access, George Lady (Acting)	633-9342
Director, Office of Energy Source Analysis, David Hulett	633-8540
Director, Short-Term Analysis Division, Vacant	
Director, Oil and Gas Analysis Division, Charles Everett	633-9108
Director, Coal and Electric Power Analysis Division, Jerry Eyster	633-8944
Director, Nuclear Energy Analysis Division, Ronald G. Clark	633-8244
Director, Office of Energy Use Analysis, Howard Walton (Acting)	633-8486
Director, Demand Analysis Division, Terry H. Morlan	633-8494
Director, Regional Energy Activity Analysis Division, Frank Hopkins	633-8533
Director, Conservation and Renewable Resources Division, Howard Magnas	633-8877
Director, Office of Integrative Analysis, Kenneth Kincel	633-8733
Director, Mid-Term Analysis Division, Charles Mylander	633-9274
Director, Long-Term Analysis Division, John Pearson	633-8515
Director, International Energy Analysis Division, W. Calvin Kilgore	633-9397
Director, Regional, Socioeconomic and Environmental Analysis Division, Hugh Knox	633-8720
Director, Macroeconomic Analysis Division, Gerard Lagace (Acting)	633-8693
Director, Office of Energy Industry Analysis, David Montgomery	252-6254
Director, Regulatory and Competitive Analysis Division, David Shapiro	252-6257
Director, Financial and Industry Studies Division, Vacant	

NOTE: All telephone numbers given are in area code 202, unless otherwise noted.

Office of Energy Information Services

	Extension
Director, John E. Daniels	634-5602
Director, National Energy Information Center, W. Neal Moerschel	634-2167
Director, Energy Information Administration Clearinghouse, Thomas T. Daugherty	634-2167

Office of Program Development

Assistant Administrator (Vacant)	
Director, Office of Financial Reporting System, Arthur T. Andersen	633-8806
Director, Office of Oil and Gas Information System, Gordon K. Zareski	252-6403
Director, Special Coal Studies Project, Jerome Temchin	633-8191

ADMINISTRATOR ECONOMIC REGULATORY ADMINISTRATION

Administrator, David J. Bardin	254-8505
Deputy Administrator, Hazel R. Rollins	254-7500
Counsel to the Administrator, Terence L. O'Rourke	254-9666
Special Assistant to Administrator, Judith Saunders	254-8505
Special Assistant to Deputy Administrator, James Solit	254-7500

Office of Special Counsel for Compliance

Special Counsel, Paul L. Bloom	633-8925
Deputy Special Counsel, Avrom Landesman	633-8925
Director, Division of Planning and Resource Management, Shelley R. Kolbert (Acting)	633-9358
Solicitor, Carl A. Corrallo	633-8440
Director, Division of Audit Program Operations, Lon W. Smith	633-9410
Director, Division of Audit Systems, Diane W. Lique	633-9110
Director, Field Operations, Robert E. Boldt (Acting)	633-9012
Director, Division of Technology and Computer Science, Ernst J. Neugroschl	633-8870

Operations Planning and Evaluation

Director, Martin I. Farfel	254-3196
Deputy Director, Marvin J. Shapiro	254-8093
Acting Executive Secretariat, April Ramey	254-7500
Chief, Correspondence Control Branch, Hariold Hall	632-4964
Director, Program Management Division, James M. Brown	254-8690
Director, Management Operations Division, John D. Blackwell	254-9770
Director, Public Affairs Division, William L. Webb	634-2170

Office of Enforcement

Assistant Administrator, Barton Isenberg	254-8740
Deputy Assistant Administrator, Gordon W. Harvey	254-6990
Director, Enforcement Information, Thomas J. McGuire, Jr.	254-6414
Director, Enforcement Review, E. Dawn Thompson	254-8795
Director, Special Investigations, Jerome Weiner	254-9670
Director, Enforcement Policy and Planning, Barry E. Wagman	254-6990
Director, Enforcement Training Center, Wayne W. Porter	254-9670
Director, Enforcement Program Operations (Vacant)	254-8877

NOTE: All telephone numbers given are in area code 202, unless otherwise noted.

Regulations and Emergency Planning

	Extension
Assistant Administrator, Douglas G. Robinson	254-8675
Special Assistant, Kathleen Deutsch	254-8675
Deputy Assistant Administrator, F. Scott Bush	632-8494
Special Assistant, Vacant	632-8494
Director, Major Emergency Planning Division, Ben Massell	632-6500
Director, Petroleum Price Regulations, Edwin Mampe	254-7200
Director, Division of Coal Regulations, Stephen Stern	254-9766
Director, Natural Gas Regulations, Linda Lapin	632-4721
Director, Petroleum Allocation Regulations, Gerald P. Emmer	254-7200

Fuels Regulations

Assistant Administrator, Barton R. House	254-3905
Deputy Assistant Administrator, Doris J. Dewton	254-3236
Director, Crude Oil Supply and Allocation Division, Peter A. Antonelli	254-7434
Director, Fuel Supply and Allocation Division, George E. Hall, Jr.	254-7886
Director, Coal Utilization Division, Walter A. Romanek	254-3910
Director, Industry and Regional Operations Division, T. Wendell Butler	254-3400
Director, Import/Export Division, Finn K. Neilsen	254-9730

Office of Utility Systems

Assistant Administrator, Douglas C. Bauer	254-9782
Deputy Assistant Administrator, Jerry L. Pfeffer	254-9655
Executive Officer, Faye Redding	254-9494
Director, Division of Utility Regulatory Assistance, Howard F. Perry	254-3118
Director, Division of Regulatory Interventions, A. Grey Staples	254-8690
Director, Division of Power Supply and Reliability, Charles A. Falcone	634-5620

DIRECTOR
OFFICE OF ENERGY RESEARCH

Director, John M. Deutch	252-5430
Deputy Director, Worth Bateman	252-5434
Special Assistant, Kevin Gorman	252-5430
Associate Director, Office of Nuclear Non-Proliferation, Marvin Moss	252-5438
Coordinator, Office of International Nuclear Fuel Cycle Evaluation, Eric Beckjord	376-9356
Director, Office of Nuclear Policy Analysis, David Bodde	252-5755

Office of Basic Energy Sciences

Associate Director, Dr. James S. Kane	353-5565
Deputy Director (Vacant)	353-3081
Director, Division of Administration, H. L. Kinney	353-5451
Director, Division of Program Planning, Dr. James S. Coleman	353-4061
Director, Division of Nuclear Sciences, Dr. George L. Rogosa (Acting)	353-3613
Director, Division of Materials Sciences, Dr. Donald K. Stevens	353-3427
Director, Division of Chemical Sciences, Dr. Elliot S. Pierce	353-5804
Director, Division of Engineering, Mathematical and Geo-Sciences, Dr. Elliot S. Pierce (Acting)	353-5804
Director, Division of Advanced Energy Projects, Dr. Ryszard Gajewski	353-5995
Director, Division of Biological Energy Research, Robert Rabson	353-3571

NOTE: All telephone numbers given are in area code 202, unless otherwise noted.

Office of High Energy and Nuclear Physics

	Extension
Associate Director, Dr. James S. Kane (Acting)	353-5565
Deputy Director (Vacant)	353-3081
Director, Division of High Energy Physics, Dr. William A. Wallenmeyer	353-3624
Director, Division of Nuclear Physics, Dr. George L. Rogosa	353-3613

Office of Program Analysis

Associate Director, Roger LeGassie	252-5436
Deputy Director, Richard Williamson	376-9376
Director, Division of Technology Implementation Analysis, Bruce Robinson	376-9264
Director, Division of Policy Issue Analysis, Robert Kelly	376-9259
Director, Division of R&D Strategy Studies, John Powers	376-9268
Director, Division of R&D Program Evaluation, Roger LeGassie (Acting)	252-5436

Office of Field Operations Management

Associate Director, Antionette (Toni) Grayson Joseph	252-5447
Deputy Director, Joseph E. Machurek	376-9384
Director, Division of Laboratory Management, Alan B. Claflin	376-9271
Director, Division of Program Implementation, James G. Ling	376-9387
Director, Division of Institutional Programs, Richard Stephens	376-9387

Office of Research Policy

Associate Director, Joel Snow	252-5444
Deputy Director, William Bartley (Acting)	376-9360
Director, Division of Research and Technical Assessment, Leonard Topper (Acting)	376-9372
Director, Division of Advanced Technology Projects, Ernest Blase	376-9360
Executive Director, R&D Coordination Council Staff, Robert Summers	376-9368
Director, Division of Solar Power Satellite Project Systems, Frederick Koomanoff	376-9362
Executive Director, Energy Research Advisory Board Staff, Donald Shapero	376-9368

Office of Program Administration and Operations

Assistant Director, Ira Adler	252-5440
Director, Division of Budgets and Program Coordination, Charles Cathey	252-6641
Director, Division of Resource Management and Operations, Robert Starrett	376-4631
Director, Division of Policy and Analysis Coordination, Kristine Forsberg	376-1699

ASSISTANT SECRETARY ENERGY TECHNOLOGY

Assistant Secretary, John M. Deutch (Acting)	252-6842
Deputy Assistant Secretary, Worth Bateman	252-5434
Deputy Assistant Secretary, Programs and Operations, Donald M. Kerr (Acting)	252-6850
Deputy Assistant Secretary, Planning and Analysis, Roger W. A. LeGassie (Acting)	252-5436
Director, Resource Management and Acquisition, Keith N. Frye	252-6500
Executive Assistant, John O. McElvey	252-6846
Director, Planning and Technology Transfer, George Jordy	252-6503
Director, Field Operations Management, Ronald Cochran (Acting)	252-6660

NOTE: All telephone numbers given are in area code 202, unless otherwise noted.

	Extension
Director, Institutional Liaison and Communications, Kevin Gorman (Acting)	252-5430
Program Director, Solar and Geothermal Energy Programs, Bennett Miller	376-4102
Program Director, Fossil Energy Programs, George Fumich	353-2642
Program Director, Nuclear Energy Programs, Robert L. Ferguson	353-3465
Director, Office of Fusion Energy, Edwin Kintner	353-3347
Director, Office of Nuclear Waste Management, Sheldon Meyers	353-5645

ASSISTANT SECRETARY ENVIRONMENT

Assistant Secretary, Ruth C. Clusen	376-4185
Deputy Assistant Secretary, James L. Liverman	353-5171
Staff Assistant for Operations, Lynda L. Brothers	376-4189
Scientific Advisor to the ASEV, Hal L. Hollister	376-4189
Special Assistant for Communications, Carl A. Eifert	376-4190
Staff Assistant to the ASEV, Carol A. Jolly	376-4185
Management Analyst, Dorothy H. Fauntleroy	376-4189

Office of Environmental Compliance and Overview

Director, Robert J. Catlin (Acting)	353-3033
Deputy Director, W. Herbert Pennington (Acting)	353-3033
Director, NEPA Affairs Division, Robert J. Stern (Acting)	376-5998
Director, Operational and Environmental Safety Division, W. J. McCool (Acting)	353-3157
Medical Director, William R. Albers (Acting)	353-3333
Chief, Environmental Protection and Public Safety Branch, L. Joe Deal (Acting)	353-4093
Chief, Policy, Measurements, and Institutional Standards Branch, George P. Dix (Acting)	353-5644
Chief, Occupational Safety Branch, David E. Patterson (Acting)	353-5605
Chief, Process and Facility Safety Branch, Herbert C. Field (Acting)	353-5629
Director, Environmental Control Technology Division, William E. Mott (Acting)	353-3016
Chief, Nuclear Technologies Branch, Robert W. Ramsey (Acting)	353-5028
Chief, Fossil and Inexhaustible Technologies Branch, Myron Gottlieb (Acting)	353-5587
Director, Safety Engineering Division, Robert W. Barber (Acting)	353-3548

Office of Technology Impacts

Director, Peter W. House	376-1820
Deputy Director, Joseph A. Coleman	376-1820
Director, Policy Analysis Division, Edward R. Williams	376-9073
Chief, Regulatory Analysis Branch, David A. Litvin (Acting)	376-4449
Chief, Issues Analysis Branch, Edward R. Williams (Acting)	376-9073
Director, Technology Assessments Division, Dario R. Monti	376-4363
Chief, Fossil Technologies Branch, Arnold J. Goldberg (Acting)	376-1785
Chief, Nuclear Technologies Branch, Joseph R. Maher (Acting)	353-4655
Chief, Conservation, Solar, and Geothermal Technologies Branch, Robert P. Blaunstein (Acting)	376-4692
Director, Environmental Impacts Division, Roger D. Shull	353-3311
Chief, Applied Analysis Branch, Manikonda Sastry (Acting)	353-4969
Chief, Methods and Data Evaluation Branch, Richard H. Ball (Acting)	353-5801
Director, Regional Assessments Division, Joan C. Hock	353-4258
Chief, Regional Characterization and Mitigation Branch, Paul Cho (Acting)	353-5897
Chief, Regional Assessments Branch, Theodore R. Harris (Acting)	353-4487

NOTE: All telephone numbers given are in area code 202, unless otherwise noted.

Office of Health and Environmental Research

Director, W. W. Burr, Jr. (Acting) 353-3153
  Deputy Director, Charles W. Edington (Acting) 353-3251
  Manager, CO_2 and Climate Research Program, David H. Slade (Acting) 353-4374
Director, Human Health and Assessments Division,
  Walter W. Weyzen (Acting) 353-5355
Director, Health Effects Research Division,
  Charles E. Carter (Acting) 353-5468
Director, Ecological Research Division, Jeff Swinebroad (Acting) 353-4208
Director, Pollutant Characterization and Safety Research Division,
  Robert W. Wood (Acting) 353-3213

Office of Program Coordination

Director, George R. Shepherd (Acting) 376-4445
  Manager, R&D Coordination Program, Murray Schulman (Acting) 353-3683
Director, Environmental Liaison Division, Shelda Weinstein (Acting) 376-4445
Director, Program Planning and Analysis Division,
  W. J. Little, Jr. (Acting) 353-3285
  376-4191

Office of Management Support

Director, John C. Whitnah (Acting) 376-4189
  Deputy Director, Marilynne F. Baldwin (Acting) 353-5268
Director, Financial Services Division, Delmar D. Mayhew 353-3541
  Manager, Management Information Services, C. Edward Miller (Acting) 353-3199
  Chief, Budget and Accounting Branch (Vacant) ....
  Chief, Contract Management and Special Projects Branch (Vacant) ....
Director, Management Services Division, Donald E. Shaw (Acting) 353-4223
  Chief, Administrative Branch, Barbara W. Rosenzweig (Acting) 353-4663
  Chief, Mail and Records Branch, Carmen E. Riddle (Acting) 353-4534

ASSISTANT SECRETARY
INTERGOVERNMENTAL AND INSTITUTIONAL RELATIONS

Assistant Secretary, Phillip S. Hughes 252-5450
  Director, Indian Affairs, George Crossland 252-5311
Director, Policy and Coordination, James J. Furse 252-5452
  Policy Assistant, Elizabeth V. Buffum 252-5454
  Policy Assistant, Daniel C. Frederick 252-5454
  Policy Assistant, Emmett J. Gavin 252-5454
  Policy Assistant, Linda Rae Gregory 252-5454
  Staff Assistant, Administrative Support, Joyce A. Brady 252-5477
  Staff Assistant, Regional Coordinator, Mary L. Dunlap 252-5477
Director, Executive Office, John van Santen 252-5456
  Assistant Director, Herbert Murphy 252-5101
  Director, Manpower Services, Thelma Twigg 252-5520
  Director, Planning and Budget, Richard Brancato 252-5736
    Chief, Budget Branch, Peggy Capozzi 252-5740
  Director, IR Secretariat, Bonnie Betancourt 252-5738
    Chief, Correspondence Management, Sophie Emami 252-5738
    Chief, General Services & Records Management, Clara Nutter 252-5832
Director, Technical Information Office, John van Santen 252-5456

NOTE: All telephone numbers given are in area code 202, unless otherwise noted.

Office of Congressional Affairs

<u>Extension</u>

Director, Office of Congressional Affairs, Richard L. Wright.................. 252-5468
 Deputy Director, Paul Cyr 633-8520
 Executive Assistant, Ray Marble... 252-5468
 Deputy Director, Legislative Liaison Division (Senate), Robert Schule....... 252-5797
 Deputy Director, Legislative Liaison Division (House),
  Frederick Merrill, Jr.. 252-5795
 Director, Legislative Support Division, Wolf Repke 633-9149
 Director, Research and Analysis Division, Rita Alicks...................... 633-9117
 Director, Operations and Inquiries Division, Lawrence Atwell 633-8520

Office of Consumer Affairs

Director, Office of Consumer Affairs, Tina Hobson 252-5373
 Deputy Director, Jerry Penno 252-5871
Director, Consumer Impact Division, Polly Craighill 252-5871
Director, Citizen Participation Division, William Holmberg 252-5841
 Chief, Consumer Outreach Branch (Vacant)....................................
 Chief, Advisory Committee Management Branch, Georgia Hildreth 252-5877

Office of Education, Business and Labor Affairs

Director, Office of Education, Business and Labor Affairs,
 Lawrence Stewart 252-5377
 Deputy Director, Bart McGarry 252-6350
 Administrative Officer, Ginger King 252-6350
Director, Education Programs Division, James Kellett, Jr...................... 252-6480
 Chief, Academic Programs Branch, Don Duggan 252-6482
 Chief, Public Program Branch, Joseph Carvajal............................... 252-6480
 Chief, Institutional Programs Branch, Harold Young 252-6350
Director, Labor Affairs and Manpower Assessment Division,
 William Tucker 252-5544
 Chief, Manpower Assessment Branch, Norman Seltzer........................... 252-5931
 Chief, Labor Affairs Branch (Vacant) 252-5544
Director, Business Programs Division, Richard Pastore......................... 252-6260
 Chief, Business Relations Branch, Wesley Barnes 252-6260
 Chief, Inventions Branch, Patrick Donohoe.................................... 252-6260

Office of Intergovernmental Affairs

Director, Office of Intergovernmental Affairs, William E. Peacock 252-5370
 Special Assistant, T. J. Hopkins 252-6335
Federal and Special Relations Division
 Assistant Director for Federal Relations, Robert Ritzmann................... 252-6336
 Assistant Director for Special Relations, Tom Dennis....................... 252-6336
Director, Program Planning and Analysis Division, Jack Daly 252-5660
Director, City and County Relations Division (Vacant)........................ 252-5632
Director, State Relations Division (Vacant) 252-5834

NOTE: All telephone numbers given are in area code 202, unless otherwise noted.

Office of Public Affairs

	Extension
Director, Office of Public Affairs, James Bishop, Jr.	252-5466
Deputy Director, John A. Harris	252-5540
Assistant Director, Robert W. Newlin	252-5540
Assistant to the Director, L. Carlisle Gustin	252-5540
Director, Press Services Division, Alfred Alibrando	252-5806
Chief, News Operations Branch, Gail Bradshaw (Acting)	252-5806
Chief, Broadcast News Branch, Alfred Alibrando (Acting)	252-5806
Still Photo Files	353-5476
Director, Editorial Services Division, William Anderson	252-6440
Chief, Policy Interpretation Branch, William Anderson (Acting)	252-6440
Chief, Publications Branch, John Sullivan (Acting)	252-6441
Chief, Conference Coordination Branch, Phyllis Corbitt (Acting)	252-5644
Chief, Special Media Branch, Martin Moon (Acting)	252-6461
Director, Special Programs Division, Victor P. Keay (Acting)	252-5640
Chief, Public Inquiries Branch, Patrick F. Donnelly (Acting)	252-5565
Chief, Audiovisual Branch, Alfred Rosenthal (Acting)	353-5365
Chief, Exhibits Branch, John Bradburne (Acting)	353-5441

ASSISTANT SECRETARY INTERNATIONAL AFFAIRS

Assistant Secretary, Harry E. Bergold, Jr.	252-5800
Principal Deputy Assistant Secretary, Bruce C. Clarke, Jr.	252-5858
Special Assistant, Charles Boykin	252-5858
Deputy Assistant Secretary for International Policy Development, Sarah Jackson	252-5916
Director, Special Regions Policy (Vacant)	252-6380
Deputy Assistant Secretary for International Programs, Holsey G. Handyside	252-5921
Director, Nuclear Affairs, Harold D. Bengelsdorf	252-6175
Director, International Programs, Jack Vanderryn	252-6140
Deputy Assistant Secretary for Trade and Resources, Peter Borre	252-5918
Director, Industry Operations, David Oliver	252-6777
Director, Resource Trade, John Treat	252-6770
Deputy Assistant Secretary for International Energy Research, A. Denny Ellerman	252-5890
Director, International Market Analysis, Herman Franssen	252-5893
Director, Current Assessments, John LaBarre	252-5174
Director, International Program Support, Sal Cianella	252-5926

ASSISTANT SECRETARY POLICY AND EVALUATION

Assistant Secretary, Policy and Evaluation, Alvin L. Alm	252-5325
Principal Deputy Assistant Secretary, Policy and Evaluation, Leslie J. Goldman	252-5318
Executive Assistant to the Principal Deputy Assistant Secretary, Charlotte Chapman-Cope	252-5316
Deputy Assistant Secretary, Planning and Evaluation, James R. Janis (Acting)	252-5421
Deputy Assistant Secretary, Policy Analysis, Darius Gaskins	252-5493
Executive Assistant to the Assistant Secretary, William J. Silvey	252-5328
Executive Officer, Harold H. Brandt	633-9177
Director, Office of Competition, William C. Lane	252-5680
Senior Technical Advisor to the Assistant Secretary, Nicolai Timenes	252-5334

NOTE: All telephone numbers given are in area code 202, unless otherwise noted.

Planning and Evaluation

	Extension
Deputy Assistant Secretary, Planning and Evaluation, James R. Janis (Acting) ..	252-5421
Director, Office of Planning and Regulatory Program Evaluation, Stuart Ray (Acting)	252-5722
Director, Planning and Budget Systems Division (Vacant)	
Director, Environmental Planning and Program Evaluation Analysis, Jack Siegel (Acting)	252-6453
Director, State and Local Planning and Program Evaluation Division (Vacant)	
Director, Regulatory Planning and Program Evaluation Division, William Strauss	252-5727
Director, Information Systems Planning and Program Evaluation Division (Vacant)	
Director, Office of Technical Programs Evaluation, J. Frederick Weinhold	252-6360
Director, Energy Supply Technologies Division, Arthur Ingberman	252-5616
Director, Energy Use Technologies and Evaluation Coordination Division, Edward Blum	252-5313
Director, Office of Policy Coordination, Richard Smith	252-5320
Director, Financial Tax Policy Division, Gina Despres (Acting)	252-5686
Director, Legislation and Policy Coordination Division, William White	252-5683
Director, International and Security Policy Division, Gina Despres	252-5686
Director, Strategic and Contingency Planning Division, David Bloom	252-5404
Director, Office of Emergency Response Planning, Carlyle E. Hystad	252-6450

Policy Analysis

	Extension
Deputy Assistant Secretary for Policy Analysis, Darius Gaskins	252-5493
Director, Office of Coal, Utility and Integrated Energy Analysis, Morris Zusman (Acting)	633-9194
Director, Synthetic Fuels and Industrial Fuel Use Division, Irving Susel	252-5340
Director, Integrated Energy Analysis Division, Tom Neville	252-6421
Director, Utility and Coal Supply Policy Division, Morris Zusman (Acting)	633-9194
Director, Office of Oil and Gas Policy, Stephen E. McGregor	252-5626
Director, Oil Division, Lucian Pugliaresi	252-5667
Director, Gas Division, Henry P. Santiago	633-9052
Director, Office of Conservation and Advanced Energy Systems Policy, Clark Bullard	252-5423
Director, Conservation Division, Sandra Rennie	633-8585
Director, Advanced Systems Division, Allan Hoffman	252-6431
Director, Office of Analytical Services, Roger Naill	252-5388
Director, Economic Impact Analysis Division (Vacant)	
Director, Energy Modeling and Analysis Division, John Stanley-Miller	252-5393

ASSISTANT SECRETARY RESOURCE APPLICATIONS

	Extension
Assistant Secretary, George S. McIsaac	633-9222
Executive Assistant, Christopher K. Chapin	633-9225
Environmental Advisor, Ellison Burton	633-9353
Special Assistant, Eleanor Holmes	633-9225
Director, Policy, Planning and Analysis, Ralph Bayrer	633-8362
Director, Division of Policy & Program Evaluation, Tony Cordesman	633-8362
Director, Division of Energy Resource Programs, Myron Allen	633-9390
Director, Resource Management, Charles W. Ebbecke	633-8322

NOTE: All telephone numbers given are in area code 202, unless otherwise noted.

	Extension
Director, Division of Management, Martin Duby	633-9445
Director, Division of Administrative Support, Raymond E. Gibson	633-9244
Director, Division of Budget, Thomas A. Keegan	633-8662
Director, Division of Executive Secretariat, Edna Eagle	633-9250

Utility and Industrial Energy Applications

Deputy Assistant Secretary, Stanley I. Weiss	633-9350
Staff Assistant, Frank Lyman	633-9350
Director, Coal Resource Management, Richard Passman	633-8350
Resource Manager, High Btu Coal Gasification, Phillip Gallo	633-9195
Resource Manager, Low/Med Btu Coal Gasification, Russell Bardos	633-9195
Resource Manager, Industry Atmospheric Fluidized Bed Coal Combustion, and Director, Division of Direct Combustion Applications, Howard Feibus	633-9272
Director, Division of Coal Liquefaction, Marvin Singer	633-9102
Director, Uranium Resources and Enrichment, William R. Voigt, Jr.	633-8616
Deputy Director, J. L. Schwennesen	633-9375
Director, Division of Business and Marketing Activities (Vacant)	633-9265
Director, Division of Gas Centrifuge Enrichment Program, Donald K. Gestson	633-9492
Director, Gaseous Diffusion Operations, Donald E. Saire	633-9385
Director, Division of Resource Assessment Operations, John A. Patterson	633-8700
Director, Emerging Energy Sources, Stanley I. Weiss (Acting)	633-9350
Resource Manager, Division of Geothermal Resource Management, Rudolph A. Black	633-8756
Resource Manager, Small Hydro Applications, Richard McDonald (Acting)	633-8910
Director, Coal Supply Development, Edward J. Myerson	633-9039
Director, Division of Coal Loan Programs, Robert M. Grubenmann	633-8377
Director, Division of Anthracite, Jerry Pell	633-9058
Director, Division of Licensing and Siting, G. Curtis Jones, Jr.	633-8600
Director, Division of Coal Logistics, I-Ling Chow	633-8404
Director, Division of Coal Production Technology, George W. Sall	633-9073

Oil, Natural Gas and Shale Resources

Deputy Assistant Secretary, R. Dobie Langenkamp	633-8400
Staff Assistant, John Shanges	633-8400
Director, Naval Petroleum and Oil Shale Reserves, Captain Robert H. Nelson	633-8674
Director, Division of Operations, Lawrence Vogel	633-8641
Director, Division of Legal, Peter M. Frank	633-8683
Director, Oil and Natural Gas Supply Development, J. Lisle Reed	633-8395
Director, Division of Special Projects, Troyt York (Acting)	633-8383
Director, Division of Enhanced Oil Recovery, Ira Mayfield (Acting)	633-8395
Director, Division of Natural Gas Resource Development, Larry Dewey (Acting)	633-8383
Director, Division of Supply Initiatives, Eugene Peer (Acting)	633-9179
Director, Shale Resource Applications, Paul Petzrick (Acting)	633-8660
Director, Division of Program Development and Evaluation, Mary Halow (Acting)	633-8651
Director, Division of Technology Assessment, Brian Harney (Acting)	633-8644
Director, Leasing Policy Development, Robert J. Kalter	633-9421
Director, Division of Planning and Program Development, Robert Lawton	633-9326
Director, Division of Regulatory, Allan Weeks	633-9035
Director, Division of Economic Analysis, Stuart Edwards (Acting)	633-9035

NOTE: All telephone numbers given are in area code 202, unless otherwise noted.

Office of Power Marketing Coordination

	Extension
Director, Daniel M. Ogden, Jr.	633-8338
Director, Staff Office, Robert K. Volk, Jr.	633-8338
Director, Division of Rates, Contracts and Power Allocation, John DiNucci	633-8376
Director, Division of Power and Water Resources, Truman P. Price	633-8336
Assistant Administrator, Bonneville Power Administration (Washington, D.C. Office), George Bell	633-8330

DIRECTOR ADMINISTRATION

Director, William S. Heffelfinger	252-5940
Secretary, William R. Allen	252-5940
Deputy Director, William P. Davis	252-5942
Secretary, Rebecca Arndt	252-5942
Administrative Officer, Lenora J. Lewis	252-5947
Assistant, Myrna K. Turturro	252-5947

Office of Personnel Management

Director, Personnel, Lloyd W. Grable	252-5610
Administrative Officer, Dottie Van Steinburg	252-5612
Deputy Director, J. Merle Schulman	376-4213
Special Assistant, John Polishuk	353-5595
Consultant, Jose A. Ramirez	376-9710
Director, Division of Personnel Policies and Programs, Geraldine P. Flowers	376-4468
Chief, Employment Policies and Programs, Anita Sciacca, (Acting)	376-1786
Chief, Classification, Compensation and ADP, Douglas L. Wood	376-1789
Chief, Employee/Labor Relations, Ronald W. Knisley	376-4441
Director, Employee Development and Training Division, Helene S. Markoff	376-4232
Chief, Executive Development Branch, John Uhrlaub	376-4233
Chief, Training, Policy and Systems Branch (Vacant)	
P.O. Box E, Oak Ridge Training and Conference Center, Oak Ridge, TN	626-1363 FTS
Director, Executive Resources Division, Eugene H. Beach	633-8228
Chief, Executive Program Planning, Leon G. Harbeson	633-8232
Chief, Energy Executive Service, Bill McCormick	633-8228
Chief, Special Executive Resources, Robert L. Willyerd	633-8228
Director, Headquarters Operations Division, Lois Schutte	376-4211
Deputy Director (Vacant)	376-9390
Chief, Central Employment and Staffing, Raymond C. Cardinal (Acting)	376-4221
Personnel Management Services -	
Chief, Germantown, Gerald C. Butts	353-5116
Chief, 12th and Pennsylvania Avenue, Richard J. Yochum	633-9282
Chief, 20 Massachusetts Avenue, Paul J. Montigny	376-9690
Chief, Training (Headquarters), James P. Baker	252-4220
Chief, Employee/Labor Relations, Jay Neel	633-9372
Chief, Personnel Management Evaluation and Field Liaison Staff, Furman Layman	252-5949

Office of Organization and Management Systems

Director, Harry L. Peebles	252-6799
Deputy Director, K. Dean Helms	252-6802
Director, Manpower Resource Management Division, Charles R. Tierney	252-6804
Chief, Manpower Utilization Branch, Elwood P. Sheetz	252-6346
Chief, Manpower Control Branch, John E. Klansek	252-6807
Director, Organization Planning and Management Division, K. Dean Helms (Acting)	252-6802

NOTE: All telephone numbers given are in area code 202, unless otherwise noted.

	Extension
Director, Management Systems Analysis Division, Andrew D. Eppelmann	252-5620
Chief, Directives and Reports Branch, Jean F. McKeel	376-4337
Chief, Management System. Development and Evaluation Branch, Andrew D. Eppelman (Acting)	252-5620

Office of Construction and Facility Management

Director, Donald A. Koss (Acting)..	353-3087
Director, Engineering and Construction Support Division, Donald A. Koss	353-5574
Chief, Project Support Branch (D.C.), Richard A. Frost	252-6760
Chief, Project Support Branch (GTN) (Vacant)	353-5573
Director, Facilities and Real Estate Management Division, Edward H. Tuttle, III (Acting)	353-3706
Chief, Facility Management Branch, Frank Lewis, Jr. (Acting)	353-5426
Chief, Energy Management Branch, Edward H. Tuttle, III	353-3698

Office of Administrative Services

Director, Gene K. Fleming.......................................	252-5710
Deputy Director, Vern Witherill	252-5712
Assistant to the Director, Ronald R. Turner	252-5710
Administrative Officer, Marian B. Ruck	252-5717
Budget and Funds Control Staff (Vacant)	252-5714
Director, Division of Engineering and Facilities, Lewis J. Claytor	252-6080
Deputy Director, Robert F. Cooper (Acting)	252-6226
Chief, Buildings Operations Branch (Vacant)	252-6100
Chief, Facilities Engineering Branch, Robert F. Cooper (Acting)	252-6226
Chief, Space Management Branch (Vacant)	252-5030
Chief, Video Tape Branch (Vacant)	252-6075
Director, Division of FOI and Privacy Acts Activities, Milton Jordan	252-5955
Deputy Director (Vacant)	
Director, Division of Library Services, ...	353-4301
C. Neil Sherman	376-9015
Chief, Library Branch (D.C.), Robert F. Kimberlin 20 Mass. Ave.	376-9015
Chief, Library Branch (Germantown), Ruth E. Perks	353-2855
Chief, Library Branch (FERC), Denise B. Diggin 825 NCA	275-4303
Director, Division of Printing and Graphics, Donald R. Hunt, Sr.	376-5768
Chief, Policy and Standards Branch, John A. Schmehl	252-5041
Chief, Printing Operations Branch (Vacant)	376-9443
Chief, Graphics Branch, Danny C. Jones	376-9447
Director, Division of Property, Mail and Transportation, James B. Sampson (Acting)	252-5712
Chief, Property and Supply Branch, Robert A. Lavalle (Acting)	252-5204
Chief, Mail and Transportation Branch, William G. McDermott (Acting)	252-6086
Chief, Records Management Branch (Vacant)	252-5712

Office of ADP Management

Director, Ronald S. Schwartz	353-4720
Deputy Director, Robert E. Greeves	353-3307

NOTE: All telephone numbers given are in area code 202, unless otherwise noted.

Office of Computer Services and Telecommunications Management

	Extension
Director, John W. Polk	353-3685
Deputy Director, Charles N. Brashears	353-3057
Assistant to Director, Gerald F. Chappell	353-4319
Director, Division of Operations, Ronald L. Kaiser	353-3193
Chief, SACNET Operations Branch, David W. Rowland	353-4622
Chief, Data Processing Operations Branch, Richard G. Shook	353-4602
Chief, Message Center and ADP Services Branch, David K. Berkau	376-4279
Chief, Operational Systems Support Branch, Peter J. Grahn, Jr.	353-4653
Director, Division of Technical Support, Howard E. Lewis, Jr.	353-4640
Chief, Systems Software Branch, Howard E. Lewis, Jr. (Acting)	353-4640
Chief, Analytic Computing Branch, Joseph D. Eckard, Jr.	353-4211
Chief, Data Base Administration Branch, Alfred W. Kuebler	353-4578
Director, Division of Telecommunications, Jesse T. Pate, Jr.	353-3674
Chief, Voice Networks Engineering and Circuit Management Branch, Jesse T. Pate, Jr. (Acting)	353-4314
Chief, Radio and Electronics Engineering Branch, Robert M. Lewis	353-4627
Chief, Data Communications and COMSEC Engineering Branch, Ben B. Barnett	353-4636

Office of Contractor Industrial Relations

	Extension
Director, C. Stuart Broad	353-5083
Deputy Director (Vacant)	353-5083
Director, Division of Program Management and Assessment, Lyle E. Crews	353-5291
Chief, Labor Relations Branch, Juanita E. Smith	353-4361
Chief, Systems and Analysis Branch, Eric Tolmach	353-4825
Director, Division of Contractor Personnel Management, James R. O'Gwin	353-4818
Chief, Compensation, Pensions & Labor Management Branch, Wayne W. Rives	353-4418
Chief, Human Resource Development Branch, J. Faherty Casey	353-4810

BOARD OF CONTRACT APPEALS

Chairman, John B. Farmakides	376-4181
Vice Chairman, John G. Roberts	376-4181
Member, Carlos R. Garza	376-4181
Recorder, Kathleen A. Jones	376-4181

CONTROLLER

Controller, C. N. Mitchell (Acting)	376-4155
Deputy Controller (Vacant)	376-4155
Special Assistant, C. N. Mitchell	376-4159
Director, GAO Liaison, Donald C. Gestiehr	353-5035
Chief, Administrative Support Branch, James R. Gregg	353-5030
James R. Gregg	376-4159

Office of Budget

Director, Mary E. Tuszka (Acting)	376-4172
Deputy Director, Mary E. Tuszka	376-4172
Director, Division of Budget Operations (Vacant)	376-4163
Director, Division of Budget Analysis, Aaron D. Edmondson (Acting)	376-4163
Director, External Coordination Staff, Kendrick Wentzel (Acting)	376-4157

NOTE: All telephone numbers given are in area code 202, unless otherwise noted.

Office of Program Mangement Support

Extension

Director, Bert Greenglass	376-4048
Director, Division of Business Management Systems, Walter F. Barney (Acting)	353-4645
Director, Division of Financial Management Support, Michael W. O'Neill (Acting)	353-4686
Director, Division of DOE-Wide Systems Development and Coordination, Frederick L. Schuyler (Acting)	353-3506

Office of Program and Project Assessment and Control

Director (Vacant)...	
Director, Independent Cost Estimating Staff, Bobby R. Scarlett (Acting)	353-5499
Director, Division of Program Management Review and Control, Edward H. Schneider (Acting)	376-4161
Director, Division of Program and Project Management Assessment, Thomas R. James (Acting)	353-4411

Office of Finance and Accounting

Director, Robert L. Zanetell	353-4461
Director, Division of Accounting Policy and Procedures, L. Stowe Lenderman (Acting)	353-4466
Director, Division of Financial Programs, William R. Mitchell (Acting)	353-3401
Director, Division of Washington Financial Services, Fletcher R. Brande (Acting)..	353-5316
Director, Division of Financial Practices and Compliance, Curtis Ingraham (Acting)	353-4340

Office of Financial Policy

Director, Thomas L. Blair (Acting)	376-1776

DIRECTOR OFFICE OF EQUAL OPPORTUNITY

Director, M. A. Bowden	376-4663
Deputy Director, N. H. Pierson	376-4663
Departmental Federal Women's Program Manager, M. Anderson	376-4663
Hispanic Employment Program Coordinator, L. G. Turrietta	376-4663

Federal Equal Employment Opportunity Programs

Director, Complaints and Investigations Division, T. K. Levi.................	376-9422
Director, Federally Assisted Program Division, C. A. Ruiz	376-1923
Director, Equal Opportunity Program Division, C. K. Zane	376-1828

NOTE: All telephone numbers given are in area code 202, unless otherwise noted.

GENERAL COUNSEL

Extension

General Counsel, Lynn R. Coleman 252-5281
  Special Assistant, Erica A. Ward 252-5281
  Staff Assistant, Eleanor Rodgers 252-5281
Deputy General Counsel, Eric J. Fygi (Acting) 252-5284
  Special Assistant, Henry Garson 252-5284
Deputy General Counsel for Regulations (Vacant) 252-6732
  Assistant General Counsel for International Trade and Emergency
    Preparedness, Robert C. Goodwin (Acting) 633-9380
  Assistant General Counsel for Petroleum Regulations,
    Michael Paige (Acting) 252-6736
  Assistant General Counsel for Interpretations and Rulings,
    Everard Marseglia (Acting) 633-8969
  Assistant General Counsel for Coal and Leasing Regulations,
    James H. Heffernan (Acting) 633-9296
Deputy General Counsel for Enforcement and Litigation,
  Gaynell C. Methvin 633-9199
  Assistant General Counsel for Enforcement, Robert G. Heiss (Acting) 254-8700
  Assistant General Counsel for General Litigation,
    Joseph DiStefano (Acting) 376-4254
  Assistant General Counsel for Regulatory Litigation I,
    Thomas P. Humphrey (Acting) 633-9475
  Assistant General Counsel for Regulatory Litigation II, John P. McKenna 633-9199
  Assistant General Counsel for Administrative Litigation,
    Richard Greene (Acting) 633-8624
  Director, Special Investigations, Jerome Weiner (Acting) 254-9670
Deputy General Counsel for Programs, Robert M. Hallman 376-4250
  Assistant General Counsel for Energy Research, Technology and
    Applications, Leonard Rawicz (Acting) 376-4252
  Assistant General Counsel for International Development and Defense
    Programs, Peter Brush (Acting) 376-4258
  Assistant General Counsel for Conservation and Solar Applications,
    Charles F. Savage 376-4100
  Assistant General Counsel for Environment, Stephen Greenleigh (Acting) 376-4266
  Assistant General Counsel for Regulatory Interventions,
    Cameron Graham 633-9473
Deputy General Counsel for Legal Services, Harry M. Yohalem (Acting) 252-5246
  Assistant General Counsel for Legal Counsel, Thomas C. Newkirk 633-8613
  Assistant General Counsel for Patents, James Denny (Acting) 353-4018
  Assistant General Counsel for Procurement, Lloyd W. Sides (Acting) 252-6911
  Assistant General Counsel for Legislation, Colin D. Mathews (Acting) 252-6718
  Assistant General Counsel for Standards of Conduct,
    Ralph Goldenberg (Acting) 353-5285
Administrative Officer, William V. Ochs, Jr. (Acting) 633-8774
  Assistant Administrative Officer
    William L. Shepherd (Acting) 633-8774
  Assistant Administrative Officer, Rita Kidd, (Acting) 353-4053
  GC Librarian, Oscar Strothers (Acting) 633-8922

DIRECTOR OFFICE OF HEARINGS AND APPEALS

Director, Melvin Goldstein 254-5134
  Office of Management Operations
    Director, Robert G. Emond 254-9711
  Management Information Branch
    Chief, Otto S. Reid 254-8480
  Docket and Publications Branch
    Chief, Marcia Proctor 254-9740
  Public Docket Room 254-6306
  Office of Legal Analysis
    Director, George B. Breznay 254-9681
  Office of Financial Analysis
    Director, Richard T. Tedrow 254-8606
  Office of Economic Analysis
    Director, Thomas L. Wieker 254-9681

NOTE: All telephone numbers given are in area code 202, unless otherwise noted.

OFFICE OF INSPECTOR GENERAL

	Extension
Inspector General, J. Kenneth Mansfield	252-4073
Deputy Inspector General, Thomas S. Williamson, Jr.	252-4073
Executive Assistant, Thomas F. Dunn	252-4073
Assistant IG for Investigations, William DeSonia	252-4143
Assistant IG for Audits, Gilbert F. Stromvall	252-4079
Assistant IG for Inspections, Edward L. Heller	252-4109
Executive Director, A. F. McGuire	252-4128
Director, Internal Policy, Planning and Evaluation, M. Thomas Abruzzo	252-4128
Director, Resource Management (Vacant)	

PROCUREMENT AND CONTRACTS MANAGEMENT DIRECTORATE

Director, Michael J. Tashjian	376-9232
Deputy Director, Hilary J. Rauch	376-9235
Executive Assistant, Steven R. Morgan	376-9236
Advisor for Architect-Engineer and Construction Procurement Office, Sonny A. Caputo	376-9191

Office of Small and Disadvantaged Business Utilization

Director, Colonel C. Armstrong	376-9195
Director, Business Liaison Division, Jack D. Koser	376-1694
Director, Small Business Division, Francis C. Brda	376-9057
Director, Minority Business Division, Leonel V. Miranda	376-1850

Policy Office

Director, Berton J. Roth	376-9227
Director, Policy and Procedures Division, Robert L. Van Ness	376-1730
Director, Contract and Property Management Division, Joseph Garcia	376-9208

Procurement Management Office

Director, John E. Reid	376-9322
Director, Contract Business Clearance Division, David J. Ball	376-9322
Director, Procurement Management Review Division, James B. Boone	376-9327
Director, Procurement Management Systems and Analysis Division, George D. Conwell	376-9351
Director, Resource Control and Administration Division, George Williams	376-9343

NOTE: All telephone numbers given are in area code 202, unless otherwise noted.

Program Support Office

	Extension
Director, William B. Ferguson	376-9060
Deputy Director, Thomas Anderson	376-9063
Director, Technology & Resource Applications Support Division, James R. Higgins	376-9063
Director, Major Systems Acquisition Division, Edward G. Cumesty	376-9063
Director, Conservation, Solar Applications and Institutional Support Division, Thomas Anderson	376-9063

Operations Office

Director, Joseph P. Cappello	376-9167
Deputy Director, Thomas J. Davin	376-9168
Associate Director, G. Edward Larson	376-9167
Director, Contract Execution Division "A", Ben Goldman	376-5902
Director, Contract Execution Division "B", Edwin R. Itnyre	376-9823
Director, Contract Support Division, John C. Regan	376-9171
Director, Contract Cost and Price Division, James A. Nelson	376-4562

NOTE: All telephone numbers given are in area code 202, unless otherwise noted.

FERC Organizational Listing

Office of the Chairman

	Extension
Chairman, Charles B. Curtis	275-4152
Executive Assistant to the Chairman, Walter W. Schroeder	275-4194
Assistants to the Chairman	
J. Curtis Moffatt	275-4195
James H. Bailey	275-4193
A. Angela Lancaster	275-4192
Confidential Assistant to the Chairman	
Joanne Aldridge	275-4152
Secretarial Staff	
Anita Battese	275-4193
Rose Marie Oliver	275-4192
Charmaine R. McNiven	275-4194
Sarah V. Howard	275-4198
Janet Mullinax	275-4183

Office of Commissioner Don S. Smith

Commissioner Don S. Smith	275-4141
Legal Advisor to Commissioner, Kim Martin Clark	275-4141
Legal Advisor to Commissioner (Vacant)	275-4141
Confidential Assistant to Commissioner, Margie Passerini	275-4141
Secretary to Commissioner, Marcie Gouldman	275-4141

Office of Commissioner Sheldon

Commissioner Georgiana Sheldon	275-4112
Attorney Advisor, Daniel Shillito	275-4112
Attorney Advisory, Peter Lesch	275-4112
Confidential Assistant, Jeraline Fennell	275-4112
Secretary, Jo Pensivy	275-4112

Office of Commissioner Holden

Commissioner Matthew Holden, Jr.	275-4175
Legal Advisor, Terry Oberstone Vogel	275-4176
Technical Advisor, Gregory D. Martin	275-4176
Confidential Assistant to the Commissioner, Kathryn L. Hesaltine	275-4175
Secretary, Lorraine M. Marrow	275-4176

Office of Commissioner Hall

Commissioner George R. Hall	275-4147
Technical Assistant, Norman Pedersen	275-4147
Technical Assistant, Paul Korman	275-4147
Confidential Assistant, Martha Stringer	275-4147
Secretary, Doris Cooper	275-4147
Clerk Typist, Janice Marshall	275-4147

NOTE: All telephone numbers given are in area code 202, unless otherwise noted.

Office of the Executive Director

	Extension
Executive Director, William G. McDonald	275-4122
Secretary, Patricia M. Jones	275-4122
Deputy Executive Director (Vacant)	275-4158
Secretary, Patricia Snesrvid	275-4158
Special Assistant to the Executive Director, Paul M. Feine	275-3925
Secretary, Barbara Anderson	275-3925
Director, Division of Planning and Evaluation, Lawrence R. Anderson	275-5168
Secretary, Dolores Jean Ankrim	275-5168
Director, Division of Administrative Liaison with DOE, James Dunn	275-4797
Secretary, Norma T. McDonald	275-4797
Budget Staff	275-4797
Building Facilities Manager, Larry Langhorne	275-4021
Contracts Office, Al Smith	275-4529

Office of Opinions and Reviews

Director, Bernard Wexler	275-4200
Deputy Director, Kenneth E. Richardson	275-4200
Secretary, Bessine B. Squirewell	275-4200

Office of Enforcement

Director, Sheila S. Hollis	275-1832
	275-4890
Special Assistant, Connie Davies	275-1832
	275-1832
Acting Special Counsels to the Director	275-3763
Philip Marston	275-1832
Senior Trial Attorney - Special Projects, Lou Rosenman	275-0303
Electric Power Enforcement	
Acting Assistant Director (Vacant)	
Trial Attorney, Andrew M. Zack	275-1832
Trial Attorney (Vacant)	275-1832
Gas and Oil Pipeline and Producer Enforcement	
Acting Assistant Director, Stanley W. Balis	275-3634
Trial Attorney, Joel F. Zipp	275-3654
Trial Attorney, Maureen P. Wilkerson	275-6516
Trial Attorney	275-1832
Special Investigations	
Acting Assistant Director, Thomson vonStein	275-1832
Trial Attorney	275-4841
Technical Advisors	
Public Utilities Specialist, Jeanne M. Zabel	275-5337
	275-1832
(Vacant)	275-1832
(Vacant)	275-1832
(Vacant)	
Secretarial Staff	
Pamela S. Dowdle	275-1832
Debra Pettitt	275-1832
Virgie Clark	275-1832
Electric Power	
Herman Bluestein	275-0303
Lawrence Acker	275-0303
Gas and Oil Pipeline and Producer Enforcement	
Craig Ellis	275-1832
Charles Friedman	275-0303
Robert Hirasuna	275-1832
Stephen Melton	275-0303
Technical Advisors	
Julie Bernt	275-1832
Secretarial Staff	
Patty Casillo	275-1832
Beatrice Jordon	275-1832

NOTE: All telephone numbers given are in area code 202, unless otherwise noted.

Office of Administrative Law Judges

Extension

Chief Administrative Law Judge, Joseph Zwerdling 275-3917
 Deputy, Chief Administrative Law Judge, Curtis L. Wagner, Jr. 275-3902
 Administrative Officer, Sandra Wyvill 275-4588
Administrative Law Judges:
 Isaac D. Benkin 275-3904
 Bruce L. Birchman 275-1546
 William L. Ellis 275-3935
 Samuel Z. Gordon 275-3904
 Stephen L. Grossman 275-2529
 David I. Harfeld 275-0405
 Thomas L. Howe 275-3918
 William Jensen 275-3904
 Max L. Kane 275-3934
 Samuel Kanell 275-3934
 Sherman P. Kimball 275-3918
 Burton S. Kolko 275-1546
 Allen C. Lande 275-3918
 Michel Levant 275-3918
 George P. Lewnes 275-4542
 Ernst Liebman 275-1546
 Jon G. Lotis 275-3934
 Graham W. McGowan 275-1546
 Victor M. Mercogliano 275-1546
 David W. Miller 275-3904
 Raymond M. Zimmet 275-1546

Office of the Secretary

Secretary, Kenneth F. Plumb 275-4166
 Assistant Secretary, Lois D. Cashell 275-4166
 Assistant to the Secretary, Edward B. Frigillana 275-4166
 Secretary, Margaret K. Nelson 275-4166
Director, Division of Agenda and Minutes, Linda L. Hancock 275-4136
Director, Division of Regulatory Support Services, Durwood W. Pate 275-4970
 Supervisory Management Assistant, James Newton 275-4970
 Chief, Dockets Branch, Constance Y. Terrell 275-4897
 Chief, Mail Branch, Lewis E. Burrus 275-3950
 Chief, Records Branch (Vacant) 275-4974
 Chief, Registry and Service Branch, Mildred T. Battle 275-4936
Director, Division of Assignment and Control (Vacant) 275-4186

Office of the General Counsel

General Counsel, Robert R. Nordhaus 275-4309
 Secretary, Jane Register 275-4309
 Confidential Assistant for Legislation, Deborah Gottheil 275-4312
Deputy General Counsel, Robert L. Baum 275-4333
Deputy General Counsel,
Solicitor, Howard E. Shapiro 275-4258
Chief Trial Counsel, David Leckie 275-4264
Assistant General Counsel, Charles E. Bullock 275-4891
Assistant General Counsel, Francis J. Gilmore 275-4851
Assistant General Counsel, Mary Wright 275-3436
Assistant General Counsel, John B. O'Sullivan 275-3411
Assistant General Counsel, Howard Jack 275-4269
Assistant General Counsel, Philip Telleen 275-4305
Deputy Assistant General Counsel-Legislation, Romulo L. Diaz, Jr. 275-3771
Deputy Assistant General Counsel-General Law, Herbert Rice 275-3801
Principal Staff Attorneys
 Richard A. Azzaro 275-4223
 David Boergers 275-4240
 Donald H. Clarke 275-4346
 Joel Cockrell 275-4320
 Bernard Cromes 275-4237
 J. Paul Douglas 275-4193
 John S. Everett, Jr. 275-4927
 Raymond E. Hagenlock 275-4271
 John J. Keating, Jr. 275-3973

NOTE: All telephone numbers given are in area code 202, unless otherwise noted.

Office of the General Counsel (Cont'd)

Extension

Principal Staff Attorneys (cont.)

Scott E. Koves	275-4808
Daniel Lamke	275-4233
Linda L. Lee	275-4313
Mark G. Magnuson	275-4891
Russell B. Mamome	275-4282
Edward R. Mark	275-4344
Watt N. Martin	275-4315
Richard L. Miles	275-4245
Charles F. Reusch	275-4216
John P. Roddy	275-4343
Susan T. Shepherd	275-4788
Robert L. Winters	275-4675
Cyril S. Wofsy	275-4247
Robert A. Wolfe	275-5110
James N. Wood	275-4955

Office of Regulatory Analysis

Director, Haskell P. Wald	275-4118
Administrative Officer, Sophie K. Parkhurst	275-4118
Secretary, Patricia Enoch	275-4128
Director, Division of Economics and Finance, Gordin T. C. Taylor (Acting)	275-4173
Chief, Economic Analysis Branch (Vacant)	--
Chief, Financial Analysis Branch, Julian M. Greene (Acting)	275-4171
Director, Division of Analysis and Policy Development, Wade P. Sewell (Acting)	275-4180
Chief, Regulatory Policy Branch, Jon H. Goldstein (Acting)	275-4116
Chief, Special Projects Branch, Frederick W. Lawrence (Acting)	275-4178
Director, Division of Energy Technology and Environment, Carl N. Shuster, Jr. (Acting)	275-3662
Advisor on Environmental Quality, Jack M. Heinemann	275-6569

Office of Pipeline and Producer Regulation

Director, Barry L. Haase	275-4473
Deputy Director, Kenneth A. Williams	275-4431
Special Assistant, Robert A. Nelson	275-4444
Special Assistant (Oil Pipeline) Leon J. Slavin	275-0285
Special Assistant, Francis J. Connor	275-1525
Assistant to the Director, Louis W. Mendonsa (Acting)	275-4580
Chief, Compliance Branch, Randolph E. Mathura	275-4489
Director, Program Planning & Administrative Staff, Dean R. Koth	275-4477
Director, Alaska Gas Project Office (AGPO), John B. Adger	275-3827
Director, Trans-Alaska Pipeline System (TAPS), John J. Lahey (Staff Counsel)	275-4830
Director & Technical Advisor NEA Task Force, Frederick D. Cornelius	275-4932
Director, Division of Pipeline Certificates, Robert J. Szekely	275-4496
Chief, Pipeline Certificate I Branch, Charles W. Strotman	275-4404
Chief, Pipeline Certificate II Branch, Cristobal Hernandez (Acting)	275-4493
Chief, Environmental Branch, Michael J. Sotak (Acting)	275-4940
Director, Division of Pipeline Curtailment & Special Cases, Joseph J. Solters (Acting)	275-4389
Chief, Curtailment Branch, James M. Kiely (Acting)	275-4384
Chief, Special Cases Branch, David C. Lathom (Acting)	275-4359
Chief, Systems Analysis Branch, John E. Moriarty	275-4451
Director, Division of Pipeline Rates, Raymond A. Beirne (Acting)	275-4371
Deputy Director, Andrew W. Battese	275-4427
Chief, Allocation & Rate Design Branch, Robert E. Scarbrough (Acting)	275-4408
Chief, Cost of Service Branch, Donald Champagny (Acting)	275-4374
Chief, Finance Branch, Georgia K. LeDakis	275-7492
Chief, Depreciation Branch, Norman Deutsch	275-4896
Chief, Tariff Branch, Herbert Horton	275-4480
Chief, Valuation Branch, Joseph M. Morgan	275-7470
Director, Division of Producer Rates & Certificates, Lundy R. Wright	275-4607
Deputy Director, Louis J. Engel (Acting)	275-4602
Chief, Rate Filing Branch, John Miller (Acting)	275-4545

NOTE: All telephone numbers given are in area code 202, unless otherwise noted.

Extension

```
Chief, Certificate Applications Branch, William H. Marmura................    275-4563
Chief, Special Rates Branch (Vacant)............................................
Chief, Gas Supply Branch, Wayne Thompson.......................................    275-4562
Director, Division of NGPA Compliance, Howard Kilchris (Acting)..............    275-4539
  Chief, General Reports Branch, Marvin Hirsh (Acting)........................    275-4623
  Chief, Jurisdictional Agency Reports Branch, Jerome Knittle (Acting).......    275-4606
  Chief, Review & Compliance I Branch, Victor Zabel (Acting).................    275-4568
  Chief, Review & Compliance II Branch, William Bushey (Acting)..............    275-4616
  Chief, Review & Compliance III Branch, Marilyn Rand (Acting)...............    275-4639
```

Office of Electric Power Regulation

```
Director, William W. Lindsay .................................................    275-4777
  Deputy Director, Robert E. Cackowski .......................................    275-4779
  Assistant Director, Edward O. Savwoir ......................................    275-4783
  Assistant to the Director, Daniel G. Lewis .................................    275-4766
  Administrative Officer, James Mellom........................................    275-4750
Director, Division of Licensed Projects, James J. Stout......................    275-4868
  Deputy Director, Ronald A. Corso ...........................................    275-4863
  Chief, Project Analysis Branch, Gerald Wilson .............................    275-4820
  Chief, Applications Branch, Fred E. Springer ..............................    275-4981
  Chief, Power Site Lands Branch, Ronald A. Corso (Acting)...................    275-4863
  Chief, Inspections Branch, Harry E. Thomas ................................    275-4885
  Chief, Environmental Analysis Branch, Quentin Edson .......................    275-4875
Director, Division of Rates & Corporate Regulation, Leo T. Markey............    275-4669
  Assistant Director, Gordon E. Murdock (Acting).............................    275-4667
  Chief, Electric Rate Investigations Branch, Lawrence M. Shulman (Acting)....    275-4776
  Chief, Electric Rate Filings Branch, Irvin E. Ball ........................    275-4693
Director, Division of River Basins, Neal C. Jennings.........................    275-4768
  Chief, Basin and Project Plans Branch, Ernest E. Sligh (Acting) ...........    275-4689
Chief, Headwater Benefits Investigations Branch, Richard K. Faubel (Acting)...    275-4691
Director, Division of Interconnection and Systems Analysis,
  Bernard B. Chew ...........................................................    275-4770
  Assistant to the Director, William I. Wheelock ............................    275-4730
  Chief, Energy and Fuels Analysis Branch, Alexander Gakner .................    275-4677
  Chief, System Evaluation Branch, Gene Biggerstaff .........................    275-4721
  Chief, Interconnection and Special Investigations Branch,
      Edward J. Fowlkes .....................................................    275-4731
```

Office of Chief Accountant

```
Chief Accountant, L. H. Drennan, Jr. ........................................    275-4031
  Deputy Chief Accountant, Kent H. Crowther .................................    275-4033
  Assistant to Chief Accountant, Eugene J. Shuchart .........................    275-4035
  Administrative Assistant, Dawn W. Hughes ..................................    275-4036
Director, Division of Audits, Morris R. Fitzgerald ..........................    275-4090
  Chief, Northern Branch, William Connelly ..................................    275-4073
  Chief, Programs & Special Projects Branch, Russell E. Faudree .............    275-4044
  Chief, Southern Branch, Pasquale P. Pagliuca ..............................    275-4076
  Chief, Western Branch, Thomas C. King .....................................    *
  Chief, Valuation & Analytical Studies Branch, Ben F. Kitashima ............    275-4048
Director, Division of Accounting Systems, John Loreg, Jr. ...................    275-4050
```

* (415) 556-1500
  555 Battery Street
  Room 418
  San Francisco, California 94111

Office of Public Information

```
Director, Joyce R. Morrison (Acting) ........................................    275-0330
  Assistant Director (Vacant) ...............................................    275-0330
  Assistant to the Director, Dianne M. Novick ...............................    275-0330
  Chief, Public Inquiries Branch, William H. Zietz .........................    275-4006
  Chief, Public Reference Branch, Grace M. Plummer ..........................    275-0336
  Chief, Media Relations Branch (Vacant) ....................................    275-0333
  Secretary, Judith F. Rhodes ...............................................    275-0330
  Secretary, Rebecca K. O'Neal ..............................................    275-0330
```

NOTE: All telephone numbers given are in area code 202, unless otherwise noted.

Part 2
Energy-Saving Tips

19   Saving Energy in the Home and Automobile

Americans use more energy per person than any other people in the world. With only 6 percent of the world's population, the United States uses about one-third of all the energy consumed on earth. Where does all this energy go? Industry uses about 36 percent. Commerce uses about 11 percent for enterprises including stores, offices, schools, and hospitals. Residences take about 26 percent. And transportation accounts for another 29 percent or so.

Most of the energy used in the United States comes from petroleum (crude oil). Because domestic production falls short of the need, almost half of it has to be imported at a cost of $45 billion a year (at 1977 rates).

Expert estimates of known and potential American domestic reserves vary, but most likely the United States has a 25-to-30-year supply of oil, if its energy-use growth rate is kept at about 2 percent per year.

However, if the United States continues using energy at the current rate, it could run out of domestic oil in the year 2007 and out of natural gas even sooner. The severe winter of 1976-77 painfully dramatized the natural gas situation with its complex supply and economic problems.

The overall energy situation in the United States is not rosy: Energy demand keeps rising; energy prices keep going up; the availability and future costs of supplies remain uncertain.

What can we do about it? Conserve energy. This will help us extend our supplies and reduce our import burdens until we develop new energy technologies and resources. Without personal hardship, we could easily cut our energy use by an estimated 30 percent or more—saving energy for our country and money for ourselves.

The energy we use for our homes and automobiles—gas, oil, electricity—draws on all our energy resources. Cutting back on these uses is the simplest, most effective way to make our resources last longer. And each individual conservation effort, multiplied by millions, can serve as an "energy bank"—a supply that can be used to help balance our energy accounts.

We can conserve if we make energy thrift a way of life by adopting common-sense energy habits. This chapter contains some practical advice on how you can help.

Where does our residential energy go? Seventy percent of it is used to heat and cool our homes. An additional 20 percent goes for heating water, the second-largest home energy user and expense. The remaining 10 percent is used for lighting, cooking, and running small applicances.

We can cut our energy use and costs by making our homes energy-efficient, even if we have to spend some money to do so. The money we spend now will be returned through lower utility bills month after month. And then the savings will be all ours—as good as a taxfree raise in income.

HOT-WATER ENERGY SAVERS

Heating water accounts for about 20 percent of all the energy used in our homes. Don't waste it.

- Repair leaky faucets promptly. A faucet that drips one drop a second can waste as much as 60 gallons of hot or cold water in a week.
- Do as much household cleaning as possible with cold water.
- Insulate your hot-water storage tank and piping.

Water Heaters

Energy-efficient water heaters may cost a little more initially, but reduced operating costs over a period of time can more than make up for the higher outlay.

- Buy a water heater with thick insulation on the shell. While the initial cost may be more than one without this conservation feature, the savings in energy costs over the years will more than repay you.
- Add insulation around the water heater you now have if it's inadequately insulated, but do not block off needed air vents. That would create a safety hazard, especially with oil and gas water heaters. When in doubt, get professional help. When properly done, you should save about $15 a year in energy cost.
- Check the temperature on your water heater. Most water heaters are set for 140°F or higher, but you may not need water that hot unless you have a dishwasher. A setting of 120°F can provide adequate hot water for most families.
- If you reduce the temperature from 140°F (medium) to 120°F (low), you could save over 18 percent of

the energy you use at the higher setting. Even reducing the setting 10°F will save you more than 6 percent in water-heating energy.

- If you are uncertain about the tank water temperature, draw some water from the heater through the faucet near the bottom and test it with a thermometer.
- Don't let sediment build up in the bottom of your hot water heater; it lowers the heater's efficiency and wastes energy. About once a month, flush the sediment out by drawing several buckets of water from the tank through the water-heater drain faucet.

ENERGY SAVERS IN THE KITCHEN, LAUNDRY, AND BATH

- Use cold water rather than hot to operate your food disposal. This saves the energy needed to heat the water, is recommended for the appliance, and aids in getting rid of grease. Grease solidifies in cold water and can be ground up and washed away.
- Install an aerator in the kitchen-sink faucet. By reducing the amount of water in the flow, you save the energy required to heat it. The lower flow pressure is hardly noticeable.
- If you are purchasing a gas oven or range, consider one with an automatic (electronic) ignition system instead of a pilot light. You'll save an average of up to 47 percent of your gas use—41 percent in the oven and 53 percent on the top burners.
- If you have a gas stove, make sure the pilot light is burning efficiently, with a blue flame. A yellowish flame indicates an adjustment is needed.
- Never boil water in an open pan. Water comes to a boil faster and uses less energy in a kettle or covered pan.
- Keep range-top burners and reflectors clean. They will reflect the heat better, and you will save energy.
- Match the size of pan to the heating element. More heat will get to the pan; less will be lost to surrounding air.
- If you cook with electricity, turn off the burners several minutes before the allotted cooking time. The heating element will stay hot long enough to finish the cooking for you without using more electricity.
- When using the oven, cook as many foods as you can at one time. Prepare dishes that can be stored or frozen for later use, or make all oven-cooked meals.
- Watch the clock or use a timer; don't open the oven door continually to check food. Every time you open the door, heat escapes and the cooking uses more energy.
- Use small electric pans or ovens for small meals rather than the kitchen range or oven. They use less energy.
- Use pressure cookers and microwave ovens if you have them. They can save energy by reducing cooking time.

SAVING ENERGY WITH YOUR DISHWASHER

About four times as much energy is used to heat the water that goes into a dishwasher than to operate the dishwasher mechanically. So the best way to save energy with your dishwasher is to use less hot water. You will probably use *less* energy using a dishwasher than if you do dishes by hand. Most people think a dishwasher fills completely with water. Actually, about 2 to 3 gallons of hot water are used for each "fill," which fills the machine tub only just below the bottom rack. What cleans the dishes is not the *amount* of hot water but its force and direction.

Purchasing Considerations

Look for these energy-saving features:

- Short-cycle selections. These shorter cycles use less hot water and are used for lightly soiled dishes. You can probably save up to 25 percent of water-heating energy cost.
- Less hot-water usage. Models vary as to the number of gallons of hot water used in a cycle. Check specification sheets, buying guides, and other product literature for this information.
- An "air dry" selector. This feature automatically shuts off the heat during the dry cycle. This can save 10 percent of the electricity. You'll find it on most newer models.

Use and Care

Don't stop using your dishwasher. You'll probably use more hot water doing dishes by hand.

Wash only *full* loads. Running two half-loads will take twice as much energy as a full load.

If you wash dishes at least once a day, don't waste water prerinsing dishes. Newer models require only that dishes be scraped off and liquids emptied.

Don't use the drying cycle. Instead, use the "no-heat" selector, or turn off the dishwasher after the last rinse. Normal air convection will dry the dishes. Although this takes longer, the savings are significant.

Use the special cycles on your dishwasher. Always use the shortest cycle that will satisfactorily wash the load.

Keep your dishwasher away from your refrigerator-freezer. Dishwashers produce moisture and heat, which will make your refrigerator-freezer use more energy.

Try not to use your dishwasher during peak utility hours.

If your dishwasher has a filter, check it often to be sure it is not clogged with food particles. Some filters are self-cleaning.

Always follow the manufacturer's instructions in the use and care booklet provided with a new dishwasher.

If you wash a load of dishes but aren't satisfied with the results, you'll more than likely run the load again. This wastes energy. The obvious solution is to make sure you will get satisfactory results. If you're not getting satisfactory results, then:

- Load the dishwasher according to the manufacturer's instructions. Every model loads a little differently. Some models have a diagram inside the door to show you how.
- Make sure the water coming into the dishwasher is no lower than 140°F.
- Use only automatic dishwasher detergent. Check your use and care booklet for the amount to use. You may have to experiment a little.
- Don't store dishwasher detergent for long periods of time. It will cake and become stale and won't perform satisfactorily. It should be kept in a dry place—not under the sink.
- If your dishwasher has a rinse-agent dispenser, keep it filled. It will help you get more spot-free drying. Rinse agents are available in both liquid and solid form.

SAVING ELECTRICITY

During late afternoon and early evening hours, the load on the nation's electrical systems usually reaches its peak. To meet the heavy demand, electric utilities often must use backup generating equipment that is not energy-efficient. Thus, try to use energy-intensive appliances such as dishwashers, clothes washers and dryers, and electric ovens in the early morning or late evening hours to help reduce that peak load. If everyone scheduled household chores during off-peak hours, the utilities' daily fuel use would be reduced, and the nation's energy would be conserved.

SAVING ENERGY WITH YOUR RANGE

Purchasing Considerations

1. A *self-cleaning oven* is more efficient than conventional ovens. Less energy is used because of the extra insulation needed for the very high temperatures in the special cleaning cycle. Assuming that you avoid unnecessary spillovers and use the cleaning cycle infrequently, you can save some energy with this oven.
2. A *convection oven* is relatively new on the market and available in both gas and electric models. A blower forces heated air into the oven directly to the food. In some models, no preheating is needed. The forced heated air is recirculated through the system with only about 10 percent of the heat exhausted through the oven vent. Cooking temperatures can be reduced from 25°F to 100°F, and cooking times, in some instances, are shorter.
3. *Electric ignition* is now available on gas ranges. It eliminates the need for oven and surface-burner pilot lights. Constantly burning pilot lights waste energy.
4. *Microwave ovens* are a form of electromagnetic energy having a specific frequency and wavelength. They are similar to visible light and high-frequency radio waves. In microwave cooking, the energy is absorbed by the food. Microwaves do not heat the surrounding air, oven walls, or recommended food containers. As the microwave energy enters the food, it causes the water, fat, and sugar molecules to rub against each other millions of times each second. This friction, similar to that caused by rubbing your hands together, creates the heat that cooks the food.

Because of this unique action, microwave ovens can save time. Microwave ovens can help you save cooking energy depending on the type and amount of food cooked. In tests conducted by the Association of Home Appliance Manufacturers, those homes using both a standard range and a microwave oven used about 14 percent less energy to cook than homes with only a standard range.

Use and Care

How much energy you save with your range depends on how you use and care for it. Always follow the manufacturer's instructions in the use and care booklet provided with your new appliance.

Don't use your range to heat the kitchen.

When cooking, follow the package or recipe instructions carefully, especially for time and temperature settings. When you've finished, always check that the indicator lights and all controls are turned off. When you have to thaw frozen food before cooking, use the refrigerator or microwave oven. Generally, frozen foods need longer cooking time.

For the oven. Plan ahead. When you turn on the oven, the entire oven is heated so make full use of it. Try to plan a complete oven or broiler meal.

After the oven has been turned off, use leftover heat to warm plates or heat rolls.

Don't be an "oven peeker." Each time you open the oven door during cooking, a significant amount of heat escapes. Food then takes longer to bake, and the loss of heat can affect browning and baking. Learn to depend on thermostats and timers instead and to look through the window in the oven door if there is one.

Double recipes, and freeze half for later use. You'll reduce your oven's energy use.

Make sure the oven door closes tightly. Check the seal for wear, tears, or cracks. Keep it clean and in good repair.

If you have a self-cleaning oven, use the special cleaning cycle right after baking or roasting while the oven is still hot. It won't have to heat up that much more to get to the cleaning temperature.

To baste or check internal temperature, or add vegetables to a roast, remove the pan from the oven and close the door. Return the pan to the oven when finished. Before turning on the oven, arrange the racks to accommodate baking pans.

In some instances, preheating is unnecessary. When you do preheat, keep it to the shortest possible time.

Using aluminum foil as an oven liner may reduce the efficiency of your oven. Be sure to check the use and care manual to see what the manufacturer recommends.

For surface units. Always fit a pot or pan to the size of the surface burner. A 6-inch pan on an 8-inch unit wastes energy. Pots and pans should have flat bottoms to make the best contact with the surface unit. This is especially important with the ceramic/smooth-top cooking surfaces.

Pots and pans should have straight sides and tight-fitting lids. Avoid peeking under the lids or stirring during cooking. Rely instead on specific time and temperature settings.

Always place a pot or pan containing food or liquid on the surface unit or burner *before* turning it on. Generally, start most foods on a high heat setting, and reduce to a lower heat as soon as foods reach boiling. Once the food is boiling, or cooking, it won't cook any faster on high. On electric ranges, the unit can be turned off minutes ahead of time so the food will cook with the retained heat.

Cook with as little liquid as possible. The food, especially vegetables, will be more nutritious and look and taste better.

Always cook foods at the lowest possible setting. Do not overcook. (Most people do!)

If your range has thermostatically controlled surface units or burners, use them. These units have a heat-sensing element in the center, which cycles on and off to maintain a certain temperature. Check the manufacturer's instructions on their use.

Keep reflector drip pans under surface units clean.

For gas ranges. If your range has pilot lights, make sure they are adjusted properly so they don't use more gas than necessary.

Check the gas flame. It should be pure blue. If the flame is yellow, it needs to be adjusted by a qualified technician.

SAVING ENERGY WITH YOUR REFRIGERATOR-FREEZER

Refrigerators (single door), combination refrigerator-freezers, and separate freezers (both chest and upright) are available. The term *refrigerator-freezer* is used here to apply to all three types. The amount of energy these appliances use varies significantly among brands and models. Their energy use depends mainly on (1) the overall size, (2) the freezer size and temperature, and (3) the type of defrost system. Generally, larger refrigerator-freezers (especially larger freezer compartments), colder freezer temperatures, and automatic defrost systems will use more energy.

Purchasing Considerations

Style. Single-door refrigerators have a small freezing section of about 15°F for ice cubes and a few days' food storage only. Other refrigerator-freezers have separate freezer compartments, which maintain a "zero zone" temperature and

are for longer-term food storage. These are available in various door styles—top or bottom-mount freezers (two-door), side-by-side, three-door, and a door-within-a-door.

Size. It is very important to choose the right size refrigerator-freezer for your needs. A "too large" or "too small" refrigerator-freezer wastes energy. The "too large" model refrigerates more space than your household needs. If your refrigerator is "too small," you will have to make extra trips to the grocery or fill the unit too full for adequate air circulation.

If you answer "yes" to most of the following questions, you probably should consider a larger refrigerator-freezer.

- Do you have a large number of persons in your household?
- Do you live quite a distance from a grocery store?
- Do you usually buy some foods in large quantities?
- Do you have a garden, or do you live near seasonal fresh-food suppliers and buy such foods in quantity for freezing?
- Do you prepare meals for several days in advance and freeze them?
- Do you have leftovers often?
- Do you entertain often?

On the other hand, if you live alone, seldom eat at home, and only open your refrigerator for ice cubes or a quick snack, the smaller single-door refrigerator may be right for you.

Because family sizes and life-styles differ, it is difficult to know exactly what size refrigerator-freezer a particular family will need. However, the following estimates may be helpful to you.

A family of two generally needs at least 8 cubic feet of space in the refrigerator section plus 1 cubic foot of space for each additional person. Add 2 cubic feet if you entertain often. The freezer space should provide about 2 cubic feet per person.

The largest combination refrigerator-freezers currently available provide 17 cubic feet of fresh-food compartment space and 10 cubic feet of freezer space. If your family needs more space than this, you may want to consider a separate freezer.

Defrost System. Refrigerator-freezers are available in three different defrost systems.

1. *Manual.* You have to turn the refrigerator control to "off" or "defrost" so that the freezer will warm up and the ice will loosen from the surfaces. The defrost water and ice from both the refrigerator and freezer sections must then be removed by hand and the refrigerator restarted by setting the temperature control to the right setting. (This system is available in the single-door refrigerator-freezer.)

2. *Partial automatic (cycle defrost).* The cooling surface in the fresh-food (refrigerator) section defrosts

automatically each time the unit stops running. Normal refrigeration is not interrupted. The freezer section, however, must be defrosted manually. This system is available in the two-door (bottom- or top-mount freezer) refrigerator-freezer.

3. *Automatic defrost.* Frost is removed automatically from the surfaces and packages in both the refrigerator and freezer sections. (This system is available in the top- and bottom-freezer mount two-door models, the side-by-side, and the three-door or door-within-a-door models.)

Generally, the more automatic defrost systems consume more energy. However, you can make up for some of this added energy use in two ways: you won't have frost building up on the surface, which, if more than ¼ inch thick, will make the refrigerator-freezer run longer and work harder; and you won't have to use outside heat to defrost and thus use extra energy to lower the temperature again.

Efficiency. Once you decide on the size and type of refrigerator-freezer you need and any special features you want, look for the most efficient one available. The efficiency will vary from model to model. The more efficient, the less it will cost you to operate.

Use and Care

Save energy by *using* your refrigerator-freezer. A refrigerator-freezer works day and night. If you don't use it to its fullest extent, you waste energy. With a little imagination and planning, you can make up for some of your refrigerator-freezer's energy use. In other words, your refrigerator-freezer can "pay you back," and you will be saving energy, too.

For example, prepare several meals at one time, and freeze them for later use. This will save energy by making the most use of your cooking heat and eliminate heating up the oven at different times. It also provides convenient and speedy meals for surprise guests or especially busy days.

Buy a larger-capacity refrigerator-freezer. This will enable you to buy foods at bargain prices and freeze them for later use. You can also buy foods in larger quantities and cut down on shopping trips. Unnecessary trips waste both valuable time and, if you drive to the store, fuel.

Preparing food—whether you bake, cook, broil, fry, or roast it—uses energy. That's why throwing away good food is another way to waste energy. If you have a larger freezer space, you can freeze leftovers and eliminate unnecessary food waste.

Always follow the manufacturer's instructions in the use and care booklet provided with your new appliance.

Place your refrigerator-freezer in a level, dry, cool, well-ventilated location. Make certain there is enough space behind and above the unit to allow enough air circulation to the condenser.

Set your refrigerator and freezer temperature controls to the medium range as indicated on the dials or in the use

and care booklet. After several days' use, you may have to move the dials to colder or warmer settings as necessary. Do avoid unnecessary colder settings that waste energy.

Check the door seals occasionally. Be sure they're clean and seal properly. Check the seal at various positions on both sides, top, and bottom of the door. Also check for tears, holes, and gaps. A deteriorated seal should be replaced immediately.

Instruct all household members—especially children—to always check carefully to make sure the door is completely closed after being opened.

Regularly defrost manual or partial automatic models. Don't let more than ¼ inch of frost buildup. Frost acts as an insulator, making the refrigerator use more energy to keep temperature levels.

Keep the refrigerator (fresh-food) section filled to almost full capacity but keep enough space between food items so that air can circulate freely. Be careful not to block the air vents.

Keep the freezer section filled to near capacity. Less warm air enters the freezer during door openings because the large amount of frozen foods helps keep the cold in.

Instruct all the members of your family to open the refrigerator door as few times as possible. Most of the food needed for one meal can be removed or replaced at one time. Keep an accurate, up-to-date list of the food in your freezer and where it is located so you can find particular items quickly.

Let hot dishes cool to near room temperature before they are placed in the refrigerator or freezer. Do be careful, however, not to leave food out too long or it will spoil. One way to avoid this is to cool the food quickly in cold water.

Clean your refrigerator condenser coils at least twice a year with the crevice nozzle attachment of your vacuum cleaner. Be sure to turn the refrigerator-freezer off first. Check the use and care booklet for condenser location and cleaning instructions.

If your model has an antisweat heater switch (sometimes referred to as a "power-saver" or "power economizer" by the manufacturer), use it according to the manufacturer's instructions. Generally, the heater should be turned "off" when the weather is dry and "on" when the weather is humid.

If you keep your old refrigerator for extra, backup storage, plug it in only when actually in use. However, put a lock on your unused, empty refrigerator to protect children.

During long vacations, empty your refrigerator-freezer of foods that will spoil, and set the temperature dials slightly warmer. Or, if possible, empty all the food from the refrigerator-freezer, unplug it, and leave the door open.

SAVING ENERGY WITH YOUR WASHER AND DRYER

Washers

About 90 percent of the energy used for washing clothes is for hot water. So the most important way to save energy and money is to use less hot water.

Purchasing Considerations. Look for these energy-saving features:

- Controls to select various water levels so you can match the amount of water to the size of the wash load.
- Controls to select water temperature. Warm and cold options are important.
- Suds saver. With this feature, you can reuse wash-water for additional loads.

Select the right tub capacity for your household's needs. Tub capacity, of course, affects the amount of water used.

Use and Care. Always use a cold-water rinse. The temperature of the rinsewater does *not* affect cleaning.

Most laundry loads can be washed in warm water. Generally, you need to use hot water only for heavily soiled loads. Some lightly soiled loads can even be washed in cold water.

Match the water level to the size of the load. Don't use more water than necessary. Use low water settings for small loads.

Use your washer only during off-peak utility hours. They are usually in the early morning and late evening, but check with your utility to be certain.

Always follow the manufacturer's instructions in the use and care booklet provided with your new washer. Operating your washer correctly will help you obtain satisfactory results. If you're not satisfied with your wash load, you'll probably wash again. This wastes energy money and your time! It also causes needless wear on your clothes. To get satisfactory results:

- Sort articles carefully by color, amount of soil, type of fabric. Separate lint "givers" (towels, chenille) from lint "takers" (corduroy, permanent press, synthetic knits).
- Pretreat stains and heavily soiled areas before washing.
- Use the right amount of detergent. *Always* check the use and care booklet and the directions on the detergent box or label for the right amount to use with a particular wash load. You'll probably need to use more detergent for hard or cold water, larger loads, and heavily soiled loads.
- Don't overload the machine. Place items loosely in the tub, not packed in, so they can move around freely.
- Use the correct cycle for the particular wash load.
- Don't wash a load longer than necessary. Lightly soiled loads usually require less washing time.

Dryers

Always follow the manufacturer's instructions in the use and care booklet provided with your new appliance. The best way to save energy and money with your dryer is to shorten the drying time whenever possible.

Purchasing Considerations: Look for these energy-saving features:

- An automatic shutoff when a certain dry level is reached.
- Time settings. To select the degree of dryness wanted.

Dryer running time is controlled in three basic ways:

- Time setting. The user sets the drying time for a specific number of minutes.
- Thermostatic sensing. The dryer senses temperature increases. Moisture tends to cool air. As the moisture decreases, the temperature rises. When it gets to a certain level, the dryer shuts off.
- Moisture sensing. The dryer has electronic sensors which respond to electricity conducted by moisture in clothes. As moisture decreases, less electricity is conducted to the sensors.

Saving energy with any of these three dryer types depends on whether you know what a proper dryer load is. It is very important not to mix heavy and light articles in the same load. If you do, you defeat the purpose of each of these three dryer designs. The dryer will run longer than necessary and waste energy. To help you reduce dryer running time:

- Don't overdry. Chances are you are overdrying some of your clothes. This not only wastes energy but shortens fabric life, causes shrinkage, and generates static. Try to reduce your regular drying time by about 5 minutes or more. Remember, most fabrics have some natural moisture and should not be "bone dry." Thus, there should be at least a hint of moisture in seams and waistbands when removing articles from the dryer.
- Don't overload the dryer. Articles must be able to tumble freely. A proper wash load is usually a proper drying load.
- Remove as much water as possible from washer articles before putting them in the dryer. Spinning in the washer usually removes water sufficiently.
- Dry only full loads. Don't use the dryer for just a few articles.
- Clean the lint screen *after* each load, and check it *before* each load. If too much lint is on the screen, the air will not move freely and the dryer will run longer.
- If your dryer is vented, make sure it is exhausted properly to the outside. Improperly exhausted dryers will run longer. Use the straightest, shortest duct for venting the dryer. Keep the outside vent clean. Check it at least once a month. Also check the flapper on the outside hood to make sure it opens and closes freely. A flapper that remains open allows heated air to escape and cold drafts to enter the house.
- If possible, dry loads one right after another. In this way, you can use the heat left over from the first load.

- Locate your dryer in a warm spot. Avoid unheated locations.
- Don't add wet items to a partially dried load.
- Don't open the dryer door unnecessarily.
- Remove articles to be ironed while still damp. Some fabrics such as corduroy, quilted, and knit keep shape better if not dried completely.
- Remove items immediately after the dryer stops to prevent wrinkling and thus cut down on ironing. If you don't, you'll have to run the dryer again to fluff out wrinkles.
- Try to use the dryer only during off-peak utility hours.

IRONING

- Remove clothes that will need ironing from the dryer while they still are damp. There's no point in wasting energy to dry them thoroughly if they only have to be dampened again.
- You can save ironing time and energy by "pressing" sheets and pillow cases on the warm top of your dryer. Fold them carefully, then smooth them out on the flat surface.
- Save energy needed for ironing by hanging clothes in the bathroom while you're bathing or showering. The steam often removes the wrinkles.

BATHROOM ENERGY SAVERS

- Take showers rather than tub baths, but limit showering time and check the waterflow if you want to save energy. It takes about 30 gallons of water to fill the average tub. A shower with a flow of 4 gallons of water a minute uses only 20 gallons in 5 minutes. Assuming you use half hot and half cold water for bathing, you would save about 5 gallons of hot water every time you substituted just one shower for one bath per day, and you would save almost 2000 gallons of hot water in a year.
- Consider installing a flow restrictor in the pipe at the showerhead. These inexpensive, easy-to-install devices restrict the flow of water to an adequate 3 to 4 gallons per minute. This can save considerable amounts of hot water and the energy used to produce them over a year's time. For example, reducing the flow from 8 to 3 gallons a minute would save the average family about $24 a year.

LIGHTING ENERGY SAVERS

More than 16 percent of the electricity used in our homes is for lighting. Most Americans overlight their homes, so lowering lighting levels is an easy conservation measure.

Indoor Lighting

- Light-zone your home, and save electricity. Concentrate lighting in reading and working areas and where it's needed for safety (stairwells, for example). Reduce lighting in other areas, but avoid very sharp contrasts.
- To reduce overall lighting in nonworking spaces, remove one bulb out of three in multiple light fixtures and replace it with a burned-out bulb for safety. Replace other bulbs throughout the house with bulbs of the next lower wattage.
- Consider installing solid-state dimmers or high-low switches when replacing light switches. They make it easy to reduce lighting intensity in a room and thus save energy.
- Use one large bulb instead of several small ones in areas where bright light is needed.
- Use long-life incandescent lamps only in hard-to-reach places. They are less energy-efficient than standard bulbs.
- Need new lamps? Consider the advantages of three-way switches. They make it easy to keep lighting levels low when intense light is not necessary, and that saves electricity. Use the high switch only for reading or other activities that require brighter light.
- Always turn three-way bulbs down to the lowest lighting level when watching television. You'll reduce the glare and use less energy.
- Use low-wattage night-light bulbs. These now come in 4-watt as well as 7-watt sizes. The 4-watt bulb with a clear finish is almost as bright as the 7-watt bulb but uses about half as much energy.
- Try 50-watt reflector floodlights in directional lamps (such as pole or spot lamps). These floodlights provide about the same amount of light as the standard 100-watt bulbs but at half the wattage.
- Try 25-watt reflector bulbs in high-intensity portable lamps. They provide about the same amount of light but use less energy than the 40-watt bulbs that normally come with these lamps.
- Use fluorescent lights whenever you can; they give out more lumens per watt. For example, a 40-watt fluorescent lamp gives off 80 lumens per watt and a 6-watt incandescent gives off only 14.7 lumens per watt. The 40-watt fluorescent lamp would save about 140 watts of electricity over a 7-hour period. These savings, over a period of time, could more than pay for the fixtures you would need to use fluorescent lighting.
- Consider fluorescent lighting for the kitchen sink and countertop areas. These lights set under kitchen cabinets or over countertops are pleasant and energy-efficient.
- Fluorescent lighting also is effective for makeup and grooming areas. Use 20-watt, deluxe, warm white lamps for these areas.
- Keep all lamps and lighting fixtures clean. Dirt absorbs light.
- You can save on lighting energy through decorating. Remember, light colors for walls, rugs, draperies, and upholstery reflect light and therefore reduce the amount of artificial light required.

OUTDOOR LIGHTING

- Turn off decorative outdoor gaslamps unless they are essential for safety or convert them to electricity. Keeping just eight gas lamps burning year-round uses as much natural gas as it takes to heat an average-size home for a winter heating season. By turning off one gas lamp, you'll save from $40 to $50 a year in natural gas costs.
- Use outdoor lights only when they are needed. One way to make sure they're off during the daylight hours is to put them on a photocell unit or timer that will turn them off automatically.

APPLIANCE ENERGY SAVERS

About 8 percent of all the energy used in the United States goes into running electrical home appliances, so appliance use and selection can make a considerable difference in home utility costs. Buying energy-efficient appliances may cost a bit more initially but the expense is more than made up by reduced operating costs over the lifetime of the appliance.

Energy efficiency may vary considerably though models seem similar. In the next few years, it will be easier to judge the energy efficiency of appliances with the government's appliance labeling program. In the meantime, wise selection requires some time and effort.

Some general energy-saving tips are:

- Don't leave appliances running when they're not in use. Remember to turn off the radio, television, or record player when you leave a room.
- Keep appliances in good working order so they will last longer, work more efficiently, and use less energy.
- When buying appliances, comparison shop. Compare energy-use information and operating costs of similar models by the same and different manufacturers. The retailer should be able to help you find the wattage of the appliance. With that information, and the list of appliances, you should be able to figure out how much it will cost you to run the appliance.
- Page 211 lists the estimated annual energy use of certain household appliances. With this information, you should be able to figure your approximate energy use and cost for each item listed. You also should get a good idea of which appliances in your home use the most energy and where energy conservation practices will be the most effective in cutting utility costs.
- Before buying new appliances with special features, find out how much energy they use compared with other, perhaps less convenient, models. A frost-free refrigerator, for example, uses more energy than one you have to defrost manually. It also costs more to purchase. The energy and dollars you save with a manual-defrost model may be worth giving up the convenience.

- Use appliances wisely; use the ones that require the least amount of energy for the job. For example, toasting bread in the oven uses three times more energy than toasting it in a toaster.
- Don't use energy-consuming special features on your appliances if you have an alternative. For example, the "instant-on" television sets, especially the tube types, use energy even when the screen is dark. Use the "vacation switch," if you have one, to eliminate this waste; plug the set into an outlet that is controlled by a wall switch.

APPLIANCE LABELING PROGRAM

The federal appliance labeling program is designed to help consumers shop for energy-saving household appliances and equipment. It is being developed by the Federal Energy Administration and the Federal Trade Commission as a result of the Energy Policy and Conservation Act, signed into law on December 22, 1975. Under the law, manufacturers must place labels showing estimated annual operating costs on all models of the following appliances:

- Central air conditioners
- Clothes dryers
- Clothes washers
- Dishwashers
- Freezers
- Furnaces
- Home heating equipment, not including furnaces
- Humidifiers and dehumidifiers
- Kitchen ranges and ovens
- Refrigerators and refrigerator-freezers
- Room air conditioners
- Television sets
- Water heaters

Appliance testing, labeling, and public information procedures are currently being developed. Consumers should hear about the appliance labels as they become available through government information programs.

For further information about the appliance labeling program, write to the Federal Energy Administration, Appliance Program, Washington, D.C. 20461.

BUILDING OR BUYING A HOUSE

Energy-wasting mistakes can be avoided if you consider climate, local building codes, and energy-efficient construction when you build or buy a house. In either case, the following energy conservation ideas should help you keep down home utility bills.

When Building a House

- Consider a square floor plan. It usually is more energy-efficient than a rectangular plan.

ANNUAL ENERGY REQUIREMENTS OF ELECTRIC HOUSEHOLD APPLIANC'

APPLIANCES	ESTIMATED KILOWATT-HOURS USED ANNUALLY
Major Appliances	
Air conditioner (room)	860
(Based on 1000 hours of operation per year.	
This figure will vary widely depending on	
the geographic area and specific size of unit.)	
Clothes dryer	993
Dishwasher, including energy used to heat	
water	2100
Dishwasher only	363
Freezer (16 cubic feet)	1190
Freezer, frostless (16.5 cubic feet)	1820
Range	
with oven	700
with self-cleaning oven	730
Refrigerator (12 cubic feet)	728
Refrigerator, frostless (12 cubic feet)	1217
Refrigerator-freezer (12.5 cubic feet)	1500
Refrigerator-freezer, frostless (17.5 cubic feet)	2250
Washing machine, automatic,	2500
including energy used to heat water	
Washing machine only	103
Washing machine, nonautomatic,	2497
including energy to heat water	
Washing machine only	76
Water heater	4811
Kitchen Appliances	
Blender	15
Broiler	100
Carving knife	8
Coffee maker	140
Deep fryer	83
Egg cooker	14
Frying pan	186
Hot plate	90
Mixer	13
Oven, microwave only	190
Roaster	205
Sandwich grill	33
Toaster	39
Trash compacter	50
Waffle iron	22
Waste disposal	30

APPLIANCES	
Heating and Cooling	
Air cleaner	216
Electric blanket	147
Dehumidifier	377
Fan	
attic	291
circulating	43
Fan	
rollaway	138
window	170
Heater, portable	176
Heating pad	10
Humidifier	163
Laundry	
Iron, hand	144
Health and Beauty	
Germicidal lamp	141
Hair dryer	14
Heat lamp, infrared	13
Shaver	1.8
Sunlamp	16
Toothbrush	0.5
Vibrator	2
Home Entertainment	
Radio	86
Radio/Record player	109
Television	
black and white	
tube type	350
solid-state	120
color	
tube type	660
solid-state	440
Housewares	
Clock	17
Floor polisher	15
Sewing machine	11
Vacuum cleaner	46

Source: Edison Electric Institute.
Note: When using these figures for projections, such factors as the size of the specific appliance, the geographic area of use, and individual use should be taken into consideration.

- Insulate walls and roof to the highest specifications recommended for your area.
- Insulate floors, especially those over crawl spaces, cold basements, and garages.
- If the base of a house is exposed, as in a mobile home, build a skirt around it—i.e., plywood reaching from the underside of the home to the ground.
- Install louvered panels or wind-powered roof ventilators rather than motor-driven fans to ventilate the attic. Only use a motor-driven fan if it can be used for whole-house ventilating during cool periods.

- Consider solar heat gain when you plan your window locations. In cool climates, install fewer windows in the north wall because there's little solar heat gain there in winter. In warm climates, put the largest number of windows in the north and east walls to reduce heating from the sun.
- Install windows that can be opened so you can use natural or fan-forced ventilation in moderate weather.
- Use double-pane glass throughout the house. Windows with double-pane heat-reflecting or heat-absorbing

glass provide additional energy savings, especially in south and west exposures.

- Place the refrigerator in the coolest part of the kitchen, well away from the range and oven.
- Install the water heater as close as possible to areas of major use to minimize heat loss through the pipes; insulate the pipes.
- In a warm climate, light-colored roofing can help keep houses cooler.

When Buying a House

- Consider all the ideas just mentioned for building a house.
- Ask the owner for a description of the insulation and for data on the efficiency of space-heating, air-conditioning, and water-heating plants, or have an independent engineer advise you about the efficiency of the equipment.
- Ask to see utility bills from the previous year, but remember to adjust them for current utility rates.
- Even some new houses don't have insulation in the exterior walls. Be sure to check.
- Is additional insulation or replacement of equipment needed? If so, you may want to seek an adjustment in the purchase price to cover all, or a reasonable share, of the costs.

YARD AND WORKSHOP

- Plant deciduous trees and vines on south and west sides of a house to provide shade in the summer and sunshine in the winter.
- Do not allow gasoline-powered yard equipment to idle for long periods. Turn off the equipment when you finish one job, and restart it when you're ready to resume work.
- Use hand tools, hand lawn mowers, pruners, and clippers whenever possible.
- Maintain electrical tools in top operating condition. They should be clean and properly lubricated.
- Keep cutting edges sharp. A sharp bit or saw cuts more quickly and therefore uses less power. Oil on bits and saws reduces friction and therefore also reduces power required.
- Buy power tools with the lowest horsepower adequate for the jobs for which you'll need them.
- Turn off shop lights, soldering irons, gluepots, and all bench-heating devices right after use.

ON THE ROAD

There are more than 100 million registered automobiles in the United States. A typical car, with an average fuel economy of less than 15 miles per gallon, travels about 10,000 miles each year and uses well over 650 gallons of gasoline. Altogether, private automobiles in the United States consume some 70 billion gallons of gasoline each year. That's about 4.5 million barrels a day or about two-thirds of the amount of petroleum currently being imported into the country.

The importance of individual gasoline savings cannot be overemphasized. If, for example, the fuel used by the average car were reduced just 15 percent through fewer daily trips, better driving practices, and better maintenance, the nation's use of petroleum would fall by nearly two-thirds of a million barrels per day, or about 3.5 percent of demand.

Here are some ways to improve our conservation efforts on the road:

- To get to work, use public transportation, a motorcycle, a moped, or a bicycle, or walk.
- Share your ride. Join a car pool or a van pool. About one-third of all private automobile mileage is for commuting to and from work. If the average occupancy per commuter car (currently 1.3 people) were increased by just one person, each commuter would reduce costs, energy consumption, and driving stress. And the nationwide gasoline savings—which would reduce our reliance on more expensive imports—would be more than 600,000 barrels per day.
- Eliminate unnecessary trips. Try to find at least one driving trip per week that could be handled by telephone or combined with another trip. If every automobile took just one less 10-mile trip a week, the nation would save 3.5 billion gallons of gas a year, or nearly 5 percent of the total passenger-car demand for gas.
- If just 1 gallon of gasoline were saved each week for every automobile in the country, about 5.6 billion gallons of gasoline a year would be saved—or about 8 percent of the demand created by all passenger cars.

ENERGY-EFFICIENT DRIVING

The driving technique of the person behind the wheel is the most important element in determining the fuel economy of any car. A careful driver may get 20 percent more miles per gallon than the average driver and 50 percent more than a wasteful one.

- Observe the 55-mph speed limit on the highway. Most automobiles get about 20 percent more miles per gallon on the highway at 55 mph than they do at 70 mph.
- Accelerate smoothly and moderately. Achieve the desired speed quickly, and then keep a steady pressure on the accelerator, just enough to maintain speed.
- Drive at a steady pace. Avoid stop-and-go traffic. Frequently check the traffic situation ahead of you. Adjust your driving to avoid unnecessary wasteful accelerations and decelerations.
- Minimize braking. Anticipate speed changes. Take your foot off the accelerator as soon as you see a red light or slowed traffic ahead.
- Don't let the motor idle for more than a minute. Turn off the engine. It takes less gasoline to restart the car

than to let it idle; generally, there is no need to press the accelerator down to restart the engine.

- Don't overfill the tank. Remove the nozzle, or ask the gas station attendant to remove it when the automatic valve closes. This will eliminate any chance of spillage.
- Plan trips carefully. Select routes that will allow you to consolidate errands and avoid congested areas.
- Record your gasoline use, and try to get more miles per gallon out of your car.

MAINTAINING YOUR CAR

Good car maintenance and a wise selection of accessories can mean fuel economy and dollars saved.

- Have your car tuned as recommended by the manufacturer. Regular tune-ups extend the engine life and improve performance. A poorly tuned car could use as much as 3 to 9 percent more gasoline than a well-tuned one. The tune-up will pay for itself in gasoline savings and car reliability.
- Keep the engine filters clean. Clogged filters waste gasoline.
- Use the gasoline octane and oil grade recommended for your car. If you change the oil yourself, take the used oil to a service station for recycling.
- Check tire pressures regularly. Underinflated tires increase gas use. You can lose about 2 percent in fuel economy for every pound of pressure under the recommended pounds per square inch.
- Consider radial tires. They can improve gas mileage from 3 to 5 percent in the city, 7 percent on the highways, and 10 percent at 55 mph after the tires are warmed up for 20 minutes. And they last longer too. Never mix radials with conventional tires.
- Remove unnecessary weight from the car. The lighter the car, the less gas it uses. An extra 100 pounds decreases fuel economy about 1 percent for the average car, 1.25 percent for small cars.

BUYING A CAR

Study the Market

- Ask your dealer for a free copy of the latest *EPA/FEA Gas Mileage Guide* (or write to Fuel Economy, Pueblo, Colorado 81009). Study the fuel-economy figures and tables that compare specifications. Review mileage test results publicized by Consumers Union and motor industry magazines. Generally, the best fuel economy is associated with low vehicle weight, small engines, manual transmission, low axle ratio, and low frontal area (the width of the car times its height).
- Buy the most energy-efficient car of the size and style you want. Don't let the car price alone determine your choice. Make your decision on the basis of the combination of purchase price and your estimated fuel costs.

Choose Accessories Wisely

- Purchase only the optional equipment and accessories you really need. Items such as air conditioning, automatic transmission, and power steering require considerable energy, all of which is derived from burning gasoline. Other equipment, such as power brakes and electric motor-driven windows, seats, and radio antennas, require less energy for their operation, but all accessories add to the vehicle weight—and this reduces fuel economy.
- Don't buy car air conditioning unless you really need it. Even when you're not using it, it adds to the weight of the car.
- If you have a car air conditioner or other power-draining accessories, use them sparingly. The cooling equipment reduces fuel economy from 10 percent on the highway up to almost 20 percent in stop-and-go traffic.

TAKING VACATIONS

- Vacation at home this year. Discover nearby attractions.
- Choose a hotel or campground close to where you live. It can often provide as complete and happy a change from routine as one that is hundreds of miles away.
- If you vacation away from home, consider staying in one place. "Hopping around" takes transportation energy.
- Take a train or bus instead of the car. Save gasoline—and relax too.
- Rediscover the pleasures of walking, hiking, and bicycling during your vacation. They're the most energy-conserving means of transportation and the healthiest for most people.
- Save energy at home if you go away. Turn off lights, lower heating temperatures in winter, and turn off air conditioning in summer.

IN THE MARKETPLACE

Buy products that will last. Durable products save the energy that would be required to make replacements.

- Buy equipment on the basis of initial cost plus operating costs rather than on the basis of purchase price alone. Often, products that are energy-efficient cost more to buy. But over the lifetime of the equipment, you will more than make up the difference in lower operating costs.
- Buy products made of recycled or recyclable materials—steel, aluminum, paper, and glass among others. More energy is used in the production of products from virgin materials than from recycled or reclaimed

materials. For example, producing steel from scrap requires only one-quarter of the energy it would take when using virgin ores. Making a product from recycled aluminum requires less than 10 percent of the energy that would be needed for the same product made from the ore.

- Buy fabrics or garments that can washed in cold water and/or require little or no ironing.

- When shopping for an unusual item, telephone ahead to see if the store has it. If it doesn't, you will have saved the energy and time of traveling there and being disappointed.

- Give gifts with year-round benefits. If you buy appliances as gifts, select long-lasting models that use the least amount of energy.

- Don't buy motorized equipment or gadgets when hand-operated versions will do.

- Purchasing the right equipment for your home and needs, using it wisely, and taking good care of it can reduce energy costs considerably.

- Bigger isn't necessarily better. Don't buy a larger, more powerful piece of equipment than you need. Whether it's a furnace, air conditioner, or water heater, make sure its size and power are right for your home. Ask the dealer, a trade association, or a consumer-interest group for assistance in judging this factor.

For more information on energy-saving tips for the home and home appliances, contact:

National Home Improvement Council—Code of Ethics
11 E. 44th St.
New York, NY 10017
(212) 867-0121

Members of the National Home Improvement Council are pledged to observe the highest standards of integrity, frankness, and responsibility in dealings with the public:

1. By encouraging only those home-improvement projects that are structurally and economically sound.
2. By making, in all advertising, only those statements that are accurate and free of the capacity to mislead or deceive the consumer.
3. By requiring all sales personnel to be accurate in their description of products and services.
4. By writing all contracts so that they are unambiguous and fair to all parties concerned.
5. By promptly fulfilling all contractual obligations.
6. By performing all work in a manner compatible with recognized standards of public health, safety, and applicable laws.

In addition to the code of ethics, the NHIC, together with the Council of Better Business Bureaus, has developed a set of standards of practice for the home improvement industry.

20 Automatic Thermostat Controls

It is said that you won't save energy by turning down the thermostat at night because it takes so much energy to warm your home in the morning. But this is a myth. Setting the thermostat back for several hours at a stretch each day during the heating season (up, during the cooling season) will save energy in a centrally heated and cooled building. Depending on your geographical location, the amount of energy you can save will range from 9 to 15 percent of what you used previously (see Table 20-1).

You can accomplish temperature "setback" and "setup" either by adjusting the thermostat manually at the proper times or by installing a device that makes the adjustments automatically. The manual technique, of course, requires no special equipment, but it does demand a greater degree of time and attention than many people are willing to put forth day in and day out. An automatic control device, on the other hand, does involve some initial investment, but it is more than repaid in dependability and energy savings over a period of time.

How Much Will a Setback Device Save?

The exact energy and cost savings from a setback device depend on building design, amount of insulation, climate, temperature setting, and utility rate structures. Several studies have been conducted to estimate the fuel and cost savings that can be realized by using a set back device during both the heating and cooling seasons. Table 20-1 represents the approximate percentage of heating costs that can be saved in various cities throughout the country for an 8-hour nighttime thermostat setback of 5° and 10°F.

How to Estimate Cost Savings

From the table, you can estimate what you are likely to save by automatically setting back your thermostat during the heating season from 65°F to either 60° or 55°F at night. For example, if you live in or around Detroit and your heating bills amount to approximately $300 for the heating season, by lowering your thermostat at night from 65°F to 55°F, you could save as much as 11 percent of $300, or $33 a heating season. These estimated figures are based on an assumed daytime setting of 65°F.

Table 20-1. Heating costs saved with nighttime "setback."

CITY	APPROXIMATE PERCENTAGE SAVED WITH 8-HOUR NIGHTTIME SETBACK OF	
	5°F	10°F
Atlanta, GA	11	15
Boston, MA	7	11
Buffalo, NY	6	10
Chicago, IL	7	11
Cincinnati, OH	8	12
Cleveland, OH	8	12
Columbus, OH	7	11
Dallas, TX	11	15
Denver, CO	7	11
Des Moines, IA	7	11
Detroit, MI	7	11
Kansas City, MO	8	12
Los Angeles, CA	12	16
Louisville, KY	9	13
Madison, WI	5	9
Miami, FL	12	18
Milwaukee, WI	6	10
Minneapolis, MN	5	9
New York, NY	8	12
Omaha, NE	7	11
Philadelphia, PA	8	12
Pittsburgh, PA	7	11
Portland, OR	9	13
Salt Lake City, UT	7	11
San Francisco, CA	10	14
Seattle, WA	8	12
St. Louis, MO	8	12
Syracuse, NY	7	11
Washington, DC	9	13

Types of Automatic Controls

Two types of automatic controls are now available on the commercial market. One is a device that works with a conventional thermostat. The other requires replacing the existing thermostat.

The *converter setback* converts any existing thermostat to a timed device. Several variations are available. One is a two-component system in which a temperature control is mounted below the existing thermostat and is connected by wires to a separate timer unit plugged into a wall outlet.

If the wires carry low-voltage current, they can be concealed in the wall—unlike 110-volt power cords, which cannot be so concealed. Another is a single-unit device that is attached to the wall below the thermostat and is either plugged into a nearby wall outlet or operated by self-contained batteries.

The *replacement setback* replaces the conventional thermostat entirely and is generally wired to the building's electrical system and heating/cooling system. Several types are available, but this type of device is usually more expensive since it usually requires the additional wiring in existing walls. Its main advantage is that, having all wires hidden, it gives a neater appearance.

Automatic setback devices are sold in many hardware and department stores as well as building material outlets. In general, converter types sell for less than replacement types and can be installed by a do-it-yourselfer. Most converters retail for less than $40, whereas the initial cost plus the cost of installing a replacement unit may range from $75 to over $100 depending on the model and on the type and extent of installation labor required.

A No-Cost Alternative

Of course, you can continue to adjust your thermostat by hand, and still achieve the same energy savings as with an automatic setback device. Manual setting costs you nothing. However, since the automatic setback device pays for itself rather quickly and will continue to return savings for years to come, you might prefer the regularity and convenience of an automatic control.

21 Selecting the Right Heating System

Different types of heating equipment and systems are available for heating the home. Considerations in selecting a unit or system include heating requirements, installation and maintenance costs, and heat cost. Heating-equipment dealers and contractors can assist in determining heating requirements and in selecting the most efficient and economical unit for your house.

For safety and efficiency, have a reputable contractor install your central heating system and inspect it once a year. A less costly system correctly installed will be more satisfactory than an expensive one that is not the right size for the house or that is not properly installed.

HOW TO REDUCE HEAT REQUIREMENTS

Much can be done to reduce the heat requirements in a house. This in turn can reduce heating costs and increase personal comfort. New houses may be oriented so that the main rooms and the large windows in the rooms face south to receive maximum sunlight in the winter. (In summer, the sunlight may be shaded out by trees, wide eaves, shutters, awnings, or other natural or artificial shading.)

Tight construction also reduces heat requirements. Insulate ceilings and outside walls. Caulk and weather-strip joints. Install storm sash or double- or tripe-glazed windows to reduce heat loss. An old house should always be repaired and insulated before a new heating system is installed.

The chimney is a part of the heating plant; proper construction and maintenance are important. Chimneys should extend a minimum of 2 feet above the roof ridge. Manufacturers usually specify the size of the flue required for heating equipment. Keep flues clean and free from leaks.

WARM-AIR HEATING

Area heating units, which include stoves, circular heaters, and "pipeless" furnaces, are installed in the room or area to be heated. In central systems, the heating unit is located in the basement or other out-of-the-way place, and heat is distributed through ducts. Central heating systems are the most efficient and economical method of heating.

It is best to buy a heating unit designed specifically for the fuel to be used. Coal or wood burners can be converted to oil or gas but usually do not have sufficient heating surface for best efficiency.

Stoves are one of the simplest heating devices. Although they are cheaper than central heating systems, stoves are dirtier, require more attention, and heat less uniformly. If more than one stove is used, more than one chimney may be needed.

Wood- or coal-burning stoves without jackets heat principally by radiation. Jacketed stoves or circulator heaters heat mainly by convection and are available for burning the four common fuels—wood, coal, oil, or gas.

With proper arrangement of rooms and doors, a circulator heater can heat four or five small rooms; in many instances, however, heating will not be uniform. A small fan to aid circulation will increase efficiency. The distance from the heater to the center of each room to be heated, measured through the door opening, should not be more than about 18 feet. Doors must be left open; otherwise, grilles or louvers are needed at the top and bottom of doors or walls for air circulation.

"Pipeless" furnaces may be used in smaller houses. They discharge warm air through a single register placed directly over the furnace. Units that burn wood, coal, gas, or oil are available for houses with basements. Gas- and oil-burning units, which can be suspended beneath the floor, are available for houses without basements.

Small gas-fired vertical heaters are sometimes recessed in the walls of the various rooms. Such units may be either manually or thermostatically controlled. Heater vents are carried up through the partitions to discharge the burned gases through a common vent extending through the roof.

CENTRAL HEATING SYSTEMS

Forced warm-air heating systems are more efficient and cost less to install than gravity warm-air heating systems. Forced warm-air systems consist of a furnace, ducts, and registers. A blower in the furnace circulates the warm air to the various rooms through supply ducts and registers. Return grilles and ducts carry the cooled room air back to the furnace where it is reheated and recirculated.

Forced warm-air systems heat uniformly and respond rapidly to changes in outdoor temperatures. They can be used in houses with or without basements—the furnace need not be below the rooms to be heated nor centrally located. Some can be adapted for summer cooling by the addition of cooling coils. Combination heating and cooling systems may be installed. The same ducts can be used for both heating and cooling.

The warm air is usually filtered through inexpensive replaceable or washable filters. Electronic air cleaners can sometimes be installed in existing systems and are available on specially designed furnaces for new installations. These remove pollen, fine dust, and other irritants that pass through ordinary filters and may be better for persons with respiratory ailments. The more expensive units feature automatic washing and drying of the cleaner. A humidifer may be added to the system to add moisture to the house air and avoid the discomfort and other disadvantages of a too-dry environment.

Warm-air supply outlets are preferably located along outside walls. They should be low in the wall, in the baseboard, or in the floor where air cannot blow directly on room occupants. Floor registers tend to collect dust and trash, but may have to be used in installing a new system in an old house.

High-wall or ceiling outlets are sometimes used when the system is designed primarily for cooling. However, satisfactory cooling as well as heating can be obtained with low-wall or baseboard registers by increasing the air volume and velocity and by directing the flow properly.

Ceiling diffusers that discharge the air downward may cause drafts; those that discharge the air across the ceiling may cause smudging. Most installations have a cold-air return in each room. When supply outlets are along outside walls, return grilles should be along inside walls in the baseboard or in the floor. When supply outlets are along inside walls, return grilles should be along outside walls.

Centrally located returns work satisfactorily with perimeter-type heating systems. One return may be adequate in smaller houses. In larger or split-level houses, return grilles are generally provided for each level or group of rooms. Locations of the returns within the space are not critical. They may be located in hallways, near entrance doors, in exposed corners, or on inside walls.

In the crawl-space plenum system, the entire crawl space is used as an air-supply plenum or chamber. Warm air flows from near the ceiling to a central duct, is forced into the crawl space, and enters the rooms through perimeter outlets, usually placed beneath windows, or through continuous slots in the floor adjacent to the outside wall. With tight, well-insulated crawl-space walls, this system can provide uniform temperatures throughout the house.*

In houses without basements, horizontal furnaces that burn gas or oil may be installed in the crawl space or hung

*For additional information on this system, see PR Report 99, *An Economical and Efficient Heating System for Homes,* and PR Report 124, *Peripheral Circulation for Low Cost Central Heating in Old Houses.* Both publications are available from the Superintendent of Documents, U.S. Government Printing Office, Washington, D.C. 20402, at 40 cents and 15 cents a copy, respectively.

from ceiling joists in a utility room or adjoining garage. The gas furnaces may also be installed in attics. Allow adequate space for servicing the furnaces. Insulate attic furnaces and ducts heavily to prevent excessive heat loss.

Vertical gas or oil furnaces designed for installation in a closet or a wall recess or against a wall are popular, especially in small houses. The counterflow type discharges the hot air at the bottom to warm the floor. Some, such as the gas-fired unit, provide discharge grilles into several rooms.

Upflow-type vertical furnaces may discharge the warm air through attic ducts and ceiling diffusers. Without return-air ducts, these furnaces are less expensive, but also heat loss uniformly.

Houses built on a concrete slab may be heated by a perimeter-loop heating system. Warm air is circulated by a counterflow-type furnace through ducts cast in the outer edge of the concrete slab. The warm ducts heat the floor, and the warm air is discharged through floor registers to heat the room. To prevent excessive heat loss, the edge of the slab should be insulated from the foundation walls and separated from the ground by a vapor barrier.

HOT WATER AND STEAM HEATING

Hot-water and steam heating systems consist of a boiler, pipes, and room heating units (radiators or convectors). Hot water or steam, heated or generated in the boiler, is circulated through the pipes to the radiators or convectors where the heat is transferred to the room air.

Boilers are made of cast iron or steel and are designed for burning coal, gas, or oil. Cast-iron boilers are more resistant to corrosion than steel ones. Corrosive water can be improved with chemicals. Proper water treatment can greatly prolong the life of steel boiler tubes.

Buy only a certified boiler. Certified cast-iron boilers are stamped "I-B-R" (Institute of Boiler and Radiator Manufacturers); steel boilers are stamped "SBI" (Steel Boiler Institute). Most boilers are rated (on the nameplate) for both hot water and steam. Contractors can advise you when selecting a boiler.

Conventional radiators are set on the floor or mounted on the wall. The newer types may be recessed in the wall. Insulate behind recessed radiators with 1-inch insulation board, a sheet of reflective insulation, or both.

Radiators may be partially or fully enclosed in a cabinet. A full cabinet must have openings at both the top and the bottom for air circulation. The preferred location for a radiator is under a window.

Baseboard radiators are hollow or finned units that resemble and replace the conventional wood baseboard along the outside walls. They will heat a well-insulated room uniformly, with little temperature difference between floor and ceiling.

Convectors usually consist of finned tubes enclosed in a cabinet with openings at the top and bottom. Hot water or steam circulates through the tubes. Air comes in at the bottom of the cabinet, is heated by the tubes, and goes out the top. Some units have fans for forced-air circulation. With this type of convector, summer cooling may be provided by adding a chiller and the necessary controls to the system. Convectors are installed against an outside wall or recessed in the wall.

FORCED HOT-WATER HEATING SYSTEMS

Forced hot-water heating systems are recommended over the less efficient gravity hot-water heating systems. In a forced hot-water system, a small booster or circulating pump forces or circulates the hot water through the pipes to the room radiators or convectors. In a one-pipe system, one pipe or main serves for both supply and return. It makes a complete circuit from the boiler and back again. Two risers extend from the main to each room heating unit. A two-pipe system has two pipes or mains. One carries the heated water to the room heating units; the other returns the cooled water to the boiler.

In the one-pipe system, cooled water from each radiator mixes with the hot water flowing through the main, and each succeeding radiator receives cooler water. Allowance must be made for this in sizing the radiator—larger ones may be required further along in the system.

Because water expands when heated, an expansion tank must be provided in the system. In an "open system," the tank is located above the highest point in the system and has an overflow pipe extending through the roof. In a "closed system," the tank is placed anywhere in the system, usually near the boiler. Half of the tank is filled with air, which compresses when the heated water expands. Higher water pressure can be used in a closed system than in an open one. Higher pressures raise the boiling point of the water. Higher temperatures can therefore be maintained without steam in the radiators, and smaller radiators can be used. There is almost no difference in fuel requirements.

With heating coils installed in the boiler or in a water heater connected to the boiler, a forced hot-water system can be used to heat domestic water year-round. If you want to use your heating plant to heat domestic water, consult an experienced heating engineer about the best arrangement.

One boiler can supply hot water for several circulation heating systems. The house can be "zoned" so that temperatures of individual rooms or areas can be controlled independently. Remote areas such as a garage, workshop, or small greenhouse can be supplied with controlled heat.

Gas- and oil-fired boilers for hot-water heating are compact and are designed for installation in a closet, utility room, or similar space—on the first floor, if desired.

Electrically heated hydronic (water) systems are especially compact, and the heat exchanger, expansion tank, and controls may be mounted on the wall. Some systems have thermostatically controlled electric heating components in the hydronic baseboard units, which eliminates the central heating unit. Such a system may be a single-loop installation for circulating water by a pump, or it may be composed of individual sealed units filled with antifreeze solution.

The sealed units depend on gravity flow of the solution in the unit. Each unit may have a thermostat, or several units may be controlled from a wall thermostat. An advantage of these types of systems is that heating capacity can be increased easily if the house is enlarged.

STEAM CENTRAL-HEATING SYSTEMS

Steam heating systems are not used as much as forced hot-water or warm-air systems. For one thing, they are less responsive to rapid changes in heat demands. One-pipe steam heating systems cost about as much to install as one-pipe hot-water systems. Two-pipe systems are more expensive. The heating plant must be below the lowest room heating unit unless a pump is used to return the condensate to the boiler.

RADIANT PANEL HEATING

Radiant panel heating is another method of heating with forced hot water or steam. (It is also a method of heating with electricity.)

Hot water or steam circulates through pipes concealed in the floor, wall, or ceiling. Heat is transmitted through the pipes to the surface of the floor, wall, or ceiling and then to the room by radiation and convection. No radiators are required—the floor, wall, or ceiling, in effect, act as radiators.

With radiant panel heating, rooms can be more comfortable at lower air temperatures than with other heating systems at higher air temperatures. The reason is that the radiated heat striking the occupant reduces body heat loss and increases body comfort. Temperatures are generally uniform throughout the room.

Underfloor radiant panel heating systems are difficult to design. For instance, a carpeted or bare wood floor might be very comfortable, while a ceramic-tiled bathroom floor or plastic kitchen-floor covering might be too hot for bare feet. An experienced engineer should design the system.

Panel heating in poorly insulated ceilings is not practical unless you want to heat the space above the ceiling. Exterior wall panels require insulation behind them to reduce heat-loss.

ELECTRIC HEATING

Many types and designs of electric house-heating equipment are available. These include: (1) ceiling unit, (2) baseboard heater, (3) heat pump, (4) central furnace, (5) floor furnace, and (6) wall unit. All but the heat pump are of the resistance type. Resistance-type heaters produce heat the same way as the familiar electric radiant heater. Heat pumps are usually supplemented with resistance heaters.

Ceiling heat may be provided with electric heating cable laid back and forth on the ceiling surface and covered with plaster or a second layer of gypsum board. Other types of ceiling heaters include infrared lamps and resistance heaters with reflectors or fans. Baseboard heaters resemble ordinary wood baseboards and are often used under a large picture window in conjunction with ceiling heat.

The heat pump is a single unit that both heats and cools. In winter, it takes heat from the outdoor air to warm the house or room. In summer, it removes heat from the house or room and discharges it to the outside air. It uses less electricity to furnish the same amount of heat as the resistance-type heater. Room air conditioners of the heat pump type are especially convenient in warmer climates where continuous heating is not needed or for supplemental heat in some areas of the house.

Either heat pumps or furnaces with resistance heaters are used in forced-air central heating systems. They require ducts similar to those discussed for forced warm-air heating. Hot-water systems with resistance-type heaters are also available.

Wall units, either radiant or convection, or both, are designed for recessed or surface wall mounting. They come equipped with various types of resistance heating elements. The warm air may be circulated either by gravity or by an electric fan.

Each room heated by the equipment just described (with the exception of some central heating systems) usually has its own thermostat and can be held at any desired temperature. Thermostats should be designed for long life and should be sensitive to a change in temperature of $\pm 0.5°F$.

FUELS AND BURNERS

The four fuels commonly used for home heating are wood, coal, oil, and gas. Electricity, although not a fuel, is being used increasingly. Modern heating equipment is relatively efficient when used with the fuel for which it is designed. But, even with modern equipment, some fuels cost more than others to do the same job.

The therms of heat per dollar should not be the sole consideration in selecting the heating fuel. (A therm is 100,000 Btu.) Installation cost, the efficiency with which each unit converts fuel into useful heat, and the insulation level of the house should also be considered. For example, electrically heated houses usually have twice the insulation thickness, particularly in the ceiling and floor, and therefore may require considerably less heat input than houses heated with fuel-burning systems. To compare costs for various fuels, the efficiency of combustion and heat value of the fuel must be known.

Heating units vary in efficiency, depending upon the type, method of operation, condition, and location. Stoker-fired (coal) steam and hot-water boilers of current design, operated under favorable conditions, have 60 to 75 percent efficiency. Gas- and oil-fired boilers have 70 to 80 percent efficiency. Forced warm-air furnaces, gas-fired or oil-fired with atomizing burner, generally provide about 80 percent efficiency. Oil-fired furnaces with pot-type burners usually develop not over 70 percent efficiency.

Fuel costs vary widely in different sections of the country. However, for estimates, the data given in Table 21-1 can be used to figure the comparative costs of various fuels and electricity based on local prices. Here the efficiency of electricity, gas, oil, and coal is taken as 100, 75, 75, and 65 percent, respectively. The efficiencies may be higher (except for electricity) or lower, depending upon conditions; however, the values used are considered reasonable. The heat values are taken as 3413 Btu per kilowatt-hour of electricity for resistance heating; 1050 Btu per cubic foot of natural gas; 92,000 Btu per gallon of propane (LP) gas; 139,000 Btu per gallon of No. 2 fuel oil; and 13,000 Btu per pound of coal.

More Btu of heat per kilowatt-hour can generally be obtained with heat-pump heating than with resistance heating. The difference varies depending upon the outside temperature and other factors. In warm climates, heat-pump heating may require about half as much electricity as resistance heating. In the extreme northern states, the consumption of electric energy may approach that required for resistance heating.

Table 21-1. Figuring comparative costs of various fuels and electrical energy to supply 1 therm of usable heat.[a]

FUEL OR ENERGY	QUANTITY TO SUPPLY 1 THERM USABLE HEAT	MULTIPLY VALUES IN COLUMN (2) BY LOCAL UNIT COSTS	COMPARATIVE COSTS PER THERM OF HEAT SUPPLIED TO LIVING SPACE (¢)
(1)	(2)	(3)	(4)
Coal	11.8 lb = 0.006 ton	per ton	
Electricity	29.3 kWh	per kWh	
Fuel oil, No. 2	0.96 gal.	per gal.	
Gas, natural	127 ft³	per ft³	
Gas, LP (propane)	1.45 gal.	per gal.	

[a]Usable heat calculated on basis of heater efficiencies given in the text.

Wood

The use of wood requires more labor and more storage space than does the use of other fuels. However, wood fires are easy to start, burn with little smoke, and leave little ash. Most well-seasoned hardwoods have about half as much heat value per pound as does good coal. A cord of hickory, oak, beech, sugar maple, or rock elm weighs about 2 tons and has about the same heat value as 1 ton of coal.

Coal

Two kinds of coal are used for heating homes—anthracite (hard) and bituminous (soft). Bituminous is used more often.

Anthracite coal sizes are standardized; bituminous coal sizes are not. Heat value of the different sizes of coal varies little, but certain sizes are better suited for burning in firepots of given sizes and depths.

Both anthracite and bituminous coal are used in stoker firing. Stokers may be installed at the front, side, or rear of a furnace or boiler. Leave space for servicing the stoker and for cleaning the furnace. Furnaces and boilers with horizontal heating surfaces require frequent cleaning, because fly ash (fine powdery ash) collects on these surfaces. Follow the manufacturer's instructions for operating stokers.

Oil

Oil is a popular heating fuel. It requires little space for storing and no handling, and it leaves no ash.

Two grades of fuel oil are commonly used for home heating. No. 1 is lighter and slightly more expensive than No. 2, but No. 2 fuel oil has higher heat value per gallon. The nameplate or guidebook that comes with the oil burner indicates what grade oil should be used. In general, No. 1 is used in pot-type burners, and No. 2 in gun-and-rotary-type burners.

For the best results, a competent service representative should install and service an oil burner. Oil burners are of two kinds—vaporizing and atomizing. Vaporizing burners premix the air and oil vapors. The pot-type burner is vaporizing and consists of a pot containing a pool of oil. An automatic or handset valve regulates the amount of oil in the pot. Heat from the flame vaporizes the oil. In some heaters, a pilot flame or electric arc ignites the oil pot when heat is required; in others, the oil is ignited manually and burns continuously at any set fuel rate between high and low fire, until shut off. There are few moving parts, and operation is quiet. Some pot-type burners can be operated without electric power.

Atomizing burners are of two general types—gun (or pressure) and rotary. The gun burner is by far the more popular type for home heating. It has a pump that forces the oil through a special atomizing nozzle. A fan blows air into the oil fog and an electric spark ignites the mixture, which burns in a refractory-line firepot.

Gas

Gas is used in many urban homes and in some rural areas. It is supplied at low pressure to a burner head, where it is mixed with the right amount of air for combustion.

A room thermostat controls the gas valve. A pilot light is required. It may be lighted at the beginning of the heating season and shut off when heat is no longer required. However, if it is kept burning during nonheating seasons, condensation and rapid corrosion of the system will be prevented.

The pilot light should be equipped with a safety thermostat to keep the gas valve from opening if the pilot goes out; no gas can then escape into the room. (The pilot light of all automatic gas-burning appliances should be equipped with this safety device.)

Three kinds of gas—natural, manufactured, and bottled—are used. Bottled gas (usually propane) is sometimes called *LPG* (liquefied petroleum gas). It is becoming more popular as a heating fuel in recent years, particularly in rural areas. Different gases have different heat values when burned. A burner adjusted for one gas must be readjusted when used with another gas.

Conversion gas burners may be used in boilers and furnaces designed for coal if they have adequate heating surfaces. Furnaces must be properly gas-tight. Conversion burners, as well as all other gas burners, should be installed by competent, experienced heating contractors who follow closely the manufacturer's instructions. Gas-burning equipment should bear the seal of approval of the American Gas Association.

Vent gas-burning equipment to the outdoors. Keep chimneys, smoke pipes, and stacks free from leaks. Connect all electrical controls for gas-burning equipment on a separate switch so that the circuit can be broken in case of trouble. Gas-burning equipment should be cleaned, inspected, and correctly adjusted each year.

Bottled gas is heavier than air. If it leaks into the basement, it will accumulate at the lowest point and create an explosion hazard. When bottled gas is used, make sure that the safety-control valve is placed so that it shuts off the gas to the pilot as well as to the burner when the pilot goes out.

Electricity

Electric heating offers convenience, cleanliness, evenness of heat, safety, and freedom from odors and fumes. A chimney is not required in building a new house, unless a fireplace is desired.

For electric heating to be more competitive economically with other types of heating, houses should be well-insulated and weather-stripped, have double- or triple-glazed windows, and be vapor-sealed. The required insulation, vapor barrier, and weather-proofing can be provided easily in new houses, but may be difficult to add to old houses.

Some power suppliers will guarantee a maximum monthly or seasonal cost when the house is insulated and the heating system installed in accordance with their specifications. The heating equipment should be only large enough to handle the heat load. Oversized equipment costs more and requires heavier wiring than does properly sized equipment.

Automatic Controls

Each type of heating plant requires special features in its control system. But even the simplest control system should include high-limit controls to prevent overheating. Limit controls are usually recommended by the equipment manufacturer.

The high-limit control, which is usually a furnace or boiler thermostat, shuts down the fire before the furnace or boiler becomes dangerously or wastefully hot. In steam systems, it responds to pressure; in other systems, it responds to temperature.

The high-limit control is often combined with the fan or pump controls. In a forced warm-air or forced hot-water system, these controls are usually set to start the fan or the pump circulating when the furnace or boiler warms up and to stop it when the heating plant cools down. They are ordinarily set just high enough to insure heating without overshooting the desired temperature and can be adjusted to suit weather conditions.

Other controls insure that all operations take place in the right order. Room thermostats control the burner or stoker on forced systems. They are sometimes equipped with timing devices that can be set to change automatically the temperatures desired at night and in the daytime.

Since the thermostat controls the house temperature, it must be in the right place—usually on an inside wall. Do not put it near a door to the outside; at the foot of an open stairway; above a heat register, television, or lamp; or where it will be affected by direct heat from the sun. Check it with a good thermometer for accuracy.

Oil-Burner Controls

The oil-burner controls allow electricity to pass through the motor and ignition transformer and shut them off in the right order. They also stop the motor if the oil does not ignite or if the flame goes out. This is done by means of a stack thermostat built into the relay. The sensing element of the stack control is inserted into the smoke pipe near the furnace boiler. Some heating units are equipped with electric-eye (cadmium sulfide) flame detectors, which are used in place of a stack control.

Without the protection of the stack thermostat or electric eye, a gun- or rotary-type burner could flood the basement with oil if it failed to ignite. With such protection, the relay allows the motor to run only a short time if the oil fails to ignite; then it opens the motor circuit and keeps it open until it is reset by hand.

The boiler thermostat acts as high-limit control if the water in the boiler gets too hot.

Stoker-Fired Coal-Burner Controls

The control system for a coal stoker is much like that for an oil burner. However, an automatic timer is usually included to operate the stoker for a few minutes every hour or half hour to keep the fire alive during cool weather when little heat is required.

A stack thermostat is not always used, but in communities where electric power failures may be long enough to let the fire go out, a stack thermostat or other control device is needed to keep the stoker from filling the cold firepot with coal when the electricity comes on again. Sometimes, a light-sensitive electronic device such as an electric eye is used. In the stoker-control setup for a forced warm-air system, the furnace thermostat acts as high-limit and fan control.

Gas-Burner Controls

Controls for the gas burner are so much a part of the burner itself that they were described in the section on gas.

OTHER HEATING-SYSTEM CONTROLS

Warm-air, hot-water, or steam-heat distribution systems may be controlled in other ways. If the furnace or boiler heats domestic water, more controls are needed.

In some installations of forced hot-water systems, especially with domestic-water hookups, a mixing valve is used. The water temperature of the boiler is maintained at some high, fixed value, such as 200°F. Only a portion of this high-temperature water is circulated through the heating system. Some of the water flowing through the radiators bypasses the boiler. The amount of hot water admitted is controlled by a differential thermostat operating on the difference between outdoor and indoor temperatures. This installation is more expensive than the more commonly used system but it responds almost immediately to demands. Although it cannot anticipate temperature changes, it is in a measure regulated by outside temperatures, which change earlier than do those indoors.

The flow of hot water to each part of a building can be separately controlled. This zoning—maintaining rooms or parts of the building at different desired temperatures—can be used to maintain sleeping quarters at a lower temperature than living quarters. Electric heating is also well adapted to zoning.

Fuel savings help to offset the initial cost of the more elaborate control systems.

22 How to Understand Your Utility Bill

As a consumer, you should know how to read your electric and gas meters and utility bills. With this skill, you can figure out whether your attempts to conserve energy are effective. You can also check the accuracy of your bill and find out how much energy is used to heat and cool your home.

USEFUL DEFINITIONS

British thermal unit (Btu): The standard unit used in measuring the heat content of fuel. A Btu is the amount of heat needed to raise the temperature of 1 pound of water 1°F.

Cubic foot: The quantity of natural gas is measured by its volume, in cubic feet.

Therm: The ability of natural gas to produce heat is measured in therms. A therm is the equivalent of 100,000 Btu. The amount of heat or the number of Btu in a cubic foot of natural gas varies, but 100 cubic feet of natural gas usually contain a little more than 1 therm. Some gas companies bill their customers by the number of cubic feet used, and some bill by the number of therms used.

Watt: The unit of electric power, which indicates the electrical demand or rate of energy delivery.

Kilowatt: 1000 watts.

Kilowatt-hour: Consumption of electricity is measured by the kilowatt-hour, which is the amount of energy delivered by an hour-long flow of 1 kilowatt of electric power. Your electric bill is based on the number of kilowatt-hours you used. For example, a 100-watt bulb burning for 10 hours will use 1 kilowatt-hour of electric energy—100 watts multiplied by 10 hours equals 1000 watt-hours or 1 kilowatt-hour.

Fuel cost adjustment: A fee, also called a pass-through, added to an electric bill to compensate the company for the increased cost of the coal, gas, oil, or nuclear fuel from which it generates electricity.

Purchased gas adjustment charge (PGA): A fuel-cost adjustment charge appearing on a natural gas bill. It compensates the gas company for the increased cost of the natural gas it buys for its customers. (Public utilities are regulated by state agencies. If you have a question about a utility company's adjustment charges, contact the utility commission in your state capital.)

Rate schedule: The rules by which a utility determines the cost of each kilowatt-hour, therm, or cubic foot of gas you use. Rate schedules are not usually shown on the bills. If you want to know your rate schedule, call the utility company. Typically, the first group of kilowatt-hours or therms used in a billing period costs more than the second group; the second group costs more than the third, and so on. Some companies, however, are experimenting with different approaches.

HOW TO READ YOUR METERS

Reading your own meters is an easy way to keep tabs on the amount of energy you use.

Your Electric Meter

An electric meter records your use of kilowatt-hours of electricity. (See Figure 22-1.) To interpret it:

1. Read dials from left to right. (Note that numbers run clockwise on some dials and counterclockwise on others.)
2. The figures above each dial show how many kilowatt-hours are recorded each time the pointer makes a complete revolution.

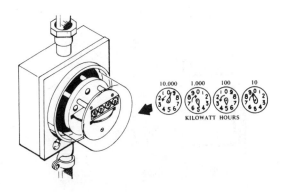

Figure 22-1

3. If the pointer is between numbers, read the smaller one. (The zero stands for 10.) If the pointer is pointed directly at a number, look at the dial to the right. If the pointer has not yet passed zero, record the smaller number; if it has passed zero, record the number the pointer is on. For example, in Figure 22-1 the pointer on the first dial is between zero and 9—read 9. The pointer on the second dial is between 5 and 4—read 4. The pointer on the third dial is almost directly on 5, but the dial on the right has not reached zero so the reading on the third dial is 4. The fourth dial is read 9. You would read the third dial as 5 after the pointer on the 10-kilowatt-hour dial reaches zero. Thus, the total reading is 9449.

4. This reading is based on a cumulative total—that is, since the meter was last set at zero, a total of 9449 kilowatt-hours of electricity have been used. To find your monthly use, take two readings 1 month apart, and subtract the earlier one from the later one.

5. Some electric meters have a constant, or multiplier, indicated on the meter. For this type follow the same steps as outlined above, and then multiply the usage reading by this number. This type of meter is primarily for high-usage customers.

Your Gas Meter

A gas meter tallies the number of cubic feet of natural gas that you have used. (See Figure 22-2.) Read it as you would an electric meter.

In Figure 22-2, from left to right, the dials register 3177. This is not the final reading, however. It must be multiplied by 100 because each time the pointer on the right-hand dial moves from one number to another, the use of 100 cubic feet of natural gas has been recorded. (The figures above each dial show how many cubic feet of natural gas are recorded each time the pointer makes a complete revolution.) Thus, the total reading is 317,700 (3177 × 100). Since the meter was last set at zero, it has recorded the use of 317,700 cubic feet of natural gas.

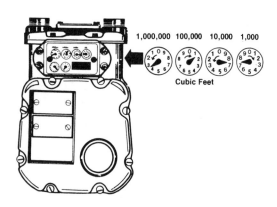

Figure 22-2

Some gas meters have more than four dials. If yours does, and you cannot decide by what number to multiply the reading, call the gas company for directions.

Many apartments are not equipped with meters. Instead, the landlord has a large master meter or meters that register the amount of gas and electricity used by all the tenants. The landlord pays the utility bills and recovers the cost through the rent charged.

HOW TO APPLY YOUR METER-READING SKILLS

Your meter-reading expertise can be used to estimate how much gas or electricity your heating system uses. Read the meter right after your family goes to bed and again before they get up. (Readings are taken while the family sleeps because it is unlikely that any lights or appliances, other than clocks or refrigerator/freezers, will be on during that time.)

Subtract the evening reading from the morning reading to find out how much energy was used for heating during the night. Divide that amount by the number of hours between readings; the answer will yield the number of either kilowatt-hours or cubic feet of natural gas used per hour. For example, assume that your natural-gas meter registered 300,000 cubic feet when you went to bed and 300,160 when you got up 8 hours later. Subtracting the evening reading from the morning one shows that your heating system used 160 cubic feet of natural gas. Dividing 160 by the number of hours between readings yields the hourly consumption: 160 divided by 8 equals 20. Your heating system uses roughly 20 cubic feet of natural gas per hour.

A similar process will tell you how much energy your air-conditioning system uses. While the air conditioning is on, take two readings 1 hour apart. Subtract the earlier reading from the later one to see how much electricity was used in that hour. Repeat the process while the air conditioner is off: subtract the amount of electricity used when the air conditioner was off from the amount used when it was on.

A Word of Caution: The answers yielded by these calculations are only very rough indications of how much energy is used. Many factors determine the amount of energy needed to heat or cool a home: the weather outside, the number of people inside, the number of appliances used, and so forth. Nevertheless, carrying out the process just described will give you an idea of how your energy is used.

HOW TO UNDERSTAND ELECTRICITY AND GAS BILLS

Next time you receive a utility bill, spend a few minutes making sure you understand what the charges are for. Compare it to bills for the previous month or for the corresponding month of the previous year, and see if your energy use has decreased or increased.

The bills shown in Figures 22-3 and 22-4 are samples. If, after reading the explanations of the samples, your

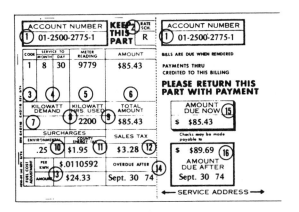

Figure 22-3

own bills still mystify you, call the consumer service department of your utility company for help.

Your Electric Bill

The following numbers identify the parts of the sample bill shown in Figure 22-3.

1. *Account number.*
2. *Rate schedule.* Code showing which rate schedule was applied.
3. *Code.* Indicates whether the bill was estimated or adjusted.
4. *Service to.* Date on which the billing period ended.
5. *Meter reading.* Number of kilowatt-hours the meter registered at end of the billing period.
6. *Amount.* In this case, the only charge shown is a figure determined by adding the price of the electricity used to the routine taxes and surcharges. However, if the customer had received some special service during this billing period, a service charge would appear in this space as a separate entry.
7. *Kilowatt demand.* Information appears in this box only when the bill is sent to a business using more than 6000 kilowatt-hours a month.
8. *Kilowatt-hours used.* Total used during the billing period.
9. *Total amount.* Total amount owed by the customer.
10. *Environmental surcharge.*
11. *County energy tax.*
12. *Sales tax.*
13. *Fuel cost adjustment.* Both the total adjustment and adjustment per kilowatt-hour are shown.
14. *Overdue after.* Date on which bill, if unpaid, become overdue.
15. *Amount due now.*
16. *Amount due after.* Amount due if the bill becomes overdue.

How much of your bill is for electricity? How much is for other charges? To answer these questions, ask your

utility company for a copy of the rate schedule that was applied to your bill. Be sure that you have the right schedule; rates sometimes vary according to the season. In most areas, the demand for electricity increases in the warm months; to meet the added burden, electric companies are forced to use spare generators that are often less efficient and consequently more expensive to run.

The sample electric bill shown in Figure 22-3 was based on the schedule that follows. Note that there is a minimum charge regardless of how little electricity you use and that the first 20 kilowatt-hours you use are covered by this flat rate. Included in the minimum, or service charge, is the cost of providing service to the customer, including metering, meter reading, billing, and various overhead expenses. (Not all rate schedules include minimum charges.)

Minimum charge (including the first 20 kWh or fraction thereof)	$2.25 per month
Next 80 kWh	.0355 per kWh
Next 100 kWh	.0321 per kWh
Next 200 kWh	.0296 per kWh
Next 400 kWh	.0265 per kWh
Consumption in excess of 800 kWh	.0220 per kWh

The sample bill shows a consumption of 2200 kilowatt-hours. The following chart shows how the schedule was applied to the bill.

20 kWh	@	$2.25 (flat rate)	=	$ 2.25
80 kWh	X	.0355	=	2.84
100 kWh	X	.0321	=	3.21
200 kWh	X	.0296	=	5.92
400 kWh	X	.0265	=	10.60
+ 1,400 kWh	X	.0220	=	30.80
2,200 kWh				$55.62

The 2200 kilowatt-hours cost $55.62. The other $29.81 ($85.43 less $55.62) is for the fuel-cost adjustment and the various taxes and surcharges listed on the face of the bill.

Your Gas Bill

The following numbers identify the parts of the sample bill in Figure 22-4.

1. *Account number.*
2. *Sales or city tax.*
3. *Overdue after.* Date on which the bill, if unpaid, becomes overdue.
4. *Code.* Indication of charges included in the bill. The key to the code is on the back of the bill. In this case, the customer is being charged only for gas, routine taxes, and surcharges. Had the customer received special services, other code indications would appear in this space.
5. *Sch.* A code indication of the rate shedule applied to the bill.

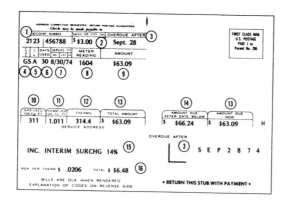

Figure 22-4

6. *Days used.* Number of days of service covered by the bill.
7. *Service to.* Date on which billing period ended.
8. *Meter reading.* Number of cubic feet of natural gas the meter registered at the end of the billing period.
9. *Amount.* In this case, the only charge shown is for the gas used and the routine taxes and surcharges. Had the customer received any special services from the gas company, service charges would appear here as separate entries.
10. *Gas used 100 cu ft.* The amount of gas—expressed in hundreds of cubic feet—that were used during the billing period.
11. *Therms per 100 cu ft.* The number of therms contained in every hundred cubic feet of gas received.
12. *Therms.* Number of therms used during the billing period.
13. *Amount due now.* Total amount owed if bill is paid on time.

14. *Amount due after date below.* Amount due if the bill becomes overdue.
15. *Interim surcharge.* Temporary surcharge granted to the company by the utility commission until the commission has had time to decide whether or not to raise the company's rates.
16. *PGA per therm* and *Total.* Purchased gas adjustment (PGA) charge.

Obtain a copy of the relevant rate schedule from your natural-gas company if you want to know either how much of your bill is for the gas you used or how much each therm costs. Following is the schedule from which the sample natural-gas bill was derived. Note that, like the electric schedule, this one also contains a minimum charge.

Minimum charge (including the first 5 therms or fraction thereof)	$2.50 per month
Next 10 therms	.1690 per therm
Next 15 therms	.1560 per therm
Next 570 therms	.1424 per therm
Next 5400 therms	.1226 per therm

The sample bill in Figure 22-4 shows a consumption of 314.4 therms. The following chart shows how to figure the cost of each therm.

5 therms	@	$2.50 (flat rate)	=	$ 2.50
10 therms	X	.1690	=	1.69
15 therms	X	.1560	=	2.34
+ 284.4 therms	X	.1424	=	40.50
314.4 therms				$47.03

The 314.4 therms cost $47.03. The other $16.06 ($69.09 less $47.03) is for the purchased gas adjustment and the various other taxes and surcharges listed on the face of the bill.

23 How to Survive a Tough Winter

As winter approaches, many people become apprehensive: Will this winter be severe? How will it affect us? Our home? Our family? This chapter will help you to prepare for winter and will show you how to handle emergencies that the worst weather can bring.

WINTERIZING THE HOME

Severe winter weather can result in serious emergencies. For example, a storm can knock down power lines, shutting off not only lights but also heating equipment, which needs electricity to operate. During a heavy snowfall, it may be difficult or even impossible for you to get out of the house. So, before winter sets in, here are some things you should do to protect your home and stay comfortable.

- *Insulate your house.* Make your house airtight to keep heat in and cold out. Caulk and weather-strip doors and windows. Install storm windows, or cover windows with plastic. Insulate walls and attics. For detailed information on how to insulate, contact your utility company or the Public Service Commission in your state. Also, see Chapter 24.
- *Have emergency heating equipment available* so you can keep at least one room warm enough to be livable if your furnace is not operating. The heat source you choose will depend on where you live and whether you own a home or rent. A fireplace with an ample wood supply is, of course, an excellent heat source. A small, well-vented wood or coal stove is an efficient, low-cost heat source that can be used in houses and some apartments. A camp stove can be used in both. Be careful. Know how to use the emergency heating equipment safely to prevent fire or dangerous fumes. Proper ventilation is essential. For detailed information on safe handling of alternative heat sources and equipment, talk to a local dealer. Check the Yellow Pages under "Heating."
- *Keep pipes from freezing.* Wrap the pipes in insulation made especially for water pipes, or in layers of old newspapers, overlapping the ends and tying them

around the pipes. Then cover the newspapers with plastic to keep out moisture.

When it is extremely cold and there is real danger of the pipes freezing, let the faucets drip a little because running water seldom freezes. Although this wastes water, it may prevent damage. Know where the valve regulating the water coming into your home is located. As a last resort, you may have to shut off this main valve and drain all the pipes to keep them from freezing and bursting.

- *If the pipes freeze despite preventive efforts, open the faucets wide to allow for expansion of the frozen water.* Remove any newspaper that may be around the pipes. Wrap pipes with rags, and pour hot water over them with the faucets still open.

NO HEAT? DON'T PANIC.

If your furnace is not operating, don't despair. You may be able to remedy the problem yourself. First, if your furnace burns oil, make sure that the fuel tank isn't empty. Second, check the electric switch that controls the blower—or some other function of the heating unit—to be sure it hasn't turned off. Also check the fuse or circuit breaker that controls the furnace.

If the unit is gas-fired, check other gas appliances to make sure the main gas supply hasn't been cut off. If that doesn't solve the problem, check the pilot light. Instructions for relighting it appear on the front of most units.

If these steps do not restore the furnace to operation, call the utility company. Or, if the unit is fueled by oil, call your fuel-oil dealer or a company that specializes in heating and cooling work. They are listed in the Yellow Pages under "Heating."

If you are a renter without heat, contact your landlord. If this is to no avail, contact your city housing authority or local community action agency to find out how to get the landlord to provide heating service.

Familiarize yourself with the basic functioning of your heating equipment. Fuel dealers and utility companies are usually willing to show customers these basics, in order to save themselves the time and expense of making emergency calls later on.

While you wait for help, to maintain a minimal heat level you should:

- Use an alternate heat source.
- Close off rooms that are not absolutely needed.
- Hang blankets over windows at night (let the sun shine in during the day). Stuff cracks around doors with rugs, newspapers, towels, etc.
- Prevent water pipes from bursting, using the techniques described on p. 227. Collect water for drinking, and store it in covered containers. Close the water inlet valve on the toilet, and then flush to prevent freezing damage.
- Don't hesitate to ask for help if the situation starts to get out of control. Call a neighbor, or a local social or emergency service agency. (Write down emergency phone numbers in advance, and keep them handy.)

KEEPING WARM IN A COLD HOUSE

When your house is without heat due to malfunctioning equipment or lack of fuel, there are ways to preserve some of the heat and stay relatively comfortable.

- Dress warmly. Layers of wool clothing, worn in direct contact with the skin, are the warmest. If you don't have any wool clothes, layers of cotton or synthetics will do. Layers of clothing protect more effectively against cold than a single layer of thick clothing, because the air between layers is warmed by body heat. The layers can be removed as needed to prevent perspiring and subsequent chill.
- Eat well-balanced, nutritional meals. This will enable the body to produce its own heat efficiently.
- Wear a hat, preferably wool, especially when sleeping. The body loses 50 to 75 percent of its heat through the head.
- For the most warmth while sleeping use several lightweight blankets rather than one very heavy blanket.

WHAT TO HAVE HANDY

Certain items and information are indispensable in an emergency. Your home emergency kit should include:

- Phone numbers you can dial for help—your neighbors, the police, the fire department, and other community service organizations.
- Emergency food and water supply. Store an adequate supply of canned and packaged food that does not require refrigeration or cooking. Keep 5 to 10 gallons of drinking water on hand.
- Battery-powered radio and extra batteries. Even if you are without electricity, you will be able to hear weather forecasts, emergency information, and advice broadcast by local authorities.

- Any medicines that may be required by family members.
- First-aid supplies, which should include:
  - Two 1-inch-wide adhesive compresses (bandages)
  - Two 2-inch bandage compresses
  - One 3-inch bandage compress
  - One 4-inch bandage compress
  - One 3-by-3-inch plain gauze pad
  - One gauze roller bandage
  - Two plain absorbent gauze, one-half square yard
  - Two plain absorbent gauze, 24-by-72-inch
  - Three 40-inch triangular bandages
  - Tourniquet, scissors, and tweezers
  - Standard first-aid or personal safety manual

  These articles can be purchased separately at most drugstores; first-aid kits can be purchased from any Red Cross unit.
- Extra blankets or sleeping bags
- Firefighting equipment, including an extinguisher, buckets of sand, a shovel, and an ax.

WHEN THE LIGHTS GO OUT

Always keep a flashlight or candles in a place that's easily accessible in the event of power failure. If the power failure seems to be affecting your house or apartment only, check the fuse box or circuit-breaker box. You may be able to correct the problem by simply replacing a fuse or resetting a circuit-breaker switch. If this doesn't work, call the electric utility that supplies your home and request assistance. If the failure is affecting your entire area, the utility company may already be aware of it and working to correct it. You may, however, wish to make sure by reporting it yourself.

While you wait for the lights to come back on:

- Turn off all but one or two light switches and the furnace switch, and unplug the freezer and refrigerator. The surge of returning electrical power can damage the motors of appliances. When the electricity returns, wait a half hour or so before turning on other lights or electrical equipment. This eases the immediate load on the electric utility system while it is trying to stabilize.
- Keep the freezer door closed as much as possible, and use the food stored in the refrigerator first. After power is restored, examine food for signs of spoilage before refreezing it. Partially thawed food can be safely refrozen only if it still contains ice crystals.

ENJOY THE GREAT OUTDOORS BUT . . .

People who are outside in low temperatures and strong winds can tire easily and become more susceptible to frostbite. A strong wind combined with a temperature slightly above freezing can have the same effect as a still-air temperature nearly 50°F lower. Following are some examples of how this wind-chill factor will dramatically lower the effective temperature:

Thermostat Temperature (°F)	Equivalent Temperature (°F)		
	At 15 mph	At 30 mph	At ≥ 40 mph
30	11	-2	-4
20	-6	-18	-22
10	-18	-33	-36
0	-33	-49	-54
-10	-45	-63	-69
-20	-60	-78	-87
-30	-70	-94	-101
-40	-85	-109	-116

IF YOU MUST GO OUTSIDE . . .

If you have to go outside during a winter weather emergency, observe the following safety measures:

- Avoid overexertion. Cold weather itself, without any physical exertion, puts an extra strain on the heart. If you add to this the strain of heavy physical activity such as shoveling snow, pushing an automobile, or even walking too fast or too far, you risk damaging your body.
- Dress warmly in loose-fitting, layered, lightweight wool clothing. Outer garments should be tightly woven and water-repellent. Wear a wool hat. Protect your face, and cover your mouth to protect your lungs from very cold air. Wear mittens instead of gloves—they allow your fingers to move freely in contact with one another and will keep your hands much warmer.
- Watch for frostbite. Frostbite causes a loss of feeling and a white or pale appearance in extremities such as fingers, toes, tip of nose, and earlobes. If such symptoms are detected, get medical attention immediately. Do not rub with snow or ice—this does not help the condition but, in fact, will make it worse. The best treatment for frostbite is the rapid rewarming of the affected tissue. Watch for symptoms of hypothermia (see below).
- Avoid alcoholic beverages. Alcohol causes the body to lose heat more rapidly, even though a person may feel warmer after drinking alcoholic beverages.
- Keep yourself and your clothes dry. Wet clothing loses all of its insulating value and transmits heat rapidly. Therefore, change wet clothing as quickly as possible to prevent loss of body heat.
- If infants or paralyzed persons must go outside in severe weather, they should be checked frequently for signs of frostbite.

COLD-WEATHER EXPOSURE

When the body begins to lose heat faster than it can produce it, a condition called *hypothermia* begins to develop. The symptoms become very apparent and include:

- Uncontrollable shivering
- Vague, slow slurred speech
- Memory lapses; incoherence
- Immobile, fumbling hands
- Frequent stumbling; lurching gait
- Drowsiness
- Apparent exhaustion; inability to get up after a rest

Often, an affected person will not realize the seriousness of the situation. If a person shows any sign of overexposure to cold or wet and windy weather, take the following measures even if the person claims to be in no difficulty:

- Get the person into dry clothing and into either a warm bed or sleeping bag with a hot-water bottle (which should actually be warm to the touch, not hot), warm towels, heating pad, or other such heat source.
- Concentrate heat on the trunk of the body first—that is, the shoulders, chest, and stomach.
- Keep the head low and the feet up to get warm blood circulating to the head.
- Give the person warm drinks.
- Never give the person alcohol, sedatives, tranquilizers, or pain relievers. They only slow down body processes even more.
- Keep the person quiet. Do not jostle, massage, or rub.
- If symptoms are extreme, call for professional medical assistance immediately.

WINTER DRIVING TIPS

If you have a car, make sure it is ready for whatever winter may bring. The following precautions can help you to avoid an unpleasant or dangerous situation while traveling during winter:

- For safety and fuel economy, keep your car in top operating condition all year long. It is especially important to winterize your car by checking that the following items are in working order:

antifreeze	defroster	lights
battery	exhaust system	snow tires
brakes	fuel system	tire treads
chains	heater	winter grade oil
cooling system	ignition system	wiper blades

- Keep your gasoline tank almost full. This will keep water out of the tank and will provide maximum advantage in case of trouble.
- A citizens band (CB) radio can be very useful in emergencies.

Carry a winter-storm car kit. The kit should contain:

- Sleeping bags, or two or more blankets. (In an emergency, a stack of newspapers can provide layers of insulation.)
- Two empty 3-pound coffee cans with lids. One may be used for sanitary purposes, the other to hold candles.

- If available, use a catalytic heater. A catalytic heater relies on a chemical reaction to produce heat. (Whether using this type or your car's own heater, leave a window slightly open for air circulation. Carbon monoxide poisoning can happen without the victim being aware of it until it's too late.)
- Matches and candles.
- Winter clothing, including wool caps, mittens, and overshoes.
- Large box of facial tissues.
- First-aid kit with pocketknife.
- Flashlight with extra batteries.
- Small sack of sand.
- One set of tire chains.
- Shovel.
- Food supply (high-calorie, nonperishable food such as canned nuts, dried fruit, candy, etc.).
- Tools (pliers, screwdriver, adjustable wrench).
- Windshield scraper.
- Transistor radio, with extra battery.
- Battery booster cables.

IF YOU MUST USE YOUR CAR DURING A STORM . . .

- Plan your travel, selecting both primary and alternate routes.
- Check latest weather information on the radio.
- Try not to travel alone. In a storm, it is preferable to have at least one other person with you.
- Travel in convoy with another vehicle, if possible.
- Always fill the gasoline tank before entering open country, even for a short distance.
- Drive carefully and defensively.

- If the storm becomes too much for you to handle, seek refuge immediately.

IF A BLIZZARD TRAPS YOU IN A CAR . . .

- Stay in the car. Do not attempt to walk in a blizzard. Disorientation comes quickly in blowing and drifting snow and being lost in open country is extremely dangerous. You are more likely to be found in your car and will at least be sheltered there.
- Avoid overexertion and exposure. Exertion from attempting to push your car, shoveling heavy drifts, and performing other difficult chores during blizzard conditions may cause a heart attack—even for persons in apparently good physical condition.
- Keep a downwind window slightly open for fresh air. Freezing rain, wet snow, and wind-driven snow can completely seal off the passenger compartment.
- Beware of carbon monoxide. Run the engine and heater sparingly, and only with a downwind window open. Make sure that snow has not blocked the exhaust pipe.
- Exercise by clapping your hands and moving your arms and legs vigorously from time to time, and don't stay in one position for long. But don't overdo it. Exercise warms you but it also increases body heat loss.
- If more than one person is in the car, don't sleep at the same time. Take turns keeping watch. If alone, stay awake as long as possible. You could freeze to death if you fall asleep.
- Turn on the dome light at night to make sure your car is more visible to working crews.
- Don't panic. Stay with the car.

24 How to Buy Insulation

Caveat emptor ("let the buyer beware") is an appropriate warning for homeowners interested in conserving energy through home insulation. They must exercise great caution when deciding on the type of insulation, selecting a contractor, and arranging financing that provides adequate protection against unfair credit practices.

There is good reason for such caution. We are faced with enormous potential for consumer losses resulting from misinformed decisions to purchase insulation. Unless decisions are made on the basis of solid information and comparison shopping, consumer losses could more than offset the savings expected from energy conservation.

Home insulation that combines quality products and good workmanship does not come cheap. Proper insulation of a three-bedroom house can cost up to $1000. So, consumers should beware of fly-by-night operators who go from door to door offering immediate low-priced installation of insulation.

Before committing themselves to high-priced insulation contracts, consumers may want to take inexpensive and effective energy-saving steps such as caulking and weatherstripping doors and windows. A more expensive but often effective measure is the installation of storm windows. All of this work can be done by the handy homeowner. Attic insulation, one of the most important energy-saving home improvements consumers can make, can also be a do-it-yourself project. Thermal clocks, which automatically turn the heat down at night and up again in the morning, are also easily installed. Although acoustical tiling, aluminum siding, and carpeting are all touted as a means to conserve energy, they have only marginal utility as energy-savers.

The major products used in home insulation are mineral wood (subdivided into glass wool and rock wool) and celulose. Other products include urea-formaldehyde (a plastic foam product that is formulated onsite and foamed into wall cavities), perlite and vermiculite (mineral-based products used primarily in masonry construction), aluminum foil, and polystyrene and polyurethane plastic foams (preformed into sheets for insertion into wall and ceiling cavities).

APPRAISALS

The amount of energy—and money—that can be saved through insulation depends on the geographic location of the house, the structure, the amount of insulation already in place, and fuel costs.

Avoid overinsulating your home.* It may be difficult to figure out how much insulation you really need without the advice of an expert. In fact, you may want to get two or three appraisals and compare them. The best insulation for the individual homeowner is that which gives the largest, long-run, *net* savings in heating and cooling costs for the investment. (Net savings are the total savings on fuel bills less the purchase price of the insulation.) Most homeowners will want to recover the cost of conservation measures in energy savings within 10 years.

JUDGING INSULATION PRODUCTS

Insulation performance is measured in *R value*. The R value is a number indicating how much resistance the insulation presents to heat flowing through it. For example, an R value of 2 means that the insulating material provides two standard units of heat resistance; an R value of 3 means the insulating material provides three standard units. The higher the material's R value, the better its insulating quality. Price the R value for any insulating material you are considering. Pay more only for more R value.

DECEPTIVE PRACTICES

Deception of consumers is likely to occur in two areas: (1) in the product's thermal resistance, either through generalizations without providing R values or through

*The Commerce Department booklet, *Making the Most of Your Energy Dollars in Home Heating and Cooling,* contains climate maps, sample costs, and worksheets for homeowners to calculate their particular needs. It is available for 70 cents a copy from Superintendent of Documents, U.S. Government Printing Office, Washington, D.C. 20402

exaggerations of the R values, and (2) in the product's flammability, either by misrepresentation of the flammability or by no reference at all to a product's unacceptable flammability.

Following are the current generally accepted maximum design standard R values for the most commonly used insulating materials:

Product	R Value per Inch of Insulation
Rock-wool batts and blankets	3.1 to 3.6
Rock-wool loose fill	2.7 to 3.2
Glass-fiber batts and blankets	2.7 to 3.7
Glass-fiber loose fill	2.1 to 2.4
Cellulose loose fill	3.1 to 3.7
Urea-formaldehyde	4.1
Polystyrene (expanded)	4.0 to 5.26
Polyurethane	6.25
Perlite	2.7
Aluminum foil	2.0

Question any claim of an R value above the maximum design standard. Some cellulose manufacturers and their distributors claim R values up in the R-4 range, and sometimes claim values as high as R-6 or even R-9. Although it is theoretically possible to achieve a higher-than-design R level with cellulose products, this is normally done by reducing the amount of flameproofing chemicals. This produces a dangerously flammable product. Therefore, an R of 3.1 to 3.7 is generally considered the maximum value for safe cellulose insulation.

There may be substantial misrepresentation of R values in connection with the sale of cellulose and (due to quality control) perhaps with rock wool. Deception may also occur in the representation of R values by the manufacturers of urea-formaldehyde insulation. The advertised R values may not take shrinkage into account, which may reduce the R value of the installed product. The National Bureau of Standards found linear shrinkage over a 28-month period to be as much as 10 percent. This produces airspace between the foam and the edge of the cavity. As a result, the R may be reduced and the cost per R raised.

FLAMMABILITY

Flammability of insulation is also an important performance characteristic, which should be carefully evaluated. Current flammability standards for insulation products, established through the American Society of Testing and Materials (ASTM), include flame spread, fuel contribution, and smoke development rates. Insulation sold to consumers, however, may not be required to meet the ASTM criteria. Moreover, no insulation should be installed near heat or exposed light fixtures. Mineral-wool flammability is less of a serious threat than the flammability of cellulose or urea-formaldehyde products. Although mineral wools themselves pose no flammability problem, their paper vapor barriers, often attached to the wool, are indeed flammable.

Cellulose is recycled paper and is inherently flammable. A flame retardant is added to cellulose to reduce flammability to accept levels. Addition of the flame retardant, however, also reduces thermal resistance. Use of the proper chemical flame retardant is also important in order to avoid a potential corrosion hazard. The ureaformaldehyde products fall within accepted levels for flammability safety, but they are not fireproof as distributors sometimes claim.

Polyurethane and polystyrene products are marketed in sheets of various dimensions and thicknesses, often through retailers directly to consumers, for retrofit installations. Their R values (over 6 per inch) and ease of installation make them an attractive choice in spite of relatively high cost. Most manufacturers offer both flame-resistant and nontreated rigid foams. The nonflameproofed products are obviously a fire hazard, but the flameproofed products also pose a risk. Once the relatively high ignition temperature for these products is reached, they emit high levels of smoke and toxic gases, which can be just as lethal as fire. Consumers who choose to use these products should know that the products cannot be used safely for insulation purposes unless they are enclosed in flame- and heat-retardant structures, such as gypsum board. While the same installation precautions are applicable to a lesser degree for mineral wool, cellulose, and urea-formaldehyde products, they are absolutely necessary for polyurethane and polystyrene products. For example, one use for which these products are being promoted is the insulation of garage walls and doors. Without application of gypsum board, these sheets would be directly exposed to heat and fire sources commonly found in garages.

INSTALLATION HAZARDS

Faulty installation is another hazardous area for consumers. At best, the practical physical problems of installation produce some loss of insulation potential. One laboratory claims that a typical loss is 20 percent. In the worst case, careless or incompetent installation can result in a major reduction of potential thermal resistance—the R value—of the product. At present, there are no tests outside the laboratory to determine the incremental change in thermal resistance as a result of installing insulation. Consumers should carefully consider the type and quantity of insulation necessary to determine the desired R value. They should then observe whether the quantity and type requested are actually installed. For example, if a certain number of bags of loose-fill insulation is needed to reach a certain R value, insist on counting the bags.

Probably the best way to avoid faulty insulation is to make sure the contractor is reputable. If the state requires licensing, find out if the contractor is licensed. If the state requires bonding, make certain the person is bonded. Once you have selected a contractor, have a specific contract written up for the job. Sign it only when you are fully satisfied that it details every thing you want done. Once you have a contract, each side knows the limits of the other's responsibility before the job begins. Consumers might also insist in their contract that the completed work be examined by a municipal building inspector. Final payment would not be made unless the inspector approved the work as satisfactory.

WARRANTY PROTECTIONS

Insulation products and services are often sold on warranty, which guarantees the replacement of defective materials and correction of defective workmanship. Warranteed products provide better protection to consumers than do products sold without warranty. Warranties should be in writing and should be part of the contract. Since consumer's may not discover that the home insulation buried behind the walls is unsatisfactory until they have scrutinized utility bills over an extended period of time, they should seek warranties that give enough time to judge whether the insulation installed has, in fact, reduced home-energy costs.

CREDIT PRACTICES

Know your rights. The federal truth-in-lending act (Consumer Credit Protection Act of 1968) obligates creditors to provide consumers with the interest rate they will be paying on an annual basis. The Fair Credit Billing Act provides a mandatory dispute resolution procedure for alleged billing errors appearing on periodic billing statements sent to consumers.

The Federal Trade Commission's "holder-in-due-course" rule requires sellers to insert a provision in the text of installment credit contracts that expressly preserves the consumer's legal rights for breach of contract, breach of warranty, misrepresentation, or fraud against the seller or against anyone who may obtain the contract and seek payment under it. It also requires sellers to include a similar provision in the text of any consumer loan contract, if the loan is used to finance the purchase and if the seller and lender work together. This means that even if a home improvement or insulating company should finance the work and then sell the note to a finance company or other credit institution, the homeowner may be able to withhold payment in the event of a dispute.

If you have reason to suspect deception or fraud, or you are having a problem, such as non-fulfillment of a contract or failure by the seller to honor a warranty, or you are being threatened with loss of your security interest on an installment-credit or vendor-related loan, contact the regional offices of the Federal Trade Commission, the neighborhood legal services offices of the Office of Economic Opportunity, or state and local consumer protection offices.

25   How to Weatherize Your Home

A QUICK LOOK AT YOUR HOME

A range of cost and savings is given for the energy-savings improvements discussed in this chapter. The cost figures given for supplies and services are only meant to be used as a general guide; they are not the actual costs of the supplies and services in today's marketplace. For comparison with your home, a single-story, 1250-square-foot house in Washington, D.C., paying $0.45 per gallon of oil, $0.16 per hundred cubic feet of gas, or $0.03 per kilowatt of electricity will fall at about the middle of the range given. Also, you should consult the Energy Checklist from time to time when making your computations (see p. 236).

You can save significantly on heating costs if you live almost anywhere in the United States and follow one of two suggested savings plans. Plan 1 is cheap and easy, and it pays for itself every year. Plan 2 saves even more, and it can cut your heating bills by as much as 50 percent. It will pay for itself within 5 years. Here's an idea of what Plan 1 costs and saves in a typical home:

Plan 1	Yearly Cost	Yearly Saving
Turn down thermostat 6°F in winter from the usual setting.	$0	$27–87
Put on plastic storm windows.	$7–9	$27–73
Service the oil furnace.	$25	$33–87
Total	$32–34	$87–247

If you already have storm windows, or if you don't have an oil furnace, then take a look at Plan 2. Note that you might not need to do all of the items listed.

Plan 2	First-Year Cost	Yearly Saving
Turn down thermostat 6°F in winter from the usual setting.	$0	$13–53
Put on plastic storm windows.	$7–9	$20–60
Service the oil furnace.	$25	$20–53
Caulk and weather-strip doors and windows.	$75–105*	$40–100
Insulate your attic.	$300–450*	$50–75
Total	$407–589	$143–341

Plan 3 can save you money on your air-conditioning bills, if you live in the southern United States.

Plan 3	First-Year Cost	Yearly Savings
Turn up thermostat 6°F in summer from the usual setting.	$0	$7–20
Insulate the attic.	$300–450*	$33–67
Caulk and weather-strip doors and windows.	$75–105*	$27–67
Total	$375–555	$67–154

*These are do-it-yourself costs. If you use a contractor, these items may cost twice as much.

If your whole house is air-conditioned and if you live in the southern part of the country, some of the energy-saving steps will save on both heating and cooling—but you only have to pay for them once. Tables 25-1 and 25-2 give an estimate of the combined costs and savings for a typical home.

Table 25-1. Plan 1 plus turning up the thermostat in summer.

	YEARLY COST	YEARLY SAVINGS
Turn down thermostat in winter	$0	$27–87
Turn up thermostat in summer	$0	$7–20
Put on plastic storm windows	$7–9	$27–73
Service oil furnace	$25	$33–87
Total	$32–34	$94–267

Table 25-2. Plan 2 and Plan 3 together.

	FIRST-YEAR COST	YEARLY SAVINGS
Turn down thermostat in winter	$0	$27–87
Turn up thermostat in summer	$0	$7–20
Put on plastic storm windows	$7–9	$20–73
Service oil furnace	$25	$20–53
Caulk and weather-strip	$75–105*	$67–167
Insulate attic	$300–450*	$80–227
Total	$407–589	$221–626

*These are do-it-yourself costs. If you use a contractor, these items may cost about twice as much.

Which Energy-Saving Steps Should You Take?

- *Should you adjust your thermostat?* Everyone can profit from turning the thermostat down in winter and up in summer.
- *Should you put on storm windows or service your oil furnace?* You should if you live in the northern part of the country. Put on storm windows if you don't have them already. Plastic ones are cheapest. Service an oil furnace each year; service a gas furnace every three years.
- *Should you add caulk or putty?* Look around the edges of the window, and check the edges of the doors. There should be some caulk or putty in all cracks (edges). If it's old, brittle, and broken, or if it's missing altogether, you should put some in.
- *Should you weather-strip?* The strips of vinyl, metal, or foam rubber around the edges of windows and doors are *weather stripping.* If it's missing or deteriorated, you should replace it.
- *Should you insulate your attic?* You should insulate if there is no, or not enough, insulation (see p. 245). Go into your attic, and find out how much insulation there is. Usually, there's a door or hatchway to an attic. If not, have a contractor check the insulation for you.

A CLOSER LOOK AT YOUR HOME

Following are valuable energy-saving steps for the home.

Step 1: Caulk and Weather-Strip Doors and Windows

Caulking and weather-stripping are good cheap ways to save energy, so you should check to see if you need caulking, putty, or weather stripping on your windows and doors.

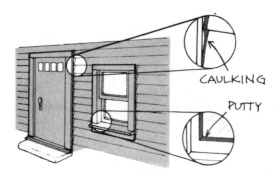

- *Are caulking or putty needed?* Look at a typical window and a typical door in your home. Check the description that best fits what you see:
  - ☐ OK... All cracks are completely filled with caulking. The putty around the windowpanes is solid and unbroken. No drafts.
  - ☐ Fair... The caulking and putty are old and cracked, and/or missing in places. Minor drafts.
  - ☐ Poor... There's no caulking at all. The putty is in poor condition. Noticeable drafts.

If you checked "Fair" or "Poor," then caulking is needed.
- *Is weather stripping needed?* Look at one or two of your typical windows. Check one:

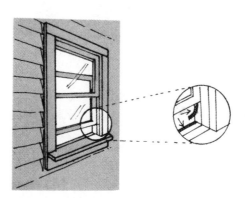

- ☐ OK... Good, unbroken weather stripping. No drafts.
- ☐ Fair... Weather stripping damaged or missing in places. Minor drafts.
- ☐ Poor... No weather stripping at all. Noticeable drafts.

If you checked "Fair" or "Poor," then your windows probably need weather stripping. Be careful; the windows could be in such poor condition that weather stripping can't be installed.

Now, apply the same criteria to your doors to determine if weather stripping is required.

To find the cost of caulking and weather-stripping doors and windows:

Multiply number of windows that need caulking and putty times the cost per window	____ X $0.90 = $ ____
Multiply number of windows that need weather stripping times the cost per window	____ X $4.00 = $ ____
Multiply number of doors that need caulking times the cost per door	____ X $0.85 = $ ____
Multiply number of doors that need weather stripping times the cost per door	____ X $6.75 = $ ____
Add these numbers to get the total	Total Cost $ ____

Enter the Total Cost on line 1 of the Energy Checklist on p. 271. This is the estimated do-it-yourself cost. If you get a contractor to do the job, it will cost at least two to four times as much.

To find the Savings Factor, fill out the lines that apply to your house:

Windows	*Number*		
Condition of caulking and putty			
Fair	____	X 0.3 =	____
Poor	____	X 1.0 =	____
Condition of weather stripping			
Fair	____	X 1.0 =	____
Poor	____	X 8.4 =	____
Doors			
Caulking			
Fair	____	X 0.3 =	____
Poor	____	X 0.9 =	____
Weather stripping			
Fair	____	X 2.0 =	____
Poor	____	X 16.8 =	____
	Savings Factor		____

Enter the Savings Factor on line 1 of the Energy Checklist on p. 271.

Caulk the Openings in Your Home. Caulking should be applied wherever two different materials or parts of the house meet; e.g., doors, windows, roof, walls. It takes no specialized skill and a minimum of tools to apply it. You will need the following tools:

Ladder
Caulking gun
Caulking cartridges
Oakum, glass fiber strips, caulking cotton, or sponge rubber
Putty knife or large screwdriver

You will need to buy caulking compound, which is available in these basic types:

- *Oil- or resin-base caulk:* readily available and will bond to most surfaces—wood, masonry, and metal; not very durable but lowest in first cost for this type of application.
- *Latex-, butyl-, or polyvinyl-based caulk:* all readily available and will bond to most surfaces; more durable, but more expensive than oil- or resin-based caulk.
- *Elastomeric caulks:* most durable and most expensive; includes silicones, polysulfides, and polyurethanes; the instructions provided on the label should be followed.
- *Filler:* includes oakum, caulking cotton, sponge rubber, and glass fiber types; used to fill extra-wide cracks or as a backup for elastomeric caulks.

Caution: Lead-base caulk is not recommended because it is toxic. Many states prohibit its use.

How much should you buy? Estimating the number of cartridges of caulking compound required is difficult since the number needed will vary greatly with the size of cracks to be filled. Rough estimates are one-half cartridge per window or door, four cartridges for the foundation sill, and two for a two-story chimney. If possible, start the job with a half-dozen cartridges and then purchase more as you need them.

You'll have to use a ladder to reach some of the areas to be caulked. Be sure you use it safely. Level and block the ladder in place. Have someone hold it if possible. Don't try to reach that extra little bit—get down and move the ladder. Carry the caulking gun with a sling so that you can use both hands while climbing the ladder.

A house needs to be caulked:

Between window drip caps (tops of windows) and siding
Between door drip caps and siding
At joints between window frames and siding
At joints between door frames and siding
Between windowsills and siding
At corners formed by siding
At sills where wood structure meets the foundation
Outside water faucets or other special breaks in the outside house surface
Where pipes and wires penetrate the ceiling below an unheated attic
Between porches and main body of house
Where chimney or masonry meets siding
Where storm windows meet the window frame, except for drain holes at windowsill
In a heated attic where the wall meets the eave at the gable ends

Before applying caulking compound, clean area of paint buildup, dirt, or deteriorated caulk with solvent and putty knife or large screwdriver.

Drawing a good bead of caulk will take a little practice.

A wide bead may be necessary to make sure caulk adheres to both sides. First attempts may be a bit messy. Make sure the bead overlaps both sides for a tight seal.

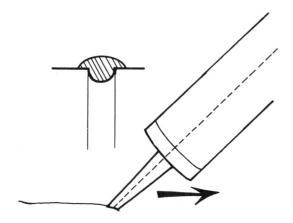

Fix extra-wide cracks such as those at the sills (where the house meets the foundation) with oakum, glass fiber, insulation strips, etc. In places where you can't quite fill the gaps, finish the job with caulk.

Caulking compound also comes in rope form. Unwind it, and force it into cracks with your fingers. You can fill extra-long cracks easily this way.

Step 2: Install Storm Windows

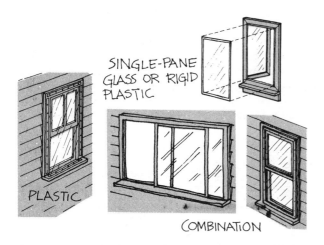

SINGLE-PANE GLASS OR RIGID PLASTIC

PLASTIC

COMBINATION

There are four kinds of storm windows:

- *Plastic* (polyethylene sheet). These come in rolls and cost about 65¢ each. You may have to replace them each year.
- *Single-pane glass or rigid plastic.* Glass storm windows cost about $25.00, and acrylic panes cost about $8.00. They are put up each fall and taken down each spring.
- *Triple-track glass (combination).* These have screens that can be opened and closed. They are for double-hung or sliding windows. They cost about $33.00 each including installation. (They are also available without screens.)

All four windows are about equally effective. However, the more expensive ones are more durable, attractive, and convenient.

Note that the following cost and savings factors are for storm windows only. They are in addition to the costs and savings for caulking and weather stripping found in the preceding section.

Install Plastic Storm Windows Without Weather Stripping

Cost: Number of windows × $0.65 = $_____ yearly cost.

Savings: Rate the weather stripping on the windows:

☐ OK   (7.9)
☐ Fair   (8.2)
☐ Poor   (10.8)

$$\frac{\text{Number in parentheses}} {} \times \frac{\text{Number of windows}} {}$$
$$= \frac{} {\text{Savings Factor}}$$

Enter yearly cost and savings factor on line 2a of the Energy Checklist on p. 271.

Install Plastic Storm Windows With Weather Stripping

Cost:

$$\frac{} {\text{Number of windows}} \times \$0.65 = \$\underline{\qquad} \text{Yearly Cost}$$

Savings:

$$\frac{} {\text{Number of windows}} \times 7.9 = \$\underline{\qquad} \text{Savings Factor}$$

Enter Yearly Cost and Savings Factor on line 2b of the Energy Checklist on p. 271.

Install glass or rigid plastic storm windows: Choose the kind of glass or rigid plastic storm windows you want, and multiply the number of windows you have by the appropriate cost.

Window	Cost
Single-pane, rigid plastic	$8.00
Single-pane, glass	$25.00
Triple-track, glass (combination), double-hung or slider	$33.00

$$\frac{} {\text{Number of windows}} \times \frac{\$} {\text{Cost}} \quad \text{Yearly Cost}$$

Savings:

$$\frac{} {\text{Number of windows}} \times 7.9 = \$\underline{\qquad} \text{Savings Factor}$$

Enter the yearly cost and savings factor on line 2c of The Energy Checklist on p. 271.

Step 3: Weather-strip Windows

Before weather-stripping, make sure that both the sash and the channels in which it slides aren't so rotted that they won't hold the small nails used with weather stripping. If they are badly rotted, consider replacing the entire window unit. Call your lumberyard or window dealer for an evaluation and cost estimate.

You will need the following tools:

Hammer and nails
Screwdriver
Tin snips
Tape measure

You'll need one of the following types of weather stripping:

- *Thin-spring metal.* Installed in the channel of a window so it is virtually invisible. Somewhat difficult to install. Very durable.
- *Rolled vinyl.* With or without metal backing. Visible when installed. Easy to install. Durable.
- *Adhesive-backed foam.* Easy to install. Breaks down and wears rather quickly. Not as effective a sealer as metal strips or rolled vinyl. Never use where friction occurs. Weather stripping can be purchased either by the running foot or in a kit. Make a list of all windows, and measure them to find the total length of weather

stripping you need. Measure the total distance around the edges of the moving parts of each window, and complete the following list:

Type	Size	Quantity	X	Length Required (inches)	=	Total (inches)
Double-hung	1	_____	X	_____	=	_____
	2	_____	X	_____	=	_____
	3	_____	X	_____	=	_____
Case-ment	1	_____	X	_____	=	_____
	2	_____	X	_____	=	_____
	3	_____	X	_____	=	_____
Tilting	1	_____	X	_____	=	_____
	2	_____	X	_____	=	_____
	3	_____	X	_____	=	_____
Sliding pane	1	_____	X	_____	=	_____
	2	_____	X	_____	=	_____
	3	_____	X	_____	=	_____

Total length of weather stripping required (inches) _____

Be sure to allow for waste. If you buy weather stripping in kit form, be sure it is intended for your window type and size.

Caution: Upper-story windows can be a problem. You should be able to do all work from inside, but avoid awkwardly leaning out of windows when tacking weather stripping into place. If you use a ladder, observe all precautions.

Thin spring metal

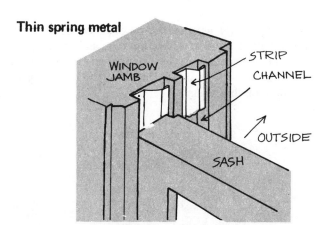

- To install *thin spring metal,* move sash to the open position and slide strip in between the sash and the channel. Tack in place into the casing. Do not cover the pulleys in the upper channels.

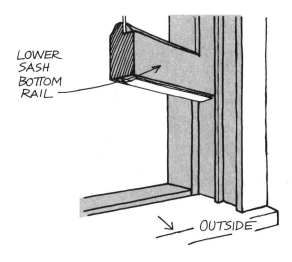

- Install strips the full width of the sash on the bottom of the lower-sash bottom rail and the top of the upper-sash top rail.

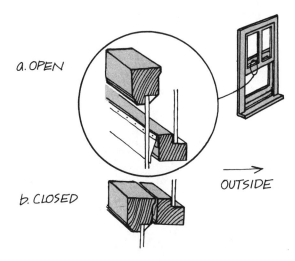

- Then attach a strip the full width of the window to the upper-sash bottom rail. Countersink the nails slightly so they won't catch on the lower-sash top rail—that is, nail the head to or below the surface.

Rolled vinyl

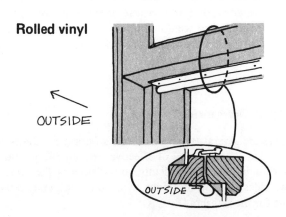

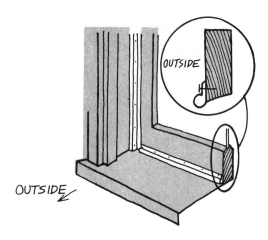

Alternative Methods and Materials

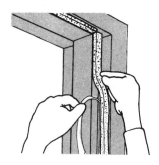

- To install *rolled vinyl,* nail vinyl strips on double-hung windows. Treat a sliding window as a double-hung window turned on its side. Casement and tilting windows should be weather-stripped with the vinyl nailed to the window casing so that as the window shuts, it compresses the roll.

- *Adhesive-backed foam.* You will need a knife or shears and a tape measure. Adhesive-backed foam is extremely easy to install, invisible when installed, and more effective on doors than windows, but not very durable. To install, stick the foam to inside face of the jamb.

Adhesive-backed foam strip

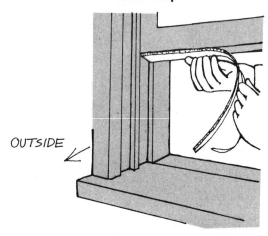

- Install *adhesive-backed foam strips* on all types of windows, but only where there is no friction. On double-hung windows, this is only on the bottom and top rails. Other types of windows can use foam strips in many more places.

- *Rolled vinyl with aluminum channel backing.* You will need a hammer, nails, tin snips, and tape measure. This material is easy to install, visible when installed and durable. To install, nail the strip snugly against door on the casing.

Step 4: Weather-strip Doors

There are several types of weather stripping for doors, and each has its own level of effectiveness, durability, and degree of installation difficulty. Installation is the same for two sides and top of a door, with a different, more durable one for the threshold.

- *Foam rubber with wood backing.* You will need hammer, nails, handsaw, and tape measure. The material is easy to install, visible when installed, and not very durable. To install, nail the strip snugly against the closed door. Space nails 8 to 12 inches apart.

- *Spring metal.* You will need tin snips, hammer, nails, and tape measure. The material is easy to install, invisible when installed, extremely durable. To install, cut strips to length, and tack in place. For better seal, lift outer edge of the strip with screwdriver after tacking.

- *Interlocking metal channels.* You will need hacksaw, hammer, nails, and tape measure. The material is difficult to install (alignment is critical), visible when installed, and durable but subject to damage because it is exposed. It has excellent seal. To install, cut and fit strips to the head of the door first (male strip on door, female on head); then hinge side of door (male strip on jamb, female on door); finally, lock side on door, with female strip on jamb.

- *Fitted interlocking metal channels (J-strips).* These are very difficult to install. They have exceptionally good weather seal, are invisible when installed, and not exposed to possible damage. They should be installed by a carpenter or accomplished handyman only.

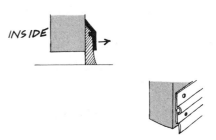

- *Sweeps.* You will need screwdriver, hacksaw, and tape measure. Sweeps are useful for flat thresholds, but may drag on carpet or rug. Models that flip up when a door is opened are available. To install, cut sweep to fit 1/16 inch from the edges of the door. Some sweeps are installed on the inside and some outside. Check the instructions for the particular type you are using.

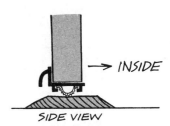

- *Doorshoes.* You will need screwdriver, hacksaw, plane, and tape measure. Doorshoes are useful with a wooden threshold that is not worn and are very durable. However, they are time-consuming to install (door must be removed). To install, remove the door from the hinges and trim the required amount off the bottom. Cut doorshoe to door width. Install by sliding the vinyl out, and fasten with screws.

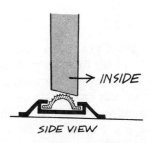

- *Vinyl bulb threshold.* You will need screwdriver, hacksaw, plane, and tape measure. This type is useful where there is no threshold or the wooden one is worn out. It is difficult to install. Vinyl will wear out, but replacements are available. To install, remove the door from the hinges, and trim required amount off the bottom. The bottom should have about 1/8-inch bevel to seal against vinyl. Be sure bevel is cut in right direction for opening.

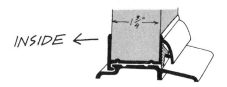

- *Interlocking threshold.* This type is very difficult to install, but has exceptionally good weather seal. It should be installed by a skilled carpenter.

Step 5: Install Plastic Storm Windows

Instead of installing permanent storm windows, you can tack plastic sheets over the outside of your windows or tape sheets over the inside. You will need the following tools and materials:

    6-mil-thick polyethylene plastic in rolls or kits
    Shears to cut and trim plastic
    2-inch-wide masking tape
    or
    Hammer and tacks
    ¼-by-1¼-inch wood slats

Measure the width of your larger windows to determine what widths of the plastic rolls to buy. Measure the length of the windows to determine how many linear feet—and therefore how many rolls or what kit size—to buy. Attach the plastic to the inside or outside of the frame so that it will block airflow that leaks around the movable parts of the window. If you attach the plastic to the outside, use

the slats and tacks. If you attach it to the inside, masking tape will work. Inside installation is easier and will provide greater protection to the plastic. Outside installation is more difficult, especially on a two-story house, and the plastic is more likely to be damaged by the elements.

Be sure to install the plastic tightly and securely, and remove all excess. Besides looking better, this will make the plastic less susceptible to deterioration during the winter.

Step 6: Install Single-Pane Storm Windows (Contractor Assembly)

- Rigid plastic storm windows. These are always installed on the inside. They are available in do-it-yourself kits. Follow the instructions.
- *Glass storm windows.* Suppliers will custom-build single-pane glass storm windows, which you then install yourself. Or, you can make your own with do-it-yourself materials available at some hardware stores. Glass storm windows can be installed either inside, if the way the window is built permits it, or on the outside. If you install them on the inside, then you won't be able to open the existing window. If you install them on the outside, they will only cover the moving part of the window and you'll save less energy; however, they will be permanently installed. With metal casement windows, exterior installation of single-pane storm windows is a job for a contractor.

Determine how you want the windows to sit in the frame. Measure the outside measurements of the storm window. Be as accurate as possible, then allow 1/8-inch along each edge for clearance. When your windows are delivered, check the actual measurements carefully against your order. Install the windows, fixing them in place with movable clips so you can take them down for cleaning.

Single-pane storm windows aren't as expensive as the double- or triple-track combination windows. Their major disadvantage is that you can't open them easily after they're installed.

When selecting glass windows, pay attention to the frame finish. A mill finish (plain aluminum) will oxidize quickly and degrade appearance. Windows with an anodized or baked enamel finish look better. In addition, check to see that the side of the storm window frame that touches the existing window frame has a permanently installed weather strip or gasket to make the joint as airtight as possible.

Step 7: Install Combination Storm Windows (Normally Contractor Installed)

Triple-track combination (windows and screen) storm windows are designed for installation over double-hung and sliding windows. They are permanently installed and can be opened at any time for ventilation. *Double-track combination* units are also available, and they cost less.

You can save 10 to 15 percent of the purchase price of your storm windows by installing the windows yourself. However, it is usually easier to have the supplier install the windows for you, although it does cost more.

The supplier will first measure all the windows where you want storm windows installed. It will take the supplier anywhere from several days to a few weeks to make up your order.

Installation should take less than one day, depending on how many windows are involved. Two very important items should be checked to make sure the installation is properly done. First, make sure that both window sashes and screen sash move smoothly and seal tightly when closed after installation. Poor installation can cause misalignment. Also, be sure there is a tightly caulked seal around the edge of each storm window. Leaks can hurt the window's performance.

Most combination units have two or three, ¼-inch diagonal holes (or other types of vents) drilled through the frame where it meets the windowsill. This is to keep winter condensation from collecting on the sill and causing rot. Keep these holes clear. If your combination units don't already have these holes, drill them yourself.

When selecting combination storm windows, keep the following criteria in mind:

- *Frame finish.* A mill finish (plain aluminum) will oxidize, reducing ease of operation and degrading appearance. An anodized or baked enamel finish is better.
- *Corner joints.* The quality of construction affects the strength and performance of storm windows. Corners are a good place to check construction. They should be strong and airtight. Normally overlapped corner

joints are better than mitered. If you can see through the joints, they will leak air.
- *Sash tracks and weather stripping.* The depth of the metal grooves (sash track) at the sides of the window and the weather-stripping quality make a big difference in how well storm windows can reduce air leakage around the windows. Compare several types before deciding.
- *Hardware quality.* The quality of locks and catches has a direct effect on durability and is a good indicator of overall construction quality.

Step 8: Install Combination Storm Doors (Normally Contractor Installed)

Combination (window and screen) storm doors are designed for installation over exterior doors. They are sold almost everywhere, with or without the cost of installation.

You can save 10 to 15 percent of the purchase price of your storm doors by installing them yourself. In most cases, however, it is easier to have the supplier install the doors.

The supplier will first measure all the doors over which you want storm doors installed. It usually takes anywhere from several days to a few weeks to make up your order. Installation should take less than one-half day.

Before the installer leaves, be sure the doors operate smoothly and close tightly. Check for cracks around the jamb, and make sure the seal is as airtight as possible. Also, remove and replace the exchangeable panels (window and screen) to make sure they fit properly and have a weathertight seal.

When selecting storm doors, use the following criteria:

- *Door finish.* A mill finish (plain aluminum) will oxidize, reducing ease of operation and degrading the appearance. An anodized or baked enamel finish is better.
- *Corner joints.* Quality of construction affects the strength and effectiveness of storm doors. Corners are a good place to check construction. They should be strong and airtight. If you can see through the joints, they will leak air.
- *Weather stripping.* Storm doors should reduce air leakage around your doors. Weather-stripping quality makes a difference in how well storm doors can do this. Compare several types before deciding.
- *Hardware quality.* The quality of locks, hinges, and catches should be evaluated since it can have a direct effect on durability and is a good indicator of overall construction quality.
- *Construction material.* Storm doors of wood or steel can be purchased within the same price range as the aluminum variety. They have the same quality differences and should be similarly evaluated. Choosing between doors of similar quality but different material is primarily a matter of your personal taste.

Step 9: Insulating

R-Value. R-value is an indicator of the amount of resistance that the insulation presents to heat flowing through it. The bigger the R-value, the better the insulation. R-value appears on insulation packaging. Thus, when choosing an insulation material, simply compare the R-value of the various brands, and get the better buy. Pay more only for greater R-value.

The following sections indicate what the R-value of the insulation should be in various situations.

Unfinished Attic, No Floor

Batts, blankets, or loose fill in the floor between the joists:

Table 25-3.

THICKNESS OF EXISTING INSULATION (INCHES)	ADD	IF YOU HAVE ELECTRIC HEAT, OR IF YOU HAVE OIL HEAT AND LIVE IN A COLD CLIMATE ADD[a]	IF YOU HAVE ELECTRIC HEAT AND LIVE IN A COLD CLIMATE, ADD[b]
0	R–38	R–38	R–38
0–2	R–22	R–30	R–38
2–4	R–11	R–11	R–30
4–6	R–11	R–11	R–19
6–8	None	None	R–11

[a]Add this much if you're doing it yourself and the Heating and Cooling Factors add up to more than 0.4, or if you're hiring a contractor and the Heating and Cooling Factors adds up to more than 0.6.

[b]Add this much if you're doing it yourself and the Heating and Cooling Factors add up to more than 0.7, or if you're hiring a contractor and your Heating and Cooling Factors add up to more than 1.0.

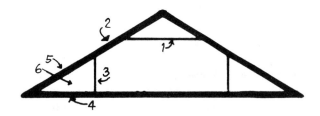

Finished Attic

1. *Attic ceiling.* See Table 19-2.
2. *Rafters.* Contractor fills completely with blow-in insulation.
3. *Knee walls.* Insulate outer attic rafters instead.
4. *Outer attic floors.* Insulate outer attic rafters.
5. *Outer attic rafters.* Add batts or blankets. If there is existing insulation in knee walls and outer attic floors, add R-11. If there is no existing insulation, add R-19.
6. *End walls.* Add R-11 batts or blankets.

Unfinished Attic With Floor
  Do-It-Yourself or Contractor-Installed

- Between the collar beams. Follow the guidelines in Table 19-2.
- *Rafters and end walls.* Buy insulation thick enough to fill the space available (usually R-19 for rafters and R-11 for end walls).

Contractor-Installed

- *Contractor blows loose-fill insulation under the floor.* Fill this space completely. Check the R-value.
- *Frame walls.* Contractor blows in insulation to fill the space inside the walls. Check the R-value.
- *Crawl space.* R-11 batts or blankets against the wall and the edge of the floor.
- *Floors.* R-11 batts or blankets between the floor joists, foiled-faced.
- *Basement walls.* R-11 batts or blankets between wall studs.

What Kind of Insulation to Buy.

Batts. Glass-fiber and rock-wool batts are used to insulate unfinished attic floors and rafters, and the underside of floors. They are best suited for standard joist or rafter spacing of 16 or 24 inches, and for spaces between joints that are relatively free of obstructions. Batts are cut in sections 15 or 23 inches wide, 1 to 7 inches thick and 4 or 8 feet long. They are available with or without a vapor barrier backing. If you need vapor barrier backing and can't get it, buy polyethylene that is used to insulate the underside of floors. Batts are easy to handle because of their relatively small size; however, their use will result in more waste from trimming sections than will the use of blankets. Batts should be fire- and moisture-resistant.

Blankets. Glass-fiber and rock-wool blankets are used to insulate unfinished attic floors and rafters, and the underside of floors. They are best suited for standard joist or rafter spacing of 16 or 24 inches, and for spaces between joists that are relatively free of obstructions. Blankets usually come in rolls 15 or 23 inches wide and 1 to 7 inches thick. They are available with or without a vapor barrier backing. Blankets are a little more difficult to handle than batts because of their size. Like batts they are fire- and moisture-resistant.

Foamed-in-Place. Urea-formaldehyde-based (foamed-in-place) insulation is used to insulate finished frame walls. It is moisture- and fire-resistant and may have higher insulating value than blown-in materials. It is more expensive than blown-in materials. The quality of application of foamed-in-place insulation has been very inconsistent, so be sure to choose a qualified contractor who will guarantee the work.

Rigid Board. Extruded polystyrene bead board (expanded polystyrene), urethane board, and glass-fiber rigid boards are used to insulate basement walls. Polystyrene and urethane rigid-board insulation should only be installed by a contractor. They must be covered with ½-inch gypsum wallboard to assure fire safety.

Extruded polystyrene and urethane are their own vapor barriers; bead board and glass fiber are not. Rigid board has a high insulating value for relatively small thicknesses, particularly urethane. Rigid board comes in 24 or 48-inch widths and in thicknesses varying from ¾ to 4 inches.

Loose Fill (poured-in). Poured-in glass-fiber, rock-wool, cellulosic-fiber, vermiculite, and perlite loose fill are used to insulate unfinished attic floors. The vapor barrier for loose fill is bought and applied separately. Loose fill is best suited for nonstandard or irregular joist spacing or when the space between joists has many obstructions.

Glass fiber and rock wool are fire- and moisture-resistant. Cellulosic-fiber loose fill is chemically treated to be fire- and moisture-resistant. However, the treatment has not yet proven to be heat-resistant, and it may break down in a hot attic. Be sure that the packaging indicates that the material meets federal specifications. Cellulosic fiber has about 30 percent more insulation value than rock wool for the same installed thickness (this can be important in walls or under attic floors).

Vermiculite is significantly more expensive, but it can be poured into smaller areas. Vermiculite and perlite have about the same insulating value. Finally, all loose fills are easy to install.

Loose Fill (blown-in). Blown-in glass-fiber, rock-wool, and cellulosic-fiber loose fill are used to insulate unfinished

and finished attic floors, finished frame walls, and the underside of floors. The vapor barrier is bought separately. They have the same physical properties as poured-in loose fill.

Because it consists of smaller tufts, cellulosic fiber gets into small nooks and corners more consistently than does rock wool or glass fiber when blown into closed spaces such as walls or joist spaces. When any of these materials are blown into a closed space, enough must be blown in to fill the whole space.

How Thick Should Insulation Be? Refer to the following table to determine the insulation thickness needed.

Step 10: Insulate Your Unfinished Attic

Should you insulate an unfinished attic? It depends on how much insulation is already there. To find out, measure the depth of the insulation in your attic. If you can't get into your attic to measure the depth of the insulation, you will need a contractor to do the work. Ask the contractor to tell you how much insulation is already there. Then ask for an estimate of how much it would cost to add the R-value recommended here. If you have 8 inches or more, you may have enough. Check Table 25-4 to be sure. If you have less than 9 inches, you may need more.

How much insulation should you add? Table 25-4 gives recommended amounts of insulation to add. For electrically heated homes or extremely cold climates, these values may not be enough. For more precise advice on how much to add, see Table 25-3. If the table recommends a greater thickness for your home than is given in Table 25-4, you can still use the latter table to estimate insulation costs, but not fuel savings.

Should you do it yourself? You can do it yourself if you can get up into the attic. If you aren't sure whether you want to do it yourself, look and see that you will need.

In Table 25-5, read across the appropriate line to find your cost and savings factor. Copy the two numbers onto line 3 of the Energy Checklist on p. 271.

Table 25-4. R-Value.

THICKNESS (INCHES)

	BATTS OR BLANKETS		LOOSE FILL (POURED-IN)			
R-VALUE	GLASS FIBER	ROCK FIBER	GLASS FIBER	ROCK FIBER	CELLULOSIC FIBER	
R-11	3½–4	3	5	4	3	R-11
R-19	6–6½	5¼	8–9	6–7	5	R-19
R-22	6½	6	10	7–8	6	R-22
R-30	9½–10½	9[a]	13–14	10–11	8	R-30
R-38	12–13[a]	10½[a]	17–18	13–14	10–11	R-38

[a]Two batts or blankets required.

How big is your attic? To find out the size of your attic, determine the area of the first floor of your home, excluding the garage, porch, and other unheated areas. The figure you get will be the same as the area of your attic. Measure its length and width to the nearest foot, and multiply them together.

Insulate an Unfinished Attic. To insulate an attic, you will need the following tools:

Temporary lighting
Temporary flooring
Duct or masking tape (2-in. wide)
Heavy-duty staple gun and staples, or hammer and tacks
Heavy-duty shears or linoleum knife to cut batts or blankets and plastic for vapor barrier

Materials you'll need are:

Batts: fiberglass or rock wool
Blankets: fiberglass or rock wool
Loose fill: rock wool, cellulosic fiber or vermiculite
Vapor barriers

How much insulating material do you need? First, determine your attic area, and then multiply the area by 0.9. The figure you obtain is the size of the area to be insulated. (To find vapor barrier area, see "How big is your attic?" above.)

Batts or blankets with vapor barrier backing may be used to fill insulation area. Polyethylene (for use with loose fill, or if backed batts or blankets are not available)—also may be used to fill insulation area, but plan on waste since the polyethylene will be installed in strips between the joists or trusses, and you may not be able to cut an even number of strips out of a roll.

If the attic requires insulation of R-30 or more, you may have to add two layers of insulation. Lay the first layer between the joists or trusses, and the second layer across them. Only the first layer should have a vapor barrier underneath it. The second layer should be an unfaced batt or blanket, loose fill, or a faced batt or blanket with the vapor barrier slashed freely.

The following safety precautions are recommended:

- Provide good lighting
- Lay boards or plywood sheets over the tops of the joists or trusses to form a walkway (the ceiling below won't support your weight).
- Beware of roofing nails protruding through roof sheathing.
- If you use glass fiber or mineral wool, wear gloves and a breathing mask, and keep the material wrapped until you're ready to put it in place.

Before you start, put in the temporary lighting and flooring, and determine whether ventilation and a vapor

Table 25-5.

			AREA (SQUARE FEET)									
			600		900		1200		1600		2000	
IF YOU HAVE THIS MUCH INSULATION	YOU NEED THIS MUCH MORE INSULATION[a]		COST	SAVING FACTOR	COST	SAVING FACTOR	COST	SAVING FACTOR	COST	SAVING FACTOR	COST	SAVING FACTOR
None	R-38 (10-18 in.)	Do-it-yourself	$282	$246	$423	$369	$564	$492	$752	$656	$ 940	$820
		Contractor	$350	$246	$525	$369	$699	$492	$932	$656	$1166	$820
Under 2 in.	R-22 (6-10 in.)	Do-it-yourself	$168	$ 56	$252	$ 86	$336	$115	$448	$154	$ 560	$206
		Contractor	$198	$ 56	$297	$ 86	$396	$115	$529	$154	$ 661	$206
2 to 4 in.	R-11 (3-5 in.)	Do-it-yourself	$ 78	$ 22	$117	$ 33	$156	$ 44	$208	$ 59	$ 260	$ 74
		Contractor	$ 86	$ 22	$129	$ 33	$172	$ 44	$229	$ 59	$ 286	$ 74
4 to 6 in.	R-11 (3-5 in.)	Do-it-yourself	$ 78	$ 12	$117	$ 18	$156	$ 24	$208	$ 32	$ 260	$ 40
		Contractor	$ 86	$ 12	$129	$ 18	$172	$ 24	$229	$ 32	$ 286	$ 40

[a]Different insulating materials require different thicknesses to achieve the same R-value.

barrier are needed. Seal all places where pipes or wires penetrate the attic floor. (Some manufacturers may recommend using polyethylene in a continuous sheet across the joists or trusses. If you aren't adding insulation that covers the tops of these framing members with at least 3½-inches of insulation, a continuous sheet may cause condensation along them; lay strips instead.)

Keep insulation in wrappers until you are ready to install. It comes wrapped in a compressed state and expands when the wrappers are removed.

Check for roof leaks by looking for water stains or marks. If you find a leak, make repairs before you insulate. Wet insulation is ineffective and can damage the structure of a house.

Install a separate vapor barrier if necessary. Lay polyethylene strips between joists or trusses, and staple or tack

them in place. Seal seams and holes with tape. (Instead of taping, seams may be overlapped 6 inches).

BAFFLE (BATT OR BLANKET INSULATION, OR CARDBOARD

TOP PLATE

If you're using loose fill, install baffles at the inside of the eave vents so that the insulation won't block the flow of air from the vents into the attic. Be sure that insulation extends out far enough to cover the top plate.

To install the insulation, either lay batts or blankets between the joists, or pour in loose fill. If you use batts or blankets with a vapor barrier, place the barrier on the side toward the living area. Batts and blankets are slightly wider than joist spacing, so they will fit snugly.

If you use blankets, cut long runs first to conserve material, and use leftovers for shorter spaces. Slide insulation under wiring wherever possible.

If you use loose fill, pour in the insulation to the depth required. If you are covering the tops of the joists, you can get uniform depth by stretching two or three strings the length of the attic at the desired height, and leveling the insulation to the strings. Use a board or garden rake. Fill all nooks and crannies, but don't cover recessed light fixtures, exhaust fans, or attic ventilation.

The space between the chimney and the wood framing should be filled with noncombustible material, preferably unfaced batts or blankets. Also, the National Electric Code requires that insulation be kept 3 inches away from light fixtures.

Cut ends of batts or blankets to fit snugly around cross bracing. Cut the next batt in a similar way to allow the ends to butt tightly together.

Insulate an Unfinished Floored Attic. *Should you insulate an unfinished floored attic?* It depends on how much insulation is already there. If there is any, it will be in either of two places. The first place to look is between the rafters and in the walls at the ends of the attic. If there's no soft, fluffy material there, it might be down under the floorboards. If so, it won't be easy to see. You'll have to look around the edges of the attic or through any large cracks in the floor. A flashlight may be handy, and also a ruler or stick that you can poke through the cracks with.

Wherever there is insulation, estimate how thick it is. If it's thicker than 4 inches, it's not economical to add more. If it's 4 inches thick or less, you might need more.

There are two basic ways to insulate an unfinished floored attic:

1. Insulate the rafters, end walls, and collar beams. This is the best way if you're doing it yourself, and/or if you think you might ever finish the attic.
2. Blow loose insulation in under the attic floor. This is a job for a contractor—you can't do it yourself. Also, this is the cheapest and most effective way. Don't insulate this way if you think you might ever finish the attic.

Table 25-6 lists three different methods of insulating an unfinished, floored attic, and lists the Cost and Savings Factors for each.

In insulating an attic, the following types of materials are used:

Blown-in insulation
Fiberglass
Rock wool
Cellulosic fiber

Table 25-6. Cost of insulating unfinished floored attic.

AREA (SQUARE FEET)

	600		900		1200		1600		2000	
	COST	SAVINGS FACTOR	COST	SAVINGS FACTOR	COST	SAVINGS FACTOR	COST	SAVINGS FACTOR	COST	SAVINGS FACTOR
Do-It-Yourself: Rafters, End Walls, Collar Beams										
No existing insulation	$190	83	$274	121	$359	165	$487	224	$ 595	284
0–2 in.	$110	27	$181	41	$236	57	$366	74	$ 444	92
2–4 in.	$ 93	11	$164	18	$219	25	$310	34	$ 388	43
Contractor Installation: Rafters, End Walls, Collar Beams										
No existing insulation	$346	83	$500	121	$657	165	$885	244	$1082	284
0–2 in.	$187	27	$314	41	$409	55	$644	72	$ 781	88
2–4 in.	$158	9	$285	15	$380	22	$549	30	$ 686	38
Contractor Installation: Under Attic Floor										
No existing insulation	$296	170	$445	256	$593	341	$790	454	$ 988	568
0–2 in.	$246	49	$369	73	$492	97	$656	130	$ 820	162
2–4 in.	$210	18	$315	27	$420	36	$560	48	$ 700	60

Once you decide to insulate, determine whether you need ventilation in your attic. (See p. 252.) Also check for roof leaks by looking for water stains or marks. If you find any, make repairs before you insulate. Wet insulation is useless and can damage the structure of a house.

When a contractor does the insulating, the insulating material is blown under air pressure through a big flexible hose into the spaces between the attic floor and the ceiling of the rooms below. Bags of insulating materials are fed into a blowing machine, which mixes the insulation with air and forces it through the hose and into place. Before starting the machine, the contractor will locate the cross bracing between the joists in the attic. The contractor will then remove the floorboards above the cross bracing and install the insulation by blowing it in on each side of the cross bracing to make sure no spaces are left unfilled. Since there's no effective way to partially fill a space, all of the spaces should be completely filled to ensure proper coverage. Normally, the job takes no longer than a day.

Be very careful about choosing a contractor (see p. 266). Before signing an agreement with a contractor, decide how much and what kind of insulation you're buying, and make sure it's included in the contract. Properly installed insulation material will achieve a single insulating value (R-value) for the depth of your joist space. You and the contractor should agree on what that insulating value is before the job begins. Next, examine a bag of the insulating material to be used. Look for the table that indicates how many square feet of attic floor the bag will cover while achieving the desired insulating value. Knowing this and the area of the attic, you can figure out how many bags you need to give you the desired R-value. You and the contractor should agree upon this number before the job is begun. While the job is in progress, have the contractor save the empty bags so that you can count them (five bags more or less than the agreed upon amount is an acceptable difference.)

After the job is finished, it's a good idea to drill ¼-in. diameter holes in the floor, about a foot apart. This will help prevent condensation from collecting under the floor in winter.

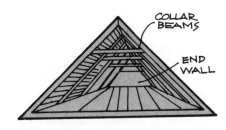

Insulating an attic yourself involves installing batts or blankets between the rafters and collar beams and the studs on the end walls. This will involve installing 2-by-4-inch beams, which span between each roof rafter at ceiling height, if the attic doesn't already have them. This provides a ventilation space above for insulation. The materials, methods, and thickness of insulation are the same whether you do the job yourself or hire a contractor. Buy either glass-fiber or rock-wool batts or blankets.

Take the following safety precautions:

- Provide good lighting.
- Beware of roofing nails protruding through the roof sheathing.
- If you use glass fiber or mineral wool, wear gloves and a breathing mask, and keep the material wrapped until you're ready to use it.

You will need the following tools:

Temporary lighting
Heavy-duty staple gun and staples
Linoleum knife or heavy-duty shears to cut the insulation
Duct or masking tape (2-in. wide)
Hammer and nails (only if you're putting in collar beams)

Do you need insulation with an attached vapor barrier? See p. 252. For the area between the collar beams, if you're laying the new insulation on top of old insulation, buy insulation without a vapor barrier if possible, or slash the vapor barrier on the new insulation.

To determine how thick the area between the collar beams should be, see p. 248. For the rafters and end walls, buy insulation that's thick enough to fill up the rafter and stud spaces. If there's some insulation in there, the combined thickness of the new and old insulation together should fill up the spaces.

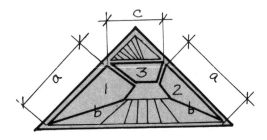

Figure out the size of the area between the rafters and collar beams that you want the insulation to cover. In general, figure each area to be covered, and add up the areas.

$$\underset{\text{distance}}{(\qquad)} \times \underset{\text{distance}}{(\qquad)} \times .9 = \underset{\text{Area 1}}{(\qquad)}$$

$$\underset{\text{distance}}{(\qquad)} \times \underset{\text{distance}}{(\qquad)} \times .9 = \underset{\text{Area 2}}{(\qquad)}$$

$$\underset{\text{distance}}{(\qquad)} \times \underset{\text{distance}}{(\qquad)} \times .9 = \underset{\text{Area 3}}{(\qquad)}$$

$$+\,\underline{\qquad\qquad}$$

TOTAL ( )
Total area of insulation needed for rafters and collar beams.

Calculate the length of 2-by-4-inch stock you'll need for collar beams. Measure the length of span needed between rafters, and count the number of collar beams you need to install. Multiply to get the length of stock needed. You can have the lumberyard cut it to length at a small charge. If you cut it yourself, allow for waste. If you plan to finish an attic, check with the lumberyard to make sure 2-by-4-inch beams are strong enough to support the ceiling you plan to install.

Figure out the area of each end wall you want to insulate. Multiply by (.9) to correct for the space taken up by the studs; then multiply by the number of end walls (area × number of end walls × area required).

Before installation check for roof leaks by looking for water stains or marks. If you find any leaks, make repairs before you insulate.

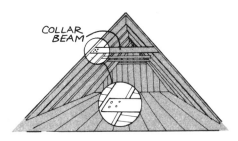

Install 2-by-4-inch collar beams spanning from rafter to rafter at the ceiling height desired. Every pair of rafters should have a collar beam spanning between them.

If you're installing new insulation over existing insulation between the rafters and between the end-wall studs, cut the old insulation loose where it has been stapled, push it in back of the cavities, and slash the old vapor barrier (if any) before you lay the new insulation over it.

If you're installing new insulation over existing insulation between the collar beams, lay the new insulation above the old. Lay it over the tops of the collar beams in an unbroken layer at right angles to the beams. Use insulation that does not have a vapor barrier for this part of the job. If you can't get insulation without a vapor barrier, slash the vapor barrier before laying it down so that moisture won't get trapped in the insulation.

Install batts or blanket sections in place between rafters and collar beams. Install with the vapor barrier on the inside (the side toward you). Don't try to use a continuous length of insulation where the collar beams meet the rafters. It will only result in gaps that are very hard to fill. Install batts in the end walls the same way. Be sure to trim carefully to fit the angles on the end walls.

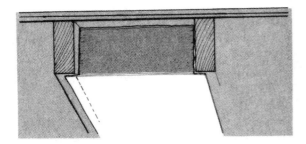

Install batts or blanket sections by stapling the facing flange to the edge of the rafter or collar beam. Don't staple insulation to the outside of the rafters; the vapor barrier will have a break at every rafter, and such stapling could compress the insulation against the sheathing, reducing its insulating value.

Insulate a Finished Attic. A finished attic is a little harder to insulate than an unfinished attic because some areas are harder to reach.

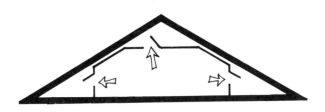

Should you insulate a finished attic? You need to find out if there's enough existing insulation. Depending on the structure of the house, you may be able to measure the insulation by getting into the unfinished spaces in the attic through a door or hatchway. If you can get in, measure the depth of insulation. If there are 9 inches or more—everywhere—you have enough. If you can't get in, have a contractor measure the insulation.

If you get into the attic you may be able to do one or more of these:

- *Insulate attic ceiling.* You can insulate an attic ceiling if there's a door to the space above the finished area. You should consider insulating if there's less than 9 inches already there.
- *Insulate outer attic rafters.* Consider insulating the outer rafters in the attic if (1) there's no insulation between the rafters and (2) there's room for more insulation in the outer attic floor and in the "knee walls" that separate the finished and unfinished parts of the attic.

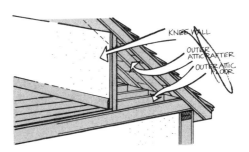

- *Insulate outer attic gables.* The "outer attic gables" are the walls of the exterior of the house; they are usually in the shape of a triangle. You should insulate them if you insulate the outer attic rafters.

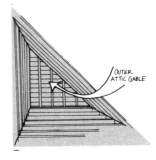

How big are the areas you want to insulate? Multiply the length times the width (in feet) of each area that you insulate:

1. attic ceiling
2. outer attic rafters
3. outer attic gables

Find your savings factor. For each part of the attic that you've measured, indicate in the following how much insulation is already there. For each row checked, multiply the area times the number given.

- *Attic ceiling*

	Area		
☐ None	(	) × .38 = (	)
☐ 0–2 inches	(	) × .09 = (	)
☐ 2–4 inches	(	) × .04 = (	)

- *Outer attic rafters* (existing insulation will be in the floor and knee walls)

	Area		
☐ None	(	) × .23 = (	)
☐ 0–2 inches	(	) × .09 = (	)
☐ 2–4 inches	(	) × .05 = (	)

- *Outer attic gables* (existing insulation will be in the floor and knee walls)

	Area		
☐ None	(	) × .16 = (	)
☐ 0–2 inches	(	) × .06 = (	)
☐ 2–4 inches	(	) × .03 = (	)
		(	) Total

Add the results from each row to get your Savings Factor. Now find your cost:

- *Attic ceiling*

  If there's no existing insulation,

  $\frac{\underline{\quad}}{\text{Area}} \times \$0.37 = \underline{\quad}$

  If there's up to 2 inches of existing insulation,

  $\frac{\underline{\quad}}{\text{Area}} \times \$0.24 = \underline{\quad}$

  If there's 2 to 4 inches of existing insulation,

  $\frac{\underline{\quad}}{\text{Area}} \times \$0.13 = \underline{\quad}$

- *Outer attic rafters*

  If there's up to 2 inches of existing insulation,

  $\frac{\underline{\quad}}{\text{Area}} \times \$0.24 = \underline{\quad}$

  If there's from 2 to 4 inches of existing insulation,

  $\frac{\underline{\quad}}{\text{Area}} \times \$0.13 = \underline{\quad}$

- *Outer attic gables*

  $\frac{\underline{\quad}}{\text{Area}} \times \$0.13 = \underline{\quad}$

Go to the Energy Checklist. On line 3, enter the Total Cost and the Savings Factor.

The cost of contractor-insulated finished attics varies among contractors. Therefore, you really can't figure out the cost by yourself; have a contractor give you an estimate.

Determine how much insulation you have already. Measure the depth of existing insulation. Usually, there's the same thickness in all parts of the attic. If there are different thicknesses, figure the average depth. Measure the length and width of the finished part of the attic. Round them off to the nearest foot, and multiply them together to find the area.

Depth	Area of Attic (square feet)				
	300	550	800	1100	1400
None	$321	$469	$587	$641	$745
	208	316	429	572	721
Under 2 in.	$233	$343	$418	$469	$555
	68	97	126	167	205
2–4 in.	$130	$214	$285	$322	$410
	31	44	56	74	90

After determining the area of your attic and the depth of the insulation, find which set of numbers on the table apply to you. Copy the top number into the Total Cost, line 3 of the Energy Checklist. Copy the bottom number into the Savings Factor on the same line.

If there's under 4 inches of insulation in your finished attic, consider one of the following insulation methods:

1. Contractor installation: insulation blown into the ceiling, sloping rafters, and outer attic floors; batts installed in the knee walls.
2. Do-it-yourself: insulation of batts, blankets, or loose fill in all accessible attic spaces.

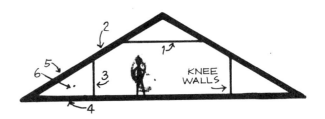

Insulation should be installed on:

1. attic ceiling
2. rafters
3. knee walls
4. outer attic floors or
5. outer attic rafters
6. end walls

How thick should the insulation be? Check the R-value. Check whether you need ventilation and a vapor barrier. Check for roof leaks by looking for water stains or marks. If you can find any leaks, make repairs before you insulate.

A contractor will blow insulation into the open joist spaces above the attic ceiling, between the rafters, and into the floor of the outer attic space, and then install batts in the knee walls. If you want to keep the outer attic spaces heated for storage or any other purpose, have the contractor install batts between the outer attic rafters instead of insulating the outer floors and knee walls.

You can insulate an attic yourself wherever you can get into the unfinished spaces. Installing insulation in an attic ceiling is the same as installing it in an unfinished attic. If you want to insulate the outer attic spaces yourself, install batts between the rafters and studs in the small triangular end walls.

Do You Need a Vapor Barrier or More Ventilation in Your Attic?

In Zone I, install a vapor barrier (unless you are blowing insulation into a finished attic). Add a ventilation area equal to 1/300 of the attic floor area if signs of condensation occur after one heating season. You can't install a vapor barrier with insulation.

In Zone II, without air conditioning, install a vapor barrier toward the living space if you are insulating a finished attic. (With other attics, a vapor barrier is optional.) Add a ventilation area equal to 1/300 of the attic floor area if signs of condensation occur after one heating season.

In Zone II, with air conditioning, install a vapor barrier toward the living space if you are insulating a finished attic (with other attics, a vapor barrier is optional). In addition, add a ventilation area equal to 1/150 of the attic floor area.

If you are installing batt or blanket insulation, and you need a vapor barrier, buy the batts or blankets with the vapor barrier attached. Install them with the vapor barrier side toward the living space. If you are installing a loose-fill insulation, lay down polyethylene (heavy, clear plastic) in strips between the joists first.

Install ventilation louvers (round or rectangular) in the eaves and gables. (Ridge vents are also available, but are more difficult and costly to install.) The total open area of these louvers should be either 1/300 or 1/150 of your attic area. Unless you are an experienced handyman, hire a carpenter to install ventilation louvers. Remember, *don't block the ventilation path with insulation.*

Insulate Walls. Should you insulate your walls? It depends on two things: the size of your energy bills and what the walls are like.

How big are your energy bills? If you have just heating—*no* whole-house air conditioning—look up the Heating Factor for your city on p. 260. The Heating Factor is a number that reflects the climate in your area and how much you pay for fuel. If your Heating Factor is 0.37 or more, you need to insulate.

If you have heating *and* whole-house air conditioning, look up both your Heating Factor and your Cooling Factor (see p. 260). Add them together. If the sum is 0.37 or more, you need to insulate.

Most houses have frame walls. They have wooden structures—usually two-by-fours—even though they may have brick or stone on the outside. Some houses have brick or block masonry walls that form the structure of the house, without a wooden backup.

If your house has frame walls, you should consider insulating them if they have no insulation at all. A contractor can fill them with insulation and cut energy waste through them by two-thirds.

You may already know whether your walls are insulated. If you don't, take the cover off a light switch on an outside wall. (Of course, turn the power off first.) Shine a flashlight into the space between the switchbox and the wall material, and see if there is any insulation. If there's no insulation there now, you may need more. If there's some there already, you don't need more. If you have uninsulated masonry walls, it may be worthwhile to insulate them. To do so is more complicated than insulating frame walls, so call a contractor to find out what's involved.

What kind of insulation do you need? Some types of wall insulation cost more than others, and some kinds work better than others. Generally, you get what you pay for. The least expensive is *mineral fiber insulation.* There are two kinds: rock wool and glass fiber. Either kind can be blown into the wall by means of a special machine. A slightly more expensive but more effective insulation is *cellulosic fiber.* Like mineral fiber, this is loose insulation that's blown in. The most expensive and perhaps the most effective insulation is *urea-formaldehyde-based foam* (not urethane foam, which is not good in walls). Quality-control problems with ureaformaldehyde-based foam require that you choose a qualified installer.

How big is your house? Measure the perimeter—the total distance around the outside—of each story of your house that has frame walls. Measure around the heated parts only. Measure in feet to the nearest 10 feet. If you have a finished heated attic, measure the width of the end walls of the attic only. Add up the width of these walls. Then carry out the following computation:

First-story perimeter	_____	feet
Second-story perimeter	_____	feet
Third-story perimeter	_____	feet
Finished-attic end walls	_____	feet
Total	_____	feet

The following table shows the Cost and Savings Factors for different kinds of insulation applied to different sizes of houses. To use the table, find the number of linear feet closest to your total. In that column, look at the estimated cost for installing each type of material. See which you can afford; remember, if the cost is higher, your savings will also be higher. When you have decided which material to use, copy the Cost and the Savings Factor on line 5 of the Energy Checklist.

Insulation	Linear Feet of Walls					
	100	150	200	250	300	400
Mineral fiber						
Cost	$397	$596	$794	$993	$1191	$1588
Savings factor	100	155	205	255	310	410
Cellulosic fiber						
Cost	$447	$671	$894	$1118	$1341	$1788
Savings factor	100	170	225	280	335	450
Urea-formaldehyde-based foam						
Cost	$596	$894	$1192	$1490	$1788	$2384
Savings factor	115	175	230	290	350	460

Insulate Wood-Frame Walls. Normally, insulating material is blown or pumped into the spaces in a wood-frame wall through holes drilled from the outside or from the inside. Note that condensation in insulated walls may be a problem. None of the insulating materials that is blown into frame walls serves as a barrier to moisture vapor, and so condensation in insulated walls may be a problem.

If you live in Zone I, and plan to insulate your walls, you have to be sure that very little moisture from the air in your house gets into the walls. If it does, it is likely to condense there in the winter, and you will run two risks: (1) that the insulation will become wet and won't insulate, and (2) that enough moisture will collect to cause rotting. Here's how to avoid these dangers:

- Seal any opening in the inside walls that could admit moisture, especially around the window and door frames.
- Paint interior walls with a low-permeability paint—that is a high-gloss enamel or other finish. Ask a paint dealer.

In Zone II, the climate is such that little condensation will form in insulated walls.

Contractors use the following types of blown-in insulation to insulate wood-frame walls: glass fiber, rock wool, and cellulosic fiber. Foam-in insulation used includes plastic foam installed as a foam under slight pressure, which hardens to form insulation.

The contractor will measure the area to be insulated to determine how much material is needed and to estimate the cost. The contractor must be able to get to all of the spaces in the wall. For each space, the contractor must drill a hole, usually in the outside wall, after removing the finished layer (usually clapboard or shingle). This amounts to a lot of holes, but a good contractor will leave no traces behind. If there is brick veneer on the exterior of the house, the procedure is much the same, except that it may be cheaper to do it from the inside.

Once the holes in the wall have been drilled, the insulation can be installed. If the insulation is blown-in, it will be done as outlined on page 249. If foam is used, it will be pumped into the wall spaces through a flexible hose with an applicator. With either method, each space should be completely filled, and the siding replaced.

Be very careful when selecting a contractor. (See p. 266.) Before signing an agreement with a contractor, make sure what you're buying is spelled out in the contract. Insulation material properly installed will add an R-value of 8 for rock wool, 10 for cellulosic fiber, or 11.5 for urea-formaldehyde in a standard wood-frame wall. You and the contractor should agree on what that R-value is before the job begins. Next, check a bag of the type of insulation to be used (unless, of course, foam will be used). It should indicate how many square feet of wall space that bag is meant to fill while giving the house the desired R-value. The information may be in different forms (number of square feet per bag, or number of bags per 1000 square feet), so you may have to do some simple division to use the number correctly. Knowing this and the area of the walls, you should be able to figure out how many bags should be installed to give the desired R-value.

This number should be agreed upon by you and the contractor before the job begins. While the job is in progress, be sure the correct amount is being installed. Have the contractor save the empty bags so you can count them—four or five bags more or less than the agreed-upon amount is an acceptable difference from the estimate.

Insulating Crawl-Space Walls and Floor, or Basement Walls. In a climate where your heating bill is big enough to be a worry, it's a good idea to insulate the underside of your house. It won't save much on air conditioning, but it will save on heating.

- *A flat concrete slab sitting on the ground.* There's not much that you can easily do to insulate this type of foundation, and, since it's hard to tell how much insulation is already there, it's hard to tell what your savings would be. Therefore, no cost and saving figures are given here for slab insulation.

- *A crawl space with walls around it.* If you have a crawl space that you can seal tightly in winter, you can insulate its walls and the ground around its outer edges.

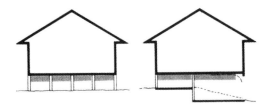

- *A floor over a garage, porch, or open crawl space.* If there's an open space under the floor that you can't seal off tightly from the outside air, the place to insulate is in the floor, between the joists.

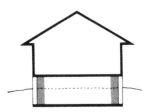

- *Walls of a heated basement that stick out of the ground.* If you have a basement that is heated and used as a living area, you might want to insulate the basement walls down to a depth of 2 feet below the ground.

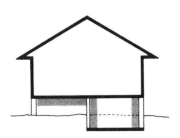

- *A combination of the preceding types.* Your house may be part heated basement and part crawl space, or some other combination. To estimate costs and savings, treat each of the parts separately. There are three individual lines on the Energy Checklist:

Insulate Crawl Space Walls
Insulate Floor
Insulate Basement Walls

Fill out as many of these as apply to you, and see which are most important for you to do.

Insulating Crawl-Space Walls

- If your house (or part of it) sits on top of a crawl space that can be tightly sealed off from the outside air in the winter, the cheapest and best place to insulate it is around the outside walls and on the adjacent ground inside the space.

- If there is no insulation at all around the crawl space walls or under the floor and if you can get into the crawl space to do the work, you should insulate.

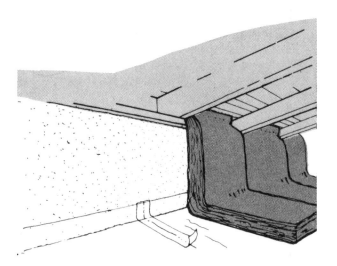

- How much will it cost? It makes a difference whether you do the work yourself or hire a contractor. Doing it yourself is hard work, but you'll save a lot of money. If you're not sure which route you want to take, read the following to see what doing-it-yourself involves.

  To estimate the cost if you do-it-yourself, first measure the distance around the outside of the heated part of your crawl space (don't include areas underneath porches, or other unheated areas). Then multiply the total distance around the crawl space by $0.80 (the cost per running foot). The result is the do-it-yourself cost. *To estimate the cost if a contractor does the work,* multiply the total distance around the crawl space by $1.10 per running foot. The result is the cost if you hire a contractor.

  How much will you save? To find the Savings Factor, multiply the distance around the crawl space by $0.54. Turn to the Energy Checklist. Go to line 4a, "Insulate Crawl Space Walls." Enter the Cost on that line and the Savings Factor next to it.

  To insulate a crawl space, you have two options available:

1. *Do-it-yourself:* Install batt or blanket insulation around the walls and perimeter of the crawl space. Lay a plastic vapor barrier down on the crawl space earth.
2. *Contractor-installed:* If the crawl space presents access or working-space problems, you should consider having a contractor to do the work for you. The contractor will probably follow a method similar to the do-it-yourself method. But if the contractor suggests something different, have him price the methods and show you which is better.

Note: The insulation method that follows should not be used in Alaska, Minnesota, and northern Maine. The extreme frost penetration in these areas can cause heaving of the foundation. Residents of these areas should contact local U.S. Department of Housing and Urban Development (Federal Housing Authority) field offices for advice.

You will need the following tools:
Staple gun
Heavy-duty shears or linoleum knife
Temporary lighting
Portable fan or blower to provide ventilation
Tape measure
Duct or masking tape (2-in.-wide)

Observe these safety measures:

- Provide adequate temporary lighting.
- Wear gloves and a breathing mask when working with glass fiber or rock wool.
- Provide adequate ventilation.
- Keep lights, fan, and all wires well off wet ground.

Materials you'll need are:

- Six-mil polyethylene plastic to lay on earth for vapor barrier.
- R-11 (3-to-3½-in. thick) blankets of rock wool or glass fiber, without a vapor barrier.

How much will you need? Determine the area to be insulated: measure the length and average height of the wall to be insulated; add 3 feet to the height (for perimeter insulation); and multiply the two to find total insulation area—that is Length × (Height + 3 feet) = Area. Determine the area to be covered by the vapor barrier by finding the area of the crawl space. You may have to divide the crawl space into several rectangles—measure them and add up the areas.

Installation. Where the joists run at right angles to the wall, press short pieces of insulation against the header—they should fit snugly. Then install the wall and perimeter insulation by stapling the top of each strip to the sill. Make sure the batts fit snugly against each other and that you cut them long enough to extend 2 feet along the floor.

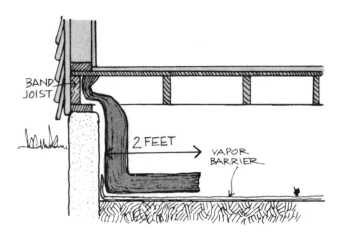

Where the joists run parallel to the wall, you don't need the short pieces of insulation, just install the wall and perimeter insulation by stapling the top of each strip to the band joist.

When all batts have been installed, lay down the polyethylene vapor barrier, and tuck it under the batts all the way to the foundation wall. Turn it up at least 6 inches at the wall. Tape the joints of the vapor barrier, or lap them at least 6 inches. Plan your work to minimize stepping or crawling on the vapor barrier.

Ventilating the Crawl Space. Even with a plastic vapor barrier on the floor, the air in the crawl space will be too damp if fresh air doesn't get in there from time to time. The insulation will be wet, and it won't keep the house as warm. It will also mean that wooden members that hold up a house will be wet, and they'll rot. Proper ventilation will prevent both of these problems.

If the crawl space is part of your forced-air heating system (in other words, if air from your furnace moves through it), seal the crawl space as tightly as possible—the air moving through it from the furnace is enough ventilation in winter. If the crawl space has vents, keep them shut in winter and open in summer. If there are no vents, run the blower on the furnace three or four times during the summer to keep the air in the crawl space from getting too damp.

All other crawl spaces should have vents in them that can be opened in summer (to clear out the damp air), and closed tightly in winter to make the most of insulation. You can make a cover for these vents and install it in winter. Note: If a furnace gets its combustion air from the crawl space, some of the vents should be left open.

Insulate the Floor

It's a good idea to insulate a floor if you have a crawl space that can't be sealed off in winter—for example, your house stands on piers—or if you have a garage, porch, or other cold unheated space with heated rooms above it.

You should insulate it if your floor is uninsulated, and, if there is a crawl space, it is high enough for a person to work in.

How big is your floor? Measure the area of the floor that you plan to insulate.

On the following table, find the number of square feet that's closest to the floor area to be insulated. Go to the Energy Checklist, and enter the Cost and Savings Factor on line 4b.

Area (Ft²)	Do-it-yourself		Contractor	
	cost	savings factor	cost	savings factor
200	$24	58	$56	58
400	$48	116	$112	116
600	$76	173	$167	173
900	$106	260	$250	260
1200	$152	347	$334	347
1600	$189	462	$445	462

Install batts or blankets between floor joists by stapling wire mesh or chicken wire to the bottom of the joists and sliding the batts or blankets in on top of the wire. Place vapor barrier up. The job is quite easy to do in most cases. If you are insulating over a crawl space, you may have some problems with access or working room, but careful planning can make things go much more smoothly and easily.

Check your floor-joist spacing—the method herein will work best with standard 16- or 24-inch joist spacing. If the spacing is nonstandard or irregular, it will entail more cutting and fitting and some waste of material.

To do it yourself, you will need certain tools:

Heavy-duty shears or linoleum knife

Temporary lighting with waterproof wiring and connectors
Portable fan or blower to provide ventilation
Tape measure
Heavy-duty staple gun and staples

Observe the following safety precautions:

- Provide adequate temporary lighting.
- Wear gloves and a breathing mask when working with glass fiber or rock wool.
- Provide adequate ventilation.
- Keep lights and all wires off wet ground.

Materials you'll need are:

- R-11 (3-to-3½-inch) batts or blankets of rock wool or glass fiber, preferably with foil facing.
- Wire mesh or chicken wire of convenient width for handling in a tight space.

Determine the area to be insulated by measuring the length and width and multiplying to get the area. You may find it necessary to first divide the floor into smaller areas and then add them. To find the area of insulation, multiply the total area by 0.9.

Installation. Start at a wall at one end of the joists, and work out. Staple the wire to the bottom of the joists and at right angles to them. Slide batts in on top of the wire. Work with short sections of wire and batts so that it won't be too difficult to get the insulation in place. Plan sections to begin and end at obstructions such as cross bracing.

Buy insulation with a vapor barrier, and install the vapor barrier facing up (next to the warm side), leaving an airspace between the vapor barrier and the floor. Get foil-faced insulation if possible, it will make the airspace insulate better. Be sure that ends of batts fit snugly up against the bottom

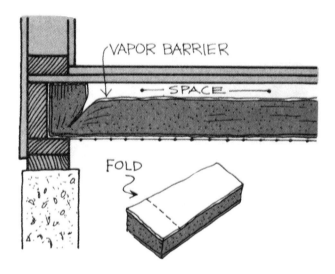

of the floor to prevent loss of heat up the end. Don't block combustion air openings for furnaces.

Insulate the Basement Walls

If you have a basement that you use as a living or work space and that is heated by air outlets, radiators, or baseboard units, you may find that it will pay to add a layer of insulation to the inside of the wall. The cost figures given here do not allow for the cost of refinishing as well as insulating.

Should you insulate? If the basement walls aren't insulated and if the basement's average height above ground is 2 feet or more, then it pays to insulate in almost any climate if you do the work yourself. If the average height above ground is less than 2 feet, then it pays to insulate these walls yourself if the Heating Factor is more than 0.7.

If you want to hire a contractor to insulate, the Heating Factor should be 0.5 or more if the basement's average height above ground is 2 feet or more. If the height is less than 2 feet, you should not have the work done.

Measure the length of each wall that sticks 2 or more feet above the ground, and add the lengths together. Referring to the following table, multiply the dollar figure that corresponds to the height of the walls above ground by the total length of the walls. The result is the estimated Total Cost.

	Average Above Ground Height (Feet)				
	0	2	4	6	8
Do-it-yourself					
Cost	$1.84	$1.84	$1.84	$1.84	$1.84
Savings Factor	0.2	1.1	2.0	2.8	3.4
Contractor					
Cost	$4.84	$4.84	$4.84	$4.84	$4.84
Savings Factor	0.2	1.1	2.0	2.8	3.4

Multiply the bottom number by the total length of the walls to get the Savings Factor. Go to the Energy Checklist, line 4c, and enter the Total Cost and the Savings Factor on the same line.

Install 2-by-4-inch studs (needed for thickness of insulation) along the walls to be insulated. Pass glass-fiber insulation between the studs. If you wish, finish with wallboard or paneling. The thickness of the finished wall material will determine the spacing of the studs.

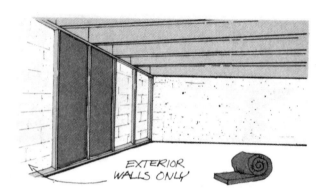

Note: The methods of insulation herein should not be used by residents of Alaska, Minnesota, and northern Maine, where the extreme frost penetration can cause heaving of the foundation. Residents of these areas should contact local offics of the U.S. Department of Housing and Urban Development (Federal Housing Authority).

You'll need the following tools:

Saw
Hammer, nails
Heavy-duty staple gun, or hammer and tacks
Tape measure
Linoleum knife or heavy-duty shears
Level
Small sledge hammer, masonry nails

Observe these safety measures:

• Provide adequate temporary lighting.
• If you use glass fiber or rock wool, wear gloves and a breathing mask, and keep the material wrapped until you are ready to use it.

Materials you'll need are:

• R-11 (3½-inch) batt or blanket insulation, glass fiber or rock wool, with a vapor barrier. (Buy polyethylene if you can't get batts or blankets with a vapor barrier.)
• 2-by-4-inch studs
• Dry-wall or paneling, if desired.
• Waterproof paint, if necessary.

To determine how much insulation to buy, measure the height and length (in feet) of the walls to be insulated. Multiply these two figures to determine how many square feet of insulation is needed.

To find the linear feet of studs you'll need, multiply the length of the walls you intend to insulate by 6.

The amount of paneling or dry-wall needed equals the basement wall height times the length of wall you intend to finish.

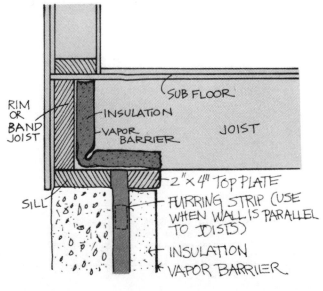

Installation. Check to see whether moisture is coming through the basement walls from the ground outside. If it is and the walls are damp, you should eliminate the cause of the dampness to prevent the insulation you're going to install from becoming wet and ineffective. To be sure, install the new studs and insulation slightly away from the wall.

Nail the bottom plate to the floor ¾ inch out from the base of the wall with a hammer and masonry nails. Install studs 16 to 24 inches apart, after the top plate is nailed to the joists above. (Whether the wall runs parallel to the joists, you may not be able to fasten the top plate in this way, but may have to fasten a ¾ inch-thick horizontal furring strip to the wall near the top and then fasten the studs to it.) Block between studs at ceiling after studs are in place if you need backing for finish wall material.

Cut blankets into sections the height of the wall. Staple them into place, with the vapor barrier toward the living space. Install another small piece of insulation above the new studs and against the sill to insulate the sill and band joist. For a more finished look, install finish wallboard or paneling over the insulation and studs.

Thermostat, Furnace, and Air Conditioner

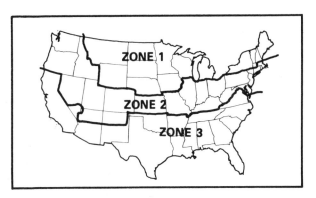

The following table tells you what percent of your heating bill you'll save by turning down your thermostat. (Look at the map above to see what zone you live in.)

Turn down thermostat	Zone 1	Zone 2	Zone 3
5°F	14%	17%	25%
8°F	19%	24%	35%

If you have whole-house air conditioning, you can save about 3 percent of your air-conditioning bill for each degree you turn up the thermostat. Usually, about a 4°F turn-up will still be comfortable; above that, the air-conditioning system will have trouble keeping the house cool during the hot part of the day. Figure out how many degrees you can turn up your thermostat; then multiply that number by 3 to get your percent savings.

The method for figuring out a heating bill depends on what kind of fuel you use. Pick the method that applies to

Table 25-7

Location	Month	Gas Electric Heat Alone **A**	Electric Heat With Electric A/C **B**	Electric A/C Alone **C**	Electric A/C With Electric Heat **D**	Location	Month	Gas Or Electric Heat Alone **A**	Electric Heat With Electric A/C **B**	Electric A/C Alone **C**	Electric A/C With Electric Heat **D**
Alabama						**Nevada**					
Montgomery	May	4.2	5.2	7.3	7.3	Elko	Sept.	6.8	6.8	2.2	*
Alaska						Las Vegas	April	4.7	4.9	4.1	5.7
Anchorage	July	7.8	7.8	*	*	**New Hampshire**					
Arizona						Concord	July	5.5	5.9	*	*
Flagstaff	July	6.4	6.6	*	*	**New Jersey**					
Phoenix	April	4.4	5.1	5.1	6.3	Atlantic City	Sept.	5.4	5.7	5.1	8.2
Arkansas						**New Mexico**					
Little Rock	May	4.3	4.7	5.6	5.9	Raton	Sept.	6.3	6.5	3.7	*
California						Silver City	Sept.	4.7	5.3	5.7	5.9
Bishop	Sept.	5.2	7.5	3.5	5.1	**New York**					
Eureka	July	17.0	17.0	*	*	New York City	Sept.	5.1	5.6	5.9	8.0
Los Angeles	Oct.	6.0	7.1	10.5	***	Rochester	Sept.	6.1	6.4	5.0	*
Bakersfield	April	4.8	5.3	4.9	8.0	**North Carolina**					
San Francisco	Sept.	6.7	7.0	*	*	Raleigh	May	4.9	5.2	5.3	6.9
Colorado						Wilmington	May	4.3	4.8	7.0	5.9
Alamosa	July	6.0	6.0	*	*	**North Dakota**					
Denver	Sept.	6.2	6.3	3.5	*	Bismarck	July	5.3	5.5	*	*
Connecticut						**Ohio**					
New Haven	Sept.	5.8	6.1	4.7	**	Youngstown	Sept.	6.1	4.8	5.2	*
Delaware						Cincinnati	May	5.4	5.4	4.0	10.0
Dover	May	5.7	5.8	3.8	9.6	**Oklahoma**					
District of Columbia						Oklahoma City	May	4.5	4.8	4.9	6.1
Washington	May	5.3	5.5	4.3	6.7	**Oregon**					
Florida						Salem	July	6.1	6.5	*	*
Miami†	Feb.	†	†	9.6	9.6	Medford	May	7.4	6.3	3.3	11.3
Tallahassee	April	4.4	4.9	5.6	6.5	**Pennsylvania**					
Georgia						Philadelphia	May	5.7	5.8	3.8	10.5
Atlanta	May	4.8	5.2	4.7	5.4	Pittsburgh	Sept.	5.9	6.2	5.4	*
Savannah	April	4.6	5.1	5.3	6.5	**Rhode Island**					
Idaho						Providence	Sept.	5.9	6.1	4.7	*
Boise	Sept.	5.9	6.0	3.1	**	**South Carolina**					
Illinois						Charleston	April	4.7	4.9	4.9	6.2
Chicago	Sept.	5.5	5.8	5.2	**	Greenville-Spartan-					
Springfield	May	5.4	5.5	3.6	**	burg	May	4.6	5.0	4.8	5.4
Cairo	May	4.7	4.9	4.2	5.3	**South Dakota**					
Indiana						Rapid City	Sept.	6.3	6.3	3.2	*
Indianapolis	Sept.	5.6	6.1	6.5	**	**Tennessee**					
Iowa						Knoxville	May	5.1	5.3	4.7	6.5
Des Moines	Sept.	5.1	5.4	11.8	**	Memphis	May	4.3	4.8	5.2	5.6
Dubuque	Sept.	5.8	6.0	4.1	*	**Texas**					
Kansas						Austin	April	4.1	4.5	5.5	7.0
Wichita	May	4.9	5.0	3.5	5.8	Dallas	April	4.6	5.0	5.0	7.7
Goodland	Sept.	5.7	5.8	3.4	9.8	Houston	April	4.0	4.8	5.9	7.0
Kentucky						Lubbock	May	4.7	5.2	6.8	9.3
Lexington	May	5.6	5.6	4.0	11.4	**Utah**					
Louisiana						Salt Lake City	Sept.	5.6	5.7	3.2	6.4
Baton Rouge	April	4.1	4.8	5.9	6.5	Milford	Sept.	5.5	5.7	2.7	*
Shreveport	April	4.6	5.0	4.9	7.7	**Vermont**					
Maine						Burlington	July	5.6	5.9	*	*
Portland	July	5.7	5.7	*	*	**Virginia**					
Maryland						Richmond	May	5.1	5.4	5.2	8.0
Baltimore	May	5.5	5.6	3.9	9.0	**Washington**					
Massachusetts						Olympia	July	6.8	7.0	*	*
Worcester	Sept.	6.2	6.4	4.4	*	Walla Walla	Sept.	5.3	5.6	3.3	8.0
Michigan						**West Virginia**					
Lansing	July	5.5	5.5	*	*	Charleston	May	5.7	5.7	4.2	9.6
Minnesota						Elkins	Sept.	6.5	5.2	4.0	*
Duluth	July	6.0	6.1	*	*	**Wisconsin**					
Minneapolis	May	6.2	6.3	2.8	*	Milwaukee	July	5.7	6.2	*	*
Mississippi						**Wyoming**					
Jackson	April	4.9	5.3	5.0	7.7	Casper	Sept.	6.7	6.8	2.8	*
Missouri											
St. Louis	May	4.8	4.8	3.4	5.9						
Springfield	May	5.6	5.7	3.7	9.1						
Montana											
Helena	July	5.8	6.0	*	*						
Nebraska											
Omaha	Sept.	5.3	5.5	4.3	*						
Scottsbluff	Sept.	6.1	6.2	3.1	*						

*Air conditioning savings not significant.

**Your air conditioning bill is about 1/10 of your electric heating bill.

***Your air conditioning bill is about 1/4 of your electric heating bill.

†Heating savings not significant.

you. (*Note:* You may heat with two fuels. For example, most of your house may be heated with oil or gas, while some newer rooms may have electric heat. In this case, *do this section once for each fuel,* and add the results together.)

Oil or Coal Heat

If you have an oil or coal furnace that heats your house but not your hot water, then all of your oil or coal bill goes to heating. Simply add up your fuel bills for last year. If your furnace also heats your hot water, add up your fuel bills for last year, and multiply the total by 0.8 to find heating cost.

Gas or Electric Heat

	$	_____	(1)
Subtract −	$	_____	(2)
	$	_____	(3)
Multiply ×	$	_____	(4)
Your Heat Bill	$	_____	(5)

If you have gas heat OR *if you have electric heat without whole-house electric air conditioning:* Write the amount of your January electric (or gas) bill on line 1. Find the city nearest you from Table 25-7. There's a month listed beside the name of that city. Enter the amount of your electric or gas bill for that month on line 2. Subtract line 2 from line 1, and write the difference on line 3. Write the number from column A Table 25-7 for the city nearest you on line 4. Multiply line 3 by line 4; write the result on line 5. That number is your estimated heating bill.
If you have electric heat AND whole-house air conditioning: Follow the preceding calculations, except for line 4, use the number from column B of Table 25-7 for your city.

Your Air-Conditioning Bill

If you have whole-house air conditioning, estimate how much it costs you each year. Use this method: Look up the city nearest you on Table 25-7. If there's an asterisk after the name of the city, your air-conditioning savings will be insignificant. If there's not an asterisk, carry out the following instructions.

	$	_____	(1)
Subtract −	$	_____	(2)
	$	_____	(3)
Multiply ×	$	_____	(4)
Your Air-Conditioning Bill	$	_____	(5)

Write the amount of your July electric bill on line 1. Find the city nearest you from Table 25-7. There's a month written beside the name of that city. Write the amount of your electric bill for that month on line 2. Subtract line 2 from line 1, and enter the difference on line 3. If you have electric heat as well as air-conditioning, write the number from column D of Table 25-7 for the city nearest you on line 4. If you have gas, oil, or coal heat, write the number from column C. Multiply line 3 by line 4; enter the result on line 5. That number is your estimated air-conditioning bill.

Your Dollar Savings

Now that you've found the amount of your heating and air-conditioning bills, you're ready to find out how much you can save on them annually.

Multiply your heating bill by either 5 or 8 percent (depending on the number of degrees you turn down your thermostat, and divide by 100. The result is your dollar savings.

If you have whole-house air conditioning, multiply your air-conditioning bill by the same percentage figure, and divide by 100. The result is your dollar savings.

Add up your thermostat savings for heating and air conditioning to find your total savings, and enter the figure on line 6 of the Energy Checklist.

If you have an oil or coal furnace that has been serviced recently, multiply your heating bill by 0.1 to find your dollar savings. Write the result on line 7 of the Energy Checklist.

If you have a central air conditioner that has been serviced recently, multiply your air-conditioning bill by 0.1 to find your dollar savings. Write the result on line 8 of the Energy Checklist.

Heating and Cooling Factors

Combine all the Savings Factors for the energy-saving home improvements you're considering with your Heating Factor and your Cooling Factor (if you have whole-house air conditioning). You'll get dollar savings. There's one Heating Factor and one Cooling Factor for a house, which are based on where it is located and how much it costs for the fuel for heating (and cooling).

Find your Heating Factor on Table 25-8. If you have whole-house air conditioning, also find your Cooling Factor.

Note: Check the fuel prices given in Table 25-8. They were collected in mid-1977 and were used to figure the Heating and Cooling Factors. Compare them with the price you now pay for fuel (see p. 223). If you find a significant difference, figure your Heating and Cooling Factors as described below.

Using Your Own Bill. You can calculate your Heating Factor (and your Cooling Factor, if you have whole-house air conditioning), using the figures from your own utility bills. To figure the exact Heating Factor, find the Heating Multiplier on Table 25-8 for your city and your fuel, and multiply it by the price you pay for heating fuel. Enter the Heating Factor on the Energy Checklist.

To figure out the exact Cooling Factor, find the Cooling Multiplier for your city, and multiply it by the price you pay for electricity in cents per kilowatt hours. Enter the Cooling Factor on the Energy Checklist.

Table 25-8

		Heating Factors				Cooling Factor	Fuel costs				Heating Multipliers				Cooling Multiplier
		Gas	Oil	Elec	Coal		Gas ¢/100 cu. ft.	Oil ¢/gal.	Elec ¢/Kwh	Coal ¢/lb.	Gas	Oil	Elec	Coal	
		A	B	C	D	E	F	G	H	I	J	K	L	M	N
ALABAMA	Montgomery	.13	.28	.34	–	.13	12.00	38.88	1.50	–	.0105	.0071	.2260	.0987	.0859
ALASKA	Anchorage	.33	.86	1.21	–	.00	12.00	53.82	2.03	–	.0275	.0160	.5956	–	.0001
ARIZONA	Flagstaff	.48	.61	.72	–	.01	20.71	37.78	1.43	–	.0233	.0162	.5040	–	.0056
	Phoenix	.07	.23	.23	–	.20	8.20	37.78	1.43	–	.0081	.0062	.1741	–	.1430
ARKANSAS	Little Rock	.12	.40	.45	–	.11	8.40	39.22	1.40	–	.0147	.0102	.3176	–	.0769
CALIFORNIA	Bishop	.29	–	.59	–	.67	17.69	–	1.69	–	.0163	–	.3515	–	.0414
	Eureka	.35	–	.84	–	.00	16.51	–	1.84	–	.0212	–	.4580	–	.0001
	Los Angeles	.09	–	.28	–	.06	12.17	–	1.69	–	.0078	–	.1682	–	.0367
	Bakersfield	.12	–	.39	–	.17	12.17	–	1.84	–	.0097	–	.2093	–	.0941
	San Francisco	.22	–	.55	–	.01	15.70	–	1.84	–	.0137	–	.2967	–	.0059
COLORADO	Alamosa	.28	.58	1.18	–	.01	10.15	38.16	1.96	–	.0278	.0153	.6010	–	.0023
	Denver	.19	.56	.71	–	.05	11.03	39.90	1.96	–	.0168	.0140	.3640	–	.0231
CONNECTICUT	New Haven	.58	.67	1.77	–	.13	30.09	46.87	4.26	–	.0192	.0143	.4155	–	.0312
DELAWARE	Dover	.28	.64	.65	–	.07	15.30	45.55	1.60	–	.0188	.0140	.4054	.1770	.0447
D.C.	Washington	.41	.55	1.12	.69	.18	25.26	45.92	3.22	4.55	.0161	.0120	.3473	.1517	.0559
FLORIDA	Miami	.10	.04	.07	–	.49	10.00	52.96	3.07	–	.0010	.0007	.0211	–	.1589
	Tallahasee	.09	.25	.37	–	.23	13.90	48.78	2.49	–	.0068	.0051	.1465	–	.0941
GEORGIA	Atlanta	.14	.40	–	.49	–	12.50	43.50	–	4.63	.0113	.0092	.2435	.1063	.0612
	Savannah	.10	.23	–	–	–	12.50	37.81	–	–	.0083	.0062	.1794	–	.0892
IDAHO	Boise	.39	.52	.36	–	.03	23.62	42.06	1.01	–	.0166	.0124	.3582	–	.0330
ILLINOIS	Chicago	.35	.60	.54	.54	.05	19.25	44.19	1.38	3.12	.0182	.0136	.3944	.1722	.0373
	Springfield	.26	.44	.77	.33	.12	19.16	42.79	2.59	2.52	.0138	.0103	.2976	.1300	.0453
	Cairo	.22	.40	.82	.24	.20	17.85	42.79	3.04	2.00	.0125	.0093	.2692	.1176	.0663
INDIANA	Indianapolis	.30	.50	.48	.50	.05	18.27	41.60	1.33	3.25	.0168	.0124	.3514	.1534	.0398
IOWA	Des Moines	.28	.58	–	–	–	14.92	41.32	–	–	.0188	.0140	.4062	–	.0406
	Dubuque	.24	.65	–	–	–	11.24	41.32	–	–	.0210	.0151	.4548	–	.0257
KANSAS	Wichita	.15	.41	.49	–	.09	9.70	36.42	1.49	–	.0151	.0112	.3255	–	.0603
	Goodland	.14	.48	.38	–	.03	8.16	36.42	1.00	–	.0175	.0133	.3780	–	.0337
KENTUCKY	Lexington	.26	.50	.63	–	.08	17.26	43.75	1.90	–	.0153	.0114	.3300	.1441	.0423
LOUISIANA	Baton Rouge	.05	–	.28	–	.19	7.10	–	1.82	–	.0071	–	.1539	–	.1046
	Shreveport	.08	–	.38	–	.16	7.50	–	1.77	–	.0100	–	.2155	–	.0914
MAINE	Portland	.65	.72	1.04	–	.03	28.00	44.70	2.25	–	.0231	.0161	.4631	–	.0138
MARYLAND	Baltimore	.45	.58	.98	.51	.12	27.12	46.74	2.71	3.25	.0167	.0124	.3603	.1573	.0461
MASSACHUSETTS	Worcester	.73	.77	1.09	–	.06	32.14	45.78	3.02	–	.0227	.0169	.4911	–	.0185
MICHIGAN	Lansing	.56	.65	1.10	.70	.05	28.47	43.86	2.60	3.75	.0197	.0147	.4260	.1860	.0200
MINNESOTA	Duluth	.60	.76	–	–	–	23.80	43.09	–	–	.0254	.0177	.5482	–	.0073
	Minneapolis	.33	.72	–	–	–	15.59	46.18	–	–	.0213	.0156	.4595	–	.0268
MISSISSIPPI	Jackson	.11	–	.57	–	.24	10.50	–	2.58	–	.0102	–	.2209	.0904	.0918
MISSOURI	St. Louis	.21	.76	.39	–	.06	14.12	45.63	1.18	–	.0153	.0167	.3306	–	.0540
	Springfield	.17	.49	.45	–	.07	10.50	41.37	1.30	–	.0160	.0119	.3453	–	.0518
MONTANA	Helena	.26	–	.69	–	.01	12.45	–	1.55	–	.0206	–	.4456	–	.0093
NEBRASKA	Omaha	.33	.58	.73	–	.08	17.65	40.90	1.80	–	.0189	.0141	.4077	–	.0446
	Scottsbluff	.25	.52	.68	–	.04	14.00	41.35	1.87	–	.0169	.0126	.3658	–	.0232
NEVADA	Elko	.33	.63	.71	–	.02	17.75	44.34	1.75	–	.0188	.0141	.4075	–	.0128
	Las Vegas	.14	.34	.46	–	.23	13.90	44.34	2.08	–	.0103	.0077	.2228	–	.1114
NEW HAMPSHIRE	Concord	.70	.73	1.05	–	.04	33.29	46.38	2.30	–	.0211	.0157	.4552	–	.0170
NEW JERSEY	Atlantic City	.62	.62	1.20	.56	.09	33.84	45.32	3.02	3.25	.0183	.0137	.3957	.1728	.0312
NEW MEXICO	Raton	.17	–	1.03	–	.09	8.60	–	2.35	–	.0203	–	.4388	–	.0391
	Silver City	.09	–	.69	–	.10	7.05	–	2.63	–	.0121	–	.2611	–	.0391
NEW YORK	New York City	.65	.68	1.72	.76	.19	35.28	49.46	4.34	4.40	.0183	.0137	.3959	.1729	.0434
	Rochester	.53	.76	1.36	.82	.07	24.05	46.53	2.85	3.95	.0220	.0164	.4755	.2076	.0259
NORTH CAROLINA	Raleigh	.30	.52	.95	.64	.12	19.41	45.10	2.35	4.40	.0155	.0115	.3347	.1462	.0500
	Wilmington	.17	.35	.54	.70	.17	15.46	43.43	2.35	4.76	.0107	.0080	.2315	.1011	.0730
NORTH DAKOTA	Bismarck	.59	–	1.21	–	.04	26.50	–	2.49	–	.0224	–	.4852	–	.0171
OHIO	Youngstown	.19	.69	1.18	.36	.05	8.88	43.95	2.62	1.80	.0209	.0156	.4522	.1974	.0204
	Cincinnati	.18	.47	.52	.53	.07	12.79	43.95	1.68	3.90	.0144	.0107	.3107	.1357	.0439
OKLAHOMA	Oklahoma City	.21	–	.74	–	.20	17.45	–	2.80	–	.0121	–	.2625	–	.0705
OREGON	Salem	.57	.72	1.02	–	.02	26.41	44.20	2.17	–	.0217	.0162	.4690	–	.0101
	Medford	.58	.75	1.05	–	.06	25.11	44.20	2.13	–	.0229	.0170	.4440	–	.0283
PENNSYLVANIA	Philadelphia	.49	.63	1.08	.61	.12	26.87	46.10	2.73	3.55	.0183	.0137	.3959	.1729	.0448
	Pittsburgh	.35	.60	1.04	.44	.03	19.25	44.75	2.68	2.60	.0180	.0134	.3890	.1698	.0120
RHODE ISLAND	Providence	.45	.68	1.37	–	.09	23.03	47.20	3.26	–	.0194	.0145	.4195	–	.0285
SOUTH CAROLINA	Charleston	.19	.27	.44	–	.20	21.58	42.17	2.34	–	.0087	.0065	.1888	–	.0869
	Greenville-Spartanburg	.18	.36	.56	–	.13	16.23	42.17	2.28	–	.0113	.0085	.2450	–	.0561
SOUTH DAKOTA	Rapid City	.22	.58	.79	–	.04	11.88	41.76	1.95	–	.0186	.0139	.4027	–	.0209
TENNESSEE	Knoxville	.19	.44	.61	.31	.12	14.20	44.20	2.13	2.50	.0133	.0099	.2873	.1255	.0557
	Memphis	.11	.38	.55	.31	.17	9.25	42.36	2.13	2.80	.0119	.0089	.2569	.1122	.0780

Table 25-8 (continued)

		Heating Factors				Cooling Factor	Fuel costs				Heating Multipliers				Cooling Multiplier
		Gas	Oil	Elec	Coal		Gas ¢/100 cu. ft.	Oil ¢/gal.	Elec ¢/Kwh	Coal ¢/lb.	Gas	Oil	Elec	Coal	
		A	B	C	D	E	F	G	H	I	J	K	L	M	N
TEXAS	Austin	.07	–	.51	–	.32	8.60	–	3.00	–	.0078	–	.1688	–	.1071
	Dallas	.07	–	.29	–	.16	7.90	–	1.49	–	.0090	–	.1943	–	.1049
	Houston	.05	–	.21	–	.16	9.00	–	1.62	–	.0061	–	.1319	–	.1000
	Lubbock	.15	–	.53	–	.13	12.46	–	2.11	–	.0117	–	.2529	–	.0617
UTAH	Salt Lake City	.23	.58	.87	–	.08	11.56	39.60	2.05	–	.0197	.0147	.4264	–	.0371
	Milford	.28	–	.90	–	.05	13.39	–	1.97	–	.0212	–	.4578	–	.0267
VERMONT	Burlington	.63	.84	1.20	–	.04	27.81	47.58	2.36	–	.0222	.0176	.5098	–	.0178
VIRGINIA	Richmond	.24	.50	.72	.61	.12	16.64	45.58	2.25	4.40	.0147	.0110	.3178	.1388	.0537
WASHINGTON	Olympia	.75	.80	.79	–	.01	32.19	44.72	1.52	–	.0239	.0178	.5165	–	.0035
	Walla Walla	.21	.46	.45	–	.05	15.40	44.72	1.51	–	.0137	.0102	.2963	–	.0357
WEST VIRGINIA	Charleston	.32	.50	.79	.13	.10	21.92	45.73	2.50	.96	.0146	.0109	.3154	.1377	.0388
	Elkins	.44	.63	1.10	.18	.06	23.82	45.73	2.76	1.05	.0185	.0138	.3999	.1746	.0208
WISCONSIN	Milwaukee	.44	.75	–	.79	–	20.29	46.10	–	3.85	.0218	.0162	–	.2055	.0216
WYOMING	Casper	.32	–	.81	–	.03	16.90	–	2.00	–	.0188	–	.4062	–	.0151

How Much Do You Really Pay for Fuel? Your true cost for 100 cubic feet of gas, a kilowatt of electricity, etc., is sometimes pretty well hidden in a bill. Call the utility company, and ask for the true cost (including all "fuel adjustment" factors and taxes) of the last unit of fuel that you buy every month. Use this cost to figure your Heating and Cooling Factors.

Saving Energy with Heating, Air Conditioning, and Water Heating

You have two options:

1. Routine servicing: A service representative should check all your heating and cooling equipment and do any needed maintenance once a year.
2. Repair or replacement: Some heating and cooling equipment may be so badly worn or outmoded that it will pay you to replace it now. The increased efficiency will allow you to recoup your investment in a few years.

Routine Servicing. Periodic checkup and maintenance of your heating and cooling equipment can reduce your fuel consumption by about 10 percent. Locating and sticking with a good heating/cooling specialist is a good way to ensure that your equipment stays in top fuel-saving condition. Your local fuel supplier or heating/cooling-system repair specialist are the people to call. You can find them in the Yellow Pages under:

Heating Contractors
Air-Conditioning Equipment
Furnaces–Heating
Electric Heating
Oil Burner–Equipment and Service

Check out the companies you contact with the Better Business Bureau and with other homeowners in your area. Once you have found a reputable outfit, it is advisable for you to take out a service contract. For an annual fee, you get a periodic tune-up of your heating/cooling system, and you're assured of regular maintenance from a company you know.

Later on in this chapter, there are lists of items that your service rep should check for each type of heating or cooling system. Some items may vary from brand to brand, but go over the list with the service rep. Also listed are service items you can probably take care of yourself, and thus save money. If you don't want to service the system yourself, make sure you add those items to the service rep's list.

Repair or replacement of equipment may be necessary. When faced with major repairs, you'll inevitably ask should I fix it or should I buy new equipment? To answer that question, do the following:

- Get several estimates–the larger the job, the more estimates. The special knowledge of the equipment dealer and installer is needed here–they'll study a house, measure the walls and windows, and should give you a written estimate.
- Check to see what your fuel costs are now.
- Ask each contractor who gives you an estimate how many years it will take before the amount you save by having the new system equals what you paid for it. Remember, fuel costs are going up.

Furnace Maintenance

Oil Burner. Every year

Adjust and clean burner unit.
Adjust fuel-to-air ratio for maximum efficiency.
Check for oil leaks.
Check electrical connections, especially on safety devices.
Clean heating elements and surfaces.
Adjust dampers and draft regulator.
Change oil filters.
Change air filter.
Change oil-burner nozzle.
Check oil pump.
Clean house thermostat contacts and adjust.

There are several tests service reps can use to check oil-furnace efficiency:

Draft test to see if heat is being lost up the chimney or if draft is not enough to properly burn the oil.
Smoke test to see if oil is being burned cleanly and completely.
Carbon dioxide test to see if fuel is being burned completely.
Stack temperature test to see if stack gases are too hot or not hot enough.

Coal Furnace. At the end of each heating season

Adjust and clean stoker.
Clean burner of all coal, ash, and clinkers.
Oil the inside of the coal screw and hopper to prevent dust.

Gas Furnace (bottled, LP, or natural). Every 3 years

Check operation of main gas valve, pressure regulator, and safety-control valve.
Adjust primary air-supply nozzle for proper combustion.
Clean thermostat contacts, and adjust for proper operation.
Perform draft test and stack temperature test (see above).

Electric Furnace. Very little maintenance required. Check the manufacturer's specifications.

Heat-Distribution Systems

You can do some of the following items yourself to keep your system at top efficiency. For the ones you can't do, call in a service rep. (Note: Unless otherwise noted, these are all once-a-year items.)

Hot Water Heating System.

Service rep:

Check pump operation.
Check operation of flow-control valve.
Check for piping leaks.

Check operation of radiator valves.
Drain and flush the boiler.
Oil pump motor.

You can do these yourself:

- Bleed air from the system. Over time, a certain amount of air will creep into the pipes in a system. It will find its way to the radiators at the top of the house, and such air keeps out hot water. There's usually a small valve at the top of each radiator. Once or twice a year, open the valve at each radiator. Hold a bucket under the valve and keep it open until the water comes out. Watch out—the water is hot.
- Draining and flushing the boiler is also something you can do yourself. Ask the service rep to show you how.

Forced Hot-Air Heating System.

Service rep:

Check blower operation.
Oil the blower motor if it doesn't have sealed bearings.
Check for duct leaks where duct is accessible.

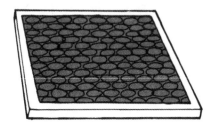

You can do these yourself:

- Clean or replace air filters—this is important, easy to do, and needs to be done more often than it pays to have a service rep do it. Every 30 to 60 days during the heating season, you should clean or replace (depending on whether they're disposable) the air filters near the furnace in the system. Ask the service rep how to do it, buy a supply, and stick to a schedule—you can save a lot of fuel this way.
- Clean the fan blade that moves the air through the system—it gets dirty easily and won't move the air well unless it's clean. Do this every year.

- Keep all registers clean—vacuum them every few weeks. Warm air coming out of the registers should have a free path, unobstructed by curtains or furniture.

Steam-Heat Systems. With steam heat, if a service rep checks the burner and the water system in the boiler, most of the work is done. There are two things you can do to save energy:

- Insulate steam pipes that are running through spaces you don't want to heat.
- Every 3 weeks during the heating season, drain a bucket of water out of the boiler (the service rep can show you how). This keeps sediment off the bottom of the boiler. If the sediment is allowed to stay there, it will actually insulate the boiler from the flame in the burner, and a lot of heat that would have heated the house will go up the chimney.

Whole-House Air Conditioning.

Service rep:

Oil bearings on fan and compressor if they are not sealed.
Measure electrical current drawn by compressor.
Check pulley-belt tension.
Check for refrigerating fluid leaks, and add fluid if needed.
Check electrical connections.
Readjust dampers—if the air conditioner uses the same ducts as the heating system, different settings are usually required for summer cooling than for winter heating.
Flush evaporator drain line.

You can do these yourself:

- Clean or replace air filters—this is important, and if you do it every 30 to 60 days, it will save you far more money in fuel than the cost of the filters.
- Clean the condenser coils of dust, grass clippings, etc. (Note: The condenser is the part of the air conditioner that sits outside the house. It should be shaded; if it has to work in the sun, it wastes a lot of fuel. When you shade it, make sure you don't obstruct the flow of air around it.)

Water Heaters

Once a year

Service rep:

Adjust damper (for gas or oil).
Adjust burner, and clean burner surfaces (for oil).
Check electrodes (for electric).
De-lime tank.

You can do these yourself:

- Once or twice a year, drain a bucket of water out of the bottom of the heater tank. This will let out any sediment that has collected. The sediment insulates the water in the tank from the burner's flame or electrode—and that wastes energy.

- Insulate the water-heater tank. This will greatly reduce the amount of fuel the heater uses to keep water hot when no hot water is being used. To insulate the heater, use 3-inch batts or blankets with a paper or foil facing, and duct tape. For a more finished-looking job, use duct-insulating blankets. Water-heater insulating kits are now sold at home-improvement centers. (Note: Do not insulate the top or bottom of an oil or gas heater. At the top, insulation may interfere with the draft of the heater's flue. At the bottom, it may cut off air from the flame. Only insulate the sides.)
- Don't set the temperature of the water heater any higher than necessary. The heater burns fuel keeping

water hot when you're not using it, and the higher you set it, the more it burns. If you have a dishwasher, 140°F is high enough, if not 120°F is sufficient. (Settings above 140°F can shorten the life of water heaters, especially those that are glass-lined.) Depending on the type of fuel you use, this will save you $5 to $45 a year. (If the heater is marked "High, Med, Low," call the dealer and ask what each setting stands for.)

Duct Insulation

If the ducts for either a heating or an air-conditioning system run exposed through an attic or garage (or any other space that is not heated or cooled), they should be insulated. Duct insulation generally comes in blankets 1 or 2 inches thick. The thicker variety is recommended, particularly if the ducts are rectangular. For air-conditioning ducts, buy the kind of insulation that has a vapor barrier (the vapor barrier goes on the outside). To avoid condensation, seal the joints of the insulation tightly with tape. Before insulating, check for leaks in the duct, and tape them tightly.

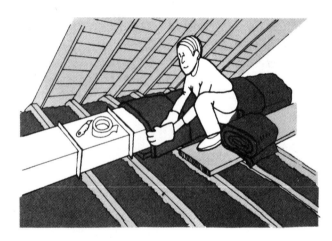

Choosing a Contractor

The large majority of contractors take pride in their business and are conscientious and honest. Nevertheless, you should still spend some time and effort in choosing a contractor and then in clearly defining the job. The following sources will help you in finding a contractor:

- Yellow Pages. See "Insulation Contractors—Cold and Heat." Don't be suspicious of small operations. Insulating is a relatively small project, and the small operation will usually do an excellent job.
- Local chapter of the National Association of Home Builders Association. They will be very helpful in recommending contractors.
- Your bank. If a bank is loaning you the money for the work, it's in their interest to recommend someone who will do a good job.
- Local government offices for government-funded or nonprofit home-improvement assistance centers.

These offices maintain files on contractors that they recommend. However, they cannot be found in all areas.

From these sources, establish a list of three or four contractors from which to select. Next, ask each contractor for a list of past customers, and check with a few to see if they were satisfied with the work. See how long each contractor has been in business—in general, the longer the better. Call the local Better Business Bureau, and ask if there have been any complaints against the contractors on your list. Get estimates from each contractor on any job you think will cost more than $200.

Once you've selected a contractor, have the contractor write up a specific contract for the job. Check the contract carefully for work content and warranty. Be sure that all the things you believe should be done in the course of the job are included.

Getting Financing

If you don't want to pay for your energy fix-up program out of your savings, and you want to get a better interest rate than either a credit-card loan or refinancing your present home mortgage will give you, try one of these:

Where to Get Financing (and Information)	What Kind of Financing	How Long to Repay
Commercial bank Savings and loan	Home improvement loan	2 to 5 years
Mutual savings bank	FHA/HUD Title 1[a]	12 years
Credit union	Depends on the credit union, but usually includes Title 1 loans; see above.	Repayment time varies with the type of loan.

[a]Lenders are not allowed to charge fees of any kind for this type of loan. Almost all of the improvements mentioned in this chapter are eligible under Title 1.

Tips on Saving Heating and Cooling Energy

Doors and windows. Keep doors and windows firmly shut and locked to cut down heat loss in winter and heat gain in summer. Check windows and door latches to see that they fit tightly; if necessary, adjust the latches, and plug any air leaks. You really don't have to open windows in winter—you usually get enough fresh air just from normal air leakage even if the house is well caulked and weather-stripped.

Use heavy or insulated draperies, keep them closed at night, and fit them tightly at the top. In the summer and in warm climates, light-colored opaque curtains will reflect the sun and help keep the house cool.

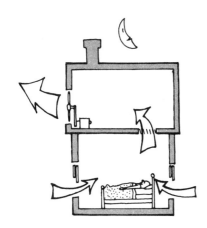

The tightest storm door in the world won't work if it's open, so try to cut down the number of times that you go in and out. A vestibule at the front and back doors also helps to keep a house warmer.

Attic and Roof. Seal any openings that might admit air between the attic and the rest of the house, such as spaces around loosely fitting attic stairway doors or pull-down stairways; penetrations of the ceiling lights or a fan; and plumbing vents, pipes, or air ducts that pass into the attic.

Instead of energy-consuming air conditioning, use an attic fan. Normally, a house holds heat so there's a lag between the time the outside air cools after sunset on a summer night and the time that the house cools. An attic fan speeds up the cooling of the house by pulling air in through open windows, up through the attic, and out. When the fan is on, air is let through to the attic either by the partial opening of the attic door or by a louver that admits air automatically. In areas that have hot days and cool nights, using an attic fan in the evening and closing the windows and curtains during the day can replace air conditioning.

You can figure out what size fan you need by finding the volume of the house: Rounding off to the nearest foot, multiply the length of the house by its width, then multiply by its height (from the ground to just below the attic). This will give you the volume in cubic feet. The capacity of all fans is marked on the fan in cfm's (cubic feet of air per minute). Divide the volume of the house by 10; this will give the cfm rating of the fan needed to change the air in the house six times an hour.

Basements. If you can't afford to insulate the exposed portions of a basement or crawl space for the winter, you can still create some barriers against wind and cold by planting shrubs around the foundation. You can also tarpaper the exposed walls and rake leaves against the foundation, covering them with a weighted tarpaulin (the tarpaper keeps moisture off the house that would otherwise come in through the leaves).

Shading the House. A good way to keep a house cool in the summer is to shade it from the sun. The east and west

sides are where the most heat comes through—shading here will result in a smaller air-conditioning bill and a cooler home. Any method of stopping the sun's rays from getting in through windows is seven times as effective at keeping a house cool as are blinds and curtains on the inside. So plant trees and vines that shade in the summer and lose their leaves to let the sun back in for the winter. If you can't shade the house with trees, concentrate on keeping the sun out of the windows—awnings, sun shutters, sunshades, or reflective foil will help do the job.

Hot water. Turn your heater down to 120°F (see p. 203). Fix all leaky faucets, particularly the hot ones. One leaky faucet can waste up to 600 gallons of water a year. You can also save by turning the water heater down before going away for a weekend or longer. Don't run the dishwasher or clothes washer when less than full, and set the clothes washer on warm wash and cold rinse. Take showers—they use less hot water than do baths. Use cold water to run the garbage disposal because, in general, you save energy every time you use cold water instead of hot.

Heating. As noted previously, you can figure out how much you can save by lowering the thermostat. For about $80, you can install a clock thermostat, which will automatically turn the heat down at night and turn it up in the morning. A time-delay thermostat, which is a windup timer wired into a thermostat, will do the same job for about $40.

If you have oil heating, check with the oil company about the new high-speed flame-retention oil burners—they can save you 10 percent on your oil bill.

If your house has been insulated since it was built, then the original furnace may be too big. In all likelihood, a smaller furnace will prove more efficient and will use less fuel. To determine if the furnace is too large, wait for a very cold night, and set the thermostat at 70°F. Once the house temperature reaches 70°F, if the furnace burner runs less than 40 minutes out of the next hour (time it only when it's running), the furnace is too big. Such a furnace turns on and off much more than it should, which wastes energy. Call the service company; depending on the type of fuel burner, they may be able to cut down the size of the burner without replacing it.

Don't overheat rooms, and don't heat or cool rooms you're not using. No room in the house should get more heat than it needs; you should be able to turn down the heating or cooling in areas of the house that you don't use. If some of the rooms get too hot before other rooms are warm enough, you are paying for fuel you don't need, and the system needs balancing—call your service rep. In a house that is "zoned," there is more than one thermostat, and the heating or cooling can be turned down in areas where they're not needed.

But if your house has only one thermostat, you can't properly adjust the temperature in rooms you're not using, and that wastes energy too. This can be corrected fairly cheaply. For example:

- *Steam radiators.* Most valves on radiators are all-on or all-off. However, valves are available that let you choose the temperature for each radiator.
- *Forced-air heating or cooling.* Many registers (the place where air comes out) are adjustable. If your registers are not adjustable, replace them so that you can balance the system.
- *Hot-water radiators.* If there are any valves on your radiators, you can use them to adjust the temperature room by room.

Air Conditioning. Closing off unused rooms is important in saving air-conditioning costs. Keep lights off during the day—most of the electricity they use makes heat, not light. You can also reduce the load on the air-conditioning system by not using heat-generating appliances such as a dishwasher during the hot part of the day (or stop the dishwasher when the drying cycle begins).

If you have central air conditioning, you may want to get an air economizer, a system that turns off the part of the air conditioner that uses a lot of electricity, and circulates outside air through the house when it's cooler outside. By using the cooler air, the system reduces its own job and saves money for you. Ask your air-conditioning dealer if one can be installed on your system.

Buying a Room Air Conditioner. Before buying a room air-conditioning unit, check its EER (energy efficiency ratio). The higher the EER, the less electricity the unit will use to cool a specified amount of air. Typical EERs range from 4 to 12; a unit with an EER of 4 will cost about three times as much to operate as one with an EER of 12. Consider the possible fuel savings when deciding how much to spend on a unit. A unit that costs more to begin with may save you enough money over the next summer to make it worth it.

Proper Installation. Inadequate electrical wiring wastes power, reduces air-conditioner efficiency, and may damage the unit. It may also be dangerous. So have the wiring checked to be sure the circuit is of the right amperage and voltage before adding a room air conditioner. Avoid lines that are already loaded with appliances. And, if a fuse blows, do not attempt to operate the unit until load and wiring have been checked.

Be sure the outlet is properly grounded. Never remove the grounding prong from the plug on a room air conditioner. Never attempt to connect a 208-volt unit to 115-volt (normal household) line nor a 115-volt unit to higher voltage lines used for an electric range or clothes dryer.

Install the unit in a shaded window if possible. However, avoid areas where excessive outside dust, odors, or pollens will be drawn into it.

Keep warm air out by weather-stripping doors and windows and keeping them tightly closed. Leave storm windows on; but be sure some can be opened so the house can be aired. Close the fireplace damper. Seal any openings to the attic or other uncooled areas.

Reduce solar heat and the heat from warm outdoor air by drawing the shades or blinds. Reflecting glass, double-pane windows, awnings, overhangs, louvered sunscreens, plantings, fences, or adjoining buildings also help. However, be sure the back of the unit is not blocked.

Six inches of ceiling insulation will reduce heat from the attic. Walls need 3 to 4 inches, which also helps keep rooms warmer in winter. To prevent problems of moisture condensation in walls, get professional advice before adding insulation.

Proper Use. Reduce air-conditioner use. Open doors and windows during mild periods, and turn off the unit. Use the air-conditioner fan to draw in cooler outside air and portable fans to increase circulation and cooling effects during periods of temporary high load due to cooking, large number of people in the room, etc.

Keep the room warmer. You can reduce the risk of power overload by using a warmer temperature setting during peak periods. Permitting the temperature to rise from 75°F to 80°F will cut power consumption by 15 percent. Use warmer settings if the room is unoccupied during the day. Don't let heat build up all day in an unoccupied room and then try to cool it quickly by turning fan and temperature to the coldest settings. Shut off units in rooms that are used only occasionally; cool them slowly before they are to be occupied.

Start units earlier in the day, before peak use periods. Spread out your use of other heat-generating appliances such as an iron or self-cleaning oven. Avoid uses that generate high heat and humidity (such as cooking, bathing, and use of laundry equipment and dishwashers) during the high-heat hours. Turn off lights, cooking equipment, and television sets when not in use.

Maintenance. Keep equipment clean. Use vacuum-cleaner attachments to clean accessible parts, being careful not to damage them. Change or clean filters every month or so, and make sure airflow is not blocked by dirty filters, leaves, bugs, outside obstructions, or frost on condenser coils. Never remove filters to get more air.

Have unit checked and cleaned at the start of the season. Shut off the unit, and call for service if the fan or cooling stops, if unusual odors or smoke are detected, or if it doesn't sound right.

Be sure the unit is properly covered and disconnected during the off season, or removed and cleaned.

The Heat Pump. A heat pump runs on electricity. It uses electricity to heat, and gets more heat out of a dollar's worth of electricity than do the resistance heaters in baseboard units and electric furnaces. How? There's heat in air outside, even when the temperature's below freezing. A heat pump can get that warmth into your house.

When should you consider installing a heat pump? If you presently have a central electric heating system, it may pay to install a heat pump in the system, next to the fur-

nace. Once the temperature drops below 20°F or so, the heat pump will need help from the furnace. Installation of a heat pump large enough for most houses should cost a little under $3000; it provides central air conditioning as well as a "furnace" that's more than twice as efficient as an electric furnace.

If you are adding on a room, consider adding a heat pump—like air conditioners, they come in room-size units. Such a unit comes with its own electric resistance coil (like a baseboard electric heater) for the times when it's too cold for the heat pump to work well by itself. Call your air-conditioner dealer for details on both central and room-size heat pumps. If the furnace runs on gas or oil, and the cost of those fuels continues to rise faster than the price of electricity, then you may want to consider a heat pump.

Fireplaces. The warm, cozy fireplace is one of the biggest energy wasters there is. Even when it's burning, it pulls more heated air up the chimney than it replaces, and pulls cold air into the house through cracks in the exterior. When a fireplace is not in use and there is no damper, or if the damper is left open or is closed but doesn't fit properly, lots of air you have paid to heat will still go up the chimney.

There are many devices on the market to make a fireplace burn wood more efficiently. These include special andirons, folding glass doors, and wood stoves that replace the fireplace. However, except in areas where wood is fairly inexpensive, wood stoves are not an economical replacement for the oil, gas, or electricity you use now. All of the devices mentioned cut the amount of fuel now wasted in a fireplace, but seldom enough to justify their cost, considering that most fireplaces are not used very often.

The most effective ways to cut the amount of fuel that fireplaces waste are free. If you don't use the fireplace, make the flue airtight by covering the front of the fireplace. If you use the fireplace, be sure that the damper is closed as soon as the fire is completely out. Normally, an open damper has the same effect on your fuel bill as a hole in the wall twice the size of the damper. A very tall chimney or high winds will increase this effect.

Lights. Reduce lighting where possible, concentrating it in work areas or reading areas where it is really needed.

Fluorescent bulbs should be used rather than the incandescent kind. A 25-watt fluorescent bulb gives off as much light as a 100-watt incandescent bulb, and only costs one-fourth as much to light.

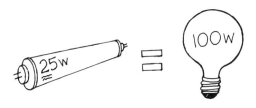

Decorative gas lanterns should be turned off (call the gas company to do this) or converted to electric lamps—they will then use much less energy to produce the same amount of light.

WHAT ARE YOUR BEST INVESTMENTS?

There are two kinds of investments here—the kind you have to make each year and the kind you only have to make once. Here's how to directly compare the two different kinds of investments and to figure out which are the best bets:

For the investments that you have to make each year (Energy Checklist, lines 2a, 2b, 7, and 8), subtract the yearly cost from the yearly savings, and write the difference in the right-hand box on the line of the checklist dealing with the measure. This number is the net savings per year for the investment.

For "one-shot" investments that you only have to pay for once (lines 1, 2c, and 3 through 5), multiply the yearly savings by 13*, subtract the cost from the result, and write the difference in the right-hand box on the line of the checklist dealing with the measure. This number is the net savings over the life of the investment.

HOW TO INTERPRET THE CHECKLIST

Now you're ready to figure out what are the best energy-saving steps for you. First, look at the cost figures. Don't consider doing anything you can't afford. But be sure you don't leave out things you can afford. Second, for the measures you can afford, look at the net savings in the right-hand column. The things to do first are those with the highest net savings. Do the highest, then the next highest, and so on, until you've done all you can afford. Don't make any improvement if the net savings are less than 0.

*Multiplying the estimated savings in the first year by 13 projects the savings (in terms of 1979 money) that a one-shot energy-saving improvement will deliver over its life. The number 13 takes into account the rate of inflation, and assumes that you can borrow money at the average available interest rate.

ENERGY CHECKLIST

	SAVINGS FACTOR FROM PART 2	YEARLY SAVINGS	COST FROM PART 2	NET SAVINGS

1. CAULK AND WEATHERSTRIP

Heating factor $\bigcirc$ + Cooling factor $\bigcirc$ = _____ Savings factor x [____] = $ _____   Total cost $ [____]   [____]

2. ADD STORM WINDOWS

a) plastic storm windows
(with no new weatherstripping)

Heating factor $\bigcirc$ x Savings factor [____] = $ _____   Yearly cost $ [____]   [____]

b) plastic storm windows
(with new weatherstripping — be sure to fill out line 1 above)

Heating factor $\bigcirc$ x Savings factor [____] = $ _____   Yearly cost $ [____]   [____]

c) glass or rigid plastic storm windows
(with new weatherstripping — be sure to fill out line 1 above)

Heating factor $\bigcirc$ x Savings factor [____] = $ _____   Total cost $ [____]   [____]

3. INSULATE ATTIC

Fill out both lines if your attic is a combination of two basic attic types (see page 11). Otherwise, fill out the top line only.

Heating factor $\bigcirc$ + Cooling factor $\bigcirc$ = _____ Savings factor x [____] = $ _____   Total cost $ [____]   [____]

Heating factor $\bigcirc$ + Cooling factor $\bigcirc$ = _____ Savings factor x [____] = $ _____   Total cost $ [____]   [____]

4. INSULATE CRAWL SPACE WALLS, FLOOR, OR BASEMENT WALLS

a) Insulate crawl space walls

Heating factor $\bigcirc$ x Savings factor [____] = $ _____   Total cost $ [____]   [____]

b) Insulate floor

Heating factor $\bigcirc$ x Savings factor [____] = $ _____   Total cost $ [____]   [____]

c) Insulate basement walls

Heating factor $\bigcirc$ x Savings factor [____] = $ _____   Total cost $ [____]   [____]

5. INSULATE FRAME WALLS

Heating factor $\bigcirc$ + Cooling factor $\bigcirc$ = _____ Savings factor x [____] = $ _____   Total cost $ [____]   [____]

6. REGULATE THERMOSTAT
Down in winter, up in summer.

Degrees turndown [____] $ [____]   Yearly cost $ 0   [____]

7. SERVICE OIL OR COAL-BURNING FURNACE

$ [____]   Yearly cost $ 30-35   [____]

8. SERVICE WHOLE-HOUSE AIR CONDITIONER

$ [____]   Yearly cost $ 30-35   [____]

26 Firewood: How to Choose It and Use It

Where can I get good firewood? How is it sold? What safety measures should I take when burning firewood? Is a fireplace efficient for home heating? These are just a few of the questions asked by the many people who plan to make greater use of the home fireplace.

A fireplace supplies radiant energy that can bring quick comfort to a cold room. In spring and fall, a fire will dispel early morning and evening chills more economically than a large heating system, since less fuel is consumed and a large volume of heat is quickly produced. In addition, should a storm or power failure interrupt the normally reliable means of heating and cooking, it is reassuring to have a fireplace for emergency use.

A fireplace, however, is not the most efficient means of producing heat for the home. An open flue draws large quantities of warm air from the home, which is replaced by cold outside air. Modern woodburning stoves with automatic draft controls, which are often sold in rural areas, are more efficient since they draw smaller quantities of air.

USING WOOD IN THE FIREPLACE

Burning wood releases much fewer irritating pollutants than do most fuels. Wood has a low ash content. It burns cleanly, leaving only a minimum of waste as ash. The ash that remains can be useful for gardening; applied to the soil, it's a valuable fertilizer.

Coal, oil, and gas are limited resources. Once used, they cannot be replaced. Wood, however, is a renewable fuel resource. New trees can be grown so that after a few years, more wood is available. Wood also has the advantage of being readily available, easily cut, and relatively inexpensive. Fossil fuels, by contrast require expensive equipment, plus considerable personnel, time, and energy, to locate, extract, and process them.

The wood that you use in your fireplace might very well be unusable in any other way. By burning it, you can help reduce the burdensome piles of wood waste riddling our environment in woodlands, urban dumps, and around wood industries. Reclaiming it for firewood provides a worthwhile environmental alternative.

Generally speaking, a standard cord of air-dry, dense hardwood weighs approximately 2 tons and provides as much heat as 1 ton of coal, or 150 to 175 gallons of No. 2 fuel oil, or 24,000 cubic feet of natural gas.

WHERE TO GET FIREWOOD

You can get wood fuel from the woods from trees that are considered undesirable. This includes trees that are poorly formed, diseased, of little-used or weed species (such as pin cherry), and genetically inferior individuals. All such trees pose problems for the forest manager. They occupy valuable growing space, which thrifty young trees might well use for continuing development. When diseased, they pose a hazard to nearby healthy trees. If genetically inferior, they may continue to reproduce and could keep the forest full of poor-quality trees for generations to come. Weeding them out often costs more than can be returned from their sale. But "fuelwood" markets may make this forest-improvement measure economical.

Contact your local forestry office for assistance in identifying trees that should be removed. Be sure to get permission to cut or remove trees from property other than your own.

Logs of elm trees killed by Dutch elm disease can be burned as firewood. Such logs should be used before the spring following the tree's death to prevent the disease-carrying beetles from emerging and infecting healthy elms. If you can't burn all the dead elm wood before spring, remove and burn the bark so that the larvae and eggs beneath it are destroyed.

You can also get fireplace wood from dumps and landfills, but be sure you have the approval of the landfill authority. Since many local ordinances forbid open burning, the quantity of dead, discarded trees keeps mounting unnecessarily on these lands, which are sorely needed for the disposal of other solid wastes. As much as 30 percent of the debris in some town and city landfills consists of reusable wood fibers, including the logs, limbs, and tops of trees toppled by storms.

Fireplace wood is also available as industrial wood scraps. Sawmills accumulate scrap materials such as slabs, trim, and edgings in their millyards. Lumber companies often offer these materials as firewood at minimal cost, since this will eliminate hauling and burning problems for them. Power companies might also offer the logs, limbs, and tree tops that result from their powerline maintenance efforts.

Firewood may be available from the national forests. Check with your nearest Forest Service District Ranger for more detailed information. State foresters, county extension agents, and county and city foresters can also provide information on local sources of firewood.

HOW TO BUY FIREWOOD

The most common measure of firewood volume is the cord. A standard can be described as a well-stacked pile of logs, 4 by 4 by 8 feet. Since few people burn wood in 4 or 8 foot lengths, most sales are a "face cord"—that is, a 4-by-8-foot, cut into desired lenghts. A face cord of 16-inch pieces is really one-third of a standard cord.

To determine the volume of firewood in cords, first stack the wood properly. Then, measure the dimensions of the stack in feet. Multiply the width by the height by the length to obtain the total cubic feet. Then divide this figure by 128, which is the number of cubic feet of wood in a standard cord. For example, if you purchase firewood in 16-inch (1.33-foot) lengths you stack it tightly to form a pile 4 feet high and 8 feet long, you have one-third of a cord: $1.33 \times 4 \times 8$ feet = 42.6 cubic feet; this divided by 128 = 0.333 cord.

In some parts of the country, firewood is more commonly sold by the ton. A ton of air-dry, dense hardwoods (such as oak, hickory, maple) is equal to approximately one-half of a cord. If you buy your wood by weight instead of volume, look for the driest wood. Don't pay for extra water.

Logs are sold in different lengths and thicknesses. The size you buy should depend on the size of your fireplace and the amount of time you want to take to get your fire going. Purchase logs that will fit when laid across your grate. Logs too large to burn readily may be split. Extra-heavy logs may be split to more manageable size. Short lengths are generally easier to split than longer logs; straight-grain knot-free wood is easier than crooked-grain. Green or wet wood splits more readily than seasoned wood, and softwoods usually split more readily than hardwoods. Elm, black gum, and locust are so difficult to split that they are rarely used as kindling. When kindling is needed, short lengths of straight-grained cottonwood, aspen, fir, and pine will split readily and prove most satisfactory. Small twigs and branches found in your yard and wood wastes found around sawmills are also good as they often do not have to be split or dried. When you buy wood, request a mixture of wood species and diameter sizes. Although the wood should be generally sound, don't worry about small pockets of rotten wood you may find in the logs.

Most wood species will not burn if freshly cut, so the wood you purchase should be reasonably dry or "seasoned." The surest way of having dry wood is to purchase it several months prior to using it. Splitting logs hastens drying. Split logs or small round logs should be stacked outside under a roof for 6 to 10 months before burning.

Gathering fuelwood and cutting and splitting it to the proper size can be a most enjoyable pastime for the entire family. For many people, such healthy outdoor exercise as sawing and splitting wood is part of the fun of having a fireplace.

BEST WOOD FOR BURNING

Choosing the kind of firewood to burn in your fireplace is much like selecting a favorite wine or cheese, since each wood species can offer something different in aroma or heat value. The fuelwood connoisseur will want to choose wood carefully and weigh both needs and taste before building a fire.

Softwoods, like pine, spruce, and fir, are easy to ignite because they are resinous. They burn rapidly with a hot flame. However, since a fire built entirely of softwoods burns out quickly, it requires frequent attention and replenishment. Thus, softwoods can be a boon if you want a quick warming fire or a short fire that will burn out before you go to bed or step out for the evening.

For a long-lasting fire, use the heavier hardwoods such as ash, beech, birch, maple, and oak. These hardwood species burn less vigorously than softwoods and with a shorter flame. Oak gives the most uniform and shortest flames and produces steady, glowing coals. With several oak logs burning in your grate, you can settle back for a steady show of flame.

Aroma is best derived from the woods of fruit trees, such as apple and cherry, and nut trees, such as beech, hickory, and pecan. Their smoke generally resembles the fragrance of the tree's fruit. Wood from fruit and nut trees often sells for more per cord than wood with greater heating values, but they generally produce a steady flame.

By mixing softwoods with hardwoods, you can achieve an easily ignited, long-lasting fire. Later, by adding some fruit or nut woods, you will add a pleasing wood-smoke aroma as well.

How much heat a fireplace log produces depends on its concentration of woody material, resin, water, and ash. Since woods have different compositions, they ignite at different temperatures and give off different heat values; therefore, it is beneficial to mix light and heavy woods to achieve the ideal fire.

Table 26-1 rates the suitability of various woods for use as firewood.

HOW TO BUILD A BETTER FIRE

There are many ways to build a fire. The basic principle is that you set a match to easily ignitable tinder, which in turn ignites the kindling, which in turn ignites the larger firewood. The following method has proved highly successful. Place

Table 26—1. Rating for firewood.

TREES	RELATIVE AMOUNT OF HEAT	EASY TO BURN?	EASY TO SPLIT?	HEAVY SMOKE?	POP, OR THROW SPARKS?	GENERAL REMARKS
Hardwoods						
Ash, red oak, white oak, beech, birch, hickory, hard maple, pecan, dogwood	High	Yes	Yes	No	No	Excellent
Soft maple, cherry, walnut	Medium	Yes	Yes	No	No	Good
Elm, sycamore, gum	Medium	Medium	No	Medium	No	Fair, contains too much water when green
Aspen, basswood, cottonwood, yellow poplar	Low	Yes	Yes	Medium	No	Fair, but good for kindling
Softwoods						
Southern yellow pine, Douglas-fir	High	Yes	Yes	Yes	No	Good, but smoky
Cypress, redwood	Medium	Medium	Yes	Medium	No	Fair
White-cedar, western red cedar, eastern red cedar	Medium	Yes	Yes	Medium	Yes	Good, excellent for kindling
Eastern white pine, western white pine, sugar pine, ponderosa pine, true firs	Low	Medium	Yes	Medium	No	Fair, good kindling
Tamarack, larch	Medium	Yes	Yes	Medium	Yes	Fair
Spruce	Low	Yes	Yes	Medium	Yes	Poor, but good for kindling

two logs on the iron grate or fire basket, and lay the tinder beneath them. (Dry scrap paper may be more readily available than such classical tinders as hemlock twigs and cedar or birch bark.) Next, above the tinder place a small handful of dry twigs or split softwood kindling. Then place small, dry logs over this base. A tepee formation of kindling and small branches will ease the fire through the early combustion stages until the logs are aglow. Place these logs close—the narrow airspaces between them promote better drafts. The heat reflected between adjacent surfaces aids in raising and maintaining combustion temperatures.

Generally, no more than four logs are needed to make a good fire. Adjust the logs, and maintain the flames by pushing the ends into the flame from time to time. Add kindling and new logs as needed to rekindle a dying glow. Rake coals toward the front of the grate before adding new logs. Add new logs at the rear of the fireplace; there, they will reflect light and heat into the room.

Only an inch or two of ashes should be left to accumulate at the bottom of the grate. A greater accumulation can ruin andirons and block the flow of air to the fire. The ashes under the grate are important, however, for they form a bed for the glowing coals that drop through the grate. They concentrate heat and direct drafts of air up to the base of the fire. By covering the burning logs, these excess ashes then can be used to check a flaming fire. A fire "banked" with ashes in this way will hold glowing coals for 8 to 10 hours, making it easier to rekindle the flames.

HOW TO BUILD A SAFE FIRE

Make sure the room is well ventilated, the damper open, and the flue unobstructed before lighting a fire. Poor ventilation will cause the fireplace to smoke. Avoid burning wet or green wood. Place a screen in front of the grate to catch any sparks that fly. Keep a fire extinguisher handy. Keep other combustibles at a distance. Never use flammable liquids indoors to light a fire.

Dry wood, when burned with abundant oxygen, produces carbon dioxide, water, and a small amount of residual ash—all of which are easily recycled by green plants. Burning green, wet, or highly resinous wood results in increased production of wood tars and several associated "smoke" products. These tars and the wood extracts may coat the chimney flue, which could cause a chimney fire if ignited.

Some modern homes, especially those with electric heating, are constructed so airtight that an air vent may have to be installed, or a window opened slightly in order to safely burn a fire. When wood or charcoal is burned without sufficient oxygen, some carbon monoxide will be released.

Glass fireplace doors offer draft control and a possible means of further reducing the amount of smoke that escapes into the room. However, if the chief reason for poor fireplace performance is faulty construction, the only safe solution is proper rebuilding.

When softwoods are used in building the fire, brief, vigorous fires result without a bed of long-lasting coals.

Within a short time when any unburned fuels have been pushed to the rear of the grate and the fireplace opening is covered with a fine mesh screen, a softwood fire can be presumed safe enough to leave unattended.

Some resinous woods are best used in stoves; they should be used with caution in the fireplace. Hemlock, larch, spruce, and juniper all contain moisture pockets in the wood. Upon heating, trapped gases and water vapor build pressure in these pockets and "pop" with great vigor. This is another reason for reducing moisture content as much as possible before burning any firewood.

With the proper size and kinds of firewood and a little practice at laying the fire, you can enjoy the warmth and beauty of your fireplace all winter long.

27 Landscaping to Conserve Energy

Only a few measurements have been made of the direct effects of landscaping on energy conservation in buildings. Nevertheless, these measurements indicate that a 25 to 35 percent reduction in the heating and cooling loads might be possible through careful use of landscape elements such as grass, paving, shrubs, trees, and fences. To achieve these savings requires careful placement and selection of landscape elements, based on an understanding of both heat gain and heat loss in buildings, and of the effects of landscape elements on the building environment. This chapter provides only a simplified view of these topics. Further information can be obtained from the various books referenced.

Heat is gained and lost through the shell of a house due to radiation, air leakage, and conductive heat transfer. Landscaping can be beneficial in reducing heat gains or losses of buildings in the following ways:

- Reducing direct solar radiation, sky reradiation, and reflected ground reradiation at windows during the summer.
- Reducing air leakage in all seasons through cracks and joints around windows and doors, at roof eaves, building corners, and at the foundation line by lowering the wind velocity at the building surfaces.
- Reducing the heat transmission of windows and, to a lesser degree, other building-shell elements, since modifying the amount of wind, sun, and rain that strikes the building's surfaces can decrease the temperature difference between indoors and outdoors.

Note: In reducing the energy requirements of a house, landscaping should not be considered as an alternative to caulking, weather stripping, insulation, good window design, or good workmanship. The choice of energy-conserving techniques for a particular home will depend on economic and design considerations.

GROUND SURFACES

Ground surfaces around houses affect the radiant and conductive heat gains and losses of windows and walls. They reflect solar radiation through windows and cause local changes in outdoor air temperatures by storing warmth and coldness in surrounding materials. Light-colored materials are usually reflective and will increase the heat load immediately by indirect radiation, while dark-colored materials will store large amounts of solar radiation and thus delay the heat load. Plants and trees, on the other hand, because of their dark color, large surface area, and evaporative cooling, neither reflect heat toward a building nor store heat for later reradiation to the building.

Several measurements have been made of variations in air temperatures due to changes in surface materials. Jeffrey E. Aronin (*Climate & Architecture*, AMS Press, 1977) reported an experiment in front of Eaton Electronic Research Laboratory in Montreal under sunny conditions in which the temperature of grass was measured at 89°F, asphalt at 106°F, and concrete at 111°F.

Aronin also reported a case in which asphalt overhung by the eave of a roof on the south side of a house reached a temperature of 120°F while the lawn nearby was 80°F. James M. Fitch, in *American Building: The Environmental Forces That Shaped It* (2nd ed., Houghton Mifflin, 1972) reported an observation made in Texas at 2 P.M. in August that showed a temperature of 125°F immediately above unshaded asphalt pavement and a temperature of 98°F above a shaded grass area 30 feet away.

It is difficult to generalize about the use of surface materials. A builder must be careful about the type and location of surface materials and examine the possibility of taking advantage of changes in winter and summer sun angles, seasonal wind directions, and defoliation of trees. As mentioned, overhanging eaves may worsen the performance of a surface material by acting as a radiant heat trap, while shade may improve its performance. Paving on the south side of a house in a warm climate can increase the air-conditioning load significantly, whereas this effect may be unimportant in a cold climate, and, in a temperate climate, paving may be an asset if exposed to the sun in the winter and shaded in the summer.

SHRUBS

Dense shrubs such as arborvitae, hemlock, or spruce, when planted close to a building, affect its outside surface temperature by blocking the wind, creating shade, and providing an insulating dead airspace between the shrub and building. Victor V. Olgyay (*Design with Climate*, Princeton University Press, 1963) reported an experiment conducted by the Lake State Forest Experimental Station in Nebraska on two identical houses. One was exposed to the winds, and the other was protected by dense shrubbery. Using the exact fuel required to maintain an indoor temperature of 70°F in each house, a savings of 23 percent was measured in the protected house. From this experiment, it was extrapolated that with good protection on three sides of the experimental house, the fuel savings could have run as high as 30 percent.

Low shrubs will have a limited benefit in conserving energy, but builders should consider planting dense shrubs 4 feet or higher on east, west, and north exposures as an energy saver.

TREES

Free-standing trees provide an effective shading device that can affect not only the walls of a building but also its roof. Deciduous trees let the sun in during the winter and provide shade during the summer, while evergreen trees provide constant shade. As the number of trees increases, their effect on the house will change. A grove of trees will not only provide shade and wind protection but will also modify outside air temperature through evaporative cooling.

William Flemer of Princeton Nursery, Princeton, New Jersey, reported that air conditioners in a fully shaded house needed to work only half as much as units in a house that has its walls and roof exposed to the sun. He also recorded a difference between shaded and unshaded outdoor wall surfaces of 8°F. This would be the equivalent to increasing the insulating value of the shaded portion of the wall by 30 percent on an instantaneous basis. Olgyay reported that shade trees will reduce solar heat gains by 40 to 80 percent depending on their density, and that even a sparse shade tree may be a better energy saver than a interior venetian blind.

For a builder, the most important thing is to consider the trees that are already on the site before cutting them, locating the house, or planting additional trees. In southern latitudes, shade trees are more effective on the east and west sides of a house, but as one moves northward into climates where south walls are exposed to the sun during the summer, shade trees are more effective on the west and south sides of a house.

WINDBREAK

A windbreak can be any barrier to wind: a row of evergreen trees spaced close together, a fence, a wall, or even a garage. A windbreak is used to control wind speed and direction. Depending on the location and design of the windbreak, it can either act as a barrier, reducing wind pressure around a house, or as a funnel, increasing wind pressure and improving ventilation through a house. When wind strikes a windbreak, it is diverted upward and to the sides, creating a large calm area on its leeward side and a small calm area on its windward side. A solid windbreak deflects the wind in such a way as to produce a wake of air that contains large organized eddies. These eddies contain high pressure, which returns to the ground downstream and often produces infiltration on the leeward side of a house.

To prevent these eddies and extend the protected area, penetrable windbreaks are preferred to solid ones.

An experiment conducted at Princeton University by George Mattingly and Eugene Peters, which modeled a group of townhouses in Twin Rivers, New Jersey, in a wind tunnel, found that a 5-foot-high wood fence would reduce air infiltration 26 to 30 percent; a single row of evergreen trees as high as the house would reduce air infiltration 40 percent; and a combination of the two would reduce air infiltration 60 percent. The study also found that a double row of staggered trees is less effective than a single row; that 59 percent greater infiltration is created by winds coming at the corner of a building, rather than perpendicular to one of the building's walls; and that the best location for a windbreak is at a distance of 1.5 to 2.5 windbreak heights upwind of the house. The results of the wind-tunnel tests are now being verified in a full-scale experiment at Twin Rivers, New Jersey. Laboratory studies reported in the *ASHRAE Guide* show that reducing the wind velocity from 15 to 5 mph will reduce the air leakage of different types of windows from 70 to 80 percent. The guide also shows that reducing the wind velocity at the surface of a single glazed window from 15 to 0 mph reduces the heat transmission by one-third.

A builder should consider placing a 5-foot-high fence or a row of 15-foot-high evergreen trees, spaced 10 feet on center, upwind of a house to protect it from winter winds. The data presented herein suggest that a 30 percent reduction in infiltration and a 12 percent reduction in heating bills can be realized.

Although the effect of landscaping is local, uneven, and varies with the season, weather, and growth, a combination of landscape elements can save energy. Table 27-1 is an overview of where landscape elements should be located in relation to the various orientations of a house.

Table 27-1. Energy-conserving locations for landscape elements.

LANDSCAPE ELEMENTS	CLIMATE		
	COLD	TEMPERATE	WARM
Ground cover or grass	Negligible effect on all sides	On south	On east, west, and south
Paving	On south	Shaded if on south	Shaded if on east, west, and south
Shrubs against house wall	On east, west, and north	On east, west, and north	On all sides
Deciduous shade trees	Negligible effect on all sides	On south and west	On east and west
Evergreen trees	On east, west, and north	On east and west	On east and west
Windbreak (trees, bushes, fences)	On sides exposed to winter winds	On sides exposed to winter winds	Undesirable effect on all sides
Windbreak used to funnel wind	Undesirable on all sides	On sides exposed to summer winds	Where cross-ventilation is possible

28 Energy Activities for Children

ELECTRICAL ENERGY

Until the oil embargo of 1973, the use of electricity in the United States had been doubling just about every 10 years. The rate slowed in 1973 and 1974, but it has speeded up again in the past couple of years.

Considering the enormous use of electricity today, it is hard to realize that by the year 2000, America will be using three to four times as much electricity as it uses today. A growth rate of 5 1/2 percent a year is projected—if certain conservation measures are taken over the next 20 years.

Electricity, unlike oil, natural gas, and coal cannot be discovered or mined. It must be manufactured, and the most prevalent means of generating it is the burning of fuels. Sixteen percent of our petroleum, 15 percent of our natural gas, and 46 percent of the coal we use are burned to make electricity. Other methods of generating electricity are hydropower, geothermal power, and nuclear fission reactors.

Some experts believe that we waste 25 percent of the energy we use. A 100-watt light bulb wastes about 1 fluid ounce of oil (or about 1 1/3 ounces of coal) for every hour it operates unnecessarily. Multiply that by the millions of needlessly lighted bulbs throughout the country on a given day, and the amount is staggering.

Another sort of waste takes place in the generating process itself. It takes three units of fossil-fuel energy to produce one unit of electricity. When coal is used to power an electrical generating plant, for example, only 38 percent of its energy is converted to electricity; the rest is lost. Consequently, heating a home electrically rather than by some other means can be expensive and wasteful. Although electricity is odorless, quiet, and easily transmitted, it does create problems of air pollution and heat emission in the immediate vicinity of a generating plant.

The experiments that follow focus on how electricity is made, how it is used, and how it is measured.

Does a 40-Watt Fluorescent Lamp Give Off More Light Than a 40-Watt Incandescent Lamp?

Materials

 Exposure meter or light meter
 Various lamps

Procedure

- Use the light meter to measure the amount of light given off at a distance of 1, 2, and 3 meters from a 40-watt fluorescent lamp, and then from a 40-watt incandescent lamp.
- Touch both lamps while they are lit. Can you explain the light-meter results?
- How much brighter is it outside in the sun compared to inside?
- How much brighter is a 100-watt lamp than a 40-watt lamp? Why?

How Much Electricity Does an LED (Light-Emitting Diode) Use Compared to a Small Electric Lamp?

Materials

 No. 40 flashlight lamp
 Masking tape
 Copper wire, 25 centimeters
 Light meter
 LED (can be purchased from an electronics supply store)
 Resistor, ½-watt 270 ohm
 Six-volt lantern battery

Note: LEDs are solid-state devices. They *must* be protected with a suitable resistor, and they only work when connected to the proper pole of a battery.

Procedure

- Set up the materials as shown in Figure 28-1. Then connect the X's (see the figure). Hook up the flashlight bulb. Record the results from the meter. Now connect the LED. Record the results. What can you do if the LED doesn't light?

How Long Does It Take a Fuse to Stop the Flow of Electricity Through a Meter?

Materials

 Piece of wood, 10 by 12 centimeters
 Two thumbtacks
 Sheet of thin aluminum foil

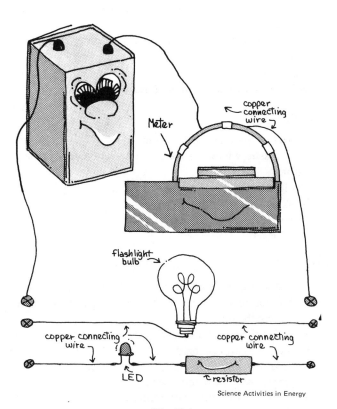

Fig. 28-1

Six-volt lantern battery
Watch with second hand
Bell wire, 25 centimeters
Light meter

Procedure (see Figure 28-2)

- Cut the aluminum foil into a bow shape as shown in Figure 28-2, and attach it to the wood with thumbtacks. Don't cut it more than 1 centimeter wide—thinner is better.
- Connect all wires but one. Then connect the last wire, and start timing. Watch the meter and the fuse. How long does it take for the fuse to blow?
- Would the fuse blow if you used a flashlight lamp instead of a meter in the circuit? Would the fuse blow faster if the middle part were wider? Does your house have fuses or circuit breakers to protect it?

SOLAR ENERGY

Anyone who has had a sunburn, walked through a greenhouse, hung clothes on the line to dry, or watched spring arrive recognizes the power of the sun. In a 1-to-2-week period, the sun transmits more energy to the earth's surface than all the energy stored in the earth's known reserves of coal, oil, and natural gas. And each year more than 500 times as much energy is radiated from the sun to the surface of the United States as is consumed by the country in all conventional forms of energy.

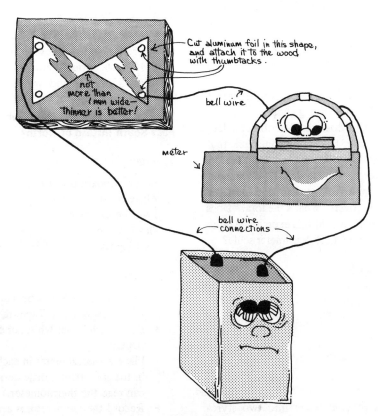

Fig. 28-2

The sun is the most inexhaustible, cleanest source of energy known. Its heat and light arrive week in and week out —free. However, the heat and light do not arrive in uniform amounts each day, or in concentrated form, or at night. Consequently, the barriers to greater use of solar energy by a world faced with dwindling energy resources are significant: mainly, they are diffusion and the inability economically to store solar energy, use it directly, or convert it to electricity.

The activities that follow cover some of the problems and some of the possible solutions.

How Much Faster Does Paper Burn with a 6-Centimeter Magnifier than with a 3-Centimeter Magnifier?

Materials

Magnifier, 6-centimeters or larger
Paper and scissors
Watch
Heavy black construction paper, one sheet
Dark or black lightweight paper, one sheet

Procedure

- Position the magnifying glass so the sun's rays focus the smallest amount of light on a piece of dark or black lightweight paper. How many seconds does it take before the paper starts to smoke? (Be careful—it gets hot!)
- Now cover the lens of the magnifying glass with a piece of black construction paper that has a 3-centimeter hole cut in the center. How long does it take this time? Try it with different colors of construction paper.

Other Ideas to Explore

- Do you think you could boil water this way?
- Solder melts at a low temperature for metal. Do you think you could melt solder?
- Do you think a glass container filled with water might do the same thing as the magnifying glass?
- How much more sunlight does the 6-centimeter lens let in than the 3-centimeter lens? Draw one circle that is 6 centimeters wide and one that is 3 centimeters wide. Draw grids over these circles on a 1-centimeter scale. Count the number of squares and parts of squares that each circle covers. What's the total? Does this explain your results? Would this exercise be more accurate if the grid were on a 0.5-centimeter scale?

How Much Warmer Do Things Get in the Sun than in the Shade?

Materials

Two styrofoam cups
Two thermometers
Watch

Procedure

- Pour equal amounts of cold water into two styrofoam cups (the colder the water the better). Place a thermometer in each cup. Set one cup in the sun and the other in the shade.
- What is the temperature of each after 5, 10, and 15 minutes?

How Much Weight Can a Grape Lose in a Week?

Materials

Grapes, approximately ¼ pound
Two plastic window screens (from hardware or variety store)
Cardboard
Masking tape
String
Food scale

Procedure

- Cut a large hole in the middle of the cardboard. Lay the plastic screen across the hole. Tape the screen to the cardboard.
- Weigh the grapes and place them on the screen.
- Tape the second plastic screen to the first screen to hold the grapes in place. (The screen protects the grapes from birds and insects.)
- Hang the structure with string in a sunny spot during the daytime. Hang it indoors at night and when it's raining.
- When the grapes look like raisins, remove them and weigh them again.

Other Ideas to Explore

- Many fruits can be dried in this way. Apricots, prunes, and peaches are especially good.
- If fresh apricots are 59¢ a pound and dried apricots are $2.00 a pound, which is a better buy?

How Much Hotter Is a House when the Windows Face South Instead of North?

Materials

Two cardboard boxes of same size
White paint or paper
Two thermometers
Plastic wrap
Masking tape

Procedure

- Cut a large hole in one side of one box, and cover it with plastic wrap. Tape tightly all the way around.
- Paint both boxes white, or cover them both with white paper.
- Place a thermometer in each box, and put both boxes in the sun. (Cut a little door in the second box so you can read the thermometer.)
- Record the temperatures after 10, 20, and 30 minutes. What do you find?

Other Ideas to Explore

- Try this experiment at different times during the day. Does it make any difference?
- Add an overhang to both boxes. Does it make any difference? How would the overhang affect the experiment at different times of the year?

What Color Absorbs the Sun's Heat Best?

Materials

White, black, green, red, and blue construction paper, all the same size
Timer
Uniformly sized ice cubes

Procedure

- Place an ice cube on top of each sheet of construction paper. Which melts first?
- Would a house with a white roof be cooler than a house with a dark roof?

Other Ideas to Explore

- Would you get the same results if the paper were on top of the ice cubes?
- Would a house with a dark roof be more or less expensive to air-condition in the summertime?

How Long Will a Magnifying Glass Focus in the Same Spot at 1 P.M.?

Materials

Magnifying glass, 6-centimeter
Chair or stool
Masking tape
Watch with second hand

Procedure

- Adjust the lens to the best focus on a piece of white paper or cardboard. Mark its position at 1 P.M. by making a circle around the light. Measure the time it takes for the light to leave the entire circle.
- How does this experiment show how complicated it is to use reflectors and lenses for the collection of solar energy?

Other Ideas to Explore

- Will the time it takes for the spot of light to move from the circle be the same at 9 A.M.? How about 3 P.M.?

Which Bakes Apples Better—A White Sun Collector or a Foil Collector?

Materials

Four paper cups, unwaxed
White paper
Aluminum foil
Black paper
Plastic food wrap
Scissors
Newspaper

Procedure

- Line two paper cups with black paper. Then place a slice of apple in each cup. Cover the cups tightly with plastic food wrap.
- Then make two large cones, one with white paper lined with aluminum foil and one with white paper only. Wrap a cone around each apple cup.
- Cover the bottom of each cone and cup with another cup to hold it together.
- Aim the cookers at the sun.
- When the apples look cooked, taste to see which cup was the better apple-baker.

CONSERVATION

It's cheaper to save a barrel of oil than to produce an additional barrel. One projection indicates that turning all thermostats in the United States down 2°F in winter and raising them 2°F in summer could save more than half a million barrels of oil a day by 1980.

Dwindling of existing resources—coal, petroleum, natural gas—and excessive financial and environmental costs of developing new sources of energy are twin problems facing the United States and some other nations of the world. Together, they make conservation—the careful use of currently available resources—essential.

Conservation has more immediate results than new resources development. It takes 5 years to bring a new coal mine to full production, 6 to 7 years for a new oil field, and 8 to 10 years for a nuclear power plant. In 1975, with 6 percent of the world's population, the United States consumed about one-third of the world's energy. On a per capita basis, West Germany used 20 percent of the gasoline the United States did, and Sweden half of the total energy of the United States. All three countries had comparable standards of living.

Personal use of energy in the United States—home heating, lighting, air conditioning, cooking, refrigeration, and transportation—accounts for 37 percent of the country's energy use. The experiments in this series on conservation all are related to such personal uses: means of effecting fuel economy in cars, insulation and other forms of home temperature regulation, and food preparation and storage. Among other things, they show how individuals can help reduce waste in the use of energy while maintaining a high standard of living.

Is Your House Drafty?

Materials

Pencil
Scotch tape
Plastic food wrap

Procedure

- Make a draftometer. Cut a 12-by-15-centimeter strip of plastic wrap. Tape it to a pencil. Blow the plastic gently, and see how freely it responds to air movement. (Note: Forced-air furnace must be off when using draftometer.)
- Test your home for air leakage by holding the draftometer near the edges of windows and doors.
- Test your fireplace with the damper open and with it closed.

Other Ideas to Explore

- Visit a hardware store, and find out what products are available to close air leaks around windows and doors.
- Why are drafts energy wasters?
- Look for dirt collected around doors and windows. What does it prove?

Does Your School Waste Heat?

Materials

Two thermometers
Draftometer
Plastic food wrap
Scotch tape
Pencil

Procedure

- Test the temperature in several rooms in your school —for example, your classroom, cafeteria, library, hallway, office, gymnasium, etc.
- Place one thermometer near the ceiling and the other near the floor in each room, and take temperature readings.

Other Ideas to Explore

- How much glass is there on the window side of your classroom? How much wall?
- Feel the glass and the wall. Which carries heat better? Would you save energy if you had fewer windows?
- How is your school situated with regard to the sun?

How Much Does a Bimetallic Strip Bend when Heated or Cooled?

Materials

Piece of metal, 4 centimeters or longer (try a strip of aluminum pie plate)
Piece of bimetallic strip, 4 centimeters or longer (a bimetallic strip is two different metals bonded together— for example, copper and iron can be purchased from a metal-work shop or perhaps a hardware store.)
Large cork or small block of wood
Pieces of wood, about 8-by-4-centimeters
Masking tape, glue, candle, ice cubes

Procedure

- Glue the cork or small wood block to the large one. Attach the bimetallic strip to the top of the cork with masking tape.
- Place the candle under the end of the strip. What happens?
- Remove the candle, and put the ice cube in its place. What happens now?
- Repeat both experiments using the metal strip. Does it react in the same way as the bimetallic strip?

Other Ideas to Explore

- Turn the bimetallic strip over. Repeat the experiments. Does the same thing happen?
- Try making your own bimetallic strip. Use epoxy glue to bond two different kinds of metal, such as a strip from an aluminum pie plate and one from a tin can.
- Can you make a thermometer from the bimetallic strip?

What's the Best Insulator?

Materials

100-watt bulb in ceramic socket
A variety of insulating and noninsulating materials such as wood, aluminum foil, fiberglass (3 or 4 inches), glass, metal, newspaper, heavy cloth, etc.
Four thermometers, cardboard box, watch
Masking tape

Procedure

- Set up the box. Cut windows in all four sides. Leave the top solid. The bottom will be open. Place the light in the center of the box.
- Tape thermometers on the outside of the box.
- Tape four insulating materials over the windows on the *inside* of the box. Then, tape a thermometer to the *outside* of each insulating material and record the temperature.
- Turn the lamp on for 5 minutes. Record the rise in temperature for each material. How much better is the best insulator compared to the worst? Record the results.

Which Gets Warmer Faster?

Materials

Two small tin cans
Two thermometers
Hot plate
Water
Sand
Spoon

Procedure

- Fill one can with water and the other with sand—both to within 3 centimeters of the top. Place the cans on the hot plate, and heat for 2 minutes. Remove the cans from the hot plate. Stir the contents, and measure their temperatures with the thermometers.

- What does this experiment reveal about the energy required to heat water in your home?
- Would a solar water heater be feasible where you live?

How Much Cooler Does Water Get if You Speed Up the Evaporation Process?

Materials

Three flat pans
Fan
Newspapers
Cooking oil
Warm water (40° to 50°C)
Three thermometers

Procedure

- Place newspaper under the pans for insulation.
- Add a few drops of oil to one pan *only*!
- Put enough water in each pan to cover the thermometers (about 2 cups).
- Let the fan blow across one of the pans containing water only.
- Which pan is the coolest? What is the purpose of the oil? How much liquid is left in each pan?
- How could you use this idea to keep things cool on a hot or humid day?

Does a Thermos Keep Liquids Hotter Longer Than Other Containers?

Materials

Two thermos bottles (if possible, one wide-mouth and one narrow-mouth)
Tin can
Newspapers
Glass jar
Water
Plastic pitcher
Measuring cup
Styrofoam cup
Thermometer

Procedure

- Fill the thermoses, tin can, glass jar, plastic pitcher, and styrofoam cup with equal amounts of hot water. Measure the temperature of each. Check the temperature again in ½ hour. Which container is hotter?
- Wrap newspaper around each container, and repeat the experiment. Compare the temperatures of the various containers after ½ hour.
- Repeat both experiments, using cold water. Compare the results to those you get with hot water. Which seems easier to conserve: heat or cold?

How Much More Energy Does It Take to Move a Heavy Object Compared to a Light One?

Materials

Shoe box
Pencils
Weights or books
Assorted rubber bands

Procedure

- Attach the rubber band to the shoe box so that it can be pulled along a tabletop. Measure how long the rubber band will stretch with the box empty.
- Now fill the box with some books or weights, and repeat the experiment. Add even more weights.
- Put some round pencils or dowels under the box. Try pulling it.

Other Ideas to Explore

- Pull an empty wagon with a rope. Then try pulling it while someone sits in it. Is there much difference?
- Why do heavy cars use more gasoline than light cars?

Will a Bicycle Coast Twice as Far if the Tires Have Twice the Pressure?

Materials

Two bicycles, similar type
Tire-pressure gauge

Procedure

- Inflate one bicycle's tires to normal pressure and another to half that amount.
- Have two children of similar weight ride side by side at the same speed. When they reach a selected line on the ground, they should coast the rest of the way. Compare how far each goes.
- Is it important to check the tire pressure on your bicycle? How about on your family car?

CHEMICAL ENERGY

The dominant source of energy in the United States is fossil fuel—coal, oil, and natural gas. The balance among these sources will undoubtedly change in the years ahead with the wider use of nuclear power; nevertheless, America's reliance on them will continue for some time to come.

Less than 12 percent of the coal and oil used in the country goes into manufacturing—of nylon, plastics, etc. Some scientists contend that these energy sources are too limited—too precious—to be wasted on heating, cooling, and electricity, and should instead be devoted almost exclusively to manufacturing. Such a possibility, though tempting, is not realizable in the immediate future simply because the nation's energy needs are so great and because wide use of alternate energy (nuclear and solar power) is still not economically plausible.

All fossil fuels are limited. Most experts believe that more than half of the world's oil has been found, and the remaining portion will be exceedingly difficult to locate. Coal, on the other hand, is more readily available, but it poses significant environmental problems such as air pollution and ash removal.

In fact, significant problems accompany the use of all the fossil fuels: locating them and recovering them at affordable costs, and using them in ways that conserve their supply and are not detrimental to the environment.

We are currently recovering less than 80 percent of the coal and 40 percent of the petroleum from our mines and wells. But new efforts are under way to increase the yields of these sources—limited though they may ultimately be.

Attempts to reduce waste of these fuels—development of cars giving better mileage and of more efficient home appliances and air conditioning, for example—are going to be important parts of American life through the remainder of this century.

Activities in this unit on chemical energy focus on comparisons of various fuels, measurements of temperatures generated by each, depletion of resources, and other facets of fossil-fuel development and use.

Which Has More Energy: Wood or Paraffin?

Materials

Small tin can
Thermometer
Wire coat hangers or stiff wire
Masking tape
Aluminum foil
Small pieces of wood
Postage scale

Procedure (See Figure 28-3)

- Bend a coat hanger to make two stands. Also bend a coat hanger to make a frame for the tin can. Put the small can on the end of the coat hanger, taping it, if necessary. The coat hanger will hold it up.
- Put the small foil-covered stand on the other end of the coat hanger. The top should be 2 centimeters from the can. Add books or cardboard to raise the stand, or bend the frame down lower.
- Fill the can three-fourths full with cold water. Record the temperature. Ignite 30 grams of wood, and record the water temperature again. Try it with 30 grams of wax. Does wood or wax raise the water temperature more?

How Long Does It Take a Birthday Candle to Use Up the Oxygen in a Glass?

Materials

Birthday candle
Drinking glass

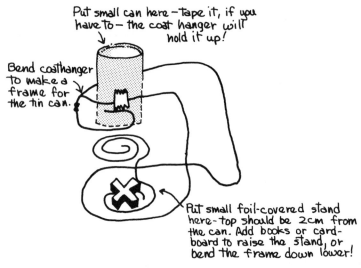

Put small can here—tape it, if you have to—the coat hanger will hold it up!

Bend coathanger to make a frame for the tin can.

Put small foil-covered stand here—top should be 2cm from the can. Add books or cardboard to raise the stand, or bend the frame down lower!

Fig. 28-3

Watch
Aluminum foil
Drinking straw
Lime or limewater*

Procedure

- Place the candle on top of the aluminum foil. Be sure that the drinking glass is dry. Place it over the candle. How long does it take for the candle to go out? What happens to the glass? Why?
- Add 10 milliliters of limewater to the glass, and shake. If it gets milky, it means the glass has carbon dioxide in it.
- Bubble your breath through 10 milliliters of fresh limewater. What happens? Why?
- Relight the candle, and hold some aluminum foil in the flame. What collects on the foil above the candle?
- What pollutants are added to the atmosphere when large amounts of oil are burned for energy?
- Can you set up an experiment to show that a piece of burning wood does the same thing as a burning candle?

How Much Methane Gas Can You Collect in 2 Days from a Small Amount of Coal?

Materials

Soft or bituminous coal
Funnel
Quart jar
Water
Test tube
Rubber band

*Make limewater by shaking 1 teaspoon of lime with 1 cup of water. Let stand until it becomes clear.

Procedure

- Hammer the lumps of coal into a coarse powder. Put the coal in the funnel, and place the funnel inside the jar. Fill the jar with water. Then fill the test tube with water, and place it over the funnel. Don't let any air get in. Mark the test tube at the waterline with a rubber band or china marker. Methane will begin to collect in the test tube. After the tube seems to be filled (this should take about 2 days), does the coal still give off methane gas (natural gas)? How can you find out?
- Will the methane burn? Remove the test tube, and put your thumb over the top. Light a match and turn the test tube upright. Let the gas out. What happens?
- Why is mining a dangerous occupation?

Which Reduces Friction More: No. 40 Oil, No. 10 Oil, or Cooking Oil?

Materials

No. 40 oil, No. 10 oil, cooking oil
Paper towels
Four pieces of wood, 25 by 8 by 2 centimeters, one side smooth
Protractor
Four blocks of wood, approximately 6 by 6 by 4 centimeters, with saw-cut (not sanded) edges

Procedure

- Coat one side of the boards and blocks with very thin layers of oil. Rub them down with paper towels. Don't leave any extra oil on them.

- Raise the board to make an inclined plane. Drop the block gently onto the board at bigger and bigger angles until the block slides quickly downhill. Record the angle at which the blocks slides best. (Measure from the same spot on each block of wood.)
- Slide the blocks down the board, and record the angles. What happens?

Are Small Amounts of Soap or Detergent Harmful to Plant Growth?

Materials

Radish or beet seeds
Soap and detergent
Three styrofoam cups
Soil
Water
Two quart jars

Procedure

- Fill the jars with the same amount of water, and then make soap and detergent solutions (¼ teaspoon detergent + water; ¼ teaspoon soap + water).
- Plant two seeds in each cup, about 1 centimeter deep, and place on a sunny windowsill. Water every day as follows:
    Cup 1, 2 tablespoons soap solution
    Cup 2, 2 tablespoons detergent solution
    Cup 3, 2 tablespoons tap water
- Chart the growth of each cup. Compare the results: Which sprouts first? Which grows fastest?

Part 3
Where to Get Help

29 Associations and Organizations

Air Conditioning and Refrigeration Institute
1815 N. Fort Myer Dr.
Arlington, VA 22209
(703) 524-8800

Alternative Sources of Energy
P.O. Box 90A
Milaca, MN 56353
(612) 983-6892

American Academy of Transportation
2783 Colony Rd.
Ann Arbor, MI 48104
(313) 665-0625

American Association for the Advancement of Science
1515 Massachusetts Ave., NW
Washington, DC 20005
(202) 467-4400

American National Standards Institute
1430 Broadway
New York, NY 10018
(212) 354-3300

American Society of Heating, Refrigerating, and Air
  Conditioning Engineers
United Engineering Center
345 E. 47th St.
New York, NY 10017
(212) 644-7953

American Society for Testing and Materials
1916 Race St.
Philadelphia, PA 19103
(215) 299-5400

American Underground Association
c/o Thomas C. Atchinson
Dept. of Civil & Mineral Engineering
University of Minnesota
Minneapolis, MN 55455
(612) 373-2851

American Wind Energy Association
55468 County Road 31
Briston, IN 46507
(219) 848-4360

Association of Home Appliance Manufacturers
20 N. Wacker Dr.
Chicago, IL 60606
(312) 236-2921

Atomic Industrial Forum, Inc.
7101 Wisconsin Ave., NW
Washington, DC 20014
(202) 833-9234

Automotive Information Council
28333 Telegraph Rd.
Southfield, MI 48076
(313) 358-0290

Building Officials and Code Administrators International
1313 E. 60th St.
Chicago, IL 60637
(312) 947-2580

Center for Energy Policy and Research
New York Institute of Technology
Old Westbury, NY 11568
(516) 686-7744

Citizens Association for Sound Energy
2125 W. Clarendon Dr.
Dallas, TX 75208
(214) 948-8489

Clean Energy Research Institute
P.O. Box 248294
Coral Gables, FL 33124
(305) 284-4666

Consumer Action Now
30 E. 68th St.
New York, NY 10021
(212) 628-2295

Consumers' Research
Bowerstown Rd.
Washington, NJ 07882
(201) 689-3300

Ectope Group
747 16th Ave. East
Seattle, WA 98112
(206) 322-3753

Electric Power Research Institute
3412 Hillview Ave.
Palo Alto, CA 94304
(415) 493-4800

Environmental Action Reprint Service
2239 E. Colfax
Denver, CO 80206
(303) 320-6537

Gas Appliance Manufacturers Association
1901 N. Fort Myer Dr.
Arlington, VA 22209
(703) 525-9565

Gas Processor Supplier's Association
1812 First National Bank Building
Tulsa, OK 74103
(918) 582-5112

The Geyser
P.O. Box 1525
Beverly Hills, CA 09213
(213) 393-4132

Independent Petroleum Association of America
1101 Sixteenth St., NW
Washington, DC 20036
(202) 466-8240

Insulation Contractors Association of America
905 Sixteenth St., NW
Washington, DC 20006
(202) 347-2791

International Association for Hydrogen Energy
P.O. Box 248266
Coral Gables, FL 33124
(305) 284-4666

Lewis Research Center
21000 Brookpark Rd.
Cleveland, OH 44135
(216) 433-4000

Major Appliance Consumer Action Panel
20 N. Wacker Dr.
Chicago, IL 60606
(312) 236-3165

Midwest Energy Alternatives (MEA)
P.O. Box 83202
Lincoln, NE 68501
(402) 477-3101

National Academy of Sciences
Office of Information
2101 Constitution Ave.
Washington, DC 20418
(202) 389-6518

National Association of Electrical Distributors
600 Summer St.
Stamford, CT 06901
(203) 327-1290

National Association of Home Builders of the U.S.
Fifteenth and M Sts., NW
Washington, DC 20005
(202) 452-0200

National Association of Home Insulation Contractors
4750 Wisconsin Ave., NW
Washington, DC 20016
(202) 363-8100

The National Center for Appropriate Technology
P.O. Box 3838
Butte, MT 59701
(406) 723-6533

National Coal Association
1130 Seventeenth St., NW
Washington, DC 20036
(202) 628-4322

National Conference of States on Building Codes and
  Standards
Building 226
National Bureau of Standards
Washington, DC 20234
(202) 533-5561

National Electric Manufacturers Association
155 E. 44th St.
New York, NY 10017
(212) 682-1500

National Energy Information Center
Federal Building
Twelfth St. and Pennsylvania Ave., NW
Washington, DC 20461
(202) 566-9820

National Home Improvement Council
11 E. 44th St.
New York, NY 10017
(212) 867-0121

National Insulation Contractors Association
1120 Nineteenth St., NW
Washington, DC 20036
(202) 233-4406

National Remodelers Association
50 E. 42nd St.
New York, NY 10017
(212) 687-5224

National Science Foundation
1800 G St., NW
Washington, DC 20550
(202) 655-4000

National Solar Heating and Cooling Information Center
P.O. Box 1607
Rockville, MD 20850
(800) 523-2929 (Maryland)
(800) 462-4983 (Pennsylvania)

Oak Ridge Associated Universities
P.O. Box 117
Oak Ridge, TN 37830
(615) 483-8411

Photovoltaic Project
Massachusetts Institute of Technology
Lincoln Laboratory
P.O. Box 73
Lexington, MA 02173
(617) 253-1000

Photovoltiac Systems Definition Project
Division 5719
Sandia Laboratories
Albuquerque, NM 87115
(505) 264-5678

Solar Energy Association
1001 Connecticut Ave., NW
Suite 632
Washington, DC 20036
(202) 293-1000

Solar Energy Society of America
P.O. Box 4264
Torrance, CA 90510
(213) 326-3283

Transportation Alternatives
20 Exchange Pl.
Rm. 5500 "F"
New York, NY 10005
(212) 221-6895

Transportation Research Board
2101 Constitution Ave., NW
Washington, DC 20418
(202) 389-6336

Total Environmental Action (TEA)
Church Hill
Harrisville, NH 03450
(603) 827-3374

U.S. National Committee of the World
Energy Conference
1620 Eye St.
Suite 808
Washington, DC 20006
(202) 331-0415

Wind Energy Society of America
1700 E. Walnut Street
Pasadena, CA 91106
(213) 433-1786

Wood Energy Institute
P.O. Box 1
Fiddlers Green
Waitsfield, VT 05673
(802) 496-2508

30 State Energy Offices

ALABAMA

Alabama Energy Management Board
Alabama Development Office
Montgomery, AL 36130
(205) 832-5010
Contact: Edwin G. Hudspeth, Staff Director

ALASKA

Alaska State Energy Office
Mackay Building
338 Denali St.
Anchorage, AK 99501
(907) 272-0527
Contact: Clarissa M. Quinlan, Acting Director

AMERICAN SAMOA

Office of Traffic Safety
Government of American Samoa
Pago Pago, American Samoa 96739
Overseas Operator: 688-9187
Contact: Marvin Leach

ARIZONA

Office of Economic Planning and Development
Capital Tower, Room 507
Phoenix, AZ 85007
(602) 271-3303
Contact: Thomas Lynch, Chief Office of Energy

ARKANSAS

Energy Conservation and Policy Office
960 Plaza West Building
Little Rock, AR 72205
(501) 371-1379
Contact: Mac B. Woodward, Director

CALIFORNIA

Energy Resources Conservation and Development
  Commission
1111 Howe Ave.
Sacramento, CA 95825
(916) 322-4523

Contact: Bob Foster, Director
Conservation Division

COLORADO

Colorado State Energy Conservation Office
State Capitol
Denver, CO 80203
(303) 892-2507
Contact: Buie Seawell, Director

CONNECTICUT

Office of Policy and Management
State Capitol
Room 308
Hartford, CT 06115
(203) 566-2800
Contact: Anthony V. Milano, Secretary

DELAWARE

Office of Management, Budget, and Planning
Townsend Building, Third Floor
P.O. Box 1401
Dover, DE 19901
(302) 678-4271
Contact: Nathan Hayward III, Director

DISTRICT OF COLUMBIA

Municipal Planning Office
Mayor's Office
Munsey Building, Room 409
1329 E. St., NW
Washington, DC 20004
(202) 629-5111
Contact: Ben Gilbert

FLORIDA

State Energy Office
108 Collins Building
Tallahassee, FL 32304
(904) 488-6764
Contact: Dr. Carlos Warren, Director

STATE ENERGY OFFICES 293

GEORGIA

Office of Energy Resources
270 Washington St., SW
Atlanta, GA 30334
(404) 656-3874
Contact: Omi Walden, Program Director

GUAM

Guam Energy Office
Box 2950
Agana, GU 96910
Overseas Operator 472-8711
Contact: Clark Jewell, Administrator

HAWAII

State Energy Office
Department of Planning and Economic Development
P.O. Box 2359
Honolulu, HI 96804
(808) 548-4080
Contact: Alfred S. Harris, Manager

IDAHO

Idaho Office of Energy
State House
Boise, ID 83720
(208) 384-3258
Contact: Kirk Hall, Director

ILLINOIS

Division of Energy
Department of Business and Economic Development
222 S. College Ave., 1st Floor
Springfield, IL 62706
(217) 782-7500
Contact: Marvin Nodiff, Director

INDIANA

Department of Commerce Energy Group
115 N. Penn Ave., 7th Floor
Indianapolis, IN 46204
(317) 633-6733
Contact: William J. Sorrels, Director

IOWA

Iowa Energy Policy Council
707 E. Locust
Des Moines, IA 30517
(515) 281-4420
Contact: Rodson L. Riggs, Director

KANSAS

State of Kansas Energy Office
503 Kansas Ave., Room 241
Topeka, KS 66603
(913) 296-2496
Contact: Steven Harris

KENTUCKY

Kentucky Department of Energy
Capitol Plaza Tower, 9th Floor
Frankfort, KY 40601
(502) 564-7416
Contact: Damon W. Harrison, Commissioner

LOUISIANA

Department of Natural Resources
P.O. Box 44396
Capitol Station
Baton Rouge, LA 70804
(504) 389-5252 or 2771
Contact: William C. Huls, Secretary

MAINE

Office of Energy Resources
55 Capitol St.
Augusta, ME 04330
(207) 289-2196
Contact: Gary Linton, Acting Director

MARYLAND

Maryland Energy Policy Office
301 W. Preston St.
Baltimore, MD 21201
(301) 383-6810
Contact: Dr. Donald E. Milsten, Director

MASSACHUSETTS

Massachusetts Energy Policy Office
McCormack Building, Room 1413
Boston, MA 02108
(617) 727-4732
Contact: Harry Lee, Director

MICHIGAN

Michigan Energy Administration
Michigan Department of Commerce
Law Building, 4th Floor
Lansing, MI 48913
(517) 374-9090
Contact: Eugene B. Hedhes

MINNESOTA

Minnesota Energy Agency
740 American Center Building
160 E. Kellogg Blvd.
St. Paul, MN 55101
(612) 296-5120
Contact: John Millhone, Director

MISSISSIPPI

Mississippi Fuel and Energy Management Commission
Woolfolk State Office Building, Room 1307
Jackson, MS 39302
(601) 354-7406
Contact: William Maddox

MISSOURI

Missouri Energy Program
Department of Natural Resources
P.O. Box 1309
Jefferson City, MO 65101
(314) 951-4000
Contact: J. Abbott, Program Manager

MONTANA

Montana Energy Office
c/o Lt. Governor's Office
Capital Building 104
Helena, MT 59601
(406) 449-3940
Contact: Jerry W. Toner, Energy Conservation Officer

NEBRASKA

Nebraska Energy Office
P.O. Box 95085
Lincoln, NE 95085
(402) 471-2867
Contact: Goerge J. Dworak, Director

NEVADA

Nevada Department of Energy
1050 E. Will, Suite 405
Carson City, NV 89710
(702) 885-5187
Contact: Noel Clark, Administrator

NEW HAMPSHIRE

Executive Vice Chairman, Governor's Council on Energy
26 Pleasant St.
Concord, NH 03301
(603) 271-2711
Contact: Marshall Cobleigh

NEW JERSEY

Department of Energy
101 Commerce St.
Newark, NJ 07102
(201) 648-3290
Contact: Joel R. Jacobson, Commissioner

NEW MEXICO

Energy Resources Board
P.O. Box 2770
Santa Fe, NM 87501
(505) 827-2146
Contact: Fred O'Cheskey, Director

NEW YORK

New York State Energy Office
Agency Building No. 2, 10th Floor
Empire State Plaza
Albany, NY 12223
Contact: James L. Larocca, Commissioner

NORTH CAROLINA

Energy Division
215 E. Lane St.
Raleigh, NC 27601
(919) 733-2230
Contact: Brian Flattery, Director

NORTH DAKOTA

North Dakota Office of Energy Management
Capitol Place Office
1533 N. 12th St.
Bismark, ND 58501
(701) 244-2250
Contact: T. Dwight Connor, Energy Conservation
 Coordinator

OHIO

Ohio Energy and Resource Development Agency
State Office Tower, 25th Floor
30 E. Broad St.
Columbus, OH 43215
(614) 466-3465
Contact: Robert S. Ryan, Director

OKLAHOMA

Oklahoma Department of Energy
4400 Lincoln Blvd., Suite 251
Oklahoma City, OK 73105
(405) 521-3941
Contact: Richard G. Hill

OREGON

Department of Energy
528 Cottage St., NE
Salem, OR 97310
(503) 378-4128
Contact: Fred Miller, Director

PENNSYLVANIA

Governor's Energy Council
State Street Building, 6th Floor
Harrisburg, PA 17120
(717) 787-8186
Contact: William B. Harral, Executive Director

PUERTO RICO

Department of Consumer Affairs
Box 41059, Minnillas Station
Santurce, PR 00940
(809) 726-6190
Contact: Dr. Carmen T. Pesquera de Busquets, Secretary

RHODE ISLAND

State Energy Coordinator
80 Dean St.
Providence, RI 02903
(401) 277-3370
Contact: Robert Coffey, State Energy Coordinator

SOUTH CAROLINA

Energy Management Office
Edgar Brown Building
1205 Pendleton St., Room 342
Columbia, SC 29201
(803) 758-2050
Contact: Robert J. Hirsch, Director

SOUTH DAKOTA

Office of Energy Policy
State Capitol
Pierre, SD 57501
(605) 224-3603
Contact: James Van Loan, Director

TENNESSEE

Tennessee Energy Office
250 Capitol Hill Building
Nashville, TN 37219
(615) 741-2994
Contact: Edward J. Spitzer, Director

TEXAS

Governor's Office
7703 N. Lamar Blvd.
Austin, TX 78752
(512) 975-5491
Contact: Alvin Askew,
  Administrative Assistant for Energy Resources

UTAH

State Energy Coordinator, Department of Natural Resources
State Capitol, Room 438
Salt Lake City, UT 84114
(801) 533-5356
Contact: Clifford R. Collings

VERMONT

Vermont State Energy Office
Pavillion Office Building
109 State St.
Montpelier, VT 05602
(802) 828-2768
Contact: Brendan Whitaker, Director

VIRGIN ISLANDS

Office of the Budget
P.O. Box 90
St. Thomas, VI 00801
(809) 774-0750
Contact: Justin Moorhead, Director

VIRGINIA

Virginia Energy Office
823 E. Maine St.
Room 300
Richmond, VA 23219
(804) 786-8451
Contact: Louis R. Lawson, Jr., Director

WASHINGTON

Washington State Energy Office
1000 S. Cherry St.
Olympia, WA 98504
(206) 753-2417
Contact: Lawrence B. Bradley, Director

WEST VIRGINIA

Fuel and Energy Division
Governor's Office of Economic and Community
  Development
1262½ Greenbrier St.
Charleston, WV 25305
(304) 343-8860
Contact: John D. Anderson III, Director

WISCONSIN

Office of State Planning and Energy
1 West Wilson St.
Room B130
Madison, WI 53702
(608) 266-8234
Contact: Craig Adams, Acting Director

WYOMING

State Capitol Building
Cheyenne, WY 82008
(307) 777-7434
Contact: Jerry Mahoney, Chief Assistant to the Governor

31   Insulation Contractors

The following list of insulation contractors was furnished by the National Association of Home Insulation Contractors. It is a roster of NAHIC members. If you are unable to locate a contractor near your home, contact the association for assistance. Call or write:

National Association of Home Insulation Contractors
4750 Wisconsin Ave., NW
Washington, DC 20016
(202) 363-8100
Contact: Gerald L. McDonald, President

Another association involved with insulation contractors is the Insulation Contractors Association of America. At the time of publication, the ICAA was not able to supply us with a list of their members. However, they too can assist you in finding an insulation contractor near your home. Call or write:

Insulation Contractors Association of America
905 16th St., NW
Washington, DC 20006
(202) 347-2791
Contact: Arthur W. Johnson, Executive Director

Before getting in touch with any insulation contractor, read "How to Buy Insulation" on p. 231. The information therein will prepare you for what lies ahead and may save you some of your hard-earned dollars.

INSULATION CONTRACTORS

ALABAMA

Bob's Insulation, Inc.
P.O. Box 152
Ozark, AL 36360
(205) 774-4944

Delta Enterprises
P.O. Box 564
Thomasville, AL 36784
(205) 636-5725

ARKANSAS

Edward's Insulating, Inc.
No. 5 Village Sq.
P.O. Box 608
Booneville, AR 72927
(501) 675-4173

CALIFORNIA

Bob Dixon Construction
P.O. Box 1356
Big Bear Lake, CA 92315
(714) 585-3301

CONNECTICUT

D. & J. Home Insulators
27 Monroe Ave.
Jewett City, CT 06351
(203) 376-0865

DISTRICT OF COLUMBIA

Capitol Insulation Co.
4750 Wisconsin Ave., NW
Washington, DC 20015
(202) 363-8100

The Insulation Co., Inc.
P.O. Box 3033
Washington, DC 20010
(202) 232-3100

FLORIDA

American Insulation Systems
P.O. Box 14725 Lake Shore Sta.
Jacksonville, FL 32210
(904) 384-1000

Quality Fiberglass Industries
Bob R. Pope
Box 641
Longwood, FL 32750
(305) 831-7666

Deep South Insulation
1102 Kissimmee St.
Tallahassee, FL 32304
(904) 385-7147

Five Hills, Inc.
P.O. Box 13905
Tampa, FL 33618
(813) 834-8241

GEORGIA

T. MacNeil, Inc.
2146 Stewart Ave., SW
Atlanta, GA 30315
(404) 768-9137

IDAHO

Bi-State Distributing Co.
1907 Burrell Ave.
Lewiston, ID 83501
(208) 746-1271

ILLINOIS

Thomason Roofing & Siding Inc.
P.O. Box 382
Blandinsville, IL 61420
(309) 652-3243

Insulation Service
Ronald H. Fairley
P.O. Box 187
Charleston, IL 61930
(217) 345-3889

All Weather Insulating & Roofing Co.
3836 W. 148th
Midlothian, IL 60445
(312) 389-6310

Leppiaho Insulation Co.
1149 Highland Rd.
Mundelein, IL 60060
(312) 566-4017

Orland Insulation, Inc.
13443 Southwest Hwy.
Orland Park, IL 60464
(312) 361-2466

Home Comfort Insulation
1314 E. Marietta
Peoria Heights, IL 61614
(309) 685-8216

Ferry Construction Co.
15 Forest Green
Springfield, IL 62702
(217) 546-0510

INDIANA

Mr. Foamy of Anderson
2444 E. Lynn
Anderson, IN 46014
(317) 649-7179

River Vista Sales & Service
410 W. Lincoln Ave.
Goshen, IN 46526
(219) 533-1194

C & D Insulation
Route 7, Box 240
Greenfield, IN 46140
(317) 899-5365

Rapco Foam Insulation
7916 Zionsville Rd.
Indianapolis, IN 46268
(317) 297-2726

Viking Blown Insulation
P.O. Box 52
Plymouth, IN 46563
(219) 936-9603

IOWA

Northern Iowa Insulation
P.O. Box 223
Badger, IA 50516
(515) 545-4549

KANSAS

Sunflower Insulation
P.O. Box 271
Newton, KS 67114
(316) 283-4755

Drywall Construction, Inc.
4830 S. Topeka
Topeka, KS 66609
(913) 862-1550

KENTUCKY

Jason & Sons, Inc.
629 Western Reserve Rd.
Crescent Springs, KY 41011
(606) 341-0066

Sunshine Energy Corp.
P.O. Box 211
Harrodsburg, KY 40330
(606) 734-5921

Ener-Check, Inc.
P.O. Box 16
Hopkinsville, KY 42240
(502) 885-7153

Becklin Insulation Co.
2338 Frankfort Ave.
Louisville, KY 40206
(502) 897-3573

MARYLAND

Aerolite Foam Home Insulators
5531 Highbridge St.
Baltimore, MD 21227
(301) 242-3521

Baltimore Insulation Systems
8845 Kelso Dr.
Baltimore, MD 21221
(301) 687-3626

James Cox & Son, Inc.
2225 Huntington Ave.
Baltimore, MD 21211
(301) 837-1512

AAA Metro Insulation, Inc.
5200 River Rd.
Bethesda, MD 20016
(301) 656-1111

Universal Specialties
4700 Webster St.
Bladensburg, MD 20710
(301) 277-1015

B. H. Improvement Co.
9243 Perfect Hour
Columbia, MD 21045
(301) 992-7844

Thermal-Acoustic Foam Insulation
6655 Dobbin Rd.
Columbia, MD 21045
(301) 995-0510

Weiger Home Improvement Co., Inc.
5808 Rolling Dr.
Derwood, MD 20855
(301) 948-9241

Air Systems
6600 Halleck
District Heights, MD 20028
(301) 735-6256

Better Home Insulation
2404 Pleasantville Rd.
Fallston, MD 21047
(301) 838-3399

Better Home Insulators
7911 Parston Dr.
Forestville, MD 20028
(301) 568-4600

Woodland Properties, Inc.
4867 Church La.
Galesville, MD 20765
(301) 867-4941

Metropolitan Foam Insulation, Inc.
150 A Penrod Court
Glen Burnie, MD 21061
(301) 761-3350

Done-Rite Assoc., Inc.
3708 Longfellow St.
Hyattsville, MD 20782
(301) 277-2478

Insulators of Maryland
3831 F. Plyers Mill Rd.
Kensington, MD 20795
(301) 933-5566

Spray Systems, Inc.
P.O. Box 349
Millersville, MD 21108
(301) 987-5930

Gaylebon Insulation Co.
16300 Oxford Ct.
Mitchellville, MD 20716
(301) 464-2333

Aerofoam Insulating System
4400 Old Court Rd.
Pikesville, MD 21208
(301) 486-7272

Thermal Contracting Corp.
3540 Carriage Hill Circle
Randallstown, MD 21133
(301) 922-0448

Energy Control, Inc.
P.O. Box 302
Reistertown, MD 21136
(301) 833-7517

EMCO Insulation & Improvement Co.
927 Maple Ave.
Rockville, MD 20851
(301) 468-2000

Savon Home Insulation, Inc.
4929 Arctic Terr.
Rockville, MD 20853
(301) 657-3322

Hearn Insulation Co.
2410 A Linden La.
Silver Spring, MD 20910
(301) 565-9300

Home Air Conditioning Co.
9112 Pennsylvania Ave.
Silver Spring, MD 20910
(301) 585-8100

Roy Barnes, Inc.
P.O. Box 9195
Suitland, MD 20023
(301) 449-3355

Weather Controller's
3018 Walnut La.
Waldorf, MD 20601
(301) 843-2336

MASSACHUSETTS

Bell Industries, Inc.
9 School St.
Dracut, MA 01826
(617) 957-1900

Foxboro Insulation & Solar Systems
6 Margaret Rd.
Foxboro, MA 02035
(617) 543-2148

All Star Insulation, Inc.
245 Stevens St.
Lowell, MA 01850
(617) 256-5226

Wolons Enterprises, Inc.
P.O. Box 97 Sterling Jct.
West Boylston, MA 01583
(617) 835-6313

MICHIGAN

Savoi Insulation Co.
9650 Dixie Hwy.
Clarkston, MI 48016
(313) 625-2601

Air-Tite Insulation, Inc.
P.O. Box 82
882 N. Holbrook St.
Plymouth, MI 48170
(313) 453-0250

Superior Insulation Co.
P.O. Box 135
407 S. Baker St.
St. Johns, MI 48879
(517) 224-7581

Rapco Foam of Kalamazoo
P.O. Box 97
Schoolcraft, MI 49087
(616) 679-4724

Morrison Insulation Co., Inc.
19516 So. Glen
Trenton, MI 48183
(313) 285-7170

MISSOURI

Henges Interiors
4133 Shoreline Dr.
Earth City, MO 63045
(314) 291-6600

Biltwell Siding Co., Inc.
P.O. Box 385
Richmond, MO 64085
(816) 776-5050

Advanced Foam Insulation, Inc.
7701 Telegraph Rd.
St. Louis, MO 63129
(314) 846-2812

MONTANA

Intermountain West Insulation
P.O. Box 1612
Billings, MT 59103
(406) 652-2852

Bob's Insulation & Garage Door Services
P.O. Box 1442
Great Falls, MT 59403
(406) 727-4011

NEBRASKA

National Insulation
3601 Calvert
Lincoln, NB 68506
(402) 489-3587

NEW HAMPSHIRE

Norm Cobb Contractor
P.O. Box 366, River Street
Charlestown, NH 03603
(603) 826-3663

Home Improvement Co.
Meriden Rd.
Lebanon, NH 03766
(603) 448-4100

United Thermal Insulators
P.O. Box 546, 2 Heather Dr.
Merrimack, NH 03054
(603) 424-2100

NEW JERSEY

Hygrade Insulators
94 Bullman St.
Phillipsburg, NJ 08865
(201) 454-1865

Penn-Jersey Foam Insulation
412 Stelko Ave.
Phillipsburg, NJ 08865
(201) 454-3000

Eastern Insulation Corp.
315 Richard Mine Rd.
Wharton, NJ 07885
(201) 328-0108

NEW YORK

Heat-Savers Insulation
141 Henry St.
Binghamton, NY 13902
(607) 723-7077

Four Seasons Insulation
805 Rein Rd.
Cheektowaga, NY 14225
(716) 632-8609

Town Insulation, Inc.
455 Cayuga Rd.
Cheektowaga, NY 14225
(716) 839-5064

Truax & Hovey, Ltd.
P.O. Box 477
Liverpool, NY 13088
(315) 652-9033

International Energy Systems
Cellutherm Insulating Division
199 Wales Ave.
Tonawanda, NY 14150
(716) 693-3990

NORTH CAROLINA

Tar Heel Insulation Co., Inc.
2710 Edmund St.
Durham, NC 27705
(919) 286-0043

Austin Insulating Co., Inc.
P.O. Box 415
Pineville, NC 28134
(704) 332-1224

Teachey Insulation
Route 1, Box 30A
Teachey, NC 28464
(919) 285-3189

B & B Insulation Service
303 N. Second St.
Wallace, NC 28466
(919) 285-5303

NORTH DAKOTA

Capitol Foamers
P.O. Box 2104
Bismark, ND 58501
(701) 258-0200

OHIO

John R. Amole, Inc.
704 Huntington La.
Chillicothe, OH 45601
(614) 775-5089

Orrville Insulation Co.
315 S. Walnut St.
Orrville, OH 44667
(216) 682-2233

Ace Insulation Co., Inc.
3633 Branda Dr.
Toledo, OH 43614
(419) 382-7561

OKLAHOMA

Pop and Sons Insulation Work
P.O. Box 417
Dill City, OK 73641
(405) 674-3538

Comfort Control Insulation
P.O. Box 3524
Enid, OK 73701
(405) 233-7300

Custom Insulators
Route 1, Box 69
Jones, OK 73049
(405) 454-2050

Hawkins Electric & Insulation
Route 3
McAlester, OK 74501
(918) 423-2925

Attic Insulation, Inc.
1311 N. Meridian No. 103
Oklahoma City, OK 73107
(405) 799-4474

High Plains Insulation
813 NW 66th
Oklahoma City, OK 73116
(405) 842-6694

High-R Home Insulators
5908 NW 83rd
Oklahoma City, OK 73132
(405) 947-1794

PENNSYLVANIA

Insul-Fil Manufacturing Co.
P.O. Box 62
Bala-Cynwyd, PA 19004
(215) 626-3466

Eugene Williams Insulation Co.
Green St.
Brockton, PA 17925
(717) 668-0197

Varbro, Inc.
63 Petrie Rd.
Coraopolis, PA 15108
(412) 264-6210

Amercian Energy Conservation
1226 W. 54th St.
Erie, PA 16509
(814) 868-3414

J-Z Insulation Co.
26 N. Lehigh Ave.
Frackville, PA 17931
(717) 874-1966

Miller's Insulating Service
R.D. No. 2
Girard, PA 16417
(814) 474-3986

Pennsylvania Weatherization Program
P.O. Box 156
Harrisburg, PA 17120
(717) 939-1396

Air-Tech Insulation, Inc.
20 W. Third St.
Hazelton, PA 18201
(717) 454-7045

Varbro Insulation Co., Inc.
Price St. and Pine St.
Holmes, PA 19043
(215) 461-1200

Tercek Electric & Insulation, Inc.
R.D. No. 3 Eisenhower Blvd.
Johnstown, PA 15904
(814) 266-3490

Mijima Conservation Systems
P.O. Box 214
Lafayette Hill, PA 19444
(215) 828-4843

Viking Insulation Systems
P.O. Box 75, Turnpike St.
Milesburg, PA 16853
(814) 355-0851

All-Seasons Insulation
1711 McKean St.
Philadelphia, PA 19145
(215) 271-5400

Energy Conservation & Insulation, Inc.
P.O. Box 14105
Pittsburgh, PA 15239
(412) 373-3584

Mur-lin Insulation Co.
324 Laurel St.
Pittston, PA 18640
(717) 654-1438

W. Rauschenberger & Co.
R.D. 1, Box 39
Uniondale, PA 18470
(717) 679-2628

Energy Saver Systems, Inc.
140 S. Main St.
Wilkes-Barre, PA 18701

RHODE ISLAND

Thermotrol Co.
390 Main St.
East Greenwich, RI 02818
(401) 884-6658

C & O Energy Conservation
108 Long Pasture Way
Tiverton, RI 02878
(401) 624-4203

SOUTH CAROLINA

Thermal Insulation Co., Inc.
511 Rhett St.
Greenville, SC 29601
(803) 271-1486

SOUTH DAKOTA

Advance Home Improvement
1815 E. 10th St.
Sioux Falls, SD 57103
(605) 339-9413

TENNESSEE

Guardian Insulation Co.
P.O. Box 11791
Knoxville, TN 37919
(615) 584-8871

Stewart Insulating Co.
Route 4, Box 180C
Livingston, TN 38570
(615) 823-5106

Chemrock Insulation of Tennessee
P.O. Box 7151
Nashville, TN 37210
(615) 254-5003

Dale Insulation Co.
1700 Eighth Ave.
Nashville, TN 37203
(615) 254-3454

Wade Electric Co., Inc.
P.O. Box 484
Trenton, TN 38382
(901) 855-2025

UTAH

Intermountain West Management
890 E. 700 North
American Fork, UT 84003
(801) 756-2696

Solar Heat & Insulation
130 E. Center St.
Midvale, UT 84047
(801) 561-1427

VIRGINIA

A P T Insulation, Inc.
358 S. Pickett St.
Alexandria, VA 22304
(703) 370-7070

Cullu-Therm Insulation, Inc.
4911 Shirely St.
Alexandria, VA 22309
(703) 780-7333

Van Note Insulation Co.
1450 Duke St.
Alexandria, VA 22314
(703) 549-0776

Hodges Home Insulation Co.
6856 Lee Hwy.
Arlington, VA 22213
(703) 532-0184

Alternative Systems, Inc.
P.O. Box 103
Bridgewater, VA 22812
(703) 828-6107

Home Insulation Co.
P.O. Box 7604
Charlottsville, VA 22906
(804) 295-2833

B & B Refrigeration Co.
2711 Dorr Ave.
Fairfax, VA 22030
(703) 560-9444

U.S. Insulation Corp.
2817 B-Dorr Ave.
Fairfax, VA 22030
(703) 698-8683

Dura Home
3010 Wallace Dr.
Falls Church, VA 22042
(703) 573-6731

People's Insulation Co., Inc.
7345 Lockport Pl.
Lorton, VA 22079
(703) 550-9885

Woodbridge Insulation
7303 Lockport Pl.
Lorton, VA 22079
(703) 550-7667

Smith Brothers, Inc.
2415 Wesley St.
Portsmouth, VA 23707
(804) 399-7543

Lakeside Insulation Co., Inc.
P.O. Box 7615
Richmond, VA 23231
(804) 746-4544

Davenport Insulation, Inc.
P.O. Box 706
Springfield, VA 22150
(703) 550-9600

Air Treatment Co.
518 Mill St., NE
Vienna, VA 22180
(703) 938-0550

WASHINGTON

American Insulation Co.
1932 First Ave., S
Seattle, WA 98134
(206) 623-5746

WEST VIRGINIA

Vaughn Insulating Co.
P.O. Box 128
Belmont, WV 26134
(304) 665-2451

WISCONSIN

Gray's Home Insulation
P.O. Box 432
Burlington, WI 53105
(414) 763-8481

Thermax Insulation Co., Ltd.
2631 S. Greeley St.
Milwaukee, WI 53207
(414) 933-1311

32 National Manufacturers and Distributors of Solar Equipment

The following codes were devised to cover general categories of equipment and accessories. The manufacturer or distributor should be contacted for more specific product information.

APPLICATIONS

AAG: Agriculture
ACA: Absorption cooling
ACD: Desiccant cooling
ACO: Space cooling
AGH: Greenhouses
AHC: Heating/Cooling
AHE: Space heating
AHW: Domestic hot water
AMH: Mobile homes
APO: Swimming pool heating

COLLECTORS

CBL: Solar heating balloons
CCM: Collector components
CCN Concentrating collectors
CFA: Air flat-plate collectors
CFE: Flat-plate evacuated collectors
CFL: Liquid flat-plate collectors
CMA: Absorber plates
CMB: Fans
CMC: Selective coatings
CMG: Glazing
CMF: Fresnel lenses
CMH: Heat transfer fluids
CMI: Corrosion inhibitors
CMM: Mounting systems
CMP: Pumps
CMR: Reflective surfaces
CMS: Sealants
CMT: Tracking devices
CTU: Tubular collectors

SOLAR CONTROLS

COC: Components
COS: Complete systems

STORAGE

STA: Air (rock bed)
STC: Components
STL: Liquid
STP: Phase change

COMPLETE SYSTEMS

HYB: Hybrid (active and passive)
SPC: Swimming-pool covers
SSF: Solar furnaces
SYA: Air
SYL: Liquid
SYP: Passive

MISCELLANEOUS

ASH: Tested according to ASHRAE standards
FEA: Listed in FEA survey of solar collector manufacturers
HEL: Heliostats
HEP: Heat pumps
HEX: Heat exchangers
INS: Instrumentation/measurement
KIT: Do-it-yourself kits
LAB: Testing equipment
NBS: Collectors tested according to NBS standards
RCE: Rankine cycle engine

ALABAMA

National Energy Systems Corp.
P.O. Box 1176
Birmingham, AL 35201
(205) 252-7726
AHE AHW APO CFL SYL COS COC STL ASH

Wolverine Tube Division
Universal Oil Products
Decatur, AL 35601
(205) 353-1310
HEX

Sun Century Systems
P.O. Box 2036
Florence, AL 35630
(205) 764-0795
AHE AHW CFL FEA APO

IBM-Federal Systems Division
Research Park
Huntsville, AL 35807
(205) 837-4000
INS

Medtherm Corp
P.O. Box 412
Huntsville, AL 35804
(205) 837-2000
INS

Solar Unlimited
4310 Governors Dr.
Huntsville, AL 35805
(205) 837-7340
HEX AHE AHW APO CFL CCM CMG CMA CMM
COS COC STL SYL

Halstead and Mitchell
P.O. Box 1110
Scottsboro, AL 35768
(205) 259-1212
AHE AHW CFL FEA

Energy Engineering Inc.
P.O. Box 1156
Tuscaloosa, AL 35401
(205) 339-5598
STL

ARIZONA

Sunshine Unlimited
900 N. Jay St.
Chandler, AZ 85224
(602) 963-3878
AHW AHE CFL

Arizona Engineering & Refrigeration
635 W. Commerce Ave.
Gilbert, AZ 85234
(602) 892-9050
AHE AHW APO CFL FEA NBS

Kreft Distributing Co.
P.O. Box 105
Lake Havasu City, AZ 86403
(602) 855-2059
AHC AHW APO CFL CCM COC STL SYL KIT CMP
CMG CMA

Matrix Inc.
537 S. 31st St.
Mesa, AZ 85204
(602) 832-1380
INS

Solar World Inc.
4449 N. 12th St., Suite 7
Phoenix, AZ 85014
(602) 266-5686
CFL STL COS SYL AHW AHE APO

Solarequip
P.O. Box 21447
Phoenix, AZ 85036
(602) 267-1166
AHE AHW APO CFL STL SYL CCM CMM CMP KIT
COC

Arizona Solar Enterprises
6719 E. Holly St.
Scottsdale, AZ 85257
(602) 945-7477
AHE AHW APO CFL

Delevan Electronics Inc.
14605 N. 73rd St.
Scottsdale, AZ 85260
(602) 948-6350
AHE CCM CMT

Diversified Natural Resources
8025 E. Roosevelt, Suite A
Scottsdale, AZ 85257
(602) 945-2330
AHC CCN FEA

Sunpower Systems Corporation
510 S. 52nd St., Suite 101
Tempe, AZ 85281
(602) 968-7425
AHW AHC APO CCN FEA SYL

Mel Kiser and Assoc.
6701 E. Kenyon Dr.
Tucson, AZ 85710
(602) 747-1552
AHW AHC CCN SYL

Hansberger Refrigeration & Electric Co.
2450 8th St.
Yuma, AZ 85364
(602) 783-3331
AHE AHW CFL FEA

CALIFORNIA

Hadbar
Division of Purosil Inc.
723 S. Fremont Ave.
Alhambra, CA 91803
(213) 283-0721
CCM CMS

Solarbeam Industries Inc.
118 N. Almansor St.
Alhambra, CA 91801
(213) 282-8451
AHE AHW CFL

Piper Hydro Inc.
2895 E. La Palma
Anaheim, CA 92806
(714) 630-4040
AHE AHW CFL FEA APO SYL

Solar Contact Systems
1415 Vernon
Anaheim, CA 92805
(714) 991-8120
AHW CFL STL SYL KIT

Sundu Company
3319 Keys La.
Anaheim, CA 92804
(714) 828-2873
APO CFL FEA AHE AHW

Vanguard Solar Systems
2727 Coronado St.
Anaheim, CA 92806
(714) 871-8181
AHW APO CFL

American Solar Systems
415 Branch St.
Arroyo Grande, CA 93420
(805) 481-1010
AHW AHC APO ACA ACD AGH AMH AAG CFL
CTU COS SYL LAB

Altenergy
P.O. Box 695
Ben Lomond, CA 95005
(408) 336-2321
AHW APO AHE AAG CTU CCN CFL COS COC HEX
RCE LAB

Sun of Man Solar Systems
Drawer W
Bethel Island, CA 94511
(415) 634-1223
AHW APO COC

Bow Jon
2829 Burton Ave.
Burbank, CA 91504
(213) 846-2620
AHW CFL

American Solar Manufacturing
P.O. Box 194
Byron, CA 94514
(415) 634-2426
AHW APO AHE CFL CCM CMP NBS

Ra-Los Inc.
559 Union Ave.
Campbell, CA 95008
(408) 371-1734
INS RCE FEA NBS LAB ASH APO AAG CCM CMG
CMH CMI CMM CMP AHC AHW CFL SYL STL
HEX COS COC

Solar King International
8577 Canoga Ave.
Canoga Park, CA 91304
(213) 998-6400
AHE AHW CFL SYL COC APO

Southwest Solar Corp.
8235 Remmet Ave.
Canoga Park, CA 91304
(213) 339-4383
AHW CFL COS STL

Swan Solar
6909 Eton St., Unit G
Canoga Park, CA 91303
(213) 884-7874
AHW AHE APO CFL HEX

Harness the Sun
P.O. Box 109
Cardiff-by-the-Sea, CA 92007
(714) 436-4822
APO CFL SYL AHE AHW

J.W. Carroll & Sons
22600 S. Bonita St.
Carson, CA 90745
(213) 775-6737
CCM CMG

Solargenics Inc.
9713 Lurline Ave.
Chatsworth, CA 91311
(213) 998-0806
AHW CFL SYL KIT AHE COS CCM CMM CMP HEX
STL NBS FEA

Unitspan Architectural Systems Inc.
9419 Mason Ave.
Chatsworth, CA 91311
(213) 998-1131
AHE AHW CFL APO KIT FEA

Grundfos Pumps Corp.
2555 Clovis Ave.
Clovis, CA 93612
(209) 299-9741
CCM CMP

Passive Solar Varient Homes by Savell
575 Birch Ct., Suite A
Colton, CA 92324
(714) 825-3394
AHC SYP

Sealed Air Corp.
2015 Saybrook Ave.
Commerce, CA 90040
(213) 685-9666
APO SPC

American Solar
1749 Pine St.
Concord, CA 94522
(415) 798-9120
AHE AHW APO CFL

Solar Energy Thermal Systems Co.
1037B Shary Circle
Concord, CA 94518
(415) 676-5392
AHE AHW APO CFL

Robertshaw Controls Co.
Uni-Line Division
P.O. Box 2000, 4190 Temescal St.
Corona, CA 91720
(714) 734-2600
AHC AHW APO COC

Solar Research Systems
3001 Red Hill Ave.
Costa Mesa, CA 92626
(714) 545-4941
APO CFL COS SYL FEA

Solar Utilities
P.O. Box 1696
2850 Mesa Verde Dr.
Costa Mesa, CA 92626
(714) 557-7125
APO CFL

SOL Power Industries Inc.
10211-C Bubb Rd.
Cupertino, CA 95014
(408) 996-3222
AHE AHW APO CFL SYL

Baker Bros. Solar Collectors
207 Cortez Ave.
Davis, CA 95616
(916) 756-4558
AHE AHW APO SYL CFL KIT

Natural Heating Systems
207 Cortez Ave.
Davis, CA 95616
(916) 756-4558
AHW AHE APO CFL

Hexcel Corp.
11711 Dublin Blvd.
Dublin, CA 94566
(415) 828-4200
AHC CCN FEA

A-1 Prototype
1288 Fayette
El Cajon, CA 92020
(714) 449-6726
AHE AHW APO CFL STL SYL KIT FEA NBS ASH
COS

Castor Development Corp.
634 Crest Dr.
El Cajon, CA 92021
(714) 280-6660
AHE AHW APO AGH AMH CFL CCM CMA COS
COC STL SYL HEX HEP HEL KIT FEA NBS LAB

Kahl Scientific Instruments Corp.
P.O. Box 1166
El Cajon, CA 92022
(714) 444-2158
INS

Natural Energy Systems
Marketing Arms Division
1632 Pioneer Way
El Cajon, CA 92020
(714) 440-6411
AHW CFL FEA APO AHC

Solar Physics Corporation
1350 Hill St.
Suite A
El Cajon, CA 92020
(714) 440-1625
AHC CFL CCN FEA AHW APO

Sunshine Utility Company
1444 Pioneer Way, Suites 9 & 10
El Cajon, CA 92020
(714) 440-3151
AHE AHW CFL APO FEA

Sunwater Energy Products
1488 Pioneer Way, Suite 17
El Cajon, CA 92020
(714) 579-0771
AHW APO CFL COS AGH AHE CCM CMA

Solar Enterprises
9803 E. Rush St.
El Monte, CA 91733
(213) 444-2551
APO CFL FEA

Vinyl-Fab Industries
10800 St. Louis Dr.
El Monte, CA 91731
(213) 575-1894
SPC

Wojick Industries Inc.
527 N. Main St.
Fallbrook, CA 92028
(714) 728-5593
APO CFL SYL COC

The Energy Factory
5622 E. Westover, Suite 105
Fresno, CA 93727
(209) 292-6622
AGH

Solar West Inc.
3636 N. Hazel No. 108, P.O. Box 892
Fresno, CA 93714
(209) 222-3455
AHW AHE APO CFL

Solar Hydro Systems Inc.
765 S. State College, Suite E
Fullerton, CA 92631
(714) 992-4470
AHW AHE APO AAG CFL COS STL SYL SPC INS

Swedlow Inc.
12122 Western Ave.
Garden Grove, CA 92645
(714) 893-7531
CCM CMF

Ying Manufacturing Corp.
1957 W. 144th St.
Gardena, CA 90249
(213) 327-8399
AHW CFL FEA AHC APO SYL CFA HEX KIT CCM
CMP CMB CMM NBS ASH STL STA

Conserdyne Corp.
4437 San Fernando Rd.
Glendale, CA 91204
(213) 246-8408
AHW AHE APO CFL SYL HEX

Era Del Sol
5960 Mandarin Ave.
Goleta, CA 93017
(805) 967-2116
AHE AHW APO CFL SYL FEA

Filon
12333 S. Van Ness Ave.
Hawthorne, CA 90250
(213) 757-5141
CCM CMG

Odin Solar Systems
26010 Eden Landing Rd., Suite 5
Hayward, CA 94545
(415) 785-2000
CCM CMA CMP INS

Solar Systems
26046 Eden Landing Rd.
Hayward, CA 94545
(415) 785-0711
AHE AHW APO CFA CFL CCM COC STL STC SYA
SYL HEP HEX KIT CMP CMB

Sunworks
1501 Felta Rd.
Healdsburg, CA 95448
(707) 443-3693
AHE AHW APC CFL

M.C. Nottingham Co.
4922 Irwindale Ave., P.O. Box 2107
Irwindale, CA 91706
(213) 283-0407
AHE STL

Technical Measurements Inc.
P.O. Box 838
La Canada, CA 91011
(213) 248-1035
INS

Spectran Instruments
P.O. Box 891
La Habra, CA 90631
(213) 694-3995
AHC AHW INS

Cushing Instruments
7911 Hershel Ave.
La Jolla, CA 92037
(714) 459-3433
INS

Helix Solar Systems
P.O. Box 2038
La Puente, CA 91746
(213) 330-3312
AHW APO CFL HEX KIT NBS

Fred Rice Productions
P.O. Box 643
48780 Eisenhower Dr.
La Quinta, CA 92253
(714) 564-4823
AHW APO CFL STL SYL CFA CTU SYP AHE AMH

Applied Sol Tech Inc.
P.O. Box 9111
Cabrillo Station
Long Beach, CA 90810
(213) 426-0127
AHC CCN CCM COS HEX AHW APO STL SYL CMP
CMH

Solarcoa
21157 E. Spring St.
Long Beach, CA 90808
(213) 426-7655
AHE AHW CFL FEA SYL APO COC CCM CMP CMM

ZZ Corp.
10806 Kaylor St.
Los Alamitos, CA 90720
(213) 598-3220
AHW AHE AGH AMH CCN CCM CMI CMM CMT
CMR COC STL SYL INS FEA NBS LAB ASH

Advanced Solar Energy Systems Inc.
3440 Wilshire Blvd.
Los Angeles, CA 90010
(213) 383-0035
AHC AHW APO CFL

Aqueduct Inc.
1934 Cotner Ave.
Los Angeles, CA 90025
(213) 447-2496
COS COC

Helio-Dynamics Inc.
327 N. Fremont St.
Los Angeles, CA 90012
(213) 624-5888
AHC AHW APO CFL SYL FEA

Highland Plating Co.
1128 N. Highland Ave.
Los Angeles, CA 90038
(213) 469-2288
CCM CMC

Sola Heat
1200 E. First St.
Los Angeles, CA 90033
(213) 263-5823
AHE AHW CFL STL SYL

Solar Concepts
12103 Washington Pl.
Los Angeles, CA 90066
(213) 398-5872
AHE AHW CFL

Solar Energy Systems Inc.
2345 Santa Fe Ave.
Los Angeles, CA 90058
(213) 472-6508
AHE AHW APO CFL COS COC SYL CCM CMA
CMM KIT

Skytherm Processes Engineering
2424 Wilshire Blvd.
Los Angeles, CA 90057
(213) 389-2300
AHC AHW SYP FEA

Advanced Energy Technology
121 Albright
Los Gatos, CA 95030
(408) 866-7686
AHW AHE APO CFL SYL

Solar Energy People
5044 Fair Grounds Dr.
Mariposa, CA 95338
(209) 966-5616
AHW APO CFL CCM CMA COS COC STL SYL KIT

Fafco
138 Jefferson Dr.
Menlo Park, CA 94025
(415) 321-6311
APO CFL FEA COS SYL HEX

Sunburst Solar Energy Inc.
P.O. Box 2799
Menlo Park, CA 94025
(415) 327-8022
AHE AHW APO CFL FEA COS STL HEX KIT CCM
CMG CMP NBS ASH

Catel Mfg. Inc.
235 West Maple Ave.
Monrovia, CA 91016
(213) 359-2593
APO SPC

Acurex Aerotherm
485 Clyde Ave.
Mountain View, CA 94042
(415) 964-3200
AHC AHW CCN CMT AAG FEA

Alten Associates Inc.
2594 Leghorn St.
Mountain View, CA 94043
(415) 969-6474
AHW CFL FEA SYL KIT AHC APO

In Solar Systems
2562 W. Middlefield Rd.
Mountain View, CA 94043
(415) 964-2801
APO CFL HEX KIT LAB

A.O. Smith Corp.
P.O. Box 484
Newark, CA 94560
(415) 792-1345
AHE AHW STL SYL

American Sun Industries
3477 Old Conejo Rd.
P.O. Box 263
Newbury Park, CA 91320
(805) 498-9700
AHE CFL SYL KIT APO FEA

Sun Water Inc.
P.O. Box 732
Northridge, CA 91324
(213) 886-3620
AHW APO CCN

Rho Sigma
11922 Valerio St.
North Hollywood, CA 91605
(213) 982-6800
AHW AHE APO COS COC INS

Wm. Lamb Co.
P.O. Box 4185
North Hollywood, CA 91607
(213) 764-6363
AHC CCC KIT

MacBall Industries
3040 Market St.
Oakland, CA 94608
(415) 658-2228
SPC

Sunglaze
P.O. Box 2634
Olympic Valley, CA 95730
(702) 831-2400
AHW AHE AAG AGH CFA SYA

Sol Ray
204 B Carleton
Orange, CA 92667
(714) 997-9431
AHW AHE APO AMH AAG CFA CCM CMG CMA
CMI CMS CMM CMR CMP CMB COS COC STA SYA
INS HEX LAB

J.G. Johnston Company
33458 Angeles Forest Hwy.
Palmdale, CA 93550
(805) 947-3791
AHE CFA FEA AHW SYA STA

Western Energy Inc.
454 Forest Ave.
Palo Alto, CA 94302
(415) 327-3371
AHE AHW APO CFL COC STL HEX CMP CCM NBS

Jacobs-Del Solar Systems Inc.
251 S. Lake Ave.
Pasadena, CA 91101
(213) 449-2171
CCN AHC AHW APO CCM CMT COC SYL FEA

Solar Pools Inc.
2200 Freed Way
Pittsburg, CA 94565
(415) 432-7344
APO CFL

Colt Inc.
71590 San Jainto
Rancho Mirage, CA 92270
(714) 346-8033
AHW AHE APO CFL CCM CMH COS COC STL SYL
INS HEX HEP KIT LAB ASH

Solar Enterprises
P.O. Box 1046
Red Bluff, CA 96080
(916) 527-0551
AHE AHW APO AMH AAG CFL CCM CMG CMA
CMM CMR COS COC STL SYL INS LAB

Helios Solar Engineering
400 Warrington Ave.
Redwood City, CA 94063
(415) 369-6414
AHW APO CFL COS STL

Solarway
P.O. Box 217
Redwood Valley, CA 95470
(707) 485-7616
AHE CCM CMG

Sunrise Solar Inc.
7359 Reseda Blvd.
Reseda, CA 91335
(213) 881-3164
AHE AHW CFL

Heliodyne Corp.
770 S. 16th
Richmond, CA 94804
(415) 237-9614
SYL APO AHW AHE CFL KIT CCM CMH COS HEX

Solar Energy Inc.
12155 Magnolia Ave., 6-E
Riverside, CA 92503
(714) 785-0610
AHW AHC APO CFL

Solar West
2711 Chicago Ave.
Riverside, CA 92507
(714) 684-1555
APO CFL

Ecosol Ltd.
3382 El Camino Ave.
Sacramento, CA 95821
(916) 485-5860
AHC AHW COS HEP INS

Weathermeasure Corp.
P.O. Box 41257
Sacramento, CA 95841
(916) 481-7565
INS

California Solar Systems Co.
421 Picadilly
Suite 12
San Bruno, CA 94066
(415) 583-4711
AHW CFL SYL

Albatross Division
Dri-Honing Corp.
975 Terminal Way
San Carlos, CA 94070
(415) 593-1465
AHW APO CFL SPC KIT ASH

Energy Systems Inc.
4570 Alvarado Canyon Rd.
Bldg. D
San Diego, CA 92120
(714) 280-6660
AHW AHE CFL FEA APO COC STL CMP CCM CMG
CMA CCN

Gamma Scientific Inc.
3777 Ruffin Rd.
San Diego, CA 92123
(714) 279-8034
INS

Solar Applications Inc.
7926 Convoy Ct.
San Diego, CA 92111
(714) 292-1857
AHE AHW APO CFL FEA CCM CMM SPC NBS

Solar Captivators Systems Inc.
7192 Clairemont Mesa Blvd.
San Diego, CA 92111
(714) 560-7454
AHW APO AHC AAG CFL

Solar Energy Digest
P.O. Box 17776
San Diego, CA 92117
(714) 277-2980
AHW CFL KIT

Solar Supply Inc.
9163 Chesapeake Dr.
San Diego, CA 92123
(714) 292-7811
COS COC AHW

Solar Tec Corp.
8250 Vickers St.
San Diego, CA 92111
(714) 560-8434
APO AHC AHW CFL FEA CCN SYL HEP NBS

Southwest Air Conditioning Inc.
7268 El Cajon Blvd.
San Diego, CA 92115
(714) 462-0512
AHW AHE APO AMH CFL STL

Sunspot Environmental Energy Systems
P.O. Box 5110
San Diego, CA 92105
(714) 264-9100
AHW AHE APC CFL SYL KIT

Vanguard Energy Systems
9133 Chesapeake Dr.
San Diego, CA 92123
(714) 292-1433
CCM CMP COS HEP

Solergy Inc.
70 Zoe St.
San Francisco, CA 94107
(415) 495-4303
AHE AHW CFL CCN NBS FEA

Burke Industries Inc.
2250 S. 10th St.
San Jose, CA 95112
(408) 297-3500
COS COC KIT APO CFL FEA SYL

David Rose Steel Co.
345 N. Montgomery
San Jose, CA 95110
(408) 295-6975
AHE AHW STL

Solar Concepts Inc.
818 Charcot Ave.
San Jose, CA 95131
(408) 263-8110
AHE AHW APO AGH AMH AAG CFL ASH

Astron Solar Industries Inc.
465 McCormick St.
San Leandro, CA 94577
(415) 632-5400
AHE AHW APO CFL COS SYL FEA

Technitrek Corp.
1999 Pike Ave.
San Leandro, CA 94577
(415) 352-0535
AHW APO CFL CCM CMP COS COC AHE STL SYL

Sunstream Environments
P.O. Box 93A
Los Osos Valley Rd.
San Luis Obispo, CA 93401
(805) 541-0760
AAG AGH CFA

Energy Sealants Inc.
1611 Borel Pl., Suite 230
San Mateo, CA 94402
(415) 574-0898
CCM CMS

Optical Sciences Group Inc.
24 Tiburon St.
San Rafael, CA 94901
(415) 453-8980
AHE CCM CMF

Aztec Solar Energy Systems of Orange City
420 Terminal St.
Santa Ana, CA 92701
(714) 558-0882
APO CFL

Sunspot
Division of Elcam Inc.
5330 Debbie La.
Santa Barbara, CA 93111
(805) 964-8676
AHE CFA FEA CCM SYL CFL COC AHW STL CMP

Troger Enterprises
2024 "A" De La Vina
Santa Barbara, CA 93105
(805) 687-6522
AHE AHW APC COS

Solar Collectors of Santa Cruz
2901 Glen Canyon Rd.
Santa Cruz, CA 95060
(408) 476-6369
AHW AHE APO AHC AGH AMH AAG CFA CFL
CCN CCM CMA CMM STA STL SYP SYL SYA KIT
HEX

Sunshine Greenhouses
109 Cooper St., Suite 5
Santa Cruz, CA 95060
(408) 425-1451
AHG AHW APO CFL

Hy-Cal Engineering
12105 Los Nietos Rd.
Santa Fe Springs, CA 90670
(213) 698-7785
INS

Solarmaster
722-D W. Betteravia Rd.
Santa Maria, CA 93454
(805) 922-0205
CFL CCN KIT AHC AHW APO AGH AAG

American Appliance Mfg. Corp.
P.O. Box 1956
2341 Michigan Ave.
Santa Monica, CA 90404
(213) 870-8541
AHW CFL STL SYL

Optical Coating Lab Inc.
2789 Giffen Ave.
Santa Rosa, CA 95403
(707) 545-6440
CCM CMG CMA CMC CMR

Solar Energy Engineering
31 Maxwell Ct.
Santa Rosa, CA 95401
(707) 542-4498
AHW AHE APO CFL

Powell Brothers Inc.
5903 Firestone Blvd.
South Gate, CA 90280
(213) 869-3307
AHE AHW CFL FEA APO

Heliotrope General
3733 Kenora Dr.
Spring Valley, CA 92077
(714) 460-3930
AHC AHW APO COC STL CCM CMP HEX

Sun Power Solar Engineering Co.
4032 Helix St.
Spring Valley, CA 92077
(714) 464-5322
AHE AHW CCN CCM CMT

Dearing Solar Energy Systems
12324 Ventura Blvd.
P.O. Box 1744
Studio City, CA 91604
(213) 769-2521
APO SPC CFL

Kasaki USA
4150 Arch Dr.
Suite 8
Studio City, CA 91604
(213) 985-9611
AHW APO CFL COS STL SYL

Molectron Corp.
177 N. Wolfe Rd.
Sunnyvale, CA 94086
(408) 738-2661
INS

Sun Power Systems Ltd.
1024 W. Maude Ave.
Suite 203
Sunnyvale, CA 94086
(408) 738-2442
AHE AHW KIT CFL SYL APO

Spectolab
12500 Gladstone Ave.
Sylmar, CA 91342
(213) 365-4611
INS

Bostik-Finch Inc.
20846 S. Normandie Ave.
Torrance, CA 90502
(213) 320-6800
CCM CMC

Western Solar Development Inc.
1236 Callen St.
Vacaville, CA 95688
(704) 446-4411
AHC AHW APO CFL KIT

The Solarshingles Co.
14532 Vanowen St.
Van Nuys, CA 91405
(213) 782-2828
AHE AHW APO CFL CFA

Raypak Inc.
31111 Agoura Rd.
Westlake Village, CA 91359
(213) 889-1500
AHE AHW APO CFL FEA SYL CCM CMM CMP NBS

Energy Absorption Systems Inc.
860 S. River Rd.
West Sacramento, CA 95691
(916) 371-3900
AHW AHE CFL CCM CMM STL SYL HEX

Solastor/Energy Absorption Systems
860 S. River Rd.
W. Sacramento, CA 95691
(916) 371-3900
AHW AHC APO CFL STL SYL HEX CCM CMP
CMM COS COC

Weathertronics
2777 Del Monte St.
West Sacramento, CA 95691
(916) 371-2660
INS

Owen Enterprises
436 N. Fries Ave.
Wilmington, CA 90744
(213) 835-7436
AHC CCN APO

Highland Manufacturing Co.
P.O. Box 563
Yucaipa, CA 92399
(714) 794-2181
APO CFL

COLORADO

Barber-Nichols Engineering
6325 W. 55th Ave.
Arvada, CO 80002
(303) 421-8111
ACO RCE

Solar Power West
709 Spruce St.
Aspen, CO 81611
(303) 925-4698
APO CFL KIT

Solar Seven Industries Inc.
3323 Moline St.
Aurora, CO 80110
(303) 364-7277
AHE AHW APO CFA SYA NBS

Colorado Sunworks Corp.
P.O. Box 455
Boulder, CO 80306
(303) 443-9199
AHW AHE AGH AAG CFL COS COC STL SYL HEX
KIT NBS LAB ASH SYP CFA CCM CMB

Entropy Ltd.
5735 Arapahoe Ave.
Boulder, CO 80303
(303) 433-5103
AHW AHC SYL STL AAG CCN

Solar Control Corp.
5595 Arapahoe Ave.
Boulder, CO 80302
(303) 449-9180
AHE COC COS

The Tub Company
P.O. Box 8
Boulder, CO 80306
(303) 449-4563
AHE AHW SYL KIT

Energy Dynamics Corporation
327 West Vermijo Rd.
Colorado Springs, CO 80903
(303) 475-0332
AHW CFL FEA APO CCM COS COC STL STC SYL
HEP AHC

First International Corp.
1354 Ford St.
Colorado Springs, CO 80915
(303) 574-4404
AHE AHW CFA SYA STA COS

American Helio Thermal Corp/Miromit
2625 S. Santa Fe Dr.
Denver, CO 80223
(303) 778-0650
AHW CFL FEA APO SYL AHC

Federal Energy Corp.
5505 E. Evans
Denver, CO 80222
(303) 753-0565
AHE AHW CFL COC HEX AHC ACA CCN STL SYL
CCM CMP

R.M. Products
5010 Cook St.
Denver, CO 80216
(303) 825-0203
AHE AHW CFL CFA SYA FEA CCM CMM APO
AGH COS COC STA STL SYL SYP

Solar Dynamics Inc.
1320 S. Lipan St.
Denver, CO 80223
(303) 777-3666
AHE SSF

Solar Technology Corp.
2160 Clay
Denver, CO 80211
(303) 455-3309
CFA SYA AHC KIT FEA AGH SSF

Solaron Corporation
300 Galleria Tower, 720 S. Colorado
Denver, CO 80222
(303) 759-0101
AHE CFA FEA AHW CCM COS STA SYA HEX CMP
NBS

Tri-State Sol-Aire Inc.
7100 Broadway, Suite 6N
Denver, CO 80221
(303) 426-4000
AHW AHE CFA COS STA SYA HEP

International Solar Industries Inc.
9555 E. Caley
Englewood, CO 80110
(303) 989-3363
AHE AHW CFA SYL

Sol-Aire/Energy Systems Inc.
2750 S. Shoshone
Englewood, CO 80110
(303) 761-4335
APO AHC AHW SYA CFA STA NBS LAB ASH

Sun-Heet Inc.
2624 S. Zuni
Englewood, CO 80110
(303) 922-6179
AHW APO AHC ACA AGH AAG CCN STL STP SYL

Future Systems Inc.
12500 W. Cedar Dr.
Lakewood, CO 80228
(303) 989-0431
AHE APO CFA SSF

Sun Power of Colorado
343 Van Gordan St., Bldg. 18/406
Lakewood, CO 80228
(303) 988-6200
AHC AHW CFA

General Solar Corp.
5575 S. Sycamore St.
Littleton, CO 80120
(303) 321-2675
AHW AHE SYL HEP HEX

Solar Energy Research Corporation
701B S. Main St.
Longmont, CO 80501
(303) 772-8406
AHC AHW APO HEP CCM CMM CMP STL KIT COS
COC CMA CMT INS HEX NBS

International Solarthermics Corp.
P.O. Box 397
Nederland, CO 80466
(303) 258-3272
AHE SSF

Design Works
P.O. Box 700
Telluride, CO 81435
(303) 728-3303
AHW CFA AHE COC KIT

CONNECTICUT

Nuclear Technology Corp.
P.O. Box 1
Amston, CT 06231
(203) 537-2387
AHE AHW CCM CMH CMI

Suntap Inc./Bross Utilities Service Inc.
42 E. Dudley Town Rd.
Bloomfield, CT 06002
(203) 243-1781
AHW AHC APO AGH AMH AAG CFL CTU CCM
CMH CMI COS COC STL SYL INS HEX FEA NBS
LAB ASH

Sun-Ray Solar Equipment
2093 Boston Ave.
Bridgeport, CT 06610
(203) 333-6264
AHE AHW APO CFL

American Solar Heat Corp.
7 National Pl.
Danbury, CT 06810
(203) 792-0077
AHW AHE COC SYL CFL INS

FTA Corp.
348 Hazard Ave.
Enfield, CT 06082
(203) 749-7054
AHW AHE CFA COS

National Solar Corp.
Novelty La.
Essex, CT 06426
(203) 767-1644
AHW CFL CCM CMM KIT

Falbel Energy Systems Corp.
P.O. Box 6
Greenwich, CT 06830
(203) 357-0626
CCN FEA AHW APO KIT COS CCM CMA CFL STL
SYL HEX AHE CMP CMG CMH CMI

International Environment Corp.
83 S. Water St.
Greenwich, CT 06830
(203) 531-4490
AHW CFL FEA AHE SYL

International Environmental Energy Inc.
275 Windsor St.
Hartford, CT 06120
(203) 549-4400
AHE AHW COS COC SYA SYL

Research Technology Corp.
151 John Downey Dr.
New Britain, CT 06051
(203) 224-8155
CCM CMH INS HEX

Solar Products Manufacturing Corp.
151 John Downey Dr.
New Britain, CT 06051
(203) 224-2164
AHE AHW APO CFL CCM CMG CMP COS SYL HEX
KIT FEA NBS

Spiral Tubing Corp.
533 John Downey Dr.
New Britain, CT 06051
(203) 244-2409
AHE AHW CCM CMA

Enthone Inc./Sunworks Division
P.O. Box 1004
New Haven, CT 06508
(203) 934-6301
AHW CFA CFL FEA CCM SYA SYL CMH AHE KIT

Kem Associates
153 East St.
New Haven, CT 06507
(203) 865-0584
CCM CMM

C & M Systems, Inc.
P.O. Box 475, Saybrook Ind. Park
Old Saybrook, CT 06475
(203) 388-3429
COC

Groundstar Energy Corp.
137 Rowayton Ave.
Rowayton, CT 06853
(203) 838-0650
AHW CCS APO AHC CFL SYL CCM CMM NBS ASH

Barnes Engineering Co.
30 Commerce Rd.
Stamford, CT 06904
(203) 348-5381
INS

Oriel Corp. of America
15 Market St.
Stamford, CT 06902
(203) 357-1600
INS LAB

Wormser Scientific Corporation
88 Foxwood Rd.
Stamford, CT 06903
(203) 322-1981
AHC CCN

Solar Kinetics Corp.
P.O. Box 17308
West Hartford, CT 06117
(203) 233-4461
AHW APO AGH AAG CCN CCM COS COC STL SYL
HEX FEA LAB ASH

Wilson Solar Kinetics Corp.
P.O. Box 17308
West Hartford, CT 06117
(203) 233-4461
CCN FEA AHW APO COS CMP STL STL AAG CCM
CMM

DELAWARE

DuPont Co.
Nemours Bldg., Rm. 24751
Wilmington, DE 19898
(302) 999-3456
AHW AHE CCM CMG CMH

Hercules Inc.
910 Market St.
Wilmington, DE 19899
(302) 575-5000
CCM SYP CMG AGH

DISTRICT OF COLUMBIA

Natural Energy Corp.
1001 Connecticut Ave., NW
Washington, DC 20036
(202) 296-7070
AHW AHC APO CFL AAG

Thomason Solar Homes Inc.
6802 Walker Mill Rd., SE
Washington, DC 20027
(301) 292-5122
AHW CFL CCM COC STL SYL AHC

FLORIDA

Aztec Solar Co.
2301 Dyan Way
Altamonte Springs, FL 32751
(305) 830-5477
AHW CFL CCM COC SYL COS CMM CMP

Solar Engineering & Mfg. Co. Inc.
P.O. Box 1358
Boca Raton, FL 33432
(305) 368-2456
INS HEL KIT FEA NBS LAB ASH AHW APO CFL
CCN CCM CMG CMA CMM CMT CMR COS COC
STL SYL

Solar Pool Heaters of SW Florida
901 SE 13th Pl.
Cape Coral, FL 33904
(813) 542-1500
APO CFL FEA

CSI Solar Systems Division
12400 49th St.
Clearwater, FL 33520
(813) 577-4228
AHE AHW CFL FEA APO COS STL SYL

General Energy Devices
1753 Ensley
Clearwater, FL 33516
(813) 586-3585
AHE AHW CFL FEA APO CCM COS COC STL SYL
CMP HEX CMH NBS ASH

Solar Heating Systems
13584 49th St. N.
Clearwater, FL 33520
(813) 577-3961
AHE AHW CFL FEA APO COS COC KIT CCM

Solar Energy Components Inc.
1605 N. Cocoa Blvd.
Cocoa, FL 32922
(305) 632-2880
AHE AHW CFL FEA CCM CMP CMG KIT COC

C.B.M. Manufacturing Inc.
621 NW Sixth Ave.
Fort Lauderdale, FL 33311
(305) 463-5810
AHW CFL COS COC STL SYL HEX FEA NBS LAB

Semco Solar Products Corp.
5701 NE 14th Ave.
Ft. Lauderdale, FL 33334
(305) 565-2516
AHE AHW CFL FEA APO SYL STL COC CCM CMP

Solarcell Corp.
1455 NE 57th St.
Fort Lauderdale, FL 33334
(305) 462-2215
AHW APO CCN COS CCM CMM KIT

Solar-Eye Products Inc.
1300 NW McNab
Bldgs. G & H
Fort Lauderdale, FL 33309
(305) 974-2500
APO AHW AHE CFL SYL COC FEA

Deko-Labs
3860 SW Archer Rd.
Gainesville, FL 32604
(904) 372-6009
COC

Solar Energy Products Inc.
1208 NW Eighth Ave.
Gainesville, FL 32601
(904) 377-6527
AHE AHW APO CFL STL SYL KIT HEX COC CCM
CMM CMP CMG NBS

Solar Dynamics Inc.
P.O. Box 3457
Hialeah, FL 33013
(305) 921-7911
AHE AHW CFL FEA SYL

D.W. Browning Contracting Co.
475 Carswell Ave.
Holly Hill, FL 32017
(904) 252-1528
AHE AHW CFL FEA APO SYL

Horizon Enterprises Inc.
P.O. Box V
Homestead, FL 33030
(305) 245-5145
AHE AHW APO AMH CFL NBS

Solar Energy Contractors
P.O. Box 17094
Jacksonville, FL 32216
(904) 641-5611
AHE AHW CFL FEA COS COC STL KIT CCM CMP
CMM

Solar Industries of Florida
P.O. Box 9013
Jacksonville, FL 32208
(904) 768-4323
AHE AHW CFL FEA

Energy Conservation Equipment Corp.
1011 Sixth Ave. S.
Lake Worth, FL 33460
(305) 586-3839
AHW APO AHC CFL COS HEX HEP NBS ASH

Solar Heater Manufacturer
1011 Sixth Ave. S.
Lake Worth, FL 33460
(305) 586-3839
AHW APO AHC CFL ASH AGH AMH COS STL SYL
INS HEX KIT NBS LAB

Universal Solar Energy Company
1802 Madrid Ave.
Lake Worth, FL 33461
(305) 586-6020
AHE AHW CFL FEA APO CCM

Solar Innovations
412 Longfellow Blvd.
Lakeland, FL 33801
(813) 688-8373
AHW SYL AHE APO CFL COS INS KIT FEA STL
CCM CMM CMP

Unit Electric Control Inc.
Sol-Ray Division
130 Atlantic Dr.
Maitland, FL 32751
(305) 831-1900
AHE AHW CFL FEA COC CCM CMP KIT

All Sunpower Inc.
10400 SW 187th St.
Miami, FL 33157
(305) 233-2224
AHC AMH CCM CMG AHW APO CFL SYL KIT COS
COC STL SPC INS HEX KIT FEA NBS LAB ASH

Beutels Solar Heating Company
7161 NW 74th St.
Miami, FL 33166
(305) 885-0122
AHW CFL FEA

Capital Solar Heating Inc.
376 NW 25th St.
Miami, FL 33127
(305) 576-2380
AHW APO AHC CFL STL SYL KIT FEA

Consumer Energy Corporation
4234 SW 75th Ave.
Miami, FL 33155
(305) 266-0124
AHW CFL FEA APO STL

Hill Bros. Inc.
Thermill Division
3501 NW 60th St.
Miami, FL 33142
(305) 693-5800
AHW AHE APO CFL CCM COS STL SYL CMP
CMM

W.R. Robbins & Son
1401 NW 20th St.
Miami, FL 33142
(305) 325-0880
AHW CFL FEA APO SYL COS COC KIT CCM
CMP CMH STL

S.P.S. Inc.
8801 Biscayne Blvd.
Miami, FL 33138
(305) 754-7766
AHC CCN CTU CCM CMT RCE

Solar Energy Resources Corporation
10639 SW 185 Terr.
Miami, FL 33157
(305) 233-0711
AHW CFL FEA AHC APO SYL HEP ACA

Solar Products Inc./Sun-Tank
614 NW 62nd St.
Miami, FL 33150
(305) 756-7609
AHE AHW CFL FEA APO KIT COC CCM
CMP

Solar Systems by Sundance
4815 SW 75th Ave.
Miami, FL 33155
(305) 264-1894
AHE AHW CFL FEA

Sun-Dance Inc.
13939 NW 60th Ave.
Miami Lakes, FL 33014
(305) 557-2882
AHE AHW APO CCM CMP STL SYL

Walter & Mart Solar Lab Inc.
3584 Progress Ave.
Naples, FL 33942
(813) 262-6257
AHE AHW APO AGH CFL STL SYL NBS

Solar Water Heaters of New Port Richey
1214 U.S. Highway 19, N.
New Port Richey, FL 33552
(813) 848-2343
AHW AHE APO AMH CFA CCM CMG CMA CMH
CMS CMM CMT CMC CMR COS STL SYL KIT FEA
NBS LAB ASH

Sun Harvesters Inc.
211 NE Fifth St.
Ocala, FL 32670
(904) 629-0687
AHW APO AMH AAG STL SYL KIT

Wilcon Corporation
3310 SW Seventh
Ocala, FL 32670
(904) 732-2550
AHE AHW FEA CFL SYL APO COC COS

Solar Electric International
2634 Taft Ave.
Orlando, FL 32804
(305) 422-8396
AHC AHW CTU SYL

Southern Lighting/Universal 100 Products
501 Elwell Ave.
Orlando, FL 32803
(305) 894-8851
AHE AHW CFL FEA COS CCM CMM

Astro Solar Corp.
457 Santa Anna Dr.
Palm Springs, FL 33460
(305) 965-0606
AHW AHC ACA AMH AAG CFL CFA CTU CFE
COC HEX FEA NBS LAB

Flagala Corporation
9700 W. Highway 98
Panama City, FL 32401
(904) 234-6559
AHE AHW CFL FEA APO COS STL

Wilcox Manufacturing Corp.
P.O. Box 455
Pinellas Park, FL 33565
(813) 531-7741
AHW AHE APO CFL FEA HEX HEP

Largo Solar Systems Inc.
991 SW 40th Ave.
Plantation, FL 33317
(305) 583-8090
AHW CFL FEA APO CCM SYL KIT COS CMP CMA

Union Correctional Institute (Florida State Agencies Only)
P.O. Box 221
Raiford, FL 32083
(904) 431-1212
AHE AHW CFL

Solar Development Inc.
3630 Reese Ave., Garden Ind. Park
Riviera Beach, FL 33404
(305) 842-8935
AHE AHW CFA CFL FEA APO CCM COC SYL CMM
CMA CMP NBS

Solar Fin Systems
140 S. Dixie Hwy.
St. Augustine, FL 32084
(904) 824-3522
AHW AHE CFL COS SYL

Chemical Processors Inc.
P.O. Box 10636
St. Petersburg, FL 33733
(813) 822-3689
AHE AHW CFL FEA SYL

Aqua Solar Inc.
1234 Zacchini Ave.
Sarasota, FL 33577
(813) 366-7080
APO CTU COS COC SYA SPC KIT FEA ASH

Gulf Thermal Corporation
P.O. Box 13124, Airgate Branch
Sarasota, FL 33578
(813) 355-9783
AHE AHW CFL FEA APO SYL

Rox International
2604 Hidden Lake Dr., Suite D
Sarasota, FL 33577
(813) 366-6053
AHW AHE CCN RCE LAB

Technology Applications Lab
1670 Highway A1A
Satellite Beach, FL 32937
(305) 777-1400
INS

Systems Technology Inc.
P.O. Box 337
Shalimar, FL 32579
(904) 863-9213
AHW APO CFL CCM SYL CMP STL HEX COS COC
FEA

J & R Simmons Construction Co.
2185 Sherwood Dr.
South Daytona, FL 32019
(904) 677-5832
AHE AHW CFL SYL FEA

Florida Solar Power Inc.
P.O. Box 5846
Tallahassee, FL 32301
(904) 224-8270
AHE AHW CFL FEA COC STL STC HEX APO CCM
COS SYL CMP CMA CMG

American Solar Power Inc.
715 Swann Ave.
Tampa, FL 33606
(813) 251-6946
AHW CFL FEA

National Solar Systems
P.O. Box 82177
Tampa, FL 33682
(813) 935-9634
AHW AHE APO CFL COS COC STL SYL HEP KIT
FEA NBS

Solarkit of Florida Inc.
1102 139th Ave.
Tampa, FL 33612
(813) 971-3934
AHE AHW CFL KIT

OEM Products Inc./Solarmatic
2701 Adamo Dr.
Tampa, FL 33605
(813) 247-5947
AHW APO CFL COC AHC FEA CMM CMP HEP
CCM CMH STC STP STL SYL HEX

Hawthorne Industries Inc.
1501 S. Dixie
West Palm Beach, FL 33401
(305) 659-5400
COS

GEORGIA

National Solar Company
2331 Adams Dr., NW
Atlanta, GA 30318
(404) 352-3478
AHE AHW CFL FEA SYL

National Solar Sales Inc.
165 W. Wieuca Rd., Suite 100
Atlanta, GA 30342
(404) 256-1660
AHW AHC APO CFL STL SYL

Scientific-Atlanta Inc.
3845 Pleasantdale Rd.
Atlanta, GA 30340
(404) 449-2000
AHE AHW CFL FEA

Solar Energy Systems of Georgia
5825 Glenridge Dr. NE, Bldg. 2
Atlanta, GA 30328
(404) 255-9588
AHW COS COC SYL CFL SYA STL CCM CMP HEX
AHE ACA AGH AMH CFA SYP HYB INS KIT

Solar Technology Inc.
3927 Oakclif Industrial Ct.
Atlanta, GA 30340
(404) 449-0900
AHE AHW CFL CCM COC COS STL SYL CMP FEA

Southeastern Solar Systems Inc.
2812 New Spring Rd., Suite 150
Atlanta, GA 30339
(404) 434-4447
AHW AHC APO CFL FEA CTU CCN CCM STC SYL
COS STL SYP HEX HEP CMA CMP NBS AAG

Southern Aluminum Finishing Co.
1581 Huber St. NW
Atlanta, GA 30318
(404) 355-1560
CCM CMC

Independent Living Inc.
2300 Peachford Rd.
Doraville, GA 30340
(404) 455-0927
AHC AHW APO COC COS SYL

Wallace Company
831 Dorsey St.
Gainsville, GA 30501
(404) 534-5971
AHE AHW CFL FEA CCM SYL APO HEP

HAWAII

Solar Enterprises Hawaii
P.O. Box 27031
Honolulu, HI 96827
(808) 922-8528
AHW CFL KIT

Solaray Corp.
2414 Makiki Heights Dr.
Honolulu, HI 96822
(808) 533-6464
AHW APO CFL SYL AHE FEA

IDAHO

Energy Alternatives Inc.
1006 E. "D" St.
Moscow, ID 83843
(208) 882-0200
AHE CFA COS STA SYA KIT

ILLINOIS

Airtex Corp.
2900 N. Western Ave.
Chicago, IL 60618
(312) 463-2500
AHW APO AHC ACA AGH AMH CFL CFE CCM
CMG CMA CMM CMC CMP COS COC STL SYL INS
HEX FEA NBS ASH

Pak-Tronics Inc.
4044 N. Rockwell Ave.
Chicago, IL 60618
(312) 478-8585
AHE AHW COS COC

Rheem Water Heater Division
City Investing
7600 S. Kedzie Ave.
Chicago, IL 60652
(312) 434-7500
AHW STL

National Industrial Sales
6501 W. 99th St.
Chicago Ridge, IL 60415
(312) 423-4924
AHE AHW CCM CMG

American Chemet Corp.
P.O. Box 165
Deerfield, IL 60015
(312) 948-0800
AHE CCM CMC

DeSoto Inc.
1700 S. Mt. Prospect Rd.
Des Plaines, IL 60018
(312) 391-9434
CCM CMC

Olin Brass Corp.
Roll-Bond Division
E. Alton, IL 62024
(618) 258-2000
AHW CCM AHC CMA

Chamberlain Manufacturing Corp.
845 Larch Ave.
Elmhurst, IL 60126
(312) 279-3600
AHE AHW CFL FEA

Sun Systems Inc.
P.O. Box 155
Eureka, IL 61530
(309) 467-3632
AHW CFA COS APO STL SYA FEA AHC AAG SSF

March Manufacturing Co. Inc.
1819 Pickwick Ave.
Glenview, IL 60025
(312) 729-5300
CCM CMP

A.O. Smith Corp.
P.O. Box 28
Kankakee, IL 60901
(815) 933-8241
AHW STL STC

Solar Dynamics Corporation
550 Frontage Rd.
Northfield, IL 60093
(312) 446-5242
AHW CFL SYL AHE

Johnson Controls Inc.
Penn Division
2221 Camden Ct.
Oak Brook, IL 60521
(312) 654-4900
AHE AHW APO COS COC

Natural Energy Systems Inc.
1117 E. Carpenter Dr.
Palatine, IL 60067
(312) 359-6760
AHW AHC APO CFL COS STL HEX HEP SYL

Illinois Solar Corp.
P.O. Box 841
Peoria, IL 61652
(309) 673-0458
AHW AHE AAG AMH AGH CFA SYA NBS FEA
COS STA SYP

S S Solar Inc.
16 Keystone Ave.
River Forest, IL 60305
(312) 771-1912
AHW AHC CCN SYL

Heliodyne Inc.
4571 Linview Dr.
Rockford, IL 61109
(815) 874-6841
AHC CFA CCM CMA CMM COS STA SYA

Airtrol Corp.
203 W. Hawick St.
Rockton, IL 61072
(815) 624-8051
AHE CCM CMB

Chicago Solar Corp.
1773 California St.
Rolling Meadows, IL 60008
(312) 358-1918
AHW CFA

ITT-Fluid Handling Division
4711 Golf Rd.
Skokie, IL 60076
(312) 677-4030
AHC AHW CCM CMP HEX COC

Shelley Radiant Ceiling Co.
8110 N. St. Louis Ave.
Skokie, IL 60076
(312) 445-2800
APO CCM

The Dexter Corp., Midland Division
East Water St.
Waukegan, IL 60085
(312) 623-4200
CCM CMC

INDIANA

C.F. Roark Welding & Engineering Co. Inc.
136 N. Green St.
Brownsburg, IN 46112
(317) 852-3163
CFL CCN AHE KIT

Arkla Industries Inc.
P.O. Box 534
Evansville, IN 47704
(812) 424-3331
ACO ACA

A/C Fabricating Corp.
P.O. Box 774
64600 U.S. Hwy. 33 East
Goshen, IN 46525
(219) 534-1415
CCM

IOWA

NRG Ltd.
901 Second Ave. E.
Coralville, IA 52241
(319) 354-2033
AHE CCN AHW

Energy King
P.O. Box 248
Creston, IA 50801
(515) 782-8566
AHE CFA

Solar-Thermics Enterprises Ltd.
Highway 34 East
P.O. Box 248
Creston, IA 50801
(515) 782-8566
AHE SSF

Decker Manufacturing
Impac Corp. Division
312 Blondeau
Keokuk, IA 52632
(319) 524-3304
AHE CFA KIT SYA

Lennox Industries Inc.
350 S. 12th Ave.
P.O. Box 280
Marshalltown, IA 50158
(515) 754-4011
AHE AHW CFL FEA SYL HEP

Iowa Solar Electronics
P.O. Box 246
North Liberty, IA 52317
(319) 626-2342
COC

Sun Saver Corp.
P.O. Box 276
North Liberty, IA 52317
(319) 626-2343
AHE APO AAG CFA CCM CMA CMM CMR COS COC
STA STL SYA SSF INS HEX KIT FEA NBS LAB ASH

United Solar
P.O. Box 67
Steamboat Rock, IA 50672
AHE CFA SYA STA COS LAB ASH

Pleiad Industries Inc.
Springdale Rd.
West Branch, IA 52358
(319) 356-2735
AHE AHW APO CFL FEA

Solar Electric Inc.
403 S. Maple
West Branch, IA 52329
(319) 643-2598
AHW AHE APO AMH AAG CCN CCM CMG CMA
CMS CMM COS COC STA SYA KIT FEA NBS ASH

KANSAS

Life Star of Kansas
N. Highway 25
Atwood, KS 67730
(913) 626-3391
AHE SYA

Sunflower Energy Works, Inc.
110 N. Main
Hillsboro, KS 67030
(316) 947-5781
CFA AHE

Salina Solar Products Inc.
620 N. Seventh
Salina, KS 6740-1
(913) 823-2131
AHW CFL STL SYL

Solar Farm Industries Inc.
P.O. Box 242
Stokton, KS 67669
(913) 425-6726
AHW AHE APO AGH AMH AAG CFA CCM CMA
CMH CMC COS COC KIT NBS ASH

Hydro-Flex Corp.
2101 NW Brickyard Rd.
Topeka, KS 66619
CCM AHC AHW APO

KENTUCKY

Swedcast Inc.
7350 Empire Dr.
Florence, KY 41042
(606) 283-1501
AHE AHW APO CCM

Midwestern Solar Systems
2235 Irvin Cobb Dr.
P.O. Box 2384
Paducah, KY 42001
AAG AHE AHW CFL SSF FEA

LOUISIANA

Seeco-Solar Engineering & Equip. Co.
3305 Metairie Rd.
Metairie, LA 70001
(504) 837-0676
AHW AHE AAG CFA COS COC STA SYA NBS ASH

MAINE

Shape Symmetry & Sun Inc.
Biddeford Industrial Park
Biddeford, ME 04005
(207) 282-6155
AHW CFL COS COC STL SYL

Dumont Industries
Main St.
Monmouth, ME 04259
(207) 933-4281
AHW SYL CFL

Thornton Sheet Metal
Waterboro, ME 04087
(207) 247-3121
AHE SSF

MARYLAND

AAI
P.O. Box 6767
Baltimore, MD 21204
(301) 666-1400
AHC CCN FEA

Belfort Instrument Co.
1600 S. Clinton St.
Baltimore, MD 21224
(301) 342-2626
INS

Solar Industries Ltd.
1727 Llewelyn
Baltimore, MD 21213
(301) 732-2072
AHE SSF

Solar Comfort Systems
Solar System Division
4853 Cordell Ave., Suite 606
Bethesda, MD 20014
(301) 652-8941
AHE AHW APO CFL SYL CFA SYA STL COC KIT
CCM CMG FEA

Solar Energy Systems & Products
500 N. Alley
Emmitsburg, MD 21727
(301) 447-6354
AHE CFL AHW COS

KTA Corporation
12300 Washington Ave.
Rockville, MD 20852
(301) 468-2066
AHE AHW FEA CCN SYL CTU

C & C Solarthermics
P.O. Box 144
Smithburg, MD 21783
(301) 631-1361
AHE SSF

N H Yates & Co. Inc.
117C Church La.
Cockeysville, MD 21030
(301) 667-6300
AHE AHW CFL CCM CMP

MASSACHUSETTS

Sunkeeper
P.O. Box 34, Shawsheen Village Sta.
Andover, MA 01801
(617) 470-0555
AHW COC INS AHC COS

Wescorp
15 Stevens St.
Andover, MA 01810
(617) 470-0520
HEP

Solar Heat Corp.
108 Summer St.
Arlington, MA 02174
(617) 646-5763
AHE AHW CFL

G.N.S.
79 Magazine St.
Boston, MA 02119
(617) 442-1000
APO AHC AAG CCN CCM CMA CMR COS COC
STL SYL INS KIT NBS LAB ASH AHW

People/Space Co.
259 Marlboro St.
Boston, MA 02109
(617) 742-8652
AHW AHE APO AGH AMH CFL COC INS KIT LAB

Solafern Ltd.
536 MacArthur Blvd.
Bourne, MA 02532
(617) 563-7181
AHE AHW CFA COS AAG STA SYA NBS

Daystar Corporation
90 Cambridge St.
Burlington, MA 01803
(617) 272-8460
AHE AHW CFL FEA SYL COC HEX CMP CMH

Acorn Structures Inc.
P.O. Box 250
Concord, MA 01742
(617) 369-4111
AHE AHW CFL COS HEX STL SYL

Solectro-Thermo Inc.
1934 Lakeview Ave.
Dracut, MA 01826
(617) 957-0028
AHW AHE APO CCN CCM CMA CMM CMT CMC
CMR CMB COS STA SYA NBS LAB

Calvin T. Frerichs
Chestnut Hill Rd.
Groton, MA 01450
(617) 448-6689
AHW APO CFL CTU COS COC STL SYL INS KIT
LAB

Dixon Energy Systems Inc.
47 East St.
Hadley, MA 01035
(413) 584-8831
AHC AHW APO AGH AMH AAG CFL STL SYL

Columbia Technical Corp.
Solar Division
55 High St.
Holbrook, MA 02343
(617) 767-0513
AHW APO CFL FEA CFA COS STL SYA SYL HEX
AHC

Itek Corp.
Optical Systems Division
10 Maguire Rd.
Lexington, MA 02117
(617) 276-5825
AHW APO CFL CCM CMF CMC CMR FEA NBS LAB

Kennecott Copper Corporation
128 Spring St.
Lexington, MA 02173
(617) 862-8268
AHE AHW CCM APO CMA

Elbart Manufacturing Co.
127 W. Main St.
Millbury, MA 01527
(617) 865-9412
AHW CFL COS COC CCM CMP STL SYL

Sun Systems Inc.
P.O. Box 347
Milton, MA 02186
(617) 268-8178
AHE AHW CFL FEA APO

Solarmaster Systems Inc.
20 Republic Rd.
N. Billerica, MA 01862
(617) 667-4668
AHW AHE APO CFA

Solargy Systems
Vaughn Corp.
386 Elm St.
Salisbury, MA 01950
(617) 462-6683
AHW SYL AHC APO

Sunsav Inc.
890 East St.
Tewksbury, MA 01876
(617) 851-5913
AHC AHW CFL FEA CCM CCN STL SYL APO ACO
ACA CMG CMA CMM CMC CMP CMB COS HEP
NBS ASH

Solar Applications Inc.
One Washington St.
Wellesley, MA 02181
(617) 237-5675
AHW CFL SYL STL

J.A. Corey Inc.
60 Woodland St.
West Boylston, MA 01583
(617) 835-3814
AHW COC

Solar Aqua Heater Corp.
15 Idlewell St.
Weymouth, MA 02188
(617) 843-7255
CCM CMR

MICHIGAN

Addison Products
Addison, MI 49220
(517) 547-6131
AHE AMH CFA COS STA SYA

Solar Research
525 N. Fifth St.
Brighton, MI 48116
(313) 227-1151
AHE AHW CFL FEA APO CCM COC STC HEX KIT
CMG

Environmental Energies Inc.
P.O. Box 73
Front St.
Copemish, MI 49625
(616) 378-2000
APO CFL

Champion Home Builders
5573 E. North St.
Dreyden, MI 48428
(313) 796-2211
AHE SSF AMH FEA

Vinyl-Fab Industries
930 E. Drayton
Ferndale, MI 48220
(313) 399-8745
APO SPC

Electric Motor Repair & Service
Lake Leelanau, MI 49653
(616) 256-9558
AHW CFL CCN CCM CMP LAB

Tranter Inc.
735 E. Hazel St.
Lansing, MI 48909
(517) 372-8410
AHE AHW CCM CMA HEX

Solarator, Inc.
P.O. Box 277
Madison Heights, MI 48071
(313) 642-9377
AAG AHC AGH AMH AHW APO CFL SPC

Dow Chemical USA
2020 Dow Center
Midland, MI 48640
(517) 636-3993
AHE AHW CCM CMH CMS CMC CMR STP

Mueller Brass Co.
Port Huron, MI 48060
(313) 987-4000
CCM

MINNESOTA

The Lord's Power Co., Inc.
726 Marshall St.
Albert Lea, MN 56007
(507) 377-1820
AHE CFA SYA KIT

ILSE Engineering
7177 Arrowhead Rd.
Duluth, MN 55811
(218) 729-6858
AHE AHW APO CFL CCM CMA STL SYL FEA

National Energy Co.
21716 Kendrick Ave.
Lakeville, MN 55044
(612) 469-3401
AHE AHW CFA SYA STA COS COC NBS FEA CCM
CMB

A to Z Solar Products
200 E. 26th St.
Minneapolis, MN 55404
(612) 870-1323
AHW CCM COC STC SYL INS KIT

Honeywell Inc.
2600 Ridgeway Rd.
Minneapolis, MN 55413
(612) 870-5200
AHC AHW APO CFL COS COC

Solar Enterprises Inc.
7830 N. Beach St.
Minneapolis, MN 55440
(612) 483-8103
AHW AHC APO AGH AMH AAG CFL CCM CMG
CMA CMM CMC CMP COS COC STL SYL HEX HEP
LAB ASH

Northern Solar Power Co.
311 S. Elm St.
Moorhead, MN 56560
(218) 233-2515
AHW AHE CFL

Sheldahl
Advanced Products Div.
Northfield, MN 55057
(507) 645-5633
AHE AHW CCM CMR

3-M Co.
P.O. Box 33331, Stop 62
St. Paul, MN 55133
(612) 733-1110
AHE AHW CCM CMC CMF

Solargizer Corporation
220 Mulberry St.
Stillwater, MN 55082
(612) 439-5734
AHE AHW CFL FEA APO SYL KIT CCM CMP STL

MISSOURI

Chemical Sealing Corp.
5401 Banks Ave.
Kansas City, MO 64130
(816) 923-8812
CCM CMS

Suncraft Solar Systems
5001 E. 59th St.
Kansas City, MO 64130
(816) 333-2100
AHE CFA FEA STA SYA COS

Weather-Made Systems Inc.
Route 2, Box 268-S
Lamar, MO 64759
(417) 682-3489
AHE AHW CFL

Midwest Solar Corp.
2359 Grissom Dr.
St. Louis, MO 63141
(314) 569-3110
AHW AGH CFA CCN STL SYL LAB

Weather-Made Systems Inc.
West Hwy. 266
Route 7, Box 300-D
Springfield, MO 65802
(417) 865-0684
AHE AHW CFL

NEBRASKA

Hot Line Solar Inc.
Box 546, 1811 Hillcrest Dr.
Bellevue, NE 68005
(402) 291-3888
AHE CCN

Solar Utilities of Nebraska
P.O. Box 387, 922 Lake St.
Gothenburg, NE 69138
(308) 537-7377
AHE CFA KIT COC AHW SSF COS CCM CMB

Lambda Instruments Corp.
P.O. Box 4425, 4421 Superior St.
Lincoln, NE 68504
(402) 467-3576
INS

Solar Inc.
P.O. Box 246
Mead, NE 68041
(402) 624-6555
AHE SYA STP COS NBS FEA

Solar America Inc.
9001 Arbor St.
Omaha, NE 68124
(402) 397-2421
AHC AGW ACA AAG CFL CFA STL SYA HEP

NEVADA

Southwest Ener-Tech Inc.
3030 S. Valley View Blvd.
Las Vegas, NV 89102
(702) 873-1975
AHE AHW CFL FEA

Richdel Inc.
P.O. Drawer A, 1851 Oregon St.
Carson City, NV 89701
(702) 882-6786
COC APO AHW

Sundog Solar
3800 N. Virginia St.
Reno, NV 89506
(702) 322-8080
AHW AHE APO AGH AMN CFL CCM CMA CMM
COS STL SYL INS HEX KIT LAB

NEW HAMPSHIRE

Hampshire Controls Corp.
P.O. Drawer M
Exeter, NH 03833
(603) 772-5442
COS COC

RDF Corp.
23 Elm Ave.
Hudson, NH 03051
(603) 882-5195
INS

Contemporary Systems Inc.
68 Charlonne St.
Jaffrey, NH 03452
(603) 532-7972
AHE CFA FEA COS STA SYA

Urethane Molding Inc.
Route 11
Laconia, NH 03246
(603) 524-7577
AHE AHW CCM

Kalwall Corp.
Solar Components Division
P.O. Box 237
Manchester, NH 03105
(603) 668-8186
AHE APO CFA CFL CCM STL STC SYA AHW CMG
SYP CMA CMS CMC COC COS CMB CMP SYL

Solarmetrics
23 Bridge St.
Manchester, NH 03101
(603) 668-3216
COS COC STA STL INS AHC AHW APO

Hollis Observatory
One Pine St.
Nashua, NH 03060
(603) 882-5017
INS

Sunhouse Inc.
6 Southgate Dr.
Nashua, NH 03060
(603) 888-0953
AHC AHW CFL SYL

Natural Power Inc.
Francestown Turnpike
New Boston, NH 03070
(603) 487-5512
AHE COS COC AHW APO AGH AMH INS

NEW JERSEY

Drew Chemical Corp.
1 Drew Chemical Plaza
Boonton, NJ 07005
(201) 263-7600
CCM CMH CMI

Burling Instrument Corp.
P.O. Box 298
Chatham, NJ 07928
(201) 635-9481
AHW AHE INS

Ominidata Inc.
16 Springdale Rd.
Cherry Hill, NJ 08003
(609) 424-4646
INS

Solar Energy Systems Inc.
One Olney Ave.
Cherry Hill, NJ 08003
(609) 424-4446
AHW CFL FEA COS STL SYL AHC APO HEP

American Solar Companies Inc.
Ford Rd., Building 4
Denville, NJ 07834
(201) 627-0021
FEA LAB AHC AGH COC STL INS HEX HEP KIT
AHE AHW APO SYL

Berry Solar Products
Woodbridge at Main
P.O. Box 327
Edison, NJ 08817
(201) 549-3800
CCM CMC

Solar Equipment Corp.
Woodbridge at Main
P.O. Box 327
Edison, NJ 08817
(201) 549-3800
CCM CMA CMC

Climatrol Corp.
Woodbridge Ave.
Edison, NJ 08812
(201) 549-7200
AHC AMH CFL COS COC STL SYL HEX HEP NBS
LAB ASH

C.W. Thornthwaite Assoc.
Route 1, Centerton
Elmer, NJ 08318
(609) 358-2350
INS

Calmac Manufacturing Corp.
P.O. Box 710E
Englewood, NJ 07631
(201) 569-0420
AHE AHW CFL FEA APO STP KIT

Solar Industries Inc.
Monmouth Airport Industrial Park
Farmingdale, NJ 07727
(201) 938-7000
APO SYL CFL COC

Universal Power Inc.
P.O. Box 339
Island Heights, NJ 08732
(201) 928-2828
AHE AMH CFA COS COC SYA

Trol-A-Temp
725 Federal Ave.
Kenilworth, NJ 07033
(201) 245-3190
AHE AHW COS

Creighton Solar Concepts
662 Whitehead Rd.
Lawrenceville, NJ 08648
(609) 587-6527
AHW CFL

American Solarize Inc.
P.O. Box 15
Martinsville, NJ 08836
(201) 356-3141
AHE CFA COS STL STA SYA STP

Allied Chemical
Fibers Division
P.O. Box 1057R
Morristown, NJ 07960
(201) 455-2000
CCM CMG

Solar Living Inc.
P.O. Box 12
Netcong, NJ 07857
(201) 691-8483
AHW CFL CCM CMA KIT

Aerco International Inc.
159 Paris Ave.
Northvale, NJ 07697
(201) 768-2400
HEX

New Jersey Aluminum
Solar Division
1007 Jersey Ave.
P.O. Box 73
North Brunswick, NJ 08902
(201) 249-6867
CCM CMA KIT AHE AHW

Solar Heating of New Jersey
811 Wynetta Pl.
Paramus, NJ 07652
(201) 652-3819
CFL COC AHW AHE APO CCM CMR

Science Associates Inc.
230 Nassau St.
P.O. Box 230
Princeton, NJ 08540
(609) 924-4470
INS

Edwards Engineering Corp.
101 Alexander Ave.
Pompton Plains, NJ 07444
(201) 835-2808
AHC AHW APO SYL

W & W Solar Systems Inc.
399 Mill St.
Rahway, NJ 07065
(201) 925-5488
AHE AHW APO CFL

Solar Life
404 Lippincott Ave.
Riverton, NJ 08077
(609) 829-7022
AHW SYL

SSP Associates
704 Blue Hill Rd.
River Vale, NJ 07675
(201) 391-4724
AHE AHW CFL CTU

HCH Associates Inc.
P.O. Box 87
Robbinsville, NJ 08691
(609) 259-9722
AHE AHW CCM CMG

Heilemann Electric
127 Mountain View Rd.
Warren, NJ 07060
(201) 757-4507
CFL HEX STL CTU COC AHW AHE APO

Cy/Ro Industries
Wayne, NJ 07470
(201) 839-4800
AGH CCM CMG SYP

NEW MEXICO

Albuquerque Western Solar Industries Inc.
612 Commanche, NE
Albuquerque, NM 87107
(505) 345-6764
AHC CCN FEA SYL

K-Line Corp.
911 Pennsylvania Ave., NE
Albuquerque, NM 87110
(505) 268-3379
AHE CFA STA SYA AHW COS INS

Sigma Energy Products
720 Rankin Rd., NE
Albuquerque, NM 87107
(505) 344-3431
AHE AHW APO CFL FEA

Soltrax Inc.
720 Rankin Rd., NE
Albuquerque, NM 87107
(505) 344-3431
AHW APO AHC ACA AAG CFL CCN COS COC STL
SYL KIT LAB

Southwest-Standard
P.O. Box 14132
Albuquerque, NM 87111
(505) 265-8871
CFL AHW APO NBS

Zomeworks Industries
P.O. Box 712
Albuquerque, NM 87103
(505) 242-5354
SYP AHC AHW SYL SYA CCM CMR CMT KIT FEA

Solar Room Company
P.O. Box 1377
Taos, NM 97511
(505) 758-9344
AHE AGH KIT

United States Solar Pillow
P.O. Box 987
Tucumcari, NM 88401
(505) 461-2608
APO CFL FEA SYP SYA SYL CFA AHC AGH AAG

NEW YORK

Sunenergy Power Ltd.
400 W. Main St.
Babylon, NY 11072
(516) 587-0611
AHE APO CFA CFL

Bi-Hex Company
P.O. Box 312
Bedford, NY 10506
(914) 764-4021
COS COC

Catalano & Sons Inc.
301 Stagg St.
Brooklyn, NY 11206
(212) 821-6100
AHE AHW CFL SYL COC

Standard Solar Collectors Inc.
1465 Gates Ave.
Brooklyn, NY 11227
(212) 456-1882
AHW AHE CFA CFL STL SYA SYL KIT FEA ASH

Sunray Solar Heat Inc.
202 Classon Ave.
Brooklyn, NY 11205
(212) 857-0193
AHW AHE APO AMH CFL COS SYL KIT

Temp-O-Matic Cooling Company
87 Luquer St.
Brooklyn, NY 11231
(212) 624-5600
AHE AHW CFL FEA

Advance Cooler Manufacturing Corp.
Route 146, Bradford Industrial Park
Clifton Park, NY 12065
(518) 371-2140
AHW CFL FEA CCM COC STL STA SYA SYL SYP
AHC

Prima Industries Inc.
P.O. Box 141
Deer Park, NY 11729
(516) 242-6347
AHW CFL SYL

Pan Solar Energy Products Inc.
50 Peabrook Ave.
Deer Park, NY 11729
(516) 586-5008
APO CFL SYL

American Acrylic Corp.
173 Marine St.
Farmingdale, NY 11735
(516) 249-1129
AHE AHW APO CCM CMG

Weksler Instruments Corp.
P.O. Box 3040
80 Mill Rd.
Freeport, NY 11520
(516) 623-0100
COC

Amprobe Instruments
630 Merrick Rd.
Lynbrook, NY 11563
(516) 593-5600
INS

Conkling Laboratories
5432 Merrick Rd.
Massapequa, NY 11758
(516) 541-1323
INS

Structured Sheets Inc.
196 E. Camp Ave.
Merrick, NY 11566
(516) 546-4868
AHE AGH CCM CMG

Ecosol Ltd.
2 W. 59th St.
The Plaza, 17th Fl.
New York, NY 10019
(212) 838-6170
AHC AHW COS HEP INS

Hitachi Chemical Co., Ltd.
437 Madison Ave.
New York, NY 10022
(212) 838-4804
AHW APO CFL STL SYL COC CMP CCM

Mechanical Mirror Works
661 Edgecombe Ave.
New York, NY 10032
(212) 795-2100
AHC CCM CMR

Union Carbide Corp.
270 Park Ave.
New York, NY 10017
(212) 551-2261
CCM CMA

Revere Copper & Brass Inc.
P.O. Box 151
Rome, NY 13440
(315) 338-2401
AHW CFL FEA APO CCM SYL COS COC AHE STL
CMP NBS

Grumman Corp.
Energy Systems Division, Dept. G-R
4175 Veterans Memorial Highway
Ronkonkoma, NY 11779
(516) 575-6205
AHE AHW CFL FEA SYL APO CCM CMH CMP

Solar Energy Systems Inc.
Concord House, Suite 2B
P.O. Box 625
Scarsdale, NY 10583
(914) 725-5570
AHW AHC APO ACA AMH CFL CCM CMH SYA
SYL SYP KIT NBS

Solar Sunstill Inc.
15 Blueberry Ridge Rd.
Setauket, NY 11733
(516) 941-4078
CCM CMC CMR

Sun Chance
P.O. Box 506
South Fallsburg, NY 12779
(914) 434-6650
AHW CTU CCM CMT CMP COS

Bio-Energy Systems Inc.
Mountaindale Rd.
Spring Glen, NY 12483
(914) 434-7858
AHE AHW APO AGH AAG CFL CFA COS STA STL
SYA SYL SYP KIT

Hudson Valley Solar
Route 9, P.O. Box 388
Valatie, NY 12184
(518) 781-4152
AHE SSF

Ford Products Corp.
Ford Products Rd.
Valley Cottage, NY 10989
(914) 358-8282
AHW STL

NORTH CAROLINA

Habitat 2000 Inc.
P.O. Box 188
Belmont, NC 28012
(704) 825-5357
AHW AHE APO AGH STL SYL HEX KIT

Carolina Aluminum
P.O. Box 2437, State Rd. 1184
Burlington, NC 27215
(919) 227-8826
CCM

McArthurs Inc.
P.O. Box 236
Forest City, NC 28043
(704) 245-7223
AHE AHW AAG CFA CFL SYP

Jenson Solar Inc.
P.O. Box 166
Goldsboro, NC 27530
(919) 566-4320
AHW CFL SYL

Solar Development & Manufacturing Co.
4000 Old Wake Forest Rd.
Raleigh, NC 27609
(919) 872-6900
AHE AHW CCN STL SYL COS

Carolina Thermal Co.
Iron Words Rd.
Route 2, Box 39
Reidsville, NC 27320
(919) 342-0352
AGH CFA

Standard Electric Co.
P.O. Box 631
Rocky Mount, NC 27801
(919) 442-1155
AHE AHW CFL FEA

Energy Applications Inc.
Route 5, Box 383
Rutherfordton, NC 28139
(704) 287-2195
AHC CCN CCM CMT COC CMM

Carolina Solar Equipment Co.
P.O. Box 2068
Salisbury, NC 28144
(704) 637-1243
AHE AHW CFL FEA SYL

Solar Comfort Inc.
Route 3, Box 139
Statesville, NC 28677
(704) 872-0753
AHE SSF

Carolina Aluminum
Metcalf Rd.
P.O. Box 177
Winton, NC 27986
(919) 358-5811
CCM

OHIO

Solar Energy Products Company
121 Miller Rd.
Avon Lake, OH 44012
(216) 933-5000
AHE AHW FEA HEX APO SYA COC STA CFA AAG
CCM CMB

Solar Usage Now Inc.
450 E. Tiffin St., Box 306
Bascom, OH 44809
(419) 937-2226
AHW CFL KIT

Solar Home Systems Inc.
12931 W. Geauga Trail
Chesterland, OH 44026
(216) 729-9350
AHW APO AHE CFL CFA COC STA SYA SYL NBS
ASH

Alpha Solarco
1014 Vine St.
Kroger Bldg., Suite 2230
Cincinnati, OH 45202
(513) 621-1243
LAB AHE CCN CFL CFE

Ohio Valley Solar Inc.
4141 Airport Rd.
Cincinnati, OH 45226
(513) 871-1961
AHE APO CFA COS STA SYA LAB SSF

Solar Sun Inc.
235 W. 12th St.
Cincinnati, OH 45210
(513) 241-4200
AHE AHW APO CFL SYL

Ferro Corp.
Coatings Division
P.O. Box 6550
4150 E. 56th St.
Cleveland, OH 44101
(216) 641-8580
CCM CMC

Fiber-Rite Products
P.O. Box 9295
Cleveland, OH 44138
(216) 228-2921
STL STC KIT AHW AHE

Glass-Lined Water Heater Co.
13000 Athens Ave.
Cleveland, OH 44107
(216) 521-1377
AHW STL

Mor-Flo Industries Inc.
18450 S. Miles Rd.
Cleveland, OH 44128
(216) 663-7300
AHW SYL STL CFL KIT CCM CMH CMI CMM HEX

Ranco Inc.
601 W. Fifth Ave.
Columbus, OH 43201
(614) 294-3511
AHE AHW COC

Mid-West Technology Inc.
P.O. Box 26238
Dayton, OH 45426
(513) 274-6020
AHE CCM CMA

Solar Glo-Thermal Energy Systems Inc.
P.O. Box 377
Dayton, OH 45459
(513) 252-6150
AHW AHC

Solar Heat Corporation
1252 French Ave.
Lakewood, OH 44107
(216) 228-2993
AHE AHW CFL FEA KIT

Solar Central
7213 Ridge Rd.
Mechanicsburg, OH 43044
(513) 828-1350
AHW CFL FEA CCM HEP AHC KIT SYL STL CMP
COS AGH ACA SYP RCE CMM CMG

Solar I
Division of Stellar Industries Inc.
7265 Commerce Dr.
Mentor, OH 44060
(216) 951-6363
AHW AHE APO CFA COS COC STA SYA INS FEA
NBS ASH

NRG Manufacturing
P.O. Box 53
Napoleon, OH 43545
(419) 559-3618
AHE SSF FEA

Solartec Inc.
250 Pennsylvania Ave.
Salem, OH 44460
(216) 332-9100
AHW AGH CFA CCM CMA CMM COC STL HEX NBS
LAB ASH

Stolle Corp.
1501 Michigan St.
Sidney, OH 45365
(513) 492-1111
AHW AHE CFL STL SYL HEX NBS ASH

Lof Solar Energy Systems
1701 Broadway
Toledo, OH 43605
(419) 247-4355
AHE AHW CFL FEA CCM CMG COC COS

Owens Illinois Inc.
P.O. Box 1035
Toledo, OH 43666
(419) 242-6543
AHC CTU FEA AHW

Van Hussel Tube Corp.
Warren, OH 44481
(216) 372-8221
CCM CMA

Sol-Era Energy Systems
P.O. Box 651
Worthington, OH 43085
(614) 846-8594
AHW AHE APO CFL STL SYL STA

Yellow Springs Instrument Co.
Yellow Springs, OH 45387
(513) 767-7241
INS

General Solar Systems
4040 Lake Park Rd.
P.O. Box 2687
Youngstown, OH 44507
(216) 783-0270
AHE CCN CCM CMM

OKLAHOMA

Professional Fiberglass Products Inc.
Ada Industrial Park
P.O. Box 1179
Ada, OK 74820
(405) 436-0223
STL

Anabil Enterprizes
525 S. Aqua Clear Dr.
Mustang, OK 73064
(405) 376-3324
COS COC

Westinghouse Electric Corp.
5005 Interstate Dr., N.
Norman, OK 73069
(405) 364-4040
HEP

Brown Manufacturing Co.
P.O. Box 14546
Oklahoma City, OK 73114
(405) 751-1323
AHW COS SYP AHC COC

Coating Laboratories
505 S. Quaker
Tulsa, OK 74120
(918) 272-1191
CCM CMR

McKim Solar Energy Systems Inc.
2627 E. Admiralty Place
Tulsa, OK 74110
(918) 936-4035
AHC AHW APO CFA CFL CTU CCN CCM CMT COS
COC STA STL STP HEX KIT NBS LAB ASH

OREGON

Tektronix Products
P.O. Box 500
Beaverton, OR 97077
(503) 644-0161
LAB

Sun Life Solar Products
12900 SE 32nd Ave.
Milwaukee, OR 97222
(503) 653-1449
AHC AHW CFL STL SYL

Solar Kits
P.O. Box 350
Philomath, OR 97370
(503) 929-6289
AHW APO CFL KIT

Solar Pre-Fab Ltd.
2625 SE Kelley St.
Portland, OR 97202
(503) 233-1652
AHW CFL KIT

Sun Power Corp.
12785 SE Hiway 212
Portland, OR 97015
(503) 655-6282
AHC AHW APO CFE

PENNSYLVANIA

Rohm & Haas
Plastics Engineering
P.O. Box 219
Bristol, PA 19007
(215) 788-5501
CCM CMG CMS

Carlisle Tile & Rubber Co.
P.O. Box 99
Carlisle, PA 17013
(717) 249-1000
AHW AHE STC

Electro-Kinetic Systems Inc.
2500 E. Ridley Ave.
Chester, PA 19013
(215) 876-6192
CCM CMC

Solar Power Inc.
201 Airport Blvd.–Cross Keys
Doylestown, PA 18901
(215) 348-9066
AHE SSF

Practical Solar Heating
209 S. Delaware Dr., Route 611
Easton, PA 18042
(215) 252-6381
AHE COC CFL STC APO KIT CCM AHW HEX STL CMP
CTU

International Environment Corp.
1400 Mill Creek Rd.
Gladwyne, PA 19035
(215) 642-3060
AHW CFL FEA AHE SYL

Sunearth Solar Products Corp.
Rd. 1, Box 337
Green Lane, PA 18054
(215) 699-7892
AHE AHW CFL FEA CCM SYL STL CMA CMG CMH
CMS CMM CMP COC COS

Solar Manufacturing Co.
40 Conneaut Lake Rd.
Greenville, PA 16125
(412) 588-2571
AHE CFA CCM STA SYA COS CMG SSF FEA

Overly Manufacturing Co.
574 W. Otterman St.
Greensburg, PA 15601
(412) 834-7300
AHE AHW APO CFL

Alco Plastic Products
266 Delray Ave.
Hanover, PA 17331
AHE CCM SYP CMG

Ametek Inc.
1 Spring Ave.
Hatfield, PA 19440
(215) 822-2971
AHW CFL FEA AHC NBS LAB

Solar Energy Associates
1063 New Jersey Ave.
Hellertown, PA 18055
(215) 838-7460
AHW CFL

Milton Roy Co.
Hartell Division
70 Industrial Dr.
Ivyland, PA 18974
(215) 322-0730
CCM CMP

Solar Shelter Engineering Co.
P.O. Box 179
Kutztown, PA 19530
(215) 683-6769
AHE SSF FEA AHW CFA COS STA SYA STP NBS ASH

Enviropane Inc.
350 N. Marshall St.
Lancaster, PA 17602
(717) 299-3737
AHE AHW CFL FEA CFA CCM CMA

Energy Systems Products Inc.
12th & Market Sts.
Lemoyne, PA 17043
(717) 761-8130
CCM CMM CMH

Heliotherm Inc.
W. Lenni Rd.
Lenni, PA 19052
(215) 459-9030
AHE AHW CFL FEA SYL

Atlas Vinyl Products
7002 Beaver Dam Rd.
Levittown, PA 19057
(215) 946-3620
APO CFL FEA

Simons Solar Environmental Systems Inc.
24 Carlisle Pike
Mechanicsburg, PA 17055
(717) 697-2778
AHW SYL AHE CFL CCM COS KIT COC CMA CMG CMP
FEA

Amicks Solar Heating
375 Aspen St.
Middletown, PA 17057
(717) 944-1842
AHE AHW APO CFL CTU CCM CMG CMA CMH CMC
CMS CMP COC STL SYL HEX

Packless Industries Inc.
P.O. Box 310
Mount Wolf, PA 17347
(717) 266-5673
AHW HEX APO AHC

Sun God
P.O. Box 54
New Britain, PA 18901
(215) 368-7719
AHW CFL COS COC HEX

Pyco
600 E. Lincoln Hwy.
Penndel, PA 19047
(215) 757-3704
INS

General Electric Co.
Bldg. 7, P.O. Box 13601
Philadelphia, PA 19101
(215) 962-2112
AHW AHC CTU SYL HEP CCM CMG

Aluminum Co. of America
Alcoa Bldg.
Pittsburgh, PA 15219
(412) 553-2321
CCM APO SYL COC CMA CMC CMM CFL AHE AHW
FEA

PPG Industries
One Gateway Center
Pittsburgh, PA 15222
(412) 434-3555
AHE AHW CFL FEA APO

Sunwall Inc.
P.O. Box 9723
Pittsburgh, PA 15229
(412) 364-5349
AHE AHW APO CFA FEA STA STL SYA SYL SYP KIT

Sundevelopment Inc.
1108 Hanover Rd.
York, PA 17404
(717) 225-5066
AHE AHW CFL

Foam Products Inc.
Gay St.
York Haven, PA 17370
(717) 266-3671
AHE CFA STA SSF FEA NBS

PUERTO RICO

Solar Devices Inc.
GPO Box 3727
San Juan, PR 00936
(809) 783-1775
AHW CFL SYL

RHODE ISLAND

Taco Inc.
1160 Cranston St.
Cranston, RI 02920
(401) 942-8000
HEX COS COC CCM CMP

Solar Homes Inc.
2707 S. County Trail
East Greenwich, RI 02818
(401) 294-2443
AHE AHW CFL CFA

Independent Energy Inc.
P.O. Box 363
Kingston, RI 02881
(401) 295-1762
COC

Eppley Laboratory Inc.
12 Sheffield Ave.
Newport, RI 02840
INS

Solar Products Inc.
12 Hylestead St.
Providence, RI 02905
(401) 467-7350
AHW AHE CFL SYL

SOUTH CAROLINA

Sunsaver Inc.
P.O. Box 21672
Columbia, SC 29221
(803) 781-4962
AHW AGH CFL STL SYL HEX KIT

Helio Thermics Inc.
110 Laurens Rd.
Greenville, SC 29601
(803) 235-8529
AHE SYA

General Solargenic Corp.
P.O. Box 307
Johns Island, SC 29455
(803) 747-4480
AHC AHW CFA COS STA SYA KIT

Solar Energy Research & Development
302 Lucas
Mt. Pleasant, SC 29464
(803) 884-0290
AHW AHE CFL COS STL SYL

SOUTH DAKOTA

The Solar Store
No. 1 Solar Lane
Parker, SD 57053
(605) 648-3465
AHE SSF

TENNESSEE

State Industries Inc.
Cumberland St.
Ashland City, TN 37015
(615) 792-4371
AHE AHW CFL FEA SYL STL KIT

Energy Converters Inc.
2501 N. Orchard Knob Ave.
Chattanooga, TN 37406
(615) 624-2608
AHE AHW CFL FEA

W. L. Jackson Manufacturing Co.
1200–26 W. 40th St.
Chattanooga, TN 37401
(615) 867-4700
AHE SYL CFL

ASG Industries
P.O. Box 929
Kingsport, TN 37662
(615) 245-0211
AHW APO CCM AHC CMG

Energy Design Corp.
P.O. Box 34294
Memphis, TN 38134
(901) 382-3000
AHW CCN AHC

TEXAS

Solar Enterprises Inc.
2816 W. Division
Arlington, TX 76012
(817) 461-5571
AHE CCM CMA CMM CMC COS COC STL HEX KIT FEA
ASH AHW APO CFL SYL

Cole Solar Systems Inc.
440A E. St. Elmo Rd.
Austin, TX 78745
(512) 444-2565
AHW APO CFL SYL CCM CMM FEA

Greenhouse Systems Corp.
P.O. Box 31407
Dallas, TX 75231
(214) 352-6174
AHE AAG CBL

Solar Kinetics Inc.
147 Parkhouse St., P.O. Box 10764
Dallas, TX 75207
(214) 747-6519
AHE AHW CCN FEA

Texas Electronics Inc.
5529 Redfield St.
Dallas, TX 75209
(214) 631-2490
INS

Wilshire Foam Products Inc.
P.O. Box 34217
Dallas, TX 75234
(214) 241-4073
SPC

Denton Greenhouse Manufacturing Inc.
3301 Fortworth
Denton, TX 76201
(817) 382-1107
AGH

Solartech Systems Corp.
P.O. Box 591
Devine, TX 78016
(512) 663-4491
AHE AHW APO CFL SYL NBS

Southwest–Standard
P.O. Box 10094
El Paso, TX 79991
(915) 533-6291
CFL AHW APO NBS

Teledyne Geotech
3401 Shiloh Rd.
Garland, TX 75041
(214) 271-2561
INS

Solar Control Corp.
P.O. Box 2201
Harker Heights, TX 76541
(817) 699-8858
AHW AHE CFA CFL CCN

Dodge Products
P.O. Box 19781
Houston, TX 77024
(713) 467-6262
INS

Exxon Company USA
P.O. Box 2180
Houston, TX 77001
(713) 656-0370
AHC AHW CCM CMH

Shell Oil Co.
1 Shell Plaza, P.O. Box 2463
Houston, TX 77001
(713) 241-6161
AHC CCM CMH

Soltex Corp.
1804 Afton St., Lock Lane
Houston, TX 77055
(713) 782-4478
AHE AHW CFL FEA

Solus, Inc.
P.O. Box 35227
Houston, TX 77035
(713) 772-6416
AHW AHC APO AAG CFL SYL HEP

Northrup Inc.
302 Nichols Dr.
Hutchins, TX 75141
(214) 225-4291
AHC AHW CFL CCN FEA APO

Butler Vent-A-Matic Corp.
P.O. Box 728
Mineral Wells, TX 76067
(800) 433-1626
APO AHW SYL AHE CFL HEX

Ace Solar Systems
Route 1, P.O. Box 50
Mission, TX 78572
(512) 585-6353
AHE AHW CFL SYP

Solar Therm
203 Point Royal Dr.
Rockwall, TX 75087
(214) 475-2201
AHE AHW APO CFL CCM CMM KIT CMA CMG STL CMP
CMI COC

Thermon Manufacturing Co.
100 Thermon Dr.
San Marcos, TX 78666
(516) 392-5801
CCM CMA AHE AHW

Friedrich Air Conditioning/Refrigeration
P.O. Box 1540
San Antonio, TX 78295
(512) 225-2000
AHE HEP

PCA
11031 Wye Dr.
San Antonio, TX 78217
(512) 656-9338
NBS STL CCM APO AHW AHE CFL SYL KIT CMP

Solarsystems Inc.
507 W. Elm St.
Tyler, TX 75702
(214) 592-5343
AHC AHW APO CFE FEA NBS

American Solar King Corp.
6801 New McGregor Hwy.
Waco, TX 76710
(817) 766-3860
AHE AHW CFL FEA

UTAH

Permaloy Corp
P.O. Box 1559
Ogden, UT 84402
(801) 731-4303
AHE CCM CMC

Griep Heating
155 E. 3600 S.
Salt Lake City, UT 84115
(801) 262-2537
AHE AHW CFA

VERMONT

Solar Alternative Inc.
30 Clark St.
Brattleboro, VT 05301
(802) 254-8221
AHE AHW CFL APO

Garden Way Laboratories
P.O. Box 66
Charlotte, VT 05445
(802) 425-2147
SYP AGH AHE

Green Mountain Homes
Royalton, VT 05068
(802) 763-8384
AHC SYP HYB

VIRGINIA

Clark's Products & Services
Route 1, P.O. Box 2138
Bluemont, VA 22012
(703) 955-3837
AHW AHE APO CFL KIT

Helios Corp.
2120 Angus Rd.
Charlottesville, VA 22901
(804) 977-3719
AHC AHW SYA SYL FEA

Puff
920 Allied St.
Charlottesville, VA 22901
(804) 977-3541
AHW AHE STL

Solar Sensor System
4220 Berritt St.
Fairfax, VA 22030
(703) 273-2683
AHE COC AHW APO COS INS

Pioneer Energy Products
Route 1, P.O. Box 189
Forest, VA 24551
(804) 239-9020
AHW AHC CFL COS SYL NBS

Martin Processing Inc.
P.O. Box 5068
Martinsville, VA 24112
(703) 629-1711
AGH AAG CCM CMG

Corillium Corp
Cronagold Division
Reston International Center
Reston, VA 22090
(703) 860-2100
AHE CCM CMC

Reynolds Metals Company
P.O. Box 27003
Richmond, VA 23261
(804) 282-3026
AHE AHW CFL FEA APO NBS

Virginia Solar Components Inc.
Route 3, Highway 29 S.
Rustburg, VA 24588
(804) 239-9523
AHW AHC APO CFL CCM CMP STL SYL COS COC INS

Solar One Ltd.
2644 Barret St.
Virginia Beach, VA 23451
(804) 340-7774
AHE STA SYA CFA FEA

Intertechnology/Solar Corp. of America
100 Main St.
Warrentown, VA 22186
(703) 347-7900
AHW CFL SYL KIT AHC APO CCN FEA

One Design Inc.
Mountain Falls Rte.
Winchester, VA 22601
(703) 662-4898
SYP

WASHINGTON

Vertrex Corp.
208 Carlson Bldg., 808 106th, NE
Bellevue, WA 98004
(206) 682-9725
AHE COC

E&K Service Company
16824 74th Ave., NE
Bothell, WA 98011
(206) 486-6660
AHE AHW CFL FEA SYL

Sun Power Northwest
16615 76th Ave., NE
Bothell, WA 98011
(206) 486-6632
AHW AHC APO CFA CCM CMC CMB COS STA SYA HEX

Sunpower Industries Inc.
10837 B6 SE 200th
Kent, WA 98031
(206) 854-0670
AHE CFA SYA KIT

Solar Northwest Corp.
Route 1, P.O. Box 114
Long Beach, WA 98631
(206) 642-2249
AHW COS

Floscan Instrument Co.
3016 NE Blakey St.
Seattle, WA 98105
(206) 524-6625
INS

Northwest Solar Systems Inc.
7700 12th NE
Seattle, WA 98115
(206) 523-3951
AHE AHW APO CFA COC STA SYA

Silverdale Fuel
P.O. Box 37
Silverdale, WA 98383
(206) 692-9221
AHW APO ACO CFA CCM SYA HEX

Mann-Russell Electronics Inc.
1401 Thorne Rd.
Tacoma, WA 98421
(206) 383-1591
AHE CCM CMM HEL KIT

WISCONSIN

Ark-Tic-Seal Systems Inc.
P.O. Box 428
Butler, WI 53007
(413) 276-0711
AHC SYP

Research Products Corp.
1015 E. Washington Ave., P.O. Box 1467
Madison, WI 53701
(608) 257-8801
AHW AHE CFA CCM CMM CMB COS LAB

Sun Stone Solar Energy Equipment
P.O. Box 941
Sheboygan, WI 53081
(414) 452-8194
AHE CFA FEA AHW CCM COS COC STA STC SYA APO
STL CMB CMM

Solaray Inc.
324 S. Kidd St.
Whitewater, WI 53190
(414) 473-2525
AHE AHW CFA CFL FEA APO STA SYA SYL COS CCM
CMA CMH CMP CMB

WYOMING

Park Energy Co.
P.O. Box SR9
Jackson, WY 83001
(307) 733-4950
CFA STA AHE AHW

33 The National Solar Heating and Cooling Information Center

The National Solar Heating and Cooling Information Center was established by the Department of Housing and Urban Development, in cooperation with the Energy Research and Development Administration, under provisions of Public Law 93-409, to help make everyone aware of the feasibility of solar energy and to encourage the public and industry to consider solar energy systems for homes and commercial buildings. If you have any questions or need any information about solar heating and cooling, contact the center at:

Box 1607
Rockville, MD 20850
(800) 535-2929

In Pennsylvania, call (800) 462-4983.

Appendix
National Energy Act

Provisions of the National Energy Act (NEA) are expected to result in reduced oil import needs by 1985, increased use of fuels other than oil and gas, and more efficient and equitable use of energy in the United States.

President Carter stated after congressional passage of the act, "We have declared to ourselves and to the world our intent to control our use of energy, and thereby to control our own destiny as a nation."

Energy Secretary James Schlesinger commented, "The NEA represents an historic turning point. The era of cheap and abundant energy is recognized to be over. For the first time, energy conservation is recognized as an indispensible ingredient in national energy policy. With the NEA, we will save 2.5 to 3 million barrels a day by 1985, compared to what we would otherwise have required for an estimated balance of payments savings of approximately $14 billion in current dollars (as much as $20 billion in 1985 dollars)."

"The purpose of the National Energy Act," Secretary Schlesinger added, "is to put into place a policy framework for decreasing oil imports by:

- replacing oil and gas with abundant domestic fuels in industry and electric utilities,
- reducing energy demand through improved efficiency,
- increasing production of conventional sources of domestic energy through more rational pricing policies, and
- building a base for the development of solar and renewable energy sources."

NEA was passed by Congress on October 15, 1978, after nearly a year and a half of deliberation. The act is composed of five bills:

- National Energy Conservation Policy Act of 1978
- Powerplant and Industrial Fuel Use Act of 1978
- Public Utilities Regulatory Policy Act
- Natural Gas Policy Act of 1978
- Energy Tax Act of 1978

ENERGY CONSERVATION

The National Energy Conservation Policy Act (EPCA) of 1978 provides for

- *Utility conservation program for residences.* A program requiring utilities to offer energy audits to their residential customers that would identify appropriate energy conservation and solar energy measures and estimate their likely costs and savings. Utilities also will be required to offer to arrange for the installation and financing of any such measures.
- *Weatherization grants for low-income families.* Extension through 1980 of the DOE weatherization grants program for insulating lower income homes at an authorized level of $200 million in fiscal year 1978 and 1980.
- *Solar energy loan program.* A $100 million program administered by HUD which will provide support for loans of up to $800 to homeowners and builders for the purchase and installation of solar heating and cooling equipment in residential units.
- *Energy-conservation loan programs.* A $5 billion program of federally supported home-improvement loans for energy-conservation measures; $3 billion for support of reduced interest loans up to $2500 for elderly or moderate-income families and $2 billion for general standby financing assistance.
- *Grant program for schools and hospitals.* Grants of $900 million over the next 3 years to improve the energy efficiency of schools and hospitals.
- *Energy audits for public buildings.* A 2-year, $65 million program for energy audits in local public buildings and public-care institutions.
- *Appliance efficiency standards.* Energy-efficiency standards for major home appliances, such as refrigerators and air-conditioning units.
- *Civil penalties relating to automobile fuel efficiency.* Authority for the Secretary of Transportation to increase the civil penalties on auto manufacturers from $5 to $10 per car for each 0.1 mile that a manufacturer's average fleet mileage fails to meet the EPCA automobile fleet average fuel economy standards.
- *Other provisions in the act include:*
  - Grants and standards for energy conservation in federally assisted housing

- Federally insured loans for conservation improvements in multifamily housing
- $100 million for a solar demonstration program in federal buildings
- Conservation requirements for federal buildings
- $98 million for solar photovoltaic systems in federal facilities
- Industrial recycling targets and reporting requirements
- Energy-efficiency labeling of industrial equipment
- A study of the energy efficiency of off-road and recreational vehicles
- An assessment of the conservation potential of bicycles.

COAL CONVERSION

The Power Plant and Industrial Fuel Use Act of 1978 provides for

- *Prohibition of new oil- and gas-fired boilers.* Prohibition against use of oil or natural gas in new electric utility generation facilities or in new industrial boilers with a fuel heat input rate of 100 million Btu per hour or greater, unless exemptions are granted by DOE.
- *Restrictions on existing coal-capable large boilers.* DOE authority to require existing coal-capable facilities, individually or by categories, to use coal and to require non-coal-capable units to use coal-oil mixtures.
- *Restrictions on users of natural gas for boiler fuel.* Limitation of natural gas use by existing utility power plants to the proportion of total fuel used during 1974 to 1976, and a requirement that there be no switches from oil to gas. There is also a requirement that natural gas use in such facilities cease by 1990 (with certain exceptions).
- *Pollution control loan program.* An $800 million loan program to assist utilities to raise necessary funds for pollution control.
- *Supplemental authority.* Supplemental authority to prohibit use of natural gas in small boilers for space heating and in decorative outdoor lighting and to allocate coal in emergencies.
- *Other provisions.* Funding of several programs to reduce negative impacts from increased coal production, energy impact assistance, and railroad rehabilitation.

PUBLIC UTILITY REGULATORY POLICIES

The Public Utility Regulatory Policies Act of 1978 provides for

- *Rate design standards.* Eleven voluntary standards on rate design and other utility practices for consideration by state regulatory authorities and nonregulated utilities—including time-of-day-rates, seasonal rates, cost of service pricing, interruptible rates, prohibition of declining block rates, and lifeline rates.
- *Consideration of rate design standards.* A requirement that state regulatory authorities and utilities consider each standard within prescribed periods and determine if they are appropriate for conservation, efficiency, and equity, as well as consistent with state laws. Voluntary guidelines with respect to the standards may be prescribed.
- *Retail policies for natural gas activities.* Consideration by gas utilities of two standards—i.e., service termination procedures and advertising expenditures. A DOE study of the best rate design for gas utilities is also required.
- *Cogeneration.* FERC rules favoring industrial cogeneration facilities, and requiring utilities to buy or sell power from qualified cogenerators at just and reasonable rates.
- *Wholesale provisions.* FERC authority to require interconnections of electric power transmission facilities, to order utilities to provide transmission services between two non-contiguous utilities, and to report anticipated power shortages; FERC review of automatic rate adjustment-clauses.
- *Aid to states and consumer representation.* Funding to assist state implementation and consumer intervention in proceedings.
- *Small hydroelectric facilities.* Loan program to aid development of small hydroelectric projects.
- *Expediting legislation for crude oil transportation systems.* Establishes a process for selecting and expediting issuance of permits for a crude oil transportation system to move oil from the west coast to northern tier inland states, as well as expediting the issuance of permits for the SOHIO pipeline in the south, running from Long Beach, California, to Midland, Texas.
- *Significant miscellaneous provisions.* Authorization funding for the National Regulatory Research Institute; establishment of three additional University Coal Research Laboratories; rules for conversion of natural-gas users to less desirable heavy fuel oils; emergency conversion of utilities and other facilities during natural gas emergencies; natural gas transportation policy, and rules for treatment of conserved natural gas.

NATURAL GAS PRICING REGULATION

The Natural Gas Policy Act of 1978 provides for

- *Price controls.* The NGPA sets a series of maximum lawful prices for various categories of natural gas, including gas sold in both the interstate and intrastate markets. This eliminates the regulatory distinction which had previously existed between the two markets, with interstate rates set on the federal level and intrastate rates largely unregulated.
- *Deregulation of certain gas.* Price controls on new gas and certain intrastate gas will be lifted as of January 1, 1985. Certain high-cost gas will be deregulated approximately one year after the NGPA's enactment. Gas

from certain new onshore wells will also be deregulated but not until July 1987. Other gas will remain under price controls indefinitely. Price controls may be reimposed by Congress or the president for one 18-month period.

- *Incremental pricing.* Protection of residential consumers by first passing through some portion of increased gas prices to industrial users. This incremental pricing to industrial users cannot result in industrial gas prices higher than the regional cost of substitute fuels as determined by the FERC.

  The average cost of natural gas to industry, even under incremental pricing, is expected to remain well below the cost of alternate fuels. Initially this rule applies only to boiler fuel users of natural gas.

- *Emergency authority.* The president may declare an emergency if a gas shortage exists or is imminent which endangers supplies for "high-priority" users. High-priority use means the use of gas in a residence or small commercial establishment, or any use, the curtailment of which would endanger life, health, or maintenance of physical property. During an emergency the president may authorize certain emergency sales of gas. If these emergency sales are not sufficient to protect high-priority users, he may allocate certain supplies of gas, as necessary.

- *Curtailment priorities.* Interstate gas supplies needed for certain agricultural and industrial uses generally will not be curtailed unless the gas is needed to serve high-priority users.

ENERGY TAX ACT

The Energy Tax Act of 1978 provides for

- *Residential insulation and conservation tax credits.* A nonrefundable income-tax credit for residential insulation and energy-conservation measures—up to $300 or 15 percent of the first $2000 expended.

- *Residential solar tax credits.* A nonrefundable income-tax credit for the residential installation of solar or wind equipment—up to a total maximum credit of $2200 covering $10,000 of expenditures.

- *Exemption of gasohol from excise tax.* Exemption of gasoline containing at least 10 percent alcohol produced from agricultural products or waste from the 4 cents per gallon federal excise tax.

- *Gas guzzler tax.* Graduated excise tax on gas guzzling cars that fall substantially below federally mandated fleetwide mileage standards for each year.

- *Geothermal energy and geopressured natural gas tax provisions.* Incentives for the development of geothermal resources through an investment tax credit, the expensing of intangible drilling costs, and a percentage depletion allowance. Geopressured natural gas is granted a special 10 percent depletion allowance.

- *Minimum tax exclusion for intangible drilling costs.* Extension of favorable minimum tax treatment of

Automobile Model Year	If the Fuel Economy of the Model Type in which the Automobile Falls is:	The Tax is
1980	At least 15	0
	At least 14 but less than 15	$200
	At least 13 but less than 14	300
	Less than 13	550
1981	At least 17	0
	At least 16 but less than 17	$200
	At least 15 but less than 16	$350
	At least 14 but less than 15	450
	At least 13 but less than 14	550
	Less than 13	650
1982	At least 18.5	0
	At least 17.5 but less than 18.5	$200
	At least 16.5 but less than 17.5	350
	At least 15.5 but less than 16.5	450
	At least 14.5 but less than 15.5	600
	At least 13.5 but less than 14.5	750
	At least 12.5 but less than 13.5	950
	Less than 12.5	1200
1983	At least 19	0
	At least 18 but less than 19	$350
	At least 17 but less than 18	500
	At least 16 but less than 17	650
	At least 15 but less than 16	800
	At least 14 but less than 15	1000
	At least 13 but less than 14	1250
	Less than 13	1550
1984	At least 19.5	0
	At least 18.5 but less than 19.5	$450
	At least 17.5 but less than 18.5	600
	At least 16.5 but less than 17.5	750
	At least 15.5 but less than 16.5	950
	At least 14.5 but less than 15.5	1150
	At least 13.5 but less than 14.5	1450
	At least 12.5 but less than 13.5	1750
	Less than 12.5	2150
1985	At least 21	0
	At least 20 but less than 21	$500
	At least 19 but less than 20	600
	At least 18 but less than 19	800
	At least 17 but less than 18	1000
	At least 16 but less than 17	1200
	At least 15 but less than 16	1500
	At least 14 but less than 15	1800
	At least 13 but less than 14	2200
	Less than 13	2650
1986	At least 22.5	0
	At least 21.5 but less than 22.5	$500
	At least 20.5 but less than 21.5	650
	At least 19.5 but less than 20.5	850
	At least 18.5 but less than 19.5	1050
	At least 17.5 but less than 18.5	1300
	At least 16.5 but less than 17.5	1500
	At least 15.5 but less than 16.5	1850
	At least 14.5 but less than 15.5	2250
	At least 13.5 but less than 14.5	2700
	At least 12.5 but less than 13.5	3200
	Less than 12.5	3850

intangible drilling cost for oil and gas into future years.

- *Business energy tax credits.* Business tax credits for industrial investment in alternative energy property (such as boilers for coal, nonboiler burners for alternate fuels, heat-conservation equipment, and recycling equipment).
- *Denial of tax benefits for new oil and gas-fired boilers.* Denial of investment tax credit and accelerated depreciation for new gas and oil boilers.

CONTACTS FOR FURTHER INFORMATION

Provisions of the National Energy Act (NEA) are carried out by a number of state and federal government units. States, for example, handle grants for energy-conservation programs in schools and hospitals. States submit plans to the Department of Energy (DOE) for utility audits and aid for residential energy conservation.

Loans and loan guarantees may fall into several categories and be handled by different units, including the Department of Housing and Urban Development (HUD), the Farmers Home Administration, and the Federal Home Loan Bank Board. Contacts, at least initially, could be made with the office of public affairs in those federal agencies or, in the case of state-administered programs, the particular state energy office.

The U.S. Internal Revenue Service sets rules for tax credits, such as for solar heating and insulation, and the local IRS offices can be contacted. Here are some contacts that may be useful:

Department of Housing and Urban Development
Washington, DC 20410
(202) 655-4000
    Conservation grants in federal housing
    Conservation loans for homeowners
    Conservation loans in multifamily housing
    Conservation standards for housing
    Solar energy loans

State Energy Offices
    Energy audits for public buildings
    Grants for schools and hospitals
    Utilities program for residential conservation
    Utility rate reform

Director
Office of Buildings and Community Systems
Department of Energy
Washington, DC 20545
(202) 252-5000
    Appliance efficiency standards
    Energy audits for public buildings
    Federal buildings conservation requirements
    Utilities program for residential conservation

Director
Office of State and Local Programs
Department of Energy
Washington, DC 20545
(202) 252-5000
    Grants for schools and hospitals
    Weatherization grants for low-income families

Office of Public Information
Federal Energy Regulatory Commission
Washington, DC 20416
(202) 275-4006 or (800) 424-5200 (toll-free)
    Natural-gas pricing regulations
    Public utility regulatory policies (cogeneration and
        wholesale provisions)

Office of Information
Economic Regulatory Administration
Washington, DC 20461
(202) 634-2170
    Coal conversion for utilities and industrial boilers
    Public utility regulatory policies (other than cogenera-
        tion and wholesale provisions)

Internal Revenue Service
1111 Constitution Ave., NW
Washington, DC 20224
(202) 488-3100 (or any local IRS office)
    Tax credits

National Solar Heating and Cooling Information Center
P.O. Box 1607
Rockville, MD 20850
(800) 523-2929
In Pennsylvania, (800) 462-4983
    Tax credits (plus information on solar equipment and
        contractors)

Director
Office of Solar Applications
Department of Energy
Washington, DC 20545
(202) 252-5000
    Solar demonstrations in federal buildings
    Photovoltaic systems in federal facilities

Director
Office of Industrial Programs
Department of Energy
Washington, DC 20545
(202) 252-5000
    Industrial equipment efficiency
    Industrial efficiency progress reporting
    Industrial recovered materials targets
    Second-law efficiency study

The National Energy Information Center of DOE's Energy Information Administration may be contacted for answers to general questions. Telephone: (202) 634-5610. Address: 1726 M St., NW, Room 210, Washington, DC 20461.

The DOE Office of Public Affairs in Washington, DC, may also be contacted for overall information or referrals.

Public Inquiries Branch
Office of Public Affairs
Department of Energy
Washington, DC 20585
(202) 252-5568

Press Services
Office of Public Affairs
Department of Energy
Washington, DC 20585
(202) 252-5806

Regional DOE offices may also be contacted for general questions or referrals. They are:

Region 1
Analex Building, Room 700
150 Causeway St.
Boston, MA 02114
(617) 223-5257

Region II
26 Federal Plaza
Room 3206
New York, NY 10007
(212) 265-0560

Region III
1421 Cherry St.
10th Floor
Philadelphia, PA 19102
(215) 597-0792

Region IV
1655 Peachtree Street, NE
8th Floor
Atlanta, GA 30309
(404) 881-2062

Region V
175 W. Jackson Blvd.
Room A-333
Chicago, IL 60604
(312) 353-5779

Region VI
2626 W. Mockingbird Lane
Box 35228
Dallas, TX 75235
(214) 749-7621

Region VII
324 E. 11th St.
Kansas City, MO 64106
(816) 374-2061

Region VIII
1075 S. Yukon St.
Box 26247—Belmar Branch
Lakewood, CO 80226
(303) 234-2420

Region IX
111 Pine St.
Third Floor
San Francisco, CA 94111
(415) 556-0418 (or 7157)

Region X
1992 Federal Building
915 Second Ave.
Seattle, WA 98174
(206) 442-7285

Glossary

Abatement. The reduction in the degree or intensity of pollution.

Absorber or **Absorber Plate.** A surface, in a solar collector, usually blackened metal, that absorbs solar radiation.

Absorber Surface. The surface of the collector plate, which absorbs solar energy and transfers it to the collector plate.

Absorptance. The ratio of the amount of radiation absorbed by a surface to the amount of radiation incident upon it.

Absorption. The penetration of one substance into or through another.

Absorption Unit. A factory-tested assembly of component parts producing refrigeration for comfort cooling by the application of heat. This definition applies to those absorption units that also produce comfort heating.

Absorption System. All of the equipment intended or installed for the purpose of heating or cooling air by an absorption unit, either by direct or indirect means, and discharging such air into any room or space.

Absorptivity. The capacity of a material to absorb radiant energy.

Accelerator. In radiation science, a device that speeds up charged particles such as electrons or protons.

Accessible. Admitting close approach. Not guarded by locked doors or other effective means.

Acclimation. The physiological and behavioral adjustments of an organism to changes in the environment.

Acclimatization. The adaption over several generations of a species to a marked change in the environment.

Activated Carbon. A highly absorbent form of carbon used to remove odors and toxic substances from gaseous emissions. In advanced waste treatment, it is used to remove dissolved organic matter from wastewater.

Activated Sludge. Sludge that has been aerated and subjected to bacterial action; used to speed breakdown of organic matter in raw sewage during secondary waste treatment.

Active Solar System. Any system that needs mechanical means, such as motors, pumps, or valves, to operate.

Acute Toxicity. Any poisonous effect produced by a single short-term exposure that results in severe biological harm or death.

Adaptation. A change in structure or habit of an organism that produces better adjustment to its surroundings.

Adhesion. Molecular attraction that holds the surfaces of two substances in contact, such as water and rock particles.

Adsorption. The attachment of the molecules of a liquid or gaseous substance to the surface of a solid.

Adulterant. Chemical impurity or substance that by law does not belong in a food, plant, animal, or pesticide formulation.

Advanced Wastewater Treatment. The tertiary stage of sewage treatment.

Aeration. To circulate oxygen through a substance, as in wastewater treatment where it aids in purification.

Aerobic. Life or processes that depend on the presence of oxygen.

Aerosol. A suspension of liquid or solid particles in a gas.

Afterburner. An air-pollution control device that removes undesirable organic gases by incineration.

Agricultural Pollution. The liquid and solid wastes from farming, including runoff from pesticides, fertilizers, and feedlots; erosion and dust from plowing; animal manure and carcasses; crop residue; and debris.

Air Change. A method of expressing the amount of air transferred out of a space, in terms of the number of space volumes transferred per hour.

Air Conditioning. The process of treating air so as to control simultaneously its temperature, humidity, cleanliness, and distribution in order to meet the comfort requirements of the occupants of the conditioned space.

Air Curtain. A method of containing oil spills in which air bubbling through a perforated pipe causes an upward waterflow that slows the spread of oil. It can also be used to prevent fish from entering polluted water.

Air Dry. Air containing no water vapor; air only.

Air Gap. An air gap in a potable water-distribution system is the unobstructed vertical distance through the free atmosphere between the lowest opening from any pipe or faucet supplying water to a tank, plumbing fixture, or other device and the flood-level rim of the receptacle.

Air Mass. A widespread body of air that gains certain characteristics, while set in one location. Its characteristics change as it moves away.

Air Monitoring. *See* Monitoring.

Air, Outdoor. Air taken from outdoors, and therefore not previously circulated through the system.

Air Pollution. The presence in the air of contaminant substances that do not disperse properly and thus interfere with human health.

Air-Pollution Episode. A period of abnormally high concentration of air pollutants, often due to low winds and temperature inversion that can cause illness and death.

Air Quality Control Region. Designated by the federal government, an area in which communities share a common air-pollution problem, sometimes involving several states.

Air-Quality Criteria. The levels of pollution and lengths of exposure above which adverse affects on health and welfare may occur.

Air-Quality Standards. The level of pollutants prescribed by law that cannot be exceeded during a specified time in a defined area.

Air, Recirculated: Return air passed through an air conditioner before being again supplied to the air-conditioned or heated space.

Air, Return. Air returned from air-conditioned or heated space to the air conditioner or heater.

Algae. Simple rootless plants that grow in bodies of water in relative proportion to the amount of nutrients available. Algal blooms, or sudden growth spurts, can affect water quality adversely.

Alpha Particle: The least penetrating type of radiation, usually not harmful to life.

Ambient Air. Any unconfined portion of the atmosphere; open air.

Anadromous Fish. Fish that swim upriver to spawn, such as salmon.

Anaerobic. Life or processes that can occur without free oxygen.

Anthracite. "Hard coal," low in volatile matter, high in carbon content, with a heat value of 6.40 million calories/ton.

Anticoagulant. A chemical that interferes with blood-clotting.

Antidegradation Clause. Clause in air- and water-quality laws that prohibits deterioration where pollution levels are within the legal limit.

Aquifer. An underground bed or layer of earth, gravel, or porous stone that contains water.

Area, Free. The total minimum area of the openings in an air outlet, or duct through which air can pass.

Area, Source. In air pollution, any small individual fuel combustion source, including vehicles. A more precise, legal definition is available in federal regulations.

Asbestos. A mineral fiber that can pollute air or water and cause cancer if inhaled or ingested.

A-Scale Sound Level. A measurement of sound approximating sensitivity of the human ear, used to note the intensity or annoyance of sounds.

Assimilation. The ability of a body of water to purify itself of pollutants.

Atmosphere. The body of air sourrounding the earth.

Atomic Pile. A nuclear reactor.

Attenuation. The sound-reduction process in which sound energy is absorbed or diminished in intensity as the result of energy conversion from sound to motion or heat.

Attractant. A chemical or agent that lures insects or other pests by stimulating their sense of smell.

Attrition. Wearing or grinding down a substance by friction. A contributing factor in air pollution, as with dust.

Audiometer. An instrument that measures hearing sensitivity.

Automatic. Self-acting; operating by its own electronic or mechanical devices rather than human operators.

Autotrophic. An organism that produces food from inorganic substances.

Autumnal Equinox. The position of the sun midway between its lowest and highest altitude during the autumn; it occurs on September 21. *See also* **Vernal Equinox.**

Auxillary Energy Subsystem. Equipment using conventional energy sources, both to supplement the output provided by a solar energy system as required by design conditions and to provide full energy backup requirements during periods when the solar systems are inoperable.

Azimuth Angle (Solar). The angular direction of the sun with respect to true south.

Backfill. The material used to refill an excavation, or the process of doing so.

Backflow. The unintentional reversal of flow in a potable water-distribution system, which may result in the transport of foreign materials or substances into the other branches of the distribution system.

Backflow Preventer. A device or means to prevent backflow.

Background Level. In air pollution, the level of pollutants present in ambient air from natural sources.

Back Pressure. Pressure created by any means in the water-distribution system on the premises, which, by being in excess of the pressure in the water supply, could cause backflow.

Back-Siphonage. The backflow of possibly contaminated water into the potable water supply system as a result of the pressure in the potable water system becoming unintentionally less than the atmospheric pressure in the plumbing fixtures, pools, tanks, or vats that may be connected to the potable water distribution piping.

Bacteria. Single-celled microorganisms that lack chlorophyll. Some cause diseases; others aid in pollution control by breaking down organic matter in air and water.

Baffle. A deflector that changes the direction of flow or velocity of water, sewage, or particulate matter. Also used to deaden sound.

Baghouse. An air-pollution abatement device used to trap particulates by filtering gas streams through larger fabric bags, which are usually made of glass fibers.

Baling. Compacting solid waste into blocks to reduce volume.

Ballistic Separator. A machine that sorts organic from inorganic matter for composting.

Band Application. In pesticides, the spreading of chemicals over or next to each row of plants in a field.

Bar Screen. In wastewater treatment, a device that removes large solids.

Barrel (bbl). A liquid measure of oil, usually crude oil, equal to 42 gallons or about 306 pounds.

Basal Application. In pesticides, the spreading of a chemical on stems or trunks just above the soil line.

Batts. Precut pieces of insulation, usually 4 to 8 feet in length. *See also* **Blankets.**

Benthic Region. The bottom layer of a body of water.

Benthos. The plants and animals that inhabit the bottom of a water body.

Beryllium. A metal that can be hazardous to human health when inhaled. It is discharged by machine shops, ceramic and propellant plants, and foundries.

Beta Particle. An elementary particle emitted by radioactive decay that may cause skin burns. It can be halted by a thin sheet of metal.

Bioassay. Using living organisms to measure the effect of a substance, factor, or condition.

Biochemical Oxygen Demand (BOD). The dissolved oxygen required to decompose organic matter in water. BOD is a measure of pollution since heavy waste loads have a high demand for oxygen.

Bioconversion. Utilization of agricultural or municipal wastes to provide fuel.

Biodegradable. Any substance that decomposes quickly through the action of microorganisms.

Biological Control. Using means other than chemicals to control pests, such as predatory organisms, sterilization, or inhibiting hormones.

Biological Magnification. The concentration of certain substances up a food chain. A very important mechanism in concentrating pesticides and heavy metals in organisms such as fish.

Biological Oxidation. The way that bacteria and microorganisms feed on and decompose complex organic materials. Used in self-purification of water bodies and activated sludge wastewater treatment.

Biomass. Plant material in any form from algae to wood.

Biomonitoring. The use of living organisms to test water quality at a discharge site or downstream.

Biosphere. The portion of earth and its atmosphere than can support life.

Biostablizer. A machine that converts solid waste into compost by grinding and aeration.

Biota. All living organisms that exist in an area.

Bituminous coal. Soft coal; coal that is high in carbonaceous and volatile matter. It is "younger" and of lower heat value than anthracite or "hard coal." Heat value: 5.92 million calories/ton.

Blankets. Insulation available in rolls of various lengths, which must be cut when installed. *See also* **Batts.**

Bloom. A proliferation of algae and/or higher aquatic plants in a body of water, often related to pollution.

BOD$_5$. The amount of dissolved oxygen consumed in 5 days by biological processes breaking down organic matter in an effluent.

Bog. Wet, spongy land that is usually poorly drained, highly acid, and rich in plant residue, as the result of lake eutrophication.

Boom. A floating device used to contain oil on a body of water.

Botanical Pesticide. A plant-produced chemical used to control pests—for example, nicotine or strychnine.

Bottoming cycle. A means of using the low-temperature heat energy exhausted from a heat engine—a steam turbine, for instance—to increase the overall efficiency. It usually employs a low-boiling point liquid as working fluid.

Brackish Water. A mixture of fresh water and salt water.

Breeder. A nuclear reactor that produces more fuel than it consumes.

Broadcast Application. In pesticides, to spread a chemical over an entire area.

BTU (British Thermal Unit). A unit of energy defined as the amount of energy required to heat 1 pound of water 1°F.

BTU/Hour. Number of Btu transferred during a period of 1 hour.

Buffer Strips. Strips of grass or other erosion-resisting vegetation between or below cultivated strips or fields.

Building. A building is a structure built, erected, and framed of component structural parts designed for the housing, shelter, enclosure, or support of persons, animals, or property of any kind.

Building Envelope. The elements of a building that enclose conditioned spaces through which energy may be transferred to or from the exterior.

Burial Ground (Graveyard). A disposal site for unwanted radioactive materials, which uses earth or water for a shield.

C. A measure of the heat flow through a given thickness of material. If you know a material's K, to find its C, divide by the thickness—for example, 3-inch-thick insulation with a K of 0.30 has a C of 0.10. The lower the K or C, the higher the insulating value.

Cadmium. A heavy metal element that accumulates in the environment.

Calorie. A metric unit of heat energy, the amount of heat that will raise the temperature of 1 kilogram of water 1°C. It is approximately equal to 4 Btu. (In scientific terminology, it is equivalent to the kilocalorie, 1000 small calories.)

Capacity. Net usable output.

Capacity, Latent. The available refrigerating capacity of an air conditioner for removing latent heat from the space to be conditioned.

Capacity, Sensible. The available refrigerating capacity of an air conditioner for removing sensible heat from the space to be conditioned.

Capacity, Total. The available refrigerating capacity of an air conditioner for removing sensible and latent heat from the space to be conditioned.

Carbon Dioxide (CO_2). A colorless, odorless, nonpoisonous gas normally part of ambient air; a result of fossil-fuel combustion.

Carbon Monoxide (CO). A colorless, odorless, poisonous gas produced by incomplete fossil-fuel combustion.

Carcinogenic. Cancer-producing.

Carrying Capacity. (1) In recreation, the amount of use a recreation area can sustain without deterioration of its quality. (2) In wildlife, the maximum number of animals an area can support during a given period of the year.

Catalyst. A substance that changes the speed of a chemical reaction without itself being changed.

Catalytic Converter. An air-pollution abatement device that removes organic contaminants by oxidizing them into carbon dioxide and water.

Cathodic Protection. The process of providing corrosion protection against electrolytic reactions that could be deleterious to the performance of the protected material or component.

Caustic Soda. Sodium hydroxide (NaOH), a strong alkaline substance used as the cleaning agent in some detergents.

Cell. In solid-waste disposal, hole where waste is dumped daily, compacted, and covered with layers of dirt.

Cellular Plastic Foam. Made of polystyrene, polyurethane, or urea formaldehyde. Polystyrene and polyurethane can be bought in preformed boards or sheets. Urea formaldehyde is foamed into place by a contractor.

Cellulose. Pulverized or shredded paper, such as newspaper, which must be chemically treated to achieve resistance to fire and vermin. It can be poured into place, or it can be "blown in."

Celsius. The metric temperature scale in which the temperature of melting ice is set at 0°, the temperature of boiling water at 100°. One degree Celsius is 9/5 of a degree Fahrenheit. The Celsius scale is also known as the centigrade scale.

Centigrade. *See* Celsius.

Centrifugal Collector. A mechanical system using centrifugal force to remove aerosals from a gas stream or to de-water sludge.

CFS. Cubic feet per second, a measure of the amount of water passing a given point.

Chain Reaction. A reaction that stimulates its own repetition. In a fission chain reaction, a fissionable nucleus absorbs a neutron and splits in two, releasing additional neutrons. These in turn can be absorbed by other fissionable nuclei, releasing still more neutrons and maintaining the reaction.

Channelization. A flood reduction or marsh drainage tactic to straighten and deepen streams so water will move faster; can interfere with waste-assimilation capacity and disturb fish habitat.

Chemical Compatibility. The ability of materials and components in contact with each other to resist mutual chemical degradation, such as that caused by electrolytic action.

Chemical Energy. A form of energy stored in the structure of atoms and molecules, and which can be released by a chemical reaction.

Chemical Oxygen Demand (COD). A measure of the oxygen required to oxidize all organic and inorganic compounds in water.

Chemosterilant. A chemical that controls pests by preventing their reproduction.

Chilling Effect. The lowering of the earth's temperature because of increased particles in the air blocking the sun's rays.

Chlorinated Hydrocarbons. A class of persistent, broad-spectrum insecticides, notably DDT, that linger in the environment and accumulate in the food chain. Other examples are aldrin, dieldrin, heptachlor, chlordane, lindane, endrin, mirex, benzene, hexachloride, and toxaphene.

Chlorination. The application of chlorine to drinking water, sewage, or industrial waste to disinfect or to oxidize undesirable compounds.

Chlorinator. A device that adds chlorine to water in gas or liquid form.

Chlorine-Contact Chamber. That part of a waste-treatment plant where effluent is disinfected by chlorine before being discharged.

Chlorosis. Discoloration of normally green plant parts that can be caused by disease, lack of nutrients, or various air pollutants.

Chromium. *See* Heavy Metals.

Chronic. Long-lasting or frequently recurring, as a disease.

Clarification. Clearing action that occurs during wastewater treatment when solids settle out, often aided by centrifugal action and chemically induced coagulation.

Clarifier. A settling tank where solids are mechanically removed from wastewater.

Clear Cut. A forest-management technique that involves harvesting all the trees in one area at one time. Under certain soil and slope conditions, a clear cut can contribute sediment to water pollution.

Climate. The prevailing or average weather conditions of a geographic area or region as shown by temperature and meterological changes over a period of years.

Coagulation. A clumping of particles in wastewater to settle out impurities, often induced by chemicals such as lime or alum.

Coastal Zone. Ocean waters and adjacent lands that exert an influence on the uses of the sea and its ecology.

Coefficient of Haze (COH). A measurement of visibility interference in the atmosphere.

Coefficient-of-Performance (COP) Cooling. The ratio of the rate of net heat removal to the rate of total energy input, expressed in consistent units and under designated rating conditions.

Coefficient-of-Performance (COP) Heat Pump, Heating. The ratio of the rate of net heat output to the rate of total energy input, expressed in consistent units and under designated rating conditions.

Coffin. A thick-walled container (usually lead) used for transporting radioactive materials.

Coliform Index. A rating of the purity of water based on a count of fecal bacteria.

Coliform Organism. Organisms found in the intestinal tract of humans and animals; their presence in water indicates pollution and potentially dangerous bacterial contamination.

Collector, or Solar Collector. A device for receiving solar radiation and converting it to heat in a fluid.

Collector Efficiency. The fraction of incoming radiation captured by a solar collector. If a system captures half of the incoming radiation, it is 50 percent efficient. Efficiency is the capability of a collector to capture Btu under various climatic conditions. Efficiency varies according to outside temperatures, whether skies are cloudy or clear, whether or not it is windy, and, of course, the quality of the collector. No collector can be 100 percent efficient; that is, no collector can capture all the Btu that fall on it. An efficiency of 55 percent under desirable weather conditions is good.

Collector Plate. The component of a solar collector that transfers the heat from solar energy to a circulating fluid.

Collector Tilt. The angle measured from the horizontal at which a solar heat collector is tilted to face the sun for better performance.

Collectors, Combined. The solar collector and storage are constructed and operated such that they functionally perform as one unit, and the thermal performance of the individual components cannot be meaningfully measured.

Collectors, Component. A collector that is not structurally integrated with the storage or building. The collector performance can be thermally characterized as an individual component with an active heat-transfer fluid.

Collectors, Integrated. The collector is constructed and operated as part of the building structure and heating system. The thermal performance is considered a part of the building heating load and solar energy provides a significant fraction of the building heat requirements.

Collector Subsystem. The assembly used for absorbing solar radiation, converting it into useful thermal energy, and transferring the thermal energy to a heat-transfer fluid.

Combined Sewers. A system that carries both sewage and stormwater runoff. In dry weather, all flow goes to the waste-treatment plant. During a storm, only part of the flow is intercepted due to overloading. The remaining mixture of sewage and stormwater overflows untreated into the receiving stream.

Combustion. Burning, or a rapid oxidation accompanied by release of energy in the form of heat and light, a basic cause of air pollution.

Comminution. Mechanical shredding or pulverizing of waste, used in solid-waste management and wastewater treatment.

Comminutor. A machine that grinds solids to make waste treatment easier.

Compaction. Reduction of the bulk of solid waste by rolling and tamping.

Components. The smallest identifiable elements of a heating subsystem, such as valves, piping, controls, and containers.

Compost. Relatively stable, decomposed organic material.

Composting. A controlled process of organic breakdown of matter. In mechanical composting, the materials are constantly mixed and aerated by a machine. The ventilated cell method mixes and aerates materials by dropping them through a vertical series of aerated chambers. Using windrows, compost is placed in piles out in the open air and mixed or turned periodically.

Concentrator. Reflector or lens designed to focus a large amount of sunshine into a small area, thus increasing the temperature.

Conductance, Thermal. A measure of the rate of heat flow for the actual thickness of a material, 1 square foot in area, at a temperature difference of $1°F$. The lower the thermal conductance, the higher the insulating value.

Conductivity. The ease with which heat will flow through a material determined by the material's physical characteristics. Copper is an excellent conductor of heat; insulating materials are poor conductors.

Connector, Flue. Class "C" pipe connecting equipment to flue.

Conservation. The protection, improvement, and use of natural resources according to principles that will assure their highest economic or social benefits.

Contact Pesticide. A chemical that kills pests when it touches them.

Contaminant. Material (solid, liquids, or gases), which when added (unintentionally or intentionally) to the potable water supply, causes it to be unfit for human consumption.

Contour Plowing. Farming methods that break ground following the shape of the land in a way that discourages erosion.

Contrails. Long narrow clouds caused when high-flying jets disturb the atmosphere.

Control Subsystem. That assembly of devices and their electrical, pneumatic, or hydraulic auxiliaries used to regulate the processes of collecting, transporting, storing, and using energy in response to the thermal, safety, and health requirements of the building occupants.

Convection. When two surfaces—one hot, the other cold—are separated by a thin layer of air, moving air currents (called *convection currents)* are established that carry heat from the hot to the cold surface.

Coolant. A liquid or gas used to reduce the heat generated by power production in nuclear reactors or electric generators.

Cooling Towers. Devices for the cooling of water used in power plants. There are two types: wet towers, in which the warm water is allowed to run over a lattice at the base of a tower and is cooled by evaporation; and dry towers, in which the water runs through a system of cooling fans and is not in contact with the air.

Core. The uranium-containing heart of a nuclear reactor, where energy is released.

Cover. Vegetation or other material providing protection.

Cover Material. Soil used to cover compacted solid waste in a sanitary landfill.

Criteria. The standards EPA has established for certain pollutants, which not only limit the concentration, but also set a limit to the number of violations per year.

Cross Connection. Any physical connection or arrangement between two otherwise separate piping systems, one of which contains potable water and the other either water of unknown or questionable safety or steam, gas, chemicals, or other substances whereby there may be a flow from one system to the other, the direction of flow depending on the pressure differential between the two systems.

Crude Oil. A mixture of hydrocarbons in liquid form found in natural underground petroleum reservoirs. It has a heat content of 1.46 million calories/barrel and is the raw material from which most refined petroleum products are made.

Cultural Eutrophication. Increasing the rate at which water bodies "die" by pollution from human activities.

Curie. A measure of radioactivity.

Curie-Pie. An instrument used to measure radiation levels.

Cyclone Collector. A device that uses centrifugal force to pull large particles from polluted air.

DDT (Dichloro-Diphenyl-Trichloro-ethane). The first chlorinated hydrocarbon insecticide. DDT has a half-life of 15 years and can collect in fatty tissues of certain animals. The EPA banned registration and interstate sale of DDT for virtually all but emergency uses in the United States in 1972 because of its persistence in the environment and accumulation in the food chain.

Decibel (dB). A unit of sound measurement.

Declining Block Rate. A method of charging for electricity wherein a certain number of kilowatt hours (the first block) is sold at a relatively high rate and succeeding blocks are sold at lower and lower rates. Thus, the charge for energy decreases as the amount consumed increases. *See* **Inverted Block Rate.**

Decomposition. The breakdown of matter by bacteria. Decomposition changes the chemical makeup and physical appearance of materials.

Degree-Day, Heating. A unit, based upon temperature difference and time, used in estimating fuel consumption and specifying nominal heating load of a building in winter. For any one day when the mean temperature is less than $65°F$, there exist as many degree-days as the difference in the number of Fahrenheit degrees between the mean temperature for the day and $65°F$.

Dehumidification. The condensation of water vapor from air by cooling below the dewpoint.

Depletion Curve (Hydraulics). A graphical representation of water depletion from storage-stream channels, surface soil, and groundwater. A depletion curve can be drawn for base flow, direct runoff, or total flow.

Dermal Toxicity. The ability of a pesticide or toxic chemical to poison people or animals by touching the skin.

DES (Diethylstilbestrol). A synthetic estrogen used as a growth stimulant in food animals. Residues in meat are believed to be carcinogenic.

Desalinization. Removing salt from ocean water or brackish water.

Desiccant. A chemical agent that dries out plants or insects causing death.

Design Life. The period of time during which a system is expected to perform its intended function without requiring major maintenance or replacement.

Desulfurization. Removal of sulfur from fossil fuels to cut pollution.

Detergent. Synthetic washing agent that helps water to remove dirt and oil. Most contain large amounts of phosphorous compounds which may kill useful bacteria and encourage algae growth in the receiving water.

Deuterium. A nonradioactive isotope of hydrogen whose nucleus contains one neutron and one proton and is therefore about twice as heavy as the nucleus of normal hydrogen, which consists of a single proton. Deuterium is often referred to as "heavy hydrogen"; it occurs in nature as 1 atom to 6500 atoms of normal hydrogen.

DHW. Domestic hot-water system.

Diatomaceous Earth (Diatomite). A chalklike material used to filter out solid wastes in wastewater treatment plants, also found in powdered pesticides.

Dielectric Fitting. An insulating or nonconducting fitting used to isolate electrochemically dissimilar materials.

Diffused Air. A type of aeration that forces oxygen into sewage by pumping air through perforated pipes inside a holding tank.

Digester. In wastewater treatment, a closed tank, sometimes heated to $95°F$, where sludge is subjected to intensified bacterial action.

Digestion. The biochemical decomposition of organic matter. Digestion of sewage sludge occurs in tanks where it breaks down into gas, liquid, and mineral matter.

Dilution Ratio. The relationship between the volume of water in a stream and the volume of incoming waste. It can affect the ability of the stream to assimilate waste.

Direct Connected. Driver and driven, as motor and fan, positively connected in line to operate at the same speed.

Disinfection. A chemical or physical process that kills organisms that cause infectious disease. Chlorine is often used to disinfect sewage-treatment effluent.

Dispersant. A chemical agent used to break up concentrations of organic material such as spilled oil.

Dissolved Oxygen (DO). A measure of the amount of oxygen available for biochemical activity in a given amount of water. Adequate levels of DO are needed to support aquatic life. Low DO concentration can result from inadequate waste treatment.

Dissolved Solids. The total of disintegrated organic and inorganic material contained in water. Excesses can make water unfit to drink or use in industrial processes.

Distillation. Purifying liquids through boiling. The steam condenses to pure water and pollutants remain in a concentrated residue.

Dose. In radiology, the quantity of energy or radiation absorbed.

Dosimeter. An instrument that measures exposure to radiation.

Draft (Comfort). A current of air that, because of high velocity, low temperature, and/or direction of flow, causes more heat to be drawn from a person's skin than is normally dissipated.

Draft (Flue). A pressure difference that causes a current of air or gases to flow through a flue, chimney, heater, or space.

Dredging. To remove earth from the bottom of water bodies using a scooping machine. This disturbs the ecosystem and causes silting that can kill aquatic life.

Dry Limestone Process. An air-pollution-control method that uses limestone to absorb the sulfur oxides in furnaces and stack gases.

Dump. A site used to dispose of solid wastes without environmental controls.

Dust. Fine grain particles light enough to be suspended in air.

Dustfall Jar. An open container used to collect large particles from air for measurement and analysis.

Dystrophic Lake. Shallow body of water containing humus and organic matter. Such a lake contains many plants but few fish and is almost eutrophic.

Ecological Impact. The total effect of an environmental change, natural or synthetic, on the community of living things.

Ecology. The relationships of living things to one another and to their environment, or the study of such relationships.

Economic Poisons. Chemicals used to control pests and to defoliate cash crops such as cotton.

Ecosphere. *See* **Biosphere.**

Ecosystem. The interacting system of a biological community and its nonliving surroundings.

Efficiency. The efficiency of an energy conversion is the ratio of the useful work or energy *output* to the total work or energy input. (This is sometimes called *First Law Efficiency.)*

Efficiency, Overall System. The ratio of the useful thermal energy (at the point of use) to the thermal energy input for a designated time period.

Effluent. Waste material discharged into the environment, an effluent can be treated or untreated. Generally refers to water pollution.

Electrodialysis. A process that uses electrical current applied to permeable membranes to remove minerals from water. Often used to desalinize salt or brackish water.

Electron. An elementary particle with a negative charge that orbits the nucleus of an atom. Its mass at rest is approximately 9 by 10^{-31} kg, and it composes only a tiny fraction of the mass of an atom. Chemical reactions consist of the transfer and rearrangement of electrons between atoms.

Electrostatic Precipitator. An air-pollution-control device that imparts an electrical charge to particles in a gas stream causing them to collect on an electrode.

Emergency Episode. *See* **Air-Pollution Episode.**

Emission. Like effluent but used in regard to air pollution.

Emission Factor. The relationship between the amount of pollution produced and the amount of raw material processed. For example, an emission factor for a blast furnace making iron would be the number of pounds of particulate per ton of raw materials.

Emission Inventory. A listing, by source, of the amounts of air pollutants discharged into the atmosphere of a community daily. It is used to establish emission standards.

Emission Standard. The maximum amount of discharge legally allowed from a single source.

Emissivity. The property of a surface that determines its ability to give off radiant energy.

Emittance. A measure of the heat reradiated back from the solar collector. Measured as fraction of the energy that would be radiated by a totally black surface at the same temperature.

Energy. A quantity having the dimensions of a force times a distance. It is conserved in all interactions within a closed system. It exists in many forms and can be converted from one form to another. Common units are calories, joules, Btu, and kilowatt-hours.

Energy Efficiency Ratio (EER). The ratio of net cooling capacity in Btu to total rate of electric input in watts under designated operating conditions.

Energy Transport Subsystem. A portion of a system, which transports energy throughout the system.

Enrichment. Sewage effluent or agricultural runoff adding nutrients (nitrogen, phosphorous, carbon compounds) to a water body, greatly increasing the growth potential for algae and acquatic plants.

Environment. The sum of all external conditions affecting the life, development, and survival of an organism.

Environmental Impact Statement. A document required of federal agencies by the National Environmental Policy Act for major projects or legislative proposals. It is used in making decisions about the positive and negative effects of the undertaking, and it lists alternatives.

Epidemiology. The study of diseases as they affect populations.

Episode (Pollution). An air-pollution incident in a given area caused by a concentration of atmospheric pollution reacting with meteorological conditions that may result in a significant increase in illness or death.

Equipment. A component of any of the energy-utilizing systems of a building.

Erosion. The wearing-away of land surface by wind or water. Erosion occurs naturally from weather or runoff but can be intensified by land-clearing practices.

Estuaries. Areas where fresh water meets salt water (bays, mouths, of rivers, salt marshes, lagoons). These brackish-water ecosystems shelter and feed marine life, birds, and widlife.

Eutrophic Lakes. Shallow murky waterbodies that have lots of algae and little oxygen.

Eutrophication. The slow aging process of a lake evolving into a marsh and eventually disappearing. During eutrophication, the lake is choked by abundant plant life. Human activities that add nutrients to a waterbody can speed up this action.

Evaporation Ponds. Areas where sewage sludge is dumped and allowed to dry out.

Evaporator. That part of a refrigeration system in which refrigerant is vaporized to produce refrigeration.

Exfiltration. Airflow outward through a wall, leak, membrane, etc.

Exterior Envelope. The elements of a building that enclose conditioned spaces through which thermal energy may be transferred to or from the exterior.

External Combustion Engine. An engine in which the fuel is burned outside the cylinders.

Fabric Filter. A cloth device that catches dust and particles from industrial emissions.

Faced Insulation. A batt or blanket with a vapor barrier on one side. Insulation without a vapor barrier is "unfaced."

Fahrenheit. A temperature scale in which the temperature of melting ice is set at 32° and the temperature of boiling water at 212° One Fahrenheit degree is equal to five-ninths of a Celsius degree.

Failure (Structural). Failure of a structure or any structural element is defined as one of the following:

1. sudden, locally increased curvature, major spalling, or structural collapse.
2. the inability of a structure to resist a further increase in load.
3. structural deflections under design loads that cause significant thermal performance degradation of the component or subsystem.

Fecal Coliform Bacteria. A group of organisms found in the intestinal tracts of people and animals. Their presence in water indicates pollution and possible dangerous bacterial contamination.

Feedlot. A relatively small, confined area for raising cattle that results in lower costs but may concentrate large amounts of animal wastes. The soil cannot absorb such large amounts of excrement, and the runoff from feedlots pollutes nearby waterways with nutrients.

Fen. Low-lying land partly covered with water.

Filling. The depositing of dirt and mud, often raised by dredging, into marshy areas to create more land for real-estate development. It can destroy the marsh ecology.

Film Badge. A piece of masked photographic film worn by nuclear workers to monitor their exposure to radiation. Nuclear radiation darkens the film.

Filtration. Removing particles of solid materials from water, usually by passing it through sand.

First Law of Thermodynamics. *Also called the* law of conservation of energy. It states that energy can neither be created nor destroyed.

Fission. The splitting of heavy nuclei into two parts (which are lighter nuclei), with the release of large amounts of energy and one or more neutrons.

Fixed Collector. A permanently oriented collector that has no provision for seasonal adjustment or tracking of the sun.

Flat-Plate Collector. A collector without external concentrators or focusing devices, usually consisting of an absorber plate, cover plates, back and side insulation, and a container.

Floc. A clump of solids formed in sewage by biological or chemical action.

Flocculation. Separation of suspended solids during wastewater treatment by the chemical creation of a clump of flocs.

Floor Area, Gross. The floor area within the perimeter of the outside walls of a building under consideration, without deduction for hallways, stairs, closets, thickness of walls, columns, or other features.

Flow Condition. The condition obtained when the heat-transfer fluid is flowing through the collector array under normal operating conditions.

Flowmeter. A gauge that shows the speed of wastewater moving through a treatment plant.

Flue. A special enclosure incorporated into a building for the removal of products of combustion to the out-of-doors. Flues are classified under three types:

Type A. A flue listed for use with oil- or coal-burning equipment.

Type B. A manufactured flue listed for use with gas-burning equipment.

Type C. A manufactured flue listed for use with oil- or coal-burning equipment.

Flue Gas. The air coming out of a chimney after combustion. It can include nitrogen, oxides, carbon oxides, water vapor, sulfur oxide, particles, and many chemical pollutants.

Fluid. Any substance such as air, water, or antifreeze used to capture heat in the collector.

Fluorides. Gaseous, solid, or dissolved compounds containing fluorine that result from industrial processes.

Fluorocarbons. A gas used as a propellant in aersols, believed to be modifying the ozone layer in the stratosphere, thereby allowing more harmful solar radiation to reach the earth's surface.

Flume. A natural or artificial channel that diverts water.

Fly Ash. Noncombustible particles carried by flue gas.

Focusing Collector. A collector using some type of focusing device (parabolic mirror, Fresnel lens, etc.) to concentrate the insolation on an absorbing element.

Fog. Suspended liquid particles formed by condensation of vapor.

Fogging. Applying a pesticide by rapidly heating the liquid chemical so that it forms very fine droplets that resemble smoke. It is used to destroy mosquitos and blackflies.

Foil Scrim. A type of blanket insulation used on heating or cooling ducts that run through an unheated space. It comes in various thicknesses of glass fiber and has a nylon-aluminum facing on one side.

Food Waste. Discarded animal and vegetable matter; garbage.

Forced Circulation. Circulation of heat-transfer fluid by a pump or fan.

Forced-Circulation Air Coil. A coil for use in an airstream whose circulation is caused by a difference in pressure produced by a fan or blower.

Forced-Circulation Air-Heating Coil. A heat exchanger, with or without extended surfaces, through which either hot water or steam is circulated for the purpose of sensible heating of a forced-circulation airstream.

Fossil Fuel. Combustibles derived from the remains of ancient plants and animals; for example, coal, oil, and natural gas.

Free Area. The total minimum area of the openings in an air inlet, outlet, or duct through which air can pass.

Freeze-up. On a coil, frost formation causing airflow to stop or to be severely restricted.

Fuel Cell. A device for combining fuel and oxygen in an electrochemical reaction to generate electricity. Chemical energy is converted directly into electrical energy without combustion.

Fume. Tiny particles trapped by vapor in a gas stream.

Fumigant. A pesticide that is vaporized to kill pests; often used in buildings or greenhouses.

Fungi. Tiny plants that lack chlorophyll. Some cause disease, others stabilize sewage and break down solid waste for compost.

Furnace. A complete heating unit for transferring heat from energy conversion to the air supplied to a heating system.

Fusion. The formation of a heavier nucleus by combining two lighter ones. In the reaction under study, as a source of energy hydrogen (or helium 3) nuclei combine to form helium 4 with a subsequent release of energy.

Galvanic Corrosion. Caused when different metals are not isolated properly, thus allowing a liquid to come in contact with both metals.

Game Fish. Species, such as trout, salmon, and bass, caught for sport. Game fish show more sensitivity to environmental changes than "rough" fish.

Gamma Ray. The most penetrating wave of radiant nuclear energy. It can be stopped by dense materials such as lead.

Garbage. *See* Food Waste.

Garbage Grinding. Use of a household disposal to crush food waste and wash it into the sewer system.

Gasification. Conversion of a solid material, such as coal, into a gas for use as fuel.

Gasoline. A petroleum product consisting primarily of light hydrocarbons. Some natural gasoline is present in crude oil but most gasoline is formed by "cracking" and refining crude oil. It has a heat value of 1.32 million calories/barrel.

Geiger Counter. An electrical device that detects the presence of radioactivity.

Generator. A device that converts mechanical energy into electrical energy.

Geothermal Energy. The heat energy in the earth's crust whose source is the earth's molten interior. When this energy occurs as steam, it can be used directly in steam turbines.

Germicide. Any compound that kills disease-carrying microorganisms. These must be registered as pesticides with the EPA.

Glazing. The cover glass for a solar collector.

Grain. A unit of weight equal to 65 milligrams or 0.002 ounce.

Grain Loading. The rate at which particles are emitted from a pollution source. Measurement is made by the numbers of grains per cubic foot of gas emitted.

Green Belt. Buffer zone created by restricting development of certain land areas.

Greenhouse Effect. The warming effect of carbon dioxide (CO_2) and water vapor in the atmosphere. These molecules are transparent to incoming sunlight but absorb and reradiate the infrared (heat) radiation from the earth.

Grille. A louvered or perforated covering for an air passage opening, which can be located in a sidewall, ceiling, or floor.

Ground Cover. Plants grown to keep soil from eroding.

Groundwater. The supply of fresh water under the earth's surface that forms a natural reservoir.

Habitat. The sum of environmental conditions in a specific place that is occupied by an organism, population, or community.

Half-Life. The time taken by certain materials to lose half their strength. For example, the half-life of DDT is 15 years; of radium, 1580 years.

Hammermill. A high-speed machine that uses hammers and cutters to crush, grind, chip, or shred waste.

Hard Water. Alkaline water containing dissolved mineral salts that interfere with some industrial processes and prevent soap from lathering.

Hazardous Air Pollutant. Substances covered by Air Quality Criteria, which may cause or contribute to illness or death; for example, asbestos, beryllium, mercury, and vinyl chloride.

Hazardous Waste. Waste materials that by nature are inherently dangerous to handle or dispose of, such as old explosives, radioactive materials, some chemicals, and some biological wastes; usually produced in industrial operations.

Heat. A form of kinetic energy that flows from one body to another because of a temperature difference between them. The effects of heat result from the motion of molecules. Heat is usually measured in calories or Btus.

Heat-Actuated Cooling. The use of thermal energy to initiate a thermodynamic cycle, which results in lowering the temperature of a heat-transfer fluid, which in turn is used to lower the indoor air temperature.

Heat Capacity. The amount of heat necessary to raise the temperature of a given mass 1°F.

Heated Space. Space, within a building, that is provided with a positive heat supply to maintain a minimum air temperature of 50°F.

Heating Season. The coldest months of the year, when pollution increases in some areas because fossil fuels are burned to provide warmth.

Heat Island Effect. A haze dome created in cities by pollutants combining with the heat trapped in the spaces between tall buildings. This haze prevents natural cooling of air, and, in the absence of strong winds, it can hold high concentrations of pollutants in one place.

Heat, Latent. The heat required to change the state of substance —that is, latent heat of vaporization is the heat required to change a liquid to a vapor. Latent heat of fusion is the heat required to change a solid to a liquid. Generally associated with moisture in the air.

Heat Pump. A device that transfers heat from cooler region to a warmer one (or vice versa) by the expenditure of mechanical or electric energy. Heat pumps work on the same general principle as refrigerators and air conditioners.

Heat, Sensible. Heat associated with a change in temperature.

Heat-Transfer Medium: A fluid used in the transport of thermal energy.

Heavy Metals. Metallic elements, such as mercury, chromium, cadmium, arsenic, and lead, with high molecular weights. Heavy metals can damage living things at low concentrations and tend to accumulate in the food chain.

Heliostat. A mirror used to reflect the sun's rays into a solar collector or furnace.

Herbicide. A chemical that controls or destroys undesirable plants.

Herbivore. An animal that feeds on plants.

Heterotrophic Organism. Humans and animals that cannot make food from inorganic chemicals.

High-Density Polyethylene. A material used to make plastic bottles that produces toxic fumes when burned.

High-Volume Sampler. A device used to measure and analyze suspended particulate pollution.

Holding Pond. A pond or reservoir, usually made of earth, built to store polluted runoff.

Horsepower. Originally, the power output of a typical working horse. Equal to three-fourths of 1 kilowatt or 0.18 calories per second.

Hot. Slang for radioactive material.

Humidifier. A device to add moisture to air.

Humidity, Relative. Approximately, it equals the ratio of the density of the water vapor in the air to the saturation density of water vapor at the same temperature.

Humus. Decomposed organic material.

Hybrid Solar System. A system that uses both active and passive methods to operate (for example, a solar system that uses pumps to heat and nocturnal cooling to cool).

Hydrocarbons. Molecules composed of carbon and hydrogen atoms in various proportions. They are usually derived from living materials.

Hydroelectric Plant. An electric power plant in which the energy of falling water is converted into electrical energy by a turbine generator.

Hydrogen Sulfide (H_2S). Gas emitted during organic decomposition that smells like rotten eggs. Also a by-product of oil-refining and burning, it can cause illness in heavy concentrations.

Hydrology. Science dealing with the properties, distribution, and circulation of water.

Impedance. The rate at which a substance absorbs and transmits sound.

Implementation Plan. An outline of steps needed to meet environmental quality standards by a set time.

Impoundment. A body of water confined by a dam, dike, floodgate, or other barrier.

Incidence, Angle of. The angle at which isolation strikes a surface.

Incineration. Disposal of solid, liquid, or gaseous wastes by burning.

Incinerator. A controlled chamber where waste substances are burned.

Indicator. In biology, an organism, species, or community that shows the presence of certain environmental conditions.

Inert Gas. A vapor that does not react with other substances under ordinary conditions.

Inertial Separator. A device that uses centrifugal force to separate waste particles.

Inflitration. The uncontrolled inward air leakage through cracks and joints in any building element and around windows and doors of a building caused by the pressure effects of wind and/or the effect of differences in the outdoor and indoor air density. The action of water moving through small openings in the earth as it seeps down into the groundwater.

In situ. In the natural or original position or location. *In situ* conversion of oil shale, for instance, is an experimental technique in which a region of shale is drilled, fractured, and set on fire. The volatile gases burn off, the oil vaporizes, then condenses and collects at the bottom of the region, from which it can be recovered by a well. There also has been some experimentation with *in situ* conversion of coal.

Insulation. Material used in ceilings, walls, and floors of buildings to reduce the loss of heat from the buildings during the winter and the gain of heat during the summer. Insulation is commonly composed of cellulose, mineral wool, cellular plastic foam, or fiberglass.

Integrated Pest Management. Combining the best of all useful techniques—biological, chemical, cultural, physical, and mechanical —into a custom-made pest control system.

Interceptor Sewers. The collection system that connects main and trunk sewers with the wastewater-treatment plant. In a combined sewer system, interceptor sewers allow some untreated wastes to flow directly into the receiving streams so the plant won't be overloaded.

Internal Combustion Engine. An engine in which power is generated within one or more cylinders by the burning of a mixture of air and fuel, and converted into mechanical work by means of a piston. The automobile engine is a common example.

Interstate Carrier Water Supply. A source of water for planes, buses, trains, and ships operating in more than one state. These sources are regulated by the federal government.

Interstate Waters. Defined by law as (1) waters that flow across or form a part of state or international boundaries, (2) the Great Lakes, and (3) coastal waters.

Inversion. An atmospheric condition caused by a layer of warm air preventing the rise of cool air trapped beneath it. An inversion holds down pollutants that might otherwise be dispersed and can cause an air-pollution episode.

Inverted Block Rate. A method of selling electricity wherein a first "block" of kilowatt hours is offered at low cost and prices increase with increased consumption.

Ionization. Removal of some or all electrons from an atom or molecule, leaving the atom or molecule with a positive charge, or the addition of one or more electrons, resulting in a negative charge.

Ionization Chamber. A device that detects ionizing radiation.

Isotope. Any of two or more species of atoms having the same number of protons in the nucleus, of the same atomic number, but with differing numbers of neutrons. All isotopes of an element have identical chemical properties, but the different nuclear masses produce different physical properties. Since nuclear stability is governed by nuclear mass, one or more isotopes of the same element may be unstable (radioactive). In the usual notation, isotopes of the same element are identified by the total of neutrons and protons in the nucleus, and the atomic number, for example, uranium 235 ($_{92}U^{235}$) and uranium 238 ($_{92}U^{238}$).

Kerosene. A petroleum distillate with a heat value of 1.43 million calories/barrel presently used in gas turbines and jet engines.

Kilowatt. One thousand watts of power; equal to about 1 1/3 horsepower.

Kilowatt-hour (kWh). The amount of energy equivalent to 1 kilowatt of power being used for 1 hour (3413 Btu).

Kinetic Energy. The energy of motion, the ability of an object to do work because of its motion.

Lagoon. A shallow pond where sunlight, bacterial action, and oxygen work to purify wastewater.

Laminated Glass. Consists of two or more sheets of glass held together by intervening layer(s) of plastic materials.

Land Subsidence. The sinking of a land surface as the result of the withdrawal of underground material. It results from underground mining and is a hazard of the development of geothermal fields.

Langley. A unit of measurement of insolation. One langley equals 1 gram-calorie per square centimeter.

Laser. A device for producing an intense beam of coherent, sharply focused light. The name is an acronym for light amplication by stimulated emission of radiation.

Lateral Sewers. Pipe running underneath city streets that collects sewage.

LC_{50}. Median lethal concentration, a standard measure of toxicity. Indicates how much of a substance is needed to kill half of a group of experimental organisms.

LDC. Less developed countries.

Leachate. Materials that pollute water as it seeps through solid waste.

Leaching. The process by which nutrient chemicals or contaminants are dissolved and carried away by water, or are moved into a lower layer of soil.

Lead. A heavy metal that may be hazardous to health if breathed or swallowed.

Life Cycle. The stages an organism passes through during its existence.

Life. In a sanitary landfill, a compacted layer of solid waste and the top layer of cover materials.

Limiting Factor. A condition whose absence, or excessive concentration, exerts some restraining influence upon a population through incompatibility with species requirements or tolerance.

Limnology. The study of the physical, chemical, meteorological, and biological aspects of fresh water.

Liquefaction. Changing a solid into a liquid form.

Lithium. The lightest metal, a silver-white alkali metal. Lithium 6 is of interest as a source of tritium for the generation of energy from a controlled fusion reaction. Molten lithium will also be the heat exchanger.

Load. The amount of heat per unit of time imposed on a refrigerating system or heating system, or the required rate of heat removal or gain.

Load Factors. Multipliers by which design loads are increased in order to obtain the loads to be used in ultimate strength design of structural elements.

Loose Fill. A type of insulation that is sold by the bag and installed by pouring or blowing it into the space to be insulated. Loose fill can be used in almost any area around the home but is most commonly used in attics. A chief advantage of loose fill is that it can provide insulation in areas that are unevenly spaced, as it will cover obstacles evenly and assure uniform efficiency. Its primary disadvantage is that it may settle over a period of time and thus reduce its effectiveness. Proper installation of the material can minimize settling.

Marsh. Wet, soft, low-lying land that provides a niche for many plants and animals. It can be destroyed by dredging and filling.

Masking. Blocking out one sight, sound, or smell with another.

Maximum Flow Temperature. The maximum temperature that will be obtained in a component when the heat transfer is flowing through the system.

Maximum No-Flow Temperature. The maximum temperature that will be obtained in a component when the heat transfer fluid is not flowing through the system.

Maximum Service Temperature. The maximum temperature to which a component will be exposed in actual service, either with or without the flow of heat transfer fluid.

Mechanical Energy. One form of energy. It is observable as the motion of an object.

Mechanical Turbulence. The eratic movement of air caused by local obstructions such as buildings.

Megawatt (MW). A unit of power. A megawatt equals 1000 kilowatts, or 1 million watts.

Mercury. A heavy metal, highly toxic if breathed or swallowed. It can accumulate in the environment.

Methane. A colorless, nonpoisonous, flammable gas emitted by marshes and dumps undergoing anaerobic decomposition.

Mgd. Millions of gallons per day. A measurement of waterflow.

Microbe. Tiny plant or animal; some microbes that cause disease are found in sewage.

Mill. A tenth of a cent. The cost of electricity is often given in mills per kilowatt hour.

Mineral Wool. Rock wool made from slag, or fibrous glass made from molten silica that has been spun like cotton candy. These materials can be installed by blowing them in or by laying the materials by hand when it is formed into batts or blankets for application. These batts or blankets can come with or without a vapor barrier.

Minimum Service Temperature. The minimum temperature to which a component will be exposed in actual service, either with or without the flow of heat-transfer fluid.

Mist. Liquid particles, measuring 500 to 40 microns, that are formed by condensation of vapor. By comparison, fog particles are smaller than 40 microns.

Mixed Liquor. Activated sludge and water containing organic matter being treated in an aeration tank.

Mobile Source. A moving producer of air pollution; mainly a form of transportation—such as a car, motorcycle, or plane.

Molecule. Atoms combined to form the smallest natural unit of a substance. For example, the water molecule is composed of two atoms of hydrogen and one atom of oxygen.

Monitoring. Periodic or continuous sampling to determine the level of pollution or radioactivity.

Muck Soil. Earth made from decaying plant materials.

Mulch. A layer of material (wood chips, straw, leaves) placed around plants to hold moisture, prevent weed growth, and enrich soil.

Multiple Use. Harmonious use of land for more than one purpose; for example, grazing of livestock, wildlife protection, recreation, watershed, and timber production. Not necessarily the combination of uses that will yield the highest economic return or greatest unit output.

Mutagen. Any substance that causes changes in the genetic structure in subsequent generations.

Natural Gas. A natural fuel containing methane and hydrocarbons that occurs in certain geologic formations.

Natural Selection. The process of survival of the fittest, by which organisms that adapt to their environment survive and those that do not adapt perish.

Necrosis. Death of cells that can discolor areas on a plant or kill the entire plant.

Neutron. An elementary particle that is present in all atomic nuclei except for the most common isotope of hydrogen. Its mass is approximately that of a proton, but it has no electric charge. Neutrons are released in fission and fusion reactions.

New Energy. Electrical or chemical energy converted to thermal or mechanical energy expressly for the purpose of heating or cooling.

Nitric Oxide (NO). A gas formed by combustion under high temperature and high pressure in an internal combustion engine. It changes into nitrogen dioxide in the ambient air and contributes to photochemical smog.

Nitrogen Dioxide (NO_2). The result of nitric oxide combining with oxygen in the atmosphere; a major component of photochemical smog.

Nitrogenous Waste. Animal or plant residue that contains large amounts of nitrogen.

NO. A notation meaning oxides of nitrogen. *See also* **Nitric Oxide.**

Nocturnal Radiation. Loss of energy by radiation to the night sky.

No-Flow Condition. The condition obtained when the heat-transfer fluid is not flowing through the collector array due to shutdown or malfunction.

Noise Reduction. A reduction or loss in sound transmission from one space to another space through one or more parallel paths of furred ceiling, through partitions, along air ducts, or door-corridor-door.

Nonpoint Source. A contributing factor to water pollution that cannot be traced to a specific spot; for example, agricultural fertilizer runoff, sediment from construction.

NTA (Nitrilotriacetic acid). a compound proposed to replace phosphates in detergents.

Nuclear Power Plant. A device that converts atomic energy into usable power; heat produced by a reactor makes steam to drive electricity-generating turbines.

Nucleus. The extremely dense, positively charged core of an atom. It contains almost the entire mass of an atom, but fills only a tiny fraction of the atomic volume.

Nutrient. Element or compound essential to growth and development of living things; for example, carbon, oxygen, nitrogen, potassium, and phosphorus.

Ocean Thermal. Providing power by harnessing the temperature differences between the surface waters and the ocean depths.

OECD. Organization for Economic Coordination and Development.

"Off-Peak" Power. Power generated during a period of low demand.

Off-Road Vehicles. Form of motorized transportation that does not require prepared surfaces; it can be used to reach remote areas.

Oil "Fingerprinting." A method that identifies oil spills so that they can be traced to their sources.

Oil Shale. A sedimentary rock containing a solid organic material called *kerogen.* When oil shale is heated at high temperatures, the oil is driven out and can be recovered.

Oil Spill. Accidental discharge of oil into bodies of water; it can be controlled by chemical dispersion, combustion, mechanical containment, and absorption.

Oligotrophic Lake. Deep clear lake, with low nutrient supplies. It contains little organic matter and has a high dissolved-oxygen level.

Oncogenic. A substance that causes both benign and malignant tumors.

Opacity. The amount of light obscured by an object or substance. A window has zero opacity; a wall has 100 percent opacity.

Opaque. Impervious radiant energy.

OPEC. The Organization of Petroleum Exporting Countries. An organization of countries in the Middle East, North Africa, and South America that aims at developing common oil-marketing policies.

Open Dump. *See* **Dump.**

Open Space. A relatively undeveloped green or wooded area, usually provided within an urban development, to minimize feelings of congested living.

Organic. Referring to or derived from living organisms. In chemistry, any compound containing carbon.

Organism. Any living thing.

Organophosphates. Pesticide chemicals that contain phosphorus, used to control insects. They are short-lived, but some can be toxic when first applied.

Osmosis. The tendency of a fluid to pass through a permeable membrane, such as the wall of a living cell, into a less concentrated solution, in order to equalize concentrations on both sides of the membrane.

Outfall. The place where an effluent is discharged into the receiving waters.

Outgassing. The emission of gases by materials and components, usually during exposure to elevated temperature, or reduced pressure.

Output. Capacity; duty; performance; net refrigeration produced by system.

Overall Coefficient of Heat Transfer. *See* **Thermal Transmittance.**

Overall System Efficiency. *See* **Efficiency, Overall Systems.**

Overfire Air. Air forced into the top of an incinerator to fan the flame.

Overturn. The period of mixing (turnover), by top to bottom circulation, of previously stratified water masses. This phenomenon may occur in spring and/or fall; the result is a uniformity of physical and chemical properties of the water at all depths.

Oxidant. A substance containing oxygen that reacts chemically in air to produce a new substance; primary source of photochemical smog.

Oxidation. Oxygen combining with other elements.

Oxidation Pond. A holding area where organic wastes are broken down by aerobic bacteria.

Ozone (O_3). A pungent, colorless, toxic gas that contributes to photochemical smog.

Package A/C Unit. Consists of a factory-made assembly, which normally includes an indoor conditioning coil, compressor(s), and condensing coil. It may also include a heating function as well.

Packaged Terminal Air Conditioner. A factory-selected combination of heating and cooling components, assemblies or sections, intended to serve a room or zone.

Packed Tower. A pollution-control device that forces dirty air through a tower packed with crushed rock or wood chips while liquid is sprayed over the packing material. The pollutants in the air-stream either dissolve or chemically react with the liquid.

Pandemic. Widespread throughout an area.

PAN (Peroxyacetyl Nitrate). A pollutant created by the action of sunlight on hydrocarbons and nitrogen oxides in the air. An ingredient of smog.

Particulate. Fine liquid or solid particle, such as dust, smoke, mist, fumes, or smog, found in the air or emissions.

Particulate Loading. The introduction of particulates into ambient air.

Passive Solar System. A system that uses gravity, heat flows, evaporation, or other acts of nature to operate without mechanical devices to collect and transfer energy (for example, south-facing window).

Pathogenic. Capable of causing disease.

PCB (Polychlorinated Biphenyls). A group of toxic, persistent chemicals used in transformers and capacitors. Further sale or new use was banned in 1979 by law.

Peak Demand Period. That time of day when the demand for electricity from a power plant is at its greatest.

Percolation. Downward flow or filtering of water through pores or spaces in rock or soil.

Persistent Pesticides. Pesticides that do not break down chemically and remain in the environment after a growing season.

Pesticide. Any substance used to control pests ranging from rats, weeds, and insects to algae and fungi. Pesticides can accumulate in the food chain and can contaminate the environment if misused.

Pesticide Tolerance. The amount of pesticide residue allowed by law to remain in or on a harvested crop. By using various safety factors, the EPA sets these levels well below the point where the chemicals might be harmful to consumers.

Petroleum (or Oil). An oily, flammable liquid that may vary from almost colorless to black and occurs in many places in the upper strata of the earth. It is a complex mixture of hydrocarbons and is the raw material for many products.

pH. A measure of the acidity of alkalinity of a material, either liquid or solid. pH is represented on a scale of 0 to 14, with 7 being a neutral state, 0 most acid, and 14 most alkaline.

Phenol. Organic compounds that are by-products of petroleum refining, tanning, textile, dye, and resin manufacture. Low concentrations can cause taste and odor problems in water, higher concentrations can kill aquatic life.

Phosphate. Chemical compound containing phosphorous.

Phosphorus. An essential food element that can contribute to the eutrophication of waterbodies.

Photochemical Oxidants. Air pollutants formed by the action of sunlight on oxides of nitrogen and hydrocarbons.

Photochemical Smog. Air pollution caused not by one pollutant but by chemical reactions of various pollutants emitted from different sources.

Photosynthesis. The manufacture by plants of carbohydrates and oxygen from carbon dioxide and water in the presence of chlorophyll, using sunlight as an energy source.

Photovoltaic. Direct conversion of the sun's energy into electricity.

Photovoltiac Conversion. Use of semiconductor or other photo-voltaic devices that convert solar radiation directly to electricity.

Physical Compatibility. The ability of materials and components in contact with each other to resist degradation such as that caused by differential coefficients of thermal expansion.

Phytotoxic. Harmful to plants.

Pig. A container, usually lead, used to ship or store radioactive materials.

Pile. A nuclear reactor.

Plasma. An electrically neutral, gaseous mixture of positive and negative ions. Sometimes called the "fourth state of matter," since it behaves differently from solids, liquids, and gases. High-temperature plasmas are used in controlled fusion experiments.

Plastic. Nonmetallic compound that results from a chemical reaction, and is molded or formed into rigid or pliable structural material.

Plenum. An air compartment connected to one or more distribution ducts.

Plume. Visible emission from a flue or chimney.

Plutonium (Pu). A heavy, radioactive, man-made, metallic element with atomic number 94. Its most important isotope is fissionable plutonium 239 ($_{94}Pu^{239}$), produced by neutron irradiation of uranium 238. It is used for reactor fuel and in weapons.

Point Source. A stationary location where pollutants, usually industrial, are discharged.

Pollen. A fine dust produced by plants; a natural or background air pollutant.

Pollutant. Any introduced substance that adversely affects the usefulness of a resource.

Pollution. The presence of matter or energy whose nature, location, or quantity produces undesired environmental effects.

Polyelectrolytes. Synthetic chemical that helps solids to clump during sewage treatment.

Polyvinyl Chloride. A plastic that releases hydrochloric acid when burned.

Potable Water. Appetizing water that is safe for drinking and use in cooking.

Power. The rate at which work is done or energy expended. It is measured in units of energy per unit of time such as calories per second, and in units such as watts and horsepower.

ppm. Parts per million; a way of expressing tiny concentrations. In air, ppm is usually a volume/volume ratio; in water, a weight/volume ratio.

Precipitate. A solid that separates from a solution because of some chemical or physical change.

Precipitator. Air-pollution control device that collects particles from an emission by mechanical or electrical means.

Pressure Design. The pressure at which a system or component is intended to operate.

Pressure Drop. Static pressure loss in fluid pressure, as from one end of conduit to the other, due to friction.

Pressure-Relief Device. Pressure-activated valve or rupture member designed to automatically relieve excessive pressure.

Pressure Static. The normal force per unit area at a small hole in the wall of the duct through which air flows.

Pressure Test. Pressure to which a system or components are subjected to prove their ability to function satisfactorily.

Pressure Total. The sum of the static pressure and the velocity pressure at the point of measurement. Also called *dynamic pressure.*

Pressure Velocity. In moving fluid, the pressure capable of causing an equivalent velocity if applied to same fluid.

Pretreatment. Processes used to reduce the amount of pollution in water before it enters the sewers or the treatment plant.

Primary Treatment. The first stage of wastewater treatment; removal of floating debris and solids by screening and sedimentation.

Process Weight. The total weight of all materials, including fuel, used in a manufacturing process. It is used to calculate the allowable rate of emission of pollutant matter from the process.

Proton. An elementary particle present in all atomic nuclei. It has a positive electric charge. Its mass is approximately 1840 times that of an electron. The nucleus of a hydrogen atom.

Psychometric Chart. A graphical representation of the thermodynamic properties of moist air.

Pulverization. The crushing or grinding of materials into small pieces.

Pumping Station. A machine installed on sewers to pull the sewage uphill. In most sewer systems, wastewater flows by gravity to the treatment plant.

Putrescible. Able to rot quickly enough to cause odors and attract flies.

Pyranometer. An instrument for measuring solar radiation.

Pyrheliometer. A measurement device to determine local values of direct insolation.

Pyrolysis. Chemical decomposition by extreme heat.

Quench Tank. A water-filled tank used to cool incinerator residue or hot material during industrial processes.

R-Value. A number assigned to insulation material referring to the ability of the material to impede the flow of heat. The higher the R-value, the more effective the insulation.

Rad. A unit of measurement of any kind of radiation absorbed by humans.

Radiation. The emission of particles or rays by the nucleus of an atom.

Radiation Standard. Regulation that governs exposure to permissible concentrations and transportation of radioactive material.

Radiation, Thermal. The transmission of heat through space by wave motion. The passage of heat from one object to another without warming the space between.

Radioactive. Substances that emit rays either naturally or as a result of scientific manipulation.

Radiobiology. The study of the principles, mechanisms, and effects of radiation on living things.

Radioecology. The study of effects of radiation on plants and animals in natural communities.

Radioisotope. Radioactive form of chemical compounds; for example, cobalt-60, used in the treatment of disease.

Rasp. A machine that grinds waste into a manageable material and helps prevent odor.

Raw Sewage. Untreated wastewater.

Readily Accessible. Means capable of being reached safely and quickly for operation, repair, or inspection.

Receiving Waters. Any body of water where untreated wastes are dumped.

Recharge. Process by which water is added to the zone of saturation, as recharge of an aquifer.

Recycling. Converting solid waste into new products by using the resources contained in discarded materials.

Red Tide. A proliferation of ocean plankton that may kill large numbers of fish. This natural phenomenon may be stimulated by the addition of nutrients.

Reflected Insolation. The portion of total solar energy reaching a surface (windows, wall, collector), which has been reflected by an adjoining surface.

Reflectivity. The property of a material that determines its ability to reflect radiant energy.

Refuge, Wildlife. An area designated for the protection of wild animals, within which hunting and fishing is either prohibited or strictly controlled.

Refuse. *See* **Solid Waste.**

Refuse Reclamation. Conversion of solid waste into useful products—for example, composting organic wastes to make a soil conditioner.

Register. A combination grille-and-damper assembly covering an air opening.

Reheat. The application of sensible heat to supply air that has been previously cooled below the temperature of the conditioned space by either mechanical refrigeration or the introduction of outdoor air to provide cooling.

Rem. Roentgen Equivalent Man. A measurement of radiation that determines the degree of biological effect on human tissue.

Rep. Roentgen Equivalent Physical. A measurement of radiation by energy development in human tissue.

Reradiation. After an object has received radiation or is otherwise heated, it often reradiates heat back. Generally speaking, matte black surfaces are good absorbers and emitters of thermal radiation, while white and metallic surfaces are not.

Reservoir. Any holding area, natural or artificial, used to store, regulate, or control water.

Reset Control. Automatic or manual adjustment of the set point of control instrument to a higher or lower value.

Resistance, Thermal (R). A measure of the ability to retard the flow of heat. The R-value is the reciprocal of a heat-transfer coefficient, as expressed by U. R = 1/U.

Resource Recovery. The process of obtaining matter or energy from materials formerly discarded—for example, solid waste, wood chips.

Retrofit, Solar. Altering an existing structure by adding solar equipment or devices.

Retrofitting. Improving the energy efficiency of an existing building by installing new insulation material.

Reverberation. The echoes of a sound that persist in an enclosed space after the sound source has stopped.

Reverse Osmosis. An advanced method of waste treatment that uses a semipermeable membrane to separate water from pollutants.

Rigid Insulation. Plastic or fiber pressed into boards.

Ringelmann Chart. A series of shaded illustrations used to measure the opacity of air-pollution emissions. The chart ranges from light grey (number 1) through black (number 5) and is used to set and enforce emission standards.

Riparian Rights. Entitlement of a landowner to the water on or bordering his property, including the right to prevent diversion or misuse of it upstream.

River Basin. The land area drained by a river and its tributaries.

Rodenticide. A chemical or agent used to destroy rats or other rodent pests, or to prevent them from damaging food, crops, etc.

Rough Fish. Those species not prized for game purposes or eating, such as gar and suckers. Most are more tolerant of changing environmental conditions than game species.

Runoff. Water from rain, snow melt, or irrigation that flows over the ground surface and returns to streams. Runoff can collect pollutants from air or land and carry them to the receiving waters.

Salinity. The degree of salt in water.

Saltwater Intrusion. The invasion of fresh surface or ground water by salt water. Saltwater intrusion from the ocean is called *seawater intrusion*.

Salvage. The utilization of waste materials.

Sanitary Landfill, Landfilling. Protecting the environment when disposing of solid waste. Waste is spread in thin layers, compacted by heavy machinery, and covered with soil daily.

Sanitary Sewers. Underground pipes that carry only domestic or commercial waste, not storm water.

Sanitation. Control of physical factors in the human environment that can harm development, health, or survival.

Scrap. Materials discarded from manufacturing operations that may be suitable for reprocessing.

Screening. Use of racks of screens to remove coarse floating and suspended solids from sewage.

Scrubber. An air-pollution control device that uses a spray of water to trap pollutants and cool emissions.

Seasonal Performance Factor. The ratio of total quantity of heat delivered by a heat pump (including supplemental resistance heaters) to the total quantity of energy input (including supplementary resistance heaters) for the total annual heating hours below 65°F.

Second Law of Thermodynamics. One of the two "limit laws" that govern the conversion of energy. Referred to sometimes as the heat tax, it can be stated in several equivalent forms, all of which describe the inevitable passage of some energy from a useful to a less useful form in any cyclic energy conversion.

Secondary Treatment. Biochemical treatment of wastewater after the primary stage, using bacteria to consume the organic wastes. Use of trickling filters or the activated sludge process removes floating and settleable solids and about 90 percent of oxygen-demanding substances and suspended solids. Disinfection with chlorine is the final stage of secondary treatment.

Sedimentation. Letting solids settle out of wastewater by gravity during wastewater treatment.

Sedimentation Tank. Holding area for wastewater where floating wastes are skimmed off and settled solids are pumped out for disposal.

Seepage. Water that flows through the soil.

Seismic. Subject to or caused by an earthquake.

Selective Pesticide. A chemical designed to affect only certain types of pests, leaving other plants and animals unharmed.

Selective Surface. A special coating sometimes applied to the absorber plate in a solar collector. The selective surface absorbs most of the incoming solar energy and reradiates very little of it.

Self-Contained A/C Unit. *See* **Packaged Air Conditioning Unit.**

Senescense. The aging process. It can refer to lakes in advanced stages of eutrophication.

Sensors. Devices (such as pressure transducers, thermocouplers, and flowmeters) used to sense individual parameters.

Septic Tank. An enclosure that stores and processes wastes where no sewer system exists, as in rural areas or on boats. Bacteria decompose the organic matter into sludge, which is pumped off periodically.

Settleable Solid. Material heavy enough to sink to the bottom of wastewater.

Settling Chamber. A series of screens placed in the way of the gases to slow the stream of air, thus helping gravity to pull particles out of the emission into a collection area.

Settling Tank. A holding area for wastewater, where heavier particles sink to the bottom and can be siphoned off.

Sewage. The organic waste and wastewater produced by residential and commercial establishments.

Sewage Lagoon. *See* **Lagoon.**

Sewer. A channel that carries wastewater and stormwater runoff from the source to a treatment plant or receiving stream. **Sanitary Sewers** carry household and commercial waste. **Storm sewers** carry runoff from rain or snow. Combined sewers are used for both purposes.

Sewerage. The entire system of sewage collection, treatment, and disposal. Also applies to all effluent carried by sewers.

Shading Loss. The loss of collector efficiency caused by the shading of the absorber plate by collector edges or components. The shading loss usually varies with the angle of incidence of the insolation.

Shield. A wall to protect people from exposure to harmful radiation.

Significant Deterioration. Pollution from a new source in previously "clean" areas.

Silt. Fine particles of soil or rock that can be picked up by air or water and deposited as sediment.

Silviculture. Management of forest land for timber. Sometimes contributes to water pollution, as in clear-cutting.

Sinking. Controlling oil spills by using an agent to trap the oil. Both sink to the bottom of the water body and biodegrade there.

Skimming. Using a machine to remove oil or scum from the surface of the water.

Sludge. The concentration of solids removed from sewage during wastewater treatment.

Slurry. A watery mixture of insoluble matter that results from some pollution-control techniques.

Smog. Air pollution associated with oxidants.

Smoke. Particles suspended in air after incomplete combustion of materials containing carbons.

SOˣ. The chemical symbol for oxides of sulfur.

Soft Detergents. Cleaning agents that break down in nature.

Soil Conditioner. An organic material such as humus or compost that helps soil absorb water, build a bacteria community, and distribute nutrients and minerals.

Solar Altitude. The angular elevation of the sun above the horizon.

Solar Building. A building that utilizes solar energy by means of an active or passive solar system.

Solar Cell. A device, usually made of silicon, that converts sunlight directly into electrical energy.

Solar Collector. Any device that collects solar energy and transforms it to another usable form of energy.

Solar Constant. The average amount of solar radiation reaching the earth's atmosphere per minute. This is just under 2 langleys, or 2 gram-calories per square centimeter. This is equivalent to 442.2 Btu/hour/square foot, 1395 watts/square meter, or 0.1395 watt/square centimeter.

Solar Degradation. The process by which exposure to sunlight deteriorates the properties of materials and components.

Solar Energy. Power collected from sunlight, used most often for heating purposes but occasionally to generate electricity.

Solar Energy Source. Thermal, chemical, or electric energy derived directly from conversion or incident solar radiation.

Solar Heating System. The complete assembly of subsystems and components necessary to convert solar energy into thermal energy for heating purposes, in combination with auxiliary energy when required.

Solar Time. The hours of the day as reckoned by the apparent position of the sun. Solar noon is that instant on any day at which the sun reaches its maximum altitude for that day.

Solid Waste. Useless, unwanted, or discarded material with insufficient liquid to be free-flowing.

Solid-Waste Disposal. The final placement of refuse that cannot be salvaged or recycled.

Solid-Waste Management. Supervised handling of waste materials from their source through recovery processes to disposal.

Sonic Boom. The thunderous noise made when shock waves reach the ground from a jet airplane exceeding the speed of sound.

Soot. Carbon dust formed by incomplete combustion.

Sorption. The action of soaking up or attracting substances; used in many pollution-control processes.

Split System. Consists of more than one factory-made assembly, which normally includes an evaporator or cooling coil, compressor, and condensor combination, and which may include a heating function as well. The separate assemblies are designed to be used together.

Spoil. Dirt or rock that has been removed from its original location, destroying the composition of the soil in the process, as with strip mining or dredging.

Sprawl. Unplanned development of open land.

Stabilization. To convert the active organic matter in sludge into inert, harmless material.

Stabilization Pond. See **Lagoon.**

Stable Air. A mass of air that is not moving normally, so that it holds rather than disperses pollutants.

Stack. A chimney or smokestack; a vertical pipe that discharges used air.

Stack Effect. Used air that moves upward, as in a chimney, because it is warmer than the surrounding atmosphere.

Stagnation. Lack of motion in a mass of air or water, which tends to hold pollutants.

Stationary Source. A pollution location that is fixed rather than moving. One point of pollution rather than widespread.

Stirling Engine. An external combustion engine in which air (or hydrogen in the newer versions) is alternately heated and cooled to drive the piston up and down. It is claimed to be nonpolluting and more efficient than the internal combustion engine.

Storage System. The assemblies used for storing thermal energy so that it can be used when required.

Storm Sewer. A system that collects and carries rain and snow runoff to a point where it can soak back into the groundwater or flow into surface waters.

Stratification. Separating into layers.

Strip Cropping. Growing crops in a systematic arrangement of strips or bands, which serve as barriers to wind and water erosion.

Strip Mining. A process that uses machines to scrape soil or rock away from mineral deposits just under the earth's surface.

Strontium 90 ($_{38}Sr^{90}$). A hazardous isotope produced in the process of nuclear fission. Strontium 90 has a "half-life" of 28 years. Thus, it takes 28 years to reduce this material to half its original amount, 56 years to one quarter, 84 years to one eighth, and so on. Strontium 90 typifies the problems of radioactive waste storage that are faced in producing power by means of nuclear fission.

Subsystem. One of several major elements of a solar system, such as collectors, thermal storage device(s), and heaters.

Sulfur Dioxide (SO_2). A heavy, pungent, colorless gas formed primarily by the combustion of fossil fuels. This major air pollutant is unhealthy for plants, animals, and people.

Sulfur Smog (Classical Smog). This smog is composed of smoke particles, sulfur oxides (SO^x), and high humidity (fog). The sulfur oxide (SO_3) reacts with water to form sulfuric acid (H_2SO_4) droplets, the major cause of damage.

Summer Solstice. The longest sunlit day of the year at which the sun is at its highest altitude; it occurs June 22.

Sump. A depression or tank, such as a cesspool, that catches liquid runoff for drainage or disposal.

Sun Tracking. Following the sun with a solar collector to make the collector more effective.

Superconductor. A material that, at very low temperatures, near absolute zero, has no electrical resistance and thus can carry large electrical currents without resistance losses.

Supersonic Transport (SST). A jet airplane that flies above the speed of sound, and that may be extremely noisy upon takeoff and landing.

Supplemental Energy. The use of fuel or electricity to provide additional energy whenever solar energy cannot satisfy the demand.

Surfactant. A surface active chemical agent, usually made up of phosphates; used in detergents to cause lathering. The phosphates may contribute to water pollution.

Surveillance System. A series of monitoring devices designed to determine environmental quality.

Suspended Solids (SS). Tiny pieces of pollutants floating in sewage that cloud the water and require special treatment for removal.

Sustained Load. A load that is considered to be applied to a structure over a projected period of time.

Synergism. A cooperative action of two substances that results in a greater effect than if both had acted independently.

System. A combination of equipment, controls, interconnecting means, and terminal elements, by which energy is transformed so as to perform a specific function, such as HVAC, service water heating, or illumination.

System Efficiency. How many Btu a system actually used in comparison with the original number collected.

Systemic Pesticide. A chemical that is taken up from the ground or absorbed through the surface and carried through the systems of the organism being protected, making it toxic to pests.

Tailings. Residue of raw materials or waste separated out during the processing of crops or mineral ores.

Temperature, Dry-Bulb. The temperature of a gas or mixture of gases indicated by an accurate thermometer after correction for radiation.

Temperature, Wet-Bulb. The temperature at which, by evaporating into air, liquid, or solid water can bring the air to saturation adiabatically at the same temperature.

Teratogenic. Substances that are suspected of causing malformations or serious deviations from the normal type, which cannot be inherited, in or on animal embryos or fetuses.

Terminal Element. The means by which the transformed energy from the system is delivered; for example, registers, diffusers, lighting fixtures, faucets.

Terracing. Building of dikes along the contour of agricultural land to hold runoff and sediment, thus reducing erosion.

Teritary Treatment. Advanced cleaning of wastewater that goes beyond the secondary or biological stage. It removes nutrients such as phosphorus and nitrogen and most suspended solids.

Therm. The quantity of heat equivalent to 100,000 Btu.

Thermal Capacity. The ability of a medium to hold heat.

Thermal Conductivity (K). A measure of the ability of a material to permit the flow of heat, K expresses the quantity of heat per hour that will pass through a 1-square-foot chunk of inch-thick material when a $1°F$ temperature difference is maintained between its two surfaces. K is measured by Btu/hour/squarefoot/°F/foot or inch.

Thermal Energy. Release of energy as the result of a heating process.

Thermal Pollution. Discharge of heated water from industrial processes that can affect the life processes of aquatic plants and animals.

Thermal Resistance (R). A measure of the ability of a substance to resist the flow of heat. R is simply the mathematical reciprocal of either C or U. Thus, R = 1/C or R = 1/U. Insulation products are typically characterized by their R-values. Thus, a specification of R-11 means the insulation displays 11 resistance units. Clearly, the higher the R-value, the better the insulating ability. R is a simple common denominator for describing all types of insulation and all kinds of dwelling construction. For example, all insulation rated R-11 has the same insulation ability, no matter what its material or thickness.

Thermal Transmittance (U). Overall coefficient of heat transmission or thermal transmittance (air to air) expressed in units of Btu per hour per square foot per degree F. It is the time rate of heat flow.

Thermal Transmittance (Uo). Overall (average) heat transmission or thermal transmittance of a gross area of the exterior building envelope, expressed in units of Btu per hour per square foot per °F. It is the time rate of heat flow.

Thermodynamics. The science and study of the relationship between heat and other forms of energy.

Thermosiphon. Circulation of a fluid by making use of the change in density of a material when it is heated and cooled. Also called *natural circulation*.

Thermostat. An instrument that measures changes in temperature and controls device(s) for maintaining a desired temperature.

Threshold Dose. The minimum application of a given substance required to produce a measurable effect.

Tidal Marsh. Low, flat marshlands traversed by interlaced channels and tidal sloughs and subject to tidal inundation; normally, salt-tolerant bushes and grasses are the only vegetation present.

Tolerance. The ability of an organism to cope with changes in its environment. Also the safe level of any chemical applied to crops that will be used as food or feed.

Ton (of Refrigeration). A useful refrigerating effect equal to 12,000 Btu per hour (200 Btu per minute).

Topography. The physical features of a surface area including relative elevations and the position of natural and man-made features.

Toxic Fluids. Gases or liquids that are poisonous, irritating, and/ or suffocating.

Toxic Substance. A chemical or mixture that may present an unreasonable risk of injury to health or the environment.

Toxicant. A chemical that controls pests by killing rather than repelling them.

Toxicity. The degree of danger posed by a substance to animal or plant life.

Tracking Collector. A solar energy collector that constantly positions itself perpendicular to the sun as the earth rotates.

Trickling Filter. A biological treatment device in which wastewater is trickled over a bed of stones covered with bacterial growth, and the bacteria break down the organic wastes in the sewage and produce cleaner water.

Tritium. A radioactive isotope of hydrogen with a half-life of 12.5 years. The nucleus contains one proton and two neutrons. It may be used as a fuel in the early fusion reactors.

Troposphere. The portion of the atmosphere between 7 and 10 miles from the earth's surface, where clouds form.

Turbidimeter. A device that measures the amount of suspended solids in a liquid.

Turbidity. Hazy air due to the presence of particles and pollutants; a similar cloudy condition in water due to suspended slit or organic matter.

Uranium. A heavy, naturally occurring, radioactive nucleus of atomic number 92. Its most common isotope is 238 ($_{92}U^{238}$), but 235 ($_{92}U^{235}$, 0.7 percent of natural ore) is used as a fission fuel. $_{92}U^{233}$, which can be made from 232 ($_{92}Th^{232}$), is also fissionable.

Urban Runoff. Storm water from city streets, usually carrying litter and organic wastes.

Utility Plan. A drawing showing location of electrical lines, gas lines, water, etc.

Vacuum. Any pressure less than that exerted by the atmosphere.

Vacuum Breaker, Pressure Type. A vacuum breaker designed to operate under conditions of static line pressure.

Vacuum Relief Valve. A device to prevent excessive vacuum in a pressure vessel.

Vapor. The gaseous phase of substances that are liquid or solid at atmospheric temperature and pressure, such as steam.

Vapor Barrier. Material that provides resistance to the transmission of water vapor under specified conditions.

Vapor Plumes. Flue gases that are visible because they contain water droplets.

Vaporization. The change of a substance from a liquid to a gas.

Variance. Government permission for a delay or exception in the application of a given law, ordinance, or regulation.

Vector. An organism, often an insect, that carries disease.

Ventilation, Air. Air supplied or removed to or from the conditioned space.

Vernal Equinox. The position of the sun midway between its lowest and highest altitude; during the spring, it occurs March 22. The sunlit period is approximately the same length as the autumnal equinox. *See also* **Autumnal Equinox.**

Vertical Penetrations. The vertical passage of a utility chase, pipe, duct, etc., through a structural assembly.

Vinyl Chloride. A chemical compound used in producing some plastics. Excessive exposure to vinyl chloride may cause cancer.

Waste. Unwanted materials left over from manufacturing processes, refuse from places of human or animal habitation.

Wastewater. Water carrying dissolved or suspended solids from homes, farms, businesses, and industries.

Water Pollution. The addition of enough harmful or objectionable material to damage water quality.

Water-Quality Criteria. The levels of pollutants that affect use of water for drinking, swimming, raising fish, farming, or industrial use.

Water-Quality Standard. A management plan that considers (1) what water will be used for, (2) setting levels to protect those uses, (3) implementing and enforcing the water-treatment plans, and (4) protecting existing high-quality waters.

Watershed. The land area that drains into a stream.

Water-Supply System. The water-service pipe, the water-distributing pipes, and the necessary connecting pipes, fittings, control valves, and all appurtenances in or adjacent to the building or premises. The collection, treatment, storage, and distribution of potable water from source to consumer.

Water Table. The level of groundwater.

Watt (W). A metric unit of power, usually used in electric measurements, which gives the rate at which work is done or energy expended. One watt equals 1 joule of work per second.

Winter Solstice. The shortest sunlit day of the year at which the sun is at its lowest altitude; it occurs December 22.

Work. Energy that is transferred from one body to another in such a way that a difference in temperature is not directly involved. The product of an external force times the distance an object moves in the direction of the force.

Zone. A space or group of spaces within a building with heating or cooling requirements sufficiently similar so that comfort conditions can be maintained throughout by a single controlling device.

Bibliography

ARTICLES AND PAMPHLETS
Pool Heaters

"For Year-round Swimming–Your Own Underground Pool," *Popular Science* 209(5): 179, November 1976.

Heat Your Home with Solar Energy from Your Swimming Pool (1977). New Allen Goodrich Enterprises, Inc., 1620 Grover Road, East Aurora, NY 14052. 25 pp., $5.00.

How to Design and Build a Solar Swimming Pool Heater (1975), F. De Winter. Copper Development Association, New York, NY. 46 pp., $1.25.

How to Heat Your Swimming Pool Using Solar Energy (1973). Brace Research Institute, Quebec, Canada. Do-It-Yourself Leaflet L-3, 4 pp., $2.00.

Solar Pool Heaters (1976). Horizon Industries, Publications Dept., Box 4518-N, Panorama City, CA 91412. $2.00.

Solar Dryers and Cookers

How to Make A Solar Cabinet/Dryer for Agricultural Produce. (1973). Brace Research Institute, Quebec, Canada. Do-It-Yourself Leaflet L-6, $2.00.

Solar Energized Food Dehydrator (1975). Solar Survival, Environmental Action Reprint Service, 2239 East Colfax, Denver, CO 80206. Four blueprints, $5.00.

"Solar Oven," C. Rose, *Alternative Sources of Energy* (22): 19–21, September 1976.

Solar Steam Cooker (1972). Brace Research Institute, Quebec, Canada. Do-It-Yourself Leaflet L-2, 13 pp., $2.00.

Solar Heaters and Collectors

Bread Box Water Heater Plans, Zomeworks Corporation, Box 712, Albuquerque, NM 87103. One blueprint, $2.50.

Build Your Own Solar Energy System (1976), C. H. Breckenridge, Solar Heating Systems, 13584 49th St., N., Clearwater, FL 33520. 40 pp., $3.95.

Build Your Own Solar Water Heaters. Florida Conservation Foundation, Inc., 935 Orange Ave., Winter Park, FL 32789. 25 pp., $2.50.

Building a Vertical Forced-Air Solar Collector (1977). San Luis Valley Solar Energy Association, Box 1284, Alamosa, CO 81101. 11 pp., $1.00 donation.

Capture the Sun: The Parabolic Curve, G. Graham. Sunway, 1301 Berkeley Way, Berkeley, CA 94702, 48 pp., $2.50.

"Concentrating Collectors for Solar Heating and Cooling," G. P. Gilmore, *Popular Science* 209(4): 96-98, October 1976.

Construction Manual: Solar Can-Type Hot Air Furnace. Northern Solar Power Company, 311 Elm St., S., Moorhead, MN 56560. February 1977, 17 pp., $2.00.

A Design Manual for Solar Water Heaters (1977). Horizon Industries, P.O. Box 4518-N, Panorama City, CA 91412. 40 pp., $5.00.

Drum Wall Plans. Zomeworks Corporation, Box 712, Albuquerque, NM 87103. Two blueprints, $5.00.

"Five Solar Water Heaters You Can Build," E. Moran, *Popular Science* 208(5): 99-103, May 1976.

The Fuel Savers: A Kit of Solar Ideas for Existing Homes (1976), D. Scully et al. Northwest New Jersey Community Action Program, Prospect and Marshall Sts., Phillipsburg, NJ 08865. 60 pp., $3.00 (add $0.25 for postage).

Home Heating Solar Heat Panels (1976). Solar Concepts, Box 462, Independence, CA 93526. $6.95.

"Hot-Line Solar Collector," *Mother Earth News* 39:108–109, May 1975.

Hot Water: Solar Water Heaters and Stack Coil Heating Systems (1975), S. Morgan et al. Hot Water, Santa Barbara, CA. 31 pp., $2.00.

How to Build a Solar Water Heater (1973). Brace Research Institute, Quebec, Canada. Do-It-Yourself Leaflet L-4, 11 pp., $2.00.

An Inexpensive Economical Solar Heating System for Homes, J.W. Allred et al. Report No. NASA TM X-3294, July 1976, Stock No. 033-000-00632-2. Superintendent of Documents, Government Printing Office, Washington, DC 20402. 55 pp., $1.15.

"Make Your Own Solar Collector," H. Kolbe, *Mechanix Illustrated* 73 (586):43–45, March 1977.

"Mother's 'Heat Grabber' (Window Collector)," *Mother Earth News* (47):101–103, September/October 1977.

"Mother's Solar Heating System," *Mother Earth News* (39): 96-98, May 1975.

Passive Solar Water Heater, Horace McCracken. Route 1, Box 417, Alpine, CA 92000. May 1976, 18 pp., $6.00.

The Nicholson Solar Energy Catalogue and Building Manual (1977), N. Nicholson. Nicholson Solar Catalogue, Box 344, Ayer's Cliff, Quebec, Canada JOB ICO. Basic components of solar air heaters, water heaters, greenhouses and storage systems, 70 pp., $9.50.

Practical Sun Power (1975), W. H. Rankins III and D. A. Wilson. Lorien House, Black Mountain, NC. 56 pp., $4.00.

"Solar Collectors You Build From Kits," R. A. Cutter, *Mechanix Illustrated* 75(58):44–45, 112-113, October, 1976.

Solar Energy: A Biased Guide (1977) by W. L. Ewers. Domus Books, 400 Anthony Trail, Northbrook, IL 60062. Illustrated guide to building and using solar collectors, in particular the Consolar collector unit, 96 pp., $4.95.

Solar Energy Notebook (1975), D. A. Wilson and W. H. Rankins III. Lorien House, Black Mountain, NC. 56 pp., $4.00.

Solar Energy Warm Air House Heater (1975), A. Meinel et al. Helio Associates, Inc., Tucson, AZ. 19 pp., $5.00.

Solar Heat: Step-by-Step Instructions on How to Use Free Solar Energy to Save on Your Utility Bills (1977), R. J. Hurley, 4701 Erie Street, College Park, MD 20740. 35 pp., $6.95.

Solar Heating Panel Instructions (1977). Astro Magnetic Labs, 457 Santa Anna Dr., Palm Springs, FL 33461. $6.00.

Solar Heating Systems (1975), Sands. Solar Systems, Box 110, Danbury, NH 03230. 46 pp., $2.50.

Solar Water Heater Plans. Zomeworks Corporation, Box 712, Albuquerque, NM 87103. Two blueprints, $5.00.

Solar Water Heating (1976) by E. T. Davey. Environmental Action Report Service, 2239 East Colfax, Denver, CO 80206. 6 pp., $.30.

Solar Water Heating for the Handyman, S. Paige (1974). Edmund Scientific Co., Edscorp Bldg., Barrington, NJ 08007. 31 pp., $4.00.

Specifications and Drawings on How to Build Your Own Solar Hot Air and Hot Water Heat Collectors by D. I. Hadley. Hadley Solar Energy Co., P.O. Box 1456, Wilmington, DE 19889. 14 pp., $7.00.

"This Solar Heater Pays for Itself Every Five Weeks," D. A. Gilman, *Mother Earth News* (41):129, September 1976.

Solar Stills

"How to Build and Use a Solar Still," D. S. Halacy, *Mother Earth News,* 19:72–74, September 1974.

How to Make a Solar Still. Brace Research Institute, Quebec, Canada. Do-It-Yourself Leaflet L-1, $1.50.

Plans for a Glass and Concrete Solar Still. Brace Research Institute, Quebec, Canada. Technical Report T 58, $3.50.

Simple Solar Still for the Production of Distilled Water. Brace Research Institute, Quebec, Canada, Technical Report T 17, $1.50.

BOOKS
Nontechnical

The Buy Wise Guide to Solar Heat (1976), F. Hickok. Hour House, St. Petersburg, FL 33700. 121 pp., $9.00.

The Coming Age of Solar Energy (1973), D. S. Halacy, Jr. Harper & Row, Inc., New York, NY 10022. 231 pp., $7.95.

Direct Use of the Sun's Energy (1964), F. Daniels. Ballantine Books, Inc., Westminister, MD 21157. 271 pp., $1.95.

Harnessing the Sun to Heat Your House (1974), J. Keyes. Morgan & Morgan, Dobbs Ferry, NY 10522. Provides basic introduction to solar heating applications in understandable language; includes comprehensive index of more technical data. 192 pp., $2.95.

Solar Homes and Sun Heating, (1976), G. Daniels. Harper & Row, New York, NY 10022. Practical guide on solar heating for the layman; presents a nontechnical description of basic principles, existing systems, and techniques for construction and installation. 178 pp., $8.95.

Technical

Applied Solar Energy: An Introduction (1976), A. B. Meinel and M. P. Meniel. Addison-Wesley Publishing Co., Reading, MA 01867. Basic textbook introduction to the theory of solar energy intended for college seniors or graduate students. 651 pp., $17.95.

Solar Energy, Technology and Applications (1974). J.R. Williams. Ann Arbor Science Publications, Ann Arbor, MI 48106. Definitive overview of the various techniques for utilizing solar energy; includes glossary and comprehensive bibliography. Supplemented by the 1976 Solar Update, which covers new developments since 1974. 120 pp., $6.95 (soft cover).

Solar Energy Thermal Processes (1974), J. A. Duffie and W.A.Beckman. John Wiley & Sons, Inc., New York, NY 10016. How to understand and predict the performance of solar collectors and solar photothermal systems for heating and cooling buildings and for heating water and air; comprehensive and coherent treatment for professionals, especially for engineers. 386 pp., $16.95.

Solar Energy Utilization for Heating and Cooling. (1974), J. I. Yellott. Chapter 59 of *ASHRAE Handbook and Product Directory,* 1974 Application Volume. Summary of current practice, including many details and a substantial section on the availability and characteristics of solar radiation at the earth's surface. ASHRAE (Sales Dept.), New York, NY. (Reprints available from American Section, ISES, c/o Dr. Jay Shelton, Physics Department, Williams College, Williamstown, MA 01276 $1.00.)

Solar Heating and Cooling: Engineering Practical Design, and Econmics (1975), J. F. Kreider and F. Kreith. McGraw-Hill, New York, NY 10026. Designed as a how-to handbook with emphasis on economically feasible heating and cooling systems; contains considerable technical detail and extensive tables of reference data. 342 pp., $22.50.

PERIODICALS

Alternative Source of Energy. Alternate Sources of Energy, Inc., Route 2, Box 90A, Milaca, MN 56353. Quarterly, $5.00/year.

The Mother Earth News. The Mother Earth News, Inc., 105 Stoney Mountain Road, Hendersonville, NC 28739. Down-to-earth descriptions of peoples' experiences with alternative lifestyles, ecology, and energy; source for what is happening in energy at the grass-roots level; source for books. Bimonthly, $10.00/year.

Solar Age, Solar Vision, Inc., Route 515, Box 288, Vernon, NJ 07462. Brief articles on developments in solar energy applications, with emphasis on solar heating and cooling. Monthly, $20.00/year.

Solar Energy, Pergamon Press Ltd., 44-01 21st St. Long Island City, NY 11101. Scientific and engineering papers on all aspects of solar energy and technology-theory and applications. Quarterly, $15.00/year.

Solar Energy Digest. CWO-4 W. B. Edmondson. P. O. Box 17776, San Diego, CA 92117. Concise summaries of solar energy developments, ongoing research, and publications, both United States and foreign. Monthly $28.50/year.

Solar Energy Intelligence Report. Business Publishers, Inc., P.O. Box 1067, Silver Spring, MD 20910. Biweekly newsletter covering the Washington beat in solar energy; also covers new developments, markets, meetings. Biweekly, $90.00/year.

Solar Engineering. Solar Engineering Publishers, Inc., 8435 N. Stemmons Freeway, Suite 880, Dallas, TX 75247. Short descriptions of activities and developments in the field of solar energy, particularly in the private sector and in the United States. Monthly, $10.00/year.

Solar Heating & Cooling. Gordon Publications, 20 Community Place, Morristown, NY 07960. Short articles on solar heating and cooling issues, developments, and equipment. Oriented to builders, developers, and manufacturers. Bimonthly, $6.00/year.

Architectural

Design Criteria for Solar-Heated Buildings (1975), E. M. Barber, Jr. and D. Watson. Sunworks, Inc., Guilford, CT 06437. Written for architects and builders; primarily concerned with new construction; outlines design criteria for solar heated buildings. 66 pp., $10.00.

Solar Home Book (1976), B. Anderson and M. Riordan. Cheshire Books, Harrisville, NH 03450. Covers all aspects of solar home heating including architectural, direct and indirect systems, do-it-yourself solar water heating and energy conservation, retrofitting, and social and cultural implications. 297 pp., $7.50.

Solar Dwelling Design Concepts (1976). A1A Research Corporation, Supt. of Documents, Government Printing Office, Washington, DC 20402. Stock No. 023-000-00334-1. Architectural analysis of solar house designs; discusses solar heating systems, climatic factors, site planning, and the impact of solar energy on the traditional design of houses. 145 pp., $2.30.

Solar Energy and Building (1975), S. V. Szokolay. John Wiley & Sons, New York, NY 10016. Provides a conceptual understanding of the problems and solutions of solar energy; contains an illustrated architectural review of "solar houses" including plans and performance data. 148 pp., $18.50.

Solar Energy and Housing Design Concepts (1975). Giffels Associates, Inc., A1A Research Corp., Washington, DC 20006. Basic orientation of designing for energy conservation and solar energy usage; specific application of these design concepts to four climatic regions. 145 pp., $15.00.

Solar Heated Buildings: A Brief Summary (1976), W. A. Shurcliff. 19 Appleton St., Cambridge, MA 02138. Twelfth edition. Descriptions of 22 buildings that are partially or fully solar heated; includes buildings that did not exist, do exist, or are expected to exist very soon. Permits comparison of characteristics and performance of a wide variety of solar heated buildings. 212 pp., $9.00 (prepaid).

Solar Primer One (1975), B. Carlson. SOLARC, Whittier, CA 90607. Written by architects; presents basic solar applications and building design with discussions of collectors, structure, heat transfer, storage, and total heating and cooling systems. 101 pp., $8.75.

Do-It-Yourself

Handbook of Homemade Power (1975). The Mother Earth News, Bantam Books, New York, NY 10019. Combination of reprint and original articles on using various energy sources including solar energy. Includes instructions on how to build and use solar cookers, ovens, heaters, and water heaters, plus an extensive bibliography. 374 pp., $2.25.

A Guide to System Sizing and Economics of Solar Water Heating in Florida Residences (1977), S. Chandra. Florida Solar Energy Center, 300 State Rd. 401, Cape Canaveral, FL 32920. Detailed guide to solar water heating systems. 56 pp., $1.25.

Other Homes and Garbage: Design for Self-Sufficient Living (1975), J. Lecki et al. Charles Scribner's Sons, New York, NY 10017. Practical approach for constructing solar panels and ovens, as well as windmills, and waterwheels; emphasizes renewable energy sources. 302 pp., $9.95.

Producing Your Own Power (1975), C. H. Stoner (ed.). Rodale Press, Inc., Book Division, Emmaus, PA 18049. How to calculate energy needs, conserve energy in existing structures, and locate required equipment. 322 pp., $8.95.

Solar Science Projects (1974), D. Halacy. Scholastic Book Services, Englewood, NJ. 07631. Easy-to-understand explanations of how to build solar-powered devices (water heaters, stills, cookers, ovens, furnaces, motors, radios). 96 pp., $.85.

Survival Scrapbook #3: Energy (1974), S. A. Szczelkum. Schocken Books, New York, NY 10017. Alternative energy for the homesteader; includes simple designs for solar collectors and stills. 112 pp., $3.95.

Your Energy-Efficient House (1975), A. Adams. Garden Way Publishing, Charlotte, VT 05445. Practical guide for every homeowner on simple, inexpensive changes to make around the home, which will make it much more energy-efficient; covers a variety of aspects including site planning, landscaping, insulating, and construction. 118 pp., $4.95

Your Home Solar Potential (1976), I. Spetgand and M. Wells. Edmund Scientific Co., Barrington, NJ 08007. Contains do-it-yourself home survey to evaluate existing energy utilization and to determine solar savings potential. 60 pp., $9.95.

General Energy

Cloudburst: A Handbook of Rural Skills and Technology (1973), V. Marks (ed). Cloudburst Press, Brackendale, British Columbia. 126 pp., $4.95.

Energy, Environment and Building (1975), P. Steadman. Cambridge University Press, New York, NY 10022. Compendium on solar energy, wind power, small scale water power; includes over 100 pages of bibliographies and appendices; lists solar hardware manufacturers. 287 pp., $5.95 (soft cover).

Energy for Survival (1974), W. Clark. Doubleday & Co., New York, NY. Energy sources in the past and present, and prospects for the future; major section on solar energy; extensive information source and guide to sources. 652 pp., $4.95 (soft cover).

Energy Primer (1974). Portola Institute, Whole Earth Truck Store, 558 Santa Cruz Ave., Menlo Park, CA 94025. A comprehensive book about renewable sources of energy—solar, water, wind, and biofuels. The focus is on small-scale systems, which can be applied to the needs of the individual, small group, or community. $5.50.

Energybook #1: Natural Sources and Backyard Applications (1975), J. Prenis (ed). Running Press, Philadelphia, PA 19103. Review of possible alternative sources of energy. Short descriptions of 55 different ideas and designs. 112 pp., $4.00.

The Poverty of Power: Energy and the Economic Crisis (1975), B. Commoner. A. Knopf, New York, NY 10022. The energy crisis as seen by an environmentalist; covers the use and misuse of fossil and nuclear energy, current government and industrial opposition to alternative energy sources, and recommendations for the future. 204 pp., $11.00.

Small Is Beautiful: Economics as if People Mattered (1974), E. Schumacher. Harper & Row, New York, NY 10022. Philosophical view of the world's social and economic problems including energy; offers some smaller-scale yet feasible solutions to these problems. 320 pp., $8.95.

DIRECTORIES

Informal Directory of the Organizations and People Involved in the Solar Heating of Buildings (1976), 2nd edition, W. A. Shurcliff. 19 Appleton St., Cambridge, MA 02138. Selective coverage of institutions and individuals involved in all aspects of solar heating of buildings; main emphasis is on United States, but some foreign groups are included. 178 pp., $7.00 (prepaid).

Solar Directory (1975), C. Peska (ed). Ann Arbor Science Publishers, Inc., Ann Arbor, MI 48106. An overall guide to solar energy activity, United States and foreign; covers information services, manufacturers, distributors, research activities, projects; and includes a bibliography. $20.00.

Index

ACES (Annual Cycle Energy System), 3
Adobe Homes, Solar Energy Applications in, 95
Air-Cooled Solar Collectors, 62
Air Conditioning, 263, 265, 268
Air Convection, 67
Air Temperature Patterns, and Solar Design, 74
Alternative Electric and Hybrid Systems, 153. See Also Electric Cars
Aluminum Foil Insulation, 231
  R Value, 232
American Society of Heating, Refrigerating, and Air Conditioning Engineers (ASHRAE), 158
American Solar Dwellings, 57
Anaerobic Digestion, and Biomass Conversion, 7
Annual Cycle Energy System (ACES), 3
Annual Energy Consumption, 47
Anthracite Coal, 221
Appliances
  and Conservation, 49
  Efficiency Standards, 334
  Energy Savers, 210
  Energy Use, 49
  Labeling of, 210
Aquatic Plants, and Biomass Fuels, 6
Architectural Design and Solar Collectors, 79-80
  Location, 80
  Orientation, 80
  Shadowing, 80
  Tilt, 80
ASHRAE (American Society of Heating, Refrigerating, and Air Conditioning Engineers), 158
Atmospheric Conditions, and Solar Design Factors, 73
Attic
  Energy Conservation, 267
  How to Insulate, 251
Attic Insulation, 245
  Conservation of, 49
Automobile
  Buying an Energy-Efficient, 213
  Fuel Efficiency Laws, 334

Balcony-Access Apartments, Solar Energy Applications in, 101
Baseboard Electric Heat, 219

Basements, 267
Bathroom Energy Savers, 209
Batteries, 3-6. See Also Fuel Cells
  Cost Savings, 5
  Primary, 4
  Secondary, 4
Batts, 244
Biomass Fuel, 6
  Production and Use
    Anaerobic Digestion, 7
    Direct Combustion, 7
    Environmental Concerns, 7
    Fermentation, 7
    Gasification and Liquefaction, 7
  Sources
    Aquatic Plants, 6
    Nonwoody Plants, 6
    Trees and Woody Plants, 6
Bituminous Coal, 221
Blankets, Glass-Fiber and Rock-Wool, 244
BOCA (Building Officials & Code Administrators), 158
Boiling Water Reactor (BWR), 144
Bottled Gas, 221
Breeder Reactors, 7
  Cost, 9
  Environmental Effects, 10
  Nuclear Power, 7
  Operation, 8
  Safety, 10
Brick Colonial Houses, Solar Energy Applications in, 95
Brick Townhouse, Solar Energy Applications in, 99
British Thermal Unit (Btu), 223
Btu Return Criteria, Solar Systems, 119
Buildings
  Construction, Conservation, 48
    Heat Exchange, 77
    Solar Design, 71, 77, 79
Building Officials & Code Administrators (BOCA), 158
Butyl-Based Caulking, 236
BWR (Boiling Water Reactor), 144

Cape Cod Houses, Solar Energy Applications in, 87
Carcinogens, Breeder Reactors, 12
Car. See Automobile
Car Pooling, 150

Caulking
  Butyl-Based, 236
  Doors and Windows, 48
  Elastomeric, 236
  Filler, 236
  Heat Loss Reduction, 217
  How To, 237
  Latex-Based, 236
  Oil-Based, 236
  Polyvinyl-Based, 236
  Resin-Based, 236
  Weatherizing, 235
Ceiling
  of Energy-Efficient Residence, 46
  Insulation, 217
Ceiling Unit, Electric Heat, 219
Cellulose Insulation, 231
  R Value, 232
Central Furnace, Electric Heat, 219
Central Heating Systems, 217
  Steam, 219
Children, Energy Activities for,
  Chemical Energy, 283
  Conservation, 281
  Electricity, 278
  Solar, 278
Circular Concentrating Collector, 63
Climate, 72
  Cool, and Solar System Use, 84
    Impact of, on Energy Usage, 129
    Solar Dwelling Design, 71, 73-75
Climatic Regions of the United States, 72
Coal, 11
  Anthracite, 221
  Bituminous, 221
  Conversion, Regulations for, 335
  Home Heating, 221
  Land Use and, 12
  Resources, 11
  Solvent Refining of, 12
Coefficient of Performance (COP), 111, 164
  Cooling, 159
  Heat Pump Heating, 159
Collector. See Also Solar Collector
  Energy Loss, 83
  Location, 82
  Orientation, 82
  Size, 82
  Shadowing, 80
  Solar, 60, 61

Solar Subsystems, 104
Storage and Distribution Components, 68
Tilt, 80, 82
Type, 81, 82
Warm-Air Flat-Plate, 69
Warm-Air Passive, 70
Warm-Water, 70
Warm-Water, Flat-Plate, 68
Combustion, Biomass Conversion, 7
Comfort, and Solar Design, 71, 75
Comfort Zone, 76
Compressed Air, and Storing Electricity, 5
Concentrating Solar Collectors, 63
Conservation
  Activities for Children, 281
  Appliances, 49
  Attic Insulation, 49
  Building Construction, 48
  Caulking Doors and Windows, 49
  Dishwasher, 204
  Electricity, 205
  Energy Policy and Conservation Act of
    1975, 158
  Gasoline, 212
  Heat Pumps, 53
  Household Appliances, 210
    Labeling Program, 210
  Hot Water, 203
  Lighting, 50
    Indoor, 209
    Outdoor, 210
  Ironing, 209
  Oven, 205
  Practice, 49
  Refrigerator-Freezer, 206
  Retrofitting, 49
  Storm Doors, 49
  Storm Windows, 49
  Strategies and Savings, 48
  Transportation, 151
  Utilities, 334
  Washer/Dryer, 207
  Weather-Stripping Doors and Windows, 49
  Yard/Workshop, 212
Conservation Codes, BOCA, 158-168
  Alternative Systems, 168
  Electrical System, 167
  Exterior Envelope, 159
  Plumbing Systems, 167
  Residential Buildings, 159
  Thermal Resistance, 159
  Thermal Transmittance, 159
Contemporary Split-Level Houses, Solar
  Energy Applications in, 95
Convection, 106
  Air, 67
  Natural, 67
Conventional Nuclear Reactors. See Nuclear
  Fission Reactors
Convector Setback, 215
Cooling, Energy-Saving, 266
  Solar, 56
Cooling Demand, Reduced, 49
Cooling System, Energy-Efficient House, 46
COP. See Coefficient of Performance
Crude Oil, 36
  Refining, 37
Cupola, 65

Dam Safety, and Hydroelectric Power, 28
Degree-Day, Heating, 159
  Solar Design Factors, 74
Department of Energy, 169
  Conservation, 172
  Energy Information Administration, 173
  Energy Regulatory Commission, 173
  Energy Research, 171
  Environment, 172
  Solar Applications, 172
  Technology, 172
Design, 71
  Solar Determinants, 75, 82
Dishwasher, 204
Distribution, Solar, 61
Doors, Energy-Efficient House, 45
  Weather-stripping, 49
Driving, in Winter, 229
Dwelling Design, Solar, 71

Earth's Axis, and Solar Design, 73
EER (Energy-Efficiency Ratio), 53, 159
EER (Energy-Efficient Residence), 44
Efficiency, Appliance Standards, 334
Elastomeric Caulking, 236
Electrical Storage, 4
  Battery, 4
  Compressed Air, 5
  Flywheels, 5
  Hydrogen, 5
Electrical Generation/Geothermal, 21
Electrical Storage, Hydrogen from Elec-
  trolysis of Water, 5
Electric Appliances, Annual Energy Re-
  quirements, 211
Electric Car, 151
Electric Heat
  Baseboard, 219
  Ceiling Unit, 219
  Central Furnace, 219
  Floor Furnace, 219
  Heat Pump, 219
  Wall Unit, 219
Electric and Hybrid Systems, Alternative,
  153. See Also Electric Cars
Electricity and Bill, 224. See Also How to
  Read Meters
  British Thermal Unit, 223
  Kilowatt-hour, 223
  Kilowatt, 223
  R value, 231
  Purchased Gas Adjustment Charge, 223
  Rate Schedule, 223
  Watt, 223
Electricity, Home Heating, 221
Electric Meter, Reading, 223
Electrolysis of Water, 5
Emergency Heating Equipment, 227
Emergency Kit, for Heat Shortage, 228
Energy Audits, for Public Buildings, 334
Energy Checklist, Home, 235
Energy Conservation, 157, 334
  Air Conditioning, 268
  Attic, 267
  Basements, 267
  BOCA Requirements, 158
  Car-Pooling, 150
  Cooling, 266

Fireplace, 269
Furnace, 264
Ground Cover, 276
Heating, 266, 268
Heat Pump, 269
Hot Water, 268
Landscaping, 267, 276, 277
Lighting, 270
Roof, 267
Shrubs, 276
Telecommunications, 152
Trees, 277
Energy Consumption
  Annual, 47
  Predictions of. See Heating and Cooling
    Degree-Days
  Transportation, 150
Energy Distribution, Thermal, 67
Energy-Efficiency
  and Cars, 213
  and Driving, 212
  and Residences, 210
Energy-Efficiency Ratio (EER), 53, 159
Energy-Efficient Residence (EER), 44
  Ceiling, 46
  Cooling System, 46
  Doors, 45
  Exterior Walls, 45
  Floor, 45
  Foundation, 45
  Heating System, 46
  Roof, 46
  Water Heating, 46
  Water-Use Appliances, 46
  Windows, 45
Energy Games, 278. See Children, Energy
  Activities for
Energy Grant Program for Schools and
  Hospitals, 334
Energy Information Administration (EIA),
  173
Energy Policy and Conservation Act of
  1975, 158
Energy Regulatory Commission, 173
Energy Requirements (Annual), Electric
  Household Appliances, 211
Energy Research, DOE, 171
Energy Savers
  in the Bath, 204
  in the Kitchen, 204
  in the Laundry, 204
Energy Sources, 126
  Geothermal, 127
  Nuclear Fission, 127
  Solar, 128
Energy Technology, 3
Energy Storage
  Batteries, 4
  Compressed Air, 5
  Flywheels, 5
  Heat, 3, 5
  Hydrogen, 29
    from Electrolysis of Water, 5
  Pumped Hydropower, 5
Energy Tax Act of 1978, 336
Energy-Transport Component, Solar System,
  61

Energy Use
  Appliances, 49
  Climatic Impact, 129
  Lighting, 50
  Population Projections, 125
Engines
  Advanced Heat, 153
  Internal Combustion, 153
Environmental Effects
  Geothermal Energy, 24
  Hydroelectric Power, 28
  Nuclear Fusion, 34
  Oil-Shale, 39
  Photovoltaics, 40
EPCA (National Energy Conservation Policy Act), 334

Farmhouse, Solar Energy Applications, 91
Federal Energy Regulatory Commission (FERC), 27
FERC (Federal Energy Regulatory Commission), 27
Fermentation, Biomass Conversion, 7
Filler Caulking, 236
Fireplace, 269
Firewood, 272-275
  Rating, 274
Flammability, of Insulation, 232
Flat-Plate Solar Collectors, 61, 62
Floor, Energy-Efficient Residence, 45
  Insulation of, 257
Floor Furnace, Electric Heat, 219
Fluorescent Light Bulbs, 157
Flywheels, Energy Storage, 16
  Storing Electricity, 5
Foamed-in-Place Insulation, 245
Forced-Air Heat Distribution, 67
Forced Hot-Water Heating Systems, 219
Forced Radiation, 68
Fossil Fuel, 126
Foundation, Energy-Efficient Residence, 45
Fraud, in Solar Systems, 120
Freezing of Pipes, Prevention, 227
Fuel Cost Adjustment, 223
Fuel Efficiency, 220
  Auto Laws, 334
Fuel Cells
  Future Fuels, 154
  Operation of, 18
Fuels, Biomass. See Also Biomass Fuels
  Coal, 11
  Fossil, 126
  Hydrogen, 28
  Nuclear, 41
  Oil, 35
Fuels, and Burners, 220
Furnace Maintenance, 264
Fusion Reactors, and Nuclear Fusion, 34
Future Fuels, 154
  Fuel Cells, 154
  Hydrogen, 154

Gas
  Bottled, 221
  Conservation of, 212
  Home Heating, 221
  Liquefied Petroleum, 221
  Manufactured, 221
  Natural, 221

Gas Bills, 225
Gas Flow Distribution, Thermal Energy, 67
Gas Guzzler Tax, 336
Gas Meter, 224
Gasahol, 7
  Exemption from Excise Tax, 336
Gasification, Biomass Conversion, 7
Geothermal Energy, 19, 127
  Environmental Effects, 24
  Generating Electricity, 21
Geothermal Heating, Direct, 20
Glass-Fiber Insulation, 232
Glass-Wool Insulation, 231
Greenhouse, Solar, 64
Ground Cover, 276

Heat Distribution
  Forced-Air, 67
  Forced Radiation, 68
  Liquid Flow, 68
  Natural Radiation, 68
  Solar, 82
Heat Engine, Advanced, 153
Heat Exchange, Solar Design Factors, 77
Heating Systems
  Electric, 219
  Radiant Panel System, 219
  Steam Central, 219
Heating
  Energy Conservation, 263, 268
  Energy Saving, 266
  Forced Hot-Water, 219
  Hot Water, 218
  Solar, 56
  Steam, 218
Heating and Cooling Degree-Days, 74
Heating Demand, Reduced, 49
Heating Equipment, Emergency, 227
Heating System
  Central, 217
  Controls, 222
  Energy-Efficient House, 46
Heat Loss. See "U" Value
  Insulated Homes, 113
  Reduction of
    Caulking, 217
    Ceiling Insulation, 217
    Storm Windows, 217
    Wall Insulation, 217
    Weather-stripping, 217
  Uninsulated Homes, 113
Heat Pumps, 53
  Conservation, 53
  Cost, 53
  Electric Heating, 219
  Energy Conservation, 269
  Solar Systems and, 110
Heat Requirements, Reducing, 217
Heat Storage
  Latent Heat, 67
  Paraffin, 67
  Rocks, 66
  Salt Hydrates, 67
  Solar, 61, 66
  Space Heating, 3
  Water, 67
Heat Traps, 64. See Also Solar Collectors
Home Buying, and Energy Efficiency, 210

Home Design
  Impact of Solar Energy, 86
  Sun Control, 79
  Temperature Control, 78
  Thermal Control, 78
  Wind Control, 79
Home Emergency Kit, 228
Home Energy Checklist, 235
Home Heating
  Coal, 221
  Electricity, 221
  Gas, 221
  Oil, 221
  Wood, 221
Home Winterizing, 227
Hot Water, Solar-Powered, 59
Hot-Water Energy Savers, 203
Hot-Water Heating, 218
Hot-Water Systems, 59
Human Comfort, and Solar Energy, 76
Humidity, and Solar Design, 74
Hybrid Solar Systems, 109
Hydroelectric Power, 26
  Dam Safety, 28
  Environmental Effects, 28
  Regulation, 26
Hydrogen
  Energy Storage, 5, 29
  Fuel of the Future, 154
  Fuels, 28
  Resources, 28
Hypothermia, 229

Incandescent Light Bulbs, 157
Indoor Lighting, 209
Inertial Confinement, 33
Insolation, 73
Insulation
  Aluminum Foil, 231
  Attic, 246, 251
  Basement Walls, 254, 258, 259
  Batts, 244
  Blankets, 244
  Cellulose, 231
  Crawl-Space Walls, 254
  Flammability, 232
  Floors, 257
  Foamed-in-Place, 245
  Glass Wool, 231
  and Heat Loss, 113, 217
  How to Judge, 231
  Loose Fill, 245
  Mineral Wool, 231
  Perlite, 231
  Polystyrene Plastic Foam, 231
  R Value, 232, 244, 246
  Rigid Board, 245
  Rock Wool, 231
  Safety, 232
  Unfinished Attic, 245
  Urea-Formaldehyde, 231
  Vermiculite, 231
  Walls, 255
  Winterizing, 227
  Wood Frame Walls, 254
In-Town Apartments, 101
Ironing, 209

Kilowatt, 223
Kilowatt-hour, 223

Land Use, and Coal, 12
Landscaping, 267, 276
Latent Heat Storage, 67
Latex-Based Caulking, 236
Legal Solar Design Factors, 71
Light Bulb, Electrodeless Fluorescent, 157
Lighting
    Conservation, 50, 270
    Energy Savers, 209
    Energy Use, 209
    Indoor, 209
    Outdoor, 210
Linear Concentrating Solar Collectors, 63
Liquefaction, 7
Liquefied Petroleum Gas (LPG), 221
Liquid-Cooled Solar Collectors, 62
Liquid Flow Heat Distribution, 68
Loan Program, Solar Energy, 334
Loose Fill, 245
LPG (Liquefied Petroleum Gas), 221

Magnetic Confinement, 33
Manufactured Gas, 221
Meters, How to Read, 223
Mid-Atlantic Garden Apartments, 101
Mineral Wool, 231
Monitors, Roof, 65

National Energy Act (NEA), 334
National Energy Conservation Policy Act
    (EPCA), 334
National Solar Heating and Cooling Infor-
    mation Center, 333
Natural Convection, 67
Natural Gas, 221
Natural Gas Policy Act of 1978, 335
Natural Radiation, 68
Northern Rowhouse, 95
Northern Townhouse, 95
Nuclear Accidents, 15
Nuclear Fission, 127, 144
    Breeder Reactors, 7, 12
    Conventional Reactors, 13, 15
    International Nuclear Safeguards, 149
    Nuclear Reactor Types and Materials, 144
    Nuclear Safeguard Systems, 148
    Uranium, 41
Nuclear Fission Reactors, Boiling Water, 144
    Breeder, 7
    Conventional, 13
    Pressurized Water, 144
    Radioactive Pollution, 14
    Radioactive Waste, 14
    Safety, 15
    Types and Materials, 144
Nuclear Fuels, 41
Nuclear Fusion, Environmental Effects of,
    34
    Fusion Reactors, 34
    Inertial Confinement, 33
    Magnetic Confinement, 33
    Plasma, 33
    Safety, 34
Nuclear Materials
    International Safeguards, 149
    Nuclear Safeguard Systems, 148

Nuclear Power, 144
Nuclear Power Plants, Safety of, 144
Nuclear Safeguard Systems, 148
    International, 149

Ocean Thermal Energy Conversion, 34
Oil, 35
    Crude, 36
    Home Heating, 221
    Petroleum, 37
    Production, 38
Oil Shale, 38
    Energy Systems, 39
    Environmental Effects, 39
Open-Water Solar Collectors, 62
Outdoor Lighting, 210
Oven, 205

Panels, Solar, 61, 62
Paraffin Heat Storage, 67
Passive Solar Collectors, 64
    Warm Air, 70
Payback, Solar Investments, 114
Perlite, Insulation, 231
    R Value, 232
Petrochemicals, 37
Petroleum, 37
PGA (Purchased Gas Adjustment Charge),
    223
Photosynthesis, 6
Photovoltaics
    Environmental Effects, 40
    Solar Cells, 40
Pipes, and Freezing, 227
Plasma, 33
Plastic Storm Windows, 238
Plumbing Systems, 167
Pollution, Thermal and Carbon Dioxide, 128
Pollution Control Loan Program, 335
Polystyrene Insulation, 231
    R Value, 232
Polyvinyl-based Caulking, 236
Ponds, Solar, 65
Population Projections, and Energy Use, 125
Power Plant
    Efficiency, 51
    and Industrial Fuel Use Act of 1978, 335
    Waste Heat, 51
Precipitation Patterns, 84
Pressurized Water Nuclear Reactor, 144
Primary Battery, 4
Production, Oil, 38
Propane, 221
Public Building, Energy Audit, 334
Public Utility Regulatory Policies Act of
    1978, 335
Pueblo, 56
Purchased Gas Adjustment Charge (PGA),
    223

R Value, 231
    Aluminum Foil Insulation, 232
    Cellulose Insulation, 232
    Glass Fiber Insulation, 232
    Insulation, 232, 244, 246
    Perlite Insulation, 232
    Polystyrene Insulation, 232
    Rock Wool Insulation, 232
    Urea-Formaldehyde Insulation, 232

Radiant Panel Heating, 219
Radiation, 106
    Solar Home Design, 73
Radioactive Pollution, 15
Radioactive Waste, 14
Rate Schedule, 223
Reactors, Nuclear. See Also Nuclear Fission
    Breeder, 7-10
    Conventional, 13
Reduced Cooling Demand, 49
Reduced Heating Demand, 49
Refrigerator-Freezer, 206
Regions, Climate, 72
Resale Value, Solar Homes, 121
Residential Buildings, 159
Resin-Based Caulking, 236
Resources
    Coal, 11
    Hydrogen, 28
    and Strategies of Transportation, 150
    Uranium, 8, 41
Retrofitting, Conservation, 49
Return on Solar Investment, 121
Rigid Board, Insulation, 245
Rigid Plastic Storm Windows, 238
Rock Heat Storage, 66
Rock Wool, 231
    R Value, 232
Rocks, Solar Storage, 108
Roof
    Energy Conservation, 267
    Energy-Efficient House, 46
Roof Monitors, 65
Room Air, Storage of, 66

Safety
    Nuclear Fission Reactors, 15
    Nuclear Fusion, 34
    Nuclear Power Plants, 144
    Wood Fires, 274
Saltbox Homes
    Old Solar Structures, 56
    Solar Energy Applications, 87
Salt Hydrates Heat Storage, 67
Solar Energy System
    Orientation, 84
    Siting, 84
Saving Energy
    With Heating, Air Conditioning, and Water
        Heating, 263
    With Range, 205
    With Refrigerator-Freezer, 206
    With Washer and Dryer, 207
Seasonal Cycles of Energy Use, 3
Secondary Battery, 4
Shadowing, Solar Collector, 80
Shrubs, 276
Site Planning for Solar Energy, 83
Site Selection, Solar Dwelling, 83
Siting, Solar Energy System, 84
Skylights, 65
Slope Orientation, 83
Sociological Solar Design Factors, 71
Solar Buildings. See Solar Dwelling
Solar Cells, 40
    Photovoltaics, 40
Solar Collector, 60, 61, 80
    Air-cooled, 62
    Architectural Design, 80

Circular Concentrating, 63
Concentrating, 63
Flat Plate, 61
Linear Concentrating, 63
Liquid-cooled, 62
Open-water, 62
Orientation, 63
Passive, 64
  Warm-air, 70
Positioning, 63
Shadowing, 80
Storage, 68
Tilt, 63
Type and Size, 81
Solar Constant, 73
Solar Design
Building Characteristics, 79
Climate, 72
Determinants, 75, 82
Factors, 71
  Air Temperature, 74
  Atmospheric Conditions, 73
  Building Characteristics, 71
  Buildings, 77
  Climate, 71, 75
  Comfort, 71, 75
  Comfort Zone, 76
  Degree-Days, 74
  Earth's Axis, 73
  Economic, 71
  Heat Exchange, 77
  Human Comfort, 76
  Humidity, 74
  Insolation, 73
  Institutional, 71
  Legal, 71
  Psychological, 71
  Sociological, 71
  Solar Constant, 73
  Solar System, 71
  Temperature, 73
  Thermal Comfort Scale, 76
Solar Distribution, 61
  Components, 68
Solar Dwelling
  American, 57
  Climate, 73
  Design, 71
  Site Selection, 83
Solar Energy, 55, 128
  Activities for Children, 279
  And the Home, 111
  Human Comfort, 76
  As an Investment, 111
  Ocean Thermal Energy, 35
  Site Planning, 83
  Suitability for Your House, 111
  Supply Projections, 128
Solar Energy Applications
  Adobe Homes, 95
  Balcony-Access Apartments, 101
  Brick Colonial Homes, 95
  Brick Townhouses, 99
  Cape Cod Homes, 87
  Contemporary Split-Level Homes, 95
  Farmhouse, 91
  In-Town Apartments, 101
  Mid-Atlantic Garden Apartments, 101

Northern Rowhouse, 95
Northern Townhouse, 95
Saltbox Homes, 87
Southern Ranch Homes, 92
  Architectural Design, 79
Solar Energy System, 109
Solar Greenhouse, 64
Solar Heat Distribution, 82
Solar Heat Storage, 66
Solar Heating and Cooling, 56
Solar Houses, 57. See Also American Solar
    Dwellings
  Resale Value, 121
Solar, Investment, 113
Solar Panels, 61, 62
Solar Ponds, 65
Solar-Powered Hot Water Systems, 59
Solar Radiation, Intensity, 73
  Patterns, and Solar Site Selection, 84
  Solar Home Design, 73
Solar Site Selection
  Geography, 83
  Precipitation Patterns, 84
  Slope Orientation, 83
  Solar Radiation Patterns, 84
  Topography, 83
  Underlying Rock, 84
  Vegetation, 84
  Wind Patterns, 84
Solar Storage, 61, 81
  Location, 81
  Rocks, 108
  Subsystem, 108
  Type and Size, 81
  Warm-Air Flat-Plate, 69
  Warm Air Passive, 70
  Warm Water, 70
  Warm Water Flat-Plate, 68
  Water, 108
Solar Storage Components, 68
Solar Subsystems, The Collector, 104
Solar System
  Depreciation, 121
  Solar Design Factors, 71
  Warranty, 107
Solar System Design
  Auxiliary Energy Component, 61
  Collector,
    Energy Loss, 83
    Location, 82
    Orientation, 82
    Size, 82
    Tilt, 82
    Type, 82
    Storage Size, 83
    Storage Type, 83
Solar System Investment, Return, 121
Solar Systems
  Combined with Heat Pumps, 110
  Hybrid, 109
  Protection Against Fraud, 120
  Questions on, 120
Solar System Use
  Cool Regions, 84
  Hot-Arid Regions, 85
  Hot-Humid Regions, 85
  Temperate Regions, 84
Solar Thermal Energy, 40

Solar Transport, 61
Solid Fuels, New, 11
Storage Size, Solar System Design, 83
Southern Ranch Homes, 92
Solvent Refining, of Coal, 12
Space Heating
  Commercial, 3
  Heat Storage, 3
  Residential, 3
  Seasonal Cycle, 3
Steam Central-Heating Systems, 219
Steam Heating, 218
Storing Electricity, 4
  Batteries, 4
  Compressed Air, 5
  Flywheels, 5
  Hydrogen from Electrolysis of Water, 5
Storm Doors
  Conservation, 49
  Installation, 243
Storm Windows
  Conservation, 49
  Heat Loss Reduction, 217
  How to Install, 242
  Plastic, 238
  Rigid Plastic, 238
  Single-Pane, 238
  Triple-Track Glass, 238
  Types, 238
Stove (Coal/Wood), 217
Sun Control, Home Design, 79
Sunlight at Earth's Surface. See Insulation
Supply and Demand, Uranium, 42
Synthetic Oils, 11

Tax Credits
  Business, 337
  Residential Insulation and Conservation,
    336
  Solar, 336
Tax Provisions, Geothermal Energy and
    Geopressured Natural Gas, 336
Telecommunications, 152
Temperate Regions, Solar System Use, 84
Temperature, 73
Temperature Control, Home Design, 78
Thermo and Carbon Dioxide Pollution, 128
Thermal Comfort Scale, 76
Thermal Control, 78
Thermal Dynamics, 21, 22
Thermal Energy Distribution, 67
Thermal Resistance, 159
Thermosiphoning, 65
Thermostat, 159
Tilt, Solar Collector, 63, 80
Tilt of Earth's Axis, 73
Topography, 83
Transport, Solar, 61
Transportation
  Conservation, 151
  Energy Consumption, 150
  Resources and Strategies, 150
Trapped Heat. See Thermosiphoning
Traps, Heat, 64. See Also Solar Collectors,
    Passive Solar Collectors
Trees, 277
  and Woody Plants, Biomass Fuels, 6
Triple-Track Glass Storm Windows, 238

Underlying Rock, Solar Site Selection, 84
Unfinished Attic, 246
Uninsulated Homes, 113
"U" Value, 112, 160
"Uo" Value, 160
Uranium, 41. *See Also* Nuclear Materials
    Reserves, 13
    Resources, 8, 41
    Supply and Demand, 42
Urea-formaldehyde, 231
    R Value, 232
U.S. Department of Energy (DOE), 169
Utility Conservation Program for Residents,
    334

Vegetation, 84
Vermiculite, 231

Wall Insulation, 253, 255
    Basement, 254, 258
    Crawl Space, 254
    Heat Loss Reduction, 217
    Wood Frame, 254
Wall Unit Electric Heat, 219
Warm-Air Flat-Plate Collector Storage,
    69

Warm-Air Heating, 217
    Coal/Wood Stoves, 217
    Ventilating, and Air Conditioning Systems
        and Equipment, 163
Warm-Air Passive Collector Storage, 70
Warm-Water Concentrating Collector
    Storage, 70
Warm-Water Flat-Plate Collector Storage, 68
Warranty, Solar System, 107
Washer/Dryer, 207
Waste-Heat Recovery, 51
Water, Hot
    Energy Conservation, 268
    Solar Storage, 108
Water Heating
    Energy Conservation, 263, 265
    Energy-Efficient House, 46
Water Heaters, 203
Water Heat Storage, 67
Water-Use Appliances, 46
Watt, 223
Weatherization Grant, for Low-Income
    Families, 334
Weatherizing
    Caulking, 235
    Weather-stripping, 235

Weather-stripping
    Adhesive-Backed Foam, 238
    Conservation, 49
    Heat Loss Reduction, 217
    How to, 239
    Rolled Vinyl, 238
    Thin-Spring Metal, 238
    Weatherizing, 235
Wind
    Control, Home Design, 79
    Energy, 42
    Patterns, 84
Windmills, 43
Windows, 238
    Energy-Efficient House, 45
    Weather-stripping, 238
Winter Driving, 229
Winterizing, 227
Wood
    Best-burning Types, 273
    Home Heating, 221
Wood Fires, Safety, 274

Yard and Workshop Energy Savers, 212